Undergraduate Texts in Mathematics

Undergraduate Texts in Mathematics

Series Editors:

Sheldon Axler
San Francisco State University, San Francisco, CA, USA

Kenneth Ribet
University of California, Berkeley, CA, USA

Advisory Board:

Colin Adams, *Williams College*
David A. Cox, *Amherst College*
Pamela Gorkin, *Bucknell University*
Roger E. Howe, *Yale University*
Michael Orrison, *Harvey Mudd College*
Lisette G. de Pillis, *Harvey Mudd College*
Jill Pipher, *Brown University*
Fadil Santosa, *University of Minnesota*

Undergraduate Texts in Mathematics are generally aimed at third- and fourth-year undergraduate mathematics students at North American universities. These texts strive to provide students and teachers with new perspectives and novel approaches. The books include motivation that guides the reader to an appreciation of interrelations among different aspects of the subject. They feature examples that illustrate key concepts as well as exercises that strengthen understanding.

More information about this series at http://www.springer.com/series/666

Charles Chapman Pugh

Real Mathematical Analysis

Second Edition

 Springer

Charles Chapman Pugh
Department of Mathematics
University of California
Berkeley, CA, USA

ISSN 0172-6056   ISSN 2197-5604   (electronic)
Undergraduate Texts in Mathematics
ISBN 978-3-319-17770-0   ISBN 978-3-319-17771-7   (eBook)
DOI 10.1007/978-3-319-17771-7

Library of Congress Control Number: 2015940438

Mathematics Subject Classification (2010): 26-xx

Springer Cham Heidelberg New York Dordrecht London
© Springer International Publishing Switzerland 2002, 2015 (Corrected at 2nd printing 2017)
This work is subject to copyright. All rights are reserved by the Publisher, whether the whole or part of the material is
concerned, specifically the rights of translation, reprinting, reuse of illustrations, recitation, broadcasting, reproduction on
microfilms or in any other physical way, and transmission or information storage and retrieval, electronic adaptation,
computer software, or by similar or dissimilar methodology now known or hereafter developed.
The use of general descriptive names, registered names, trademarks, service marks, etc. in this publication does not imply,
even in the absence of a specific statement, that such names are exempt from the relevant protective laws and regulations
and therefore free for general use.
The publisher, the authors and the editors are safe to assume that the advice and information in this book are believed to be
true and accurate at the date of publication. Neither the publisher nor the authors or the editors give a warranty, express or
implied, with respect to the material contained herein or for any errors or omissions that may have been made.

Printed on acid-free paper

Springer International Publishing AG Switzerland is part of Springer Science+Business Media (www.springer.com)

To Candida
and to the students who have encouraged me –
– especially A.W., D.H., and M.B.

Preface

Was plane geometry your favorite math course in high school? Did you like proving theorems? Are you sick of memorizing integrals? If so, real analysis could be your cup of tea. In contrast to calculus and elementary algebra, it involves neither formula manipulation nor applications to other fields of science. None. It is pure mathematics, and I hope it appeals to you, the budding pure mathematician.

This book is set out for college juniors and seniors who love math and who profit from pictures that illustrate the math. Rarely is a picture a proof, but I hope a good picture will cement your understanding of *why* something is true. Seeing is believing.

Chapter 1 gets you off the ground. The whole of analysis is built on the system of real numbers $\mathbb{R}$, and especially on its Least Upper Bound property. Unlike many analysis texts that assume $\mathbb{R}$ and its properties as axioms, Chapter 1 contains a natural construction of $\mathbb{R}$ and a natural proof of the LUB property. You will also see why some infinite sets are more infinite than others, and how to visualize things in four dimensions.

Chapter 2 is about metric spaces, especially subsets of the plane. This chapter contains many pictures you have never seen. ϵ and δ will become your friends. Most of the presentation uses sequences and limits, in contrast to open coverings. It may be less elegant but it's easier to begin with. You will get to know the Cantor set well.

Chapter 3 is about Freshman Calculus – differentiation, integration, L'Hôpital's Rule, and so on, for functions of a single variable – but this time you will find out why what you were taught before is actually true. In particular you will see that a bounded function is integrable if and only if it is continuous almost everywhere, and how this fact explains many other things about integrals.

Chapter 4 is about functions viewed en masse. You can treat a set of functions as a metric space. The "points" in the space aren't numbers or vectors – they are functions. What is the distance between two functions? What should it mean that a sequence of functions converges to a limit function? What happens to derivatives and integrals when your sequence of functions converges to a limit function? When can you approximate a bad function with a good one? What is the best kind of function? What does the typical continuous function look like? (Answer: "horrible.")

Chapter 5 is about Sophomore Calculus – functions of several variables, partial derivatives, multiple integrals, and so on. Again you will see why what you were taught before is actually true. You will revisit Lagrange multipliers (with a picture proof), the Implicit Function Theorem, etc. The main new topic for you will be differential forms. They are presented not as mysterious "multi-indexed expressions," but rather as things that assign numbers to smooth domains. A 1-form assigns to a smooth curve a number, a 2-form assigns to a surface a number, a 3-form assigns to a solid a number, and so on. Orientation (clockwise, counterclockwise, etc.) is important and lets you see why cowlicks are inevitable – the Hairy Ball Theorem. The culmination of the differential forms business is Stokes' Formula, which unifies what you know about div, grad, and curl. It also leads to a short and simple proof of the Brouwer Fixed Point Theorem – a fact usually considered too advanced for undergraduates.

Chapter 6 is about Lebesgue measure and integration. It is not about measure theory in the abstract, but rather about measure theory in the plane, where you can see it. Surely I am not the first person to have rediscovered J.C. Burkill's approach to the Lebesgue integral, but I hope you will come to value it as much as I do. After you understand a few nontrivial things about area in the plane, you are naturally led to define the integral as the area under the curve – the elementary picture you saw in high school calculus. Then the basic theorems of Lebesgue integration simply fall out from the picture. Included in the chapter is the subject of density points – points at which a set "clumps together." I consider density points central to Lebesgue measure theory.

At the end of each chapter are a great many exercises. Intentionally, there is no solution manual. You should expect to be confused and frustrated when you first try to solve the harder problems. Frustration is a good thing. It will strengthen you and it is the natural mental state of most mathematicians most of the time. Join the club! When you do solve a hard problem yourself or with a group of your friends, you will treasure it far more than something you pick up off the web. For encouragement, read Sam Young's story at http://legacyrlmoore.org/reference/young.html.

I have adopted Moe Hirsch's star system for the exercises. One star is hard, two stars is very hard, and a three-star exercise is a question to which I do not know the answer. Likewise, starred sections are more challenging.

Berkeley, California, USA   CHARLES CHAPMAN PUGH

Contents

Contents

Contents

1

Real Numbers

1 Preliminaries

Before we discuss the system of real numbers it is best to make a few general remarks about mathematical outlook.

Language

By and large, mathematics is expressed in the language of set theory. Your first order of business is to get familiar with its vocabulary and grammar. A set is a collection of elements. The elements are members of the set and are said to belong to the set. For example, $\mathbb{N}$ denotes the set of **natural numbers**, 1, 2, 3, ... The members of $\mathbb{N}$ are whole numbers greater than or equal to 1. Is 10 a member of $\mathbb{N}$? Yes, 10 belongs to $\mathbb{N}$. Is 0 a member of $\mathbb{N}$? No. We write

$$x \in A \quad \text{and} \quad y \notin B$$

to indicate that the element x is a member of the set A and y is not a member of B. Thus, $6819 \in \mathbb{N}$ and $0 \notin \mathbb{N}$.

We try to write capital letters for sets and small letters for elements of sets. Other standard sets have standard names. The set of **integers** is denoted by $\mathbb{Z}$, which stands for the German word *Zahlen*. (An integer is a positive whole number, zero, or a negative whole number.) Is $\sqrt{2} \in \mathbb{Z}$? No, $\sqrt{2} \notin \mathbb{Z}$. How about -15? Yes, $-15 \in \mathbb{Z}$.

© Springer International Publishing Switzerland 2015
C.C. Pugh, *Real Mathematical Analysis*, Undergraduate Texts
in Mathematics, DOI 10.1007/978-3-319-17771-7_1

The set of **rational numbers** is called $\mathbb{Q}$, which stands for "quotient." (A rational number is a fraction of integers, the denominator being nonzero.) Is $\sqrt{2}$ a member of $\mathbb{Q}$? No, $\sqrt{2}$ does not belong to $\mathbb{Q}$. Is π a member of $\mathbb{Q}$? No. Is 1.414 a member of $\mathbb{Q}$? Yes.

You should practice reading the notation "$\{x \in A :$" as "the set of x that belong to A such that." The **empty set** is the collection of no elements and is denoted by $\emptyset$. Is 0 a member of the empty set? No, $0 \notin \emptyset$.

A **singleton set** has exactly one member. It is denoted as $\{x\}$ where x is the member. Similarly if exactly two elements x and y belong to a set, the set is denoted as $\{x, y\}$.

If A and B are sets and each member of A also belongs to B then A is a subset of B and A is contained in B. We write[†]

$$A \subset B.$$

Is $\mathbb{N}$ a subset of $\mathbb{Z}$? Yes. Is it a subset of $\mathbb{Q}$? Yes. If A is a subset of B and B is a subset of C, does it follow that A is a subset of C? Yes. Is the empty set a subset of $\mathbb{N}$? Yes, $\emptyset \subset \mathbb{N}$. Is 1 a subset of $\mathbb{N}$? No, but the singleton set $\{1\}$ is a subset of $\mathbb{N}$. Two sets are equal if each member of one belongs to the other. Each is a subset of the other. This is how you prove two sets are equal: Show that each element of the first belongs to the second, and each element of the second belongs to the first.

The union of the sets A and B is the set $A \cup B$, each of whose elements belongs to either A, or to B, or to both A and to B. The intersection of A and B is the set $A \cap B$ each of whose elements belongs to both A and to B. If $A \cap B$ is the empty set then A and B are **disjoint**. The **symmetric difference** of A and B is the set $A \triangle B$ each of whose elements belongs to A but not to B, or belongs to B but not to A. The **difference** of A to B is the set $A \setminus B$ whose elements belong to A but not to B. See Figure 1.

A **class** is a collection of sets. The sets are members of the class. For example we could consider the class $\mathcal{E}$ of sets of even natural numbers. Is the set $\{2, 15\}$ a member of $\mathcal{E}$? No. How about the singleton set $\{6\}$? Yes. How about the empty set? Yes, each element of the empty set is even.

When is one class a subclass of another? When each member of the former belongs also to the latter. For example the class $\mathcal{T}$ of sets of positive integers divisible by 10

[†]When some mathematicians write $A \subset B$ they mean that A is a subset of B, but $A \neq B$. We do *not* adopt this convention. We accept $A \subset A$.

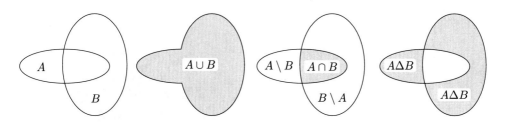

Figure 1 Venn diagrams of union, intersection, and differences

is a subclass of $\mathcal{E}$, the class of sets of even natural numbers, and we write $\mathcal{T} \subset \mathcal{E}$. Each set that belongs to the class $\mathcal{T}$ also belongs to the class $\mathcal{E}$. Consider another example. Let $\mathcal{S}$ be the class of singleton subsets of $\mathbb{N}$ and let $\mathcal{D}$ be the class of subsets of $\mathbb{N}$ each of which has exactly two elements. Thus $\{10\} \in \mathcal{S}$ and $\{2,6\} \in \mathcal{D}$. Is $\mathcal{S}$ a subclass of $\mathcal{D}$? No. The members of $\mathcal{S}$ are singleton sets and they are not members of $\mathcal{D}$. Rather they are subsets of members of $\mathcal{D}$. Note the distinction, and think about it.

Here is an analogy. Each citizen is a member of his or her country – I am an element of the USA and Tony Blair is an element of the UK. Each country is a member of the United Nations. Are citizens members of the UN? No, countries are members of the UN.

In the same vein is the concept of an **equivalence relation** on a set S. It is a relation $s \sim s'$ that holds between some members $s, s' \in S$ and it satisfies three properties: For all $s, s', s'' \in S$

(a) $s \sim s$.
(b) $s \sim s'$ implies that $s' \sim s$.
(c) $s \sim s' \sim s''$ implies that $s \sim s''$.

Figure 2 on the next page shows how the equivalence relation breaks S into disjoint subsets called **equivalence classes**[†] defined by mutual equivalence: The equivalence class containing s consists of all elements $s' \in S$ equivalent to s and is denoted $[s]$. The element s is a **representative** of its equivalence class. Think again of citizens and countries. Say two citizens are equivalent if they are citizens of the same country. The world of equivalence relations is egalitarian: I represent my equivalence class USA just as much as does the president.

[†]The phrase "equivalence class" is standard and widespread, although it would be more consistent with the idea that a class is a collection of sets to refer instead to an "equivalence set."

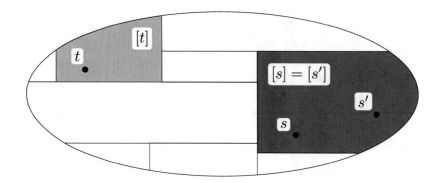

Figure 2 Equivalence classes and representatives

Truth

When is a mathematical statement accepted as true? Generally, mathematicians would answer "Only when it has a proof inside a familiar mathematical framework." A picture may be vital in getting you to believe a statement. An analogy with something you know to be true may help you understand it. An authoritative teacher may force you to parrot it. A formal proof, however, is the ultimate and only reason to accept a mathematical statement as true. A recent debate in Berkeley focused the issue for me. According to a math teacher from one of our local private high schools, his students found proofs in mathematics were of little value, especially compared to "convincing arguments." Besides, the mathematical statements were often seen as obviously true and in no need of formal proof anyway. I offer you a paraphrase of Bob Osserman's response.

But a convincing argument is not a proof. A mathematician generally wants both, and certainly would be less likely to accept a convincing argument by itself than a formal proof by itself. Least of all would a mathematician accept the proposal that we should generally replace proofs with convincing arguments.

There has been a tendency in recent years to take the notion of proof down from its pedestal. Critics point out that standards of rigor change from century to century. New gray areas appear all the time. Is a proof by computer an acceptable proof? Is a proof that is spread over many journals and thousands of pages, that is too long for any one person to master, a proof? And of course, venerable Euclid is full of flaws, some filled in by Hilbert, others possibly still lurking.

Clearly it is worth examining closely and critically the most basic notion of mathematics, that of proof. On the other hand, it is important to bear in mind that all distinctions and niceties about what precisely constitutes a proof are mere quibbles compared to the enormous gap between any generally accepted version of a proof and the notion of a convincing argument. Compare Euclid, with all his flaws to the most eminent of the ancient exponents of the convincing argument – Aristotle. Much of Aristotle's reasoning was brilliant, and he certainly convinced most thoughtful people for over a thousand years. In some cases his analyses were exactly right, but in others, such as heavy objects falling faster than light ones, they turned out to be totally wrong. In contrast, there is not to my knowledge a single theorem stated in Euclid's *Elements* that in the course of two thousand years turned out to be false. That is quite an astonishing record, and an extraordinary validation of proof over convincing argument.

Here are some guidelines for writing a rigorous mathematical proof. See also Exercise 0.

1. Name each object that appears in your proof. (For instance, you might begin your proof with a phrase, "Consider a set X, and elements x, y that belong to X," etc.)
2. Draw a diagram that captures how these objects relate, and extract logical statements from it. Quantifiers precede the objects quantified; see below.
3. Become confident that the mathematical assertion you are trying to prove is really true before trying to write down a proof of it. If there a specific function involved – say $\sin x^\alpha$ – draw the graph of the function for a few values of α before starting any ϵ, δ analysis. Belief first and proof second.
4. Proceed step by step, each step depending on the hypotheses, previously proved theorems, or previous steps in your proof.
5. Check for "rigor": All cases have been considered, all details have been tied down, and circular reasoning has been avoided.
6. Before you sign off on the proof, check for counterexamples and any implicit assumptions you made that could invalidate your reasoning.

Logic

Among the most frequently used logical symbols in math are the quantifiers $\forall$ and $\exists$. Read them always as "for each" and "there exists." Avoid reading $\forall$ as "for all," which in English has a more inclusive connotation. Another common symbol is $\Rightarrow$. Read it as "implies."

The rules of correct mathematical grammar are simple: Quantifiers appear at the beginning of a sentence, they modify only what follows them in the sentence, and assertions occur at the end of the sentence. Here is an example.

(1)   *For each integer n there is a prime number p which is greater than n.*

In symbols the sentence reads

$$\forall n \in \mathbb{Z} \quad \exists p \in P \quad \text{such that} \quad p > n,$$

where P denotes the set of prime numbers. (A **prime number** is a whole number greater than 1 whose only divisors in $\mathbb{N}$ are itself and 1.) In English, the same idea can be reexpressed as

(2)   *Every integer is less than some prime number.*

or

(3)   *A prime number can always be found which is bigger than any integer.*

These sentences are correct in English grammar, but disastrously WRONG when transcribed directly into mathematical grammar. They translate into disgusting mathematical gibberish:

(WRONG (2)) $\forall n \in \mathbb{Z} \quad n < p \quad \exists p \in P$

(WRONG (3)) $\exists p \in P \quad p > n \quad \forall n \in \mathbb{Z}.$

Moral Quantifiers first and assertions last. In stating a theorem, try to apply the same principle. Write the hypothesis first and the conclusion second. See Exercise 0.

The order in which quantifiers appear is also important. Contrast the next two sentences in which we switch the position of two quantified phrases.

(4) $(\forall n \in \mathbb{N}) \quad (\forall m \in \mathbb{N}) \quad (\exists p \in P) \quad \text{such that} \quad (nm < p).$

(5) $(\forall n \in \mathbb{N}) \quad (\exists p \in P) \quad \text{such that} \quad (\forall m \in \mathbb{N}) \quad (nm < p).$

(4) is a true statement but (5) is false. A quantifier modifies the part of a sentence that follows it but not the part that precedes it. This is another reason never to end with a quantifier.

Moral Quantifier order is crucial.

There is a point at which English and mathematical meaning diverge. It concerns the word "or." In mathematics "a or b" always means "a or b or both a and b," while in English it can mean "a or b but not both a and b." For example, Patrick Henry certainly would not have accepted both liberty and death in response to his cry of "Give me liberty or give me death." In mathematics, however, the sentence "17 is a prime or 23 is a prime" is correct even though both 17 and 23 are prime. Similarly, in mathematics $a \Rightarrow b$ means that if a is true then b is true but that b might also be true for reasons entirely unrelated to the truth of a. In English, $a \Rightarrow b$ is often confused with $b \Rightarrow a$.

Moral In mathematics "or" is inclusive. It means *and/or*. In mathematics $a \Rightarrow b$ is not the same as $b \Rightarrow a$.

It is often useful to form the negation or logical opposite of a mathematical sentence. The symbol $\sim$ is usually used for negation, despite the fact that the same symbol also indicates an equivalence relation. Mathematicians refer to this as an **abuse of notation**. Fighting a losing battle against abuse of notation, we write $\neg$ for negation. For example, if $m, n \in \mathbb{N}$ then $\neg(m < n)$ means it is not true that m is less than n. In other words

$$\neg(m < n) \quad \equiv \quad m \geq n.$$

(We use the symbol $\equiv$ to indicate that the two statements are equivalent.) Similarly, $\neg(x \in A)$ means it is not true that x belongs to A. In other words,

$$\neg(x \in A) \quad \equiv \quad x \notin A.$$

Double negation returns a statement to its original meaning. Slightly more interesting is the negation of "and" and "or." Just for now, let us use the symbols & for "and" and $\vee$ for "or." We claim

(6) $$\neg(a \ \& \ b) \quad \equiv \quad \neg a \ \vee \ \neg b.$$

(7) $$\neg(a \ \vee \ b) \quad \equiv \quad \neg a \ \& \ \neg b.$$

For if it is not the case that both a and b are true then at least one must be false. This proves (6), and (7) is similar. Implication also has such interpretations:

$$(8) \qquad\qquad a \Rightarrow b \quad \equiv \quad \neg a \Leftarrow \neg b \quad \equiv \quad \neg a \vee b.$$

$$(9) \qquad\qquad \neg(a \Rightarrow b) \quad \equiv \quad a \mathbin{\&} \neg b.$$

What about the negation of a quantified sentence such as

$$\neg(\forall n \in \mathbb{N}, \ \exists p \in P \text{ such that } n < p).$$

The rule is: change each $\forall$ to $\exists$ and vice versa, leaving the order the same, and negate the assertion. In this case the negation is

$$\exists n \in \mathbb{N}, \quad \forall p \in P, \quad n \geq p.$$

In English it reads "There exists a natural number n, and for all primes p we have $n \geq p$." The sentence has correct mathematical grammar but of course is false. To help translate from mathematics to readable English, a comma can be read as "and," "we have," or "such that."

All mathematical assertions take an implication form $a \Rightarrow b$. The hypothesis is a and the conclusion is b. If you are asked to prove $a \Rightarrow b$, there are several ways to proceed. First you may just see right away why a does imply b. Fine, if you are so lucky. Or you may be puzzled. Does a really imply b? Two routes are open to you. You may view the implication in its equivalent contrapositive form $\neg a \Leftarrow \neg b$ as in (8). Sometimes this will make things clearer. Or you may explore the possibility that a fails to imply b. If you can somehow deduce from the failure of a implying b a contradiction to a known fact (for instance, if you can deduce the existence of a planar right triangle with legs x, y but $x^2 + y^2 \neq h^2$, where h is the hypotenuse), then you have succeeded in making an **argument by contradiction**. Clearly (9) is pertinent here. It tells you what it means that a fails to imply b, namely that a is true and simultaneously b is false.

Euclid's proof that $\mathbb{N}$ contains infinitely many prime numbers is a classic example of this method. The hypothesis is that $\mathbb{N}$ is the set of natural numbers and that P is the set of prime numbers. The conclusion is that P is an infinite set. The proof of this fact begins with the phrase "Suppose not." It means to suppose, after all, that the set of prime numbers P is merely a finite set, and see where this leads you. It does not mean that we think P really is a finite set, and it is not a hypothesis of a theorem. Rather it just means that we will try to find out what awful consequences would follow from P being finite. In fact if P were[†] finite then it would consist of m numbers $p_1, \ldots, p_m$. Their product $N = 2 \cdot 3 \cdot 5 \cdots \cdot p_m$ would be evenly divisible (i.e., remainder 0 after division) by each p_i and therefore $N + 1$ would be evenly divisible by no prime (the remainder of p_i divided into $N + 1$ would always be 1), which would contradict the fact that every integer ≥ 2 can be factored as a product of primes. (The latter fact has nothing to do with P being finite or not.) Since the supposition that P is finite led to a contradiction of a known fact, prime factorization, the supposition was incorrect, and P is, after all, infinite.

Aficionados of logic will note our heavy use here of the "law of the excluded middle," to wit, that a mathematically meaningful statement is either true or false. The possibilities that it is neither true nor false, or that it is both true and false, are excluded.

Notation The symbol $\lightning$ indicates a contradiction. It is used when writing a proof in longhand.

Metaphor and Analogy

In high school English, you are taught that a metaphor is a figure of speech in which one idea or word is substituted for another to suggest a likeness or similarity. This can occur very simply as in "The ship plows the sea." Or it can be less direct, as in "His lawyers dropped the ball." What give a metaphor its power and pleasure are the secondary suggestions of similarity. Not only did the lawyers make a mistake, but it was their own fault, and, like an athlete who has dropped a ball, they could not follow through with their next legal action. A secondary implication is that their enterprise was just a game.

Often a metaphor associates something abstract to something concrete, as "Life is a journey." The preservation of inference from the concrete to the abstract in this metaphor suggests that like a journey, life has a beginning and an end, it progresses in one direction, it may have stops and detours, ups and downs, etc. The beauty of a metaphor is that hidden in a simple sentence like "Life is a journey" lurk a great many parallels, waiting to be uncovered by the thoughtful mind.

[†]In English grammar, the subjunctive mode indicates doubt, and I have written Euclid's proof in that form – "if P *were* finite" instead of "if P *is* finite," "each prime *would* divide N evenly," instead of "each prime *divides* N evenly," etc. At first it seems like a fine idea to write all arguments by contradiction in the subjunctive mode, clearly exhibiting their impermanence. Soon, however, the subjunctive and conditional language becomes ridiculously stilted and archaic. For consistency then, as much as possible, *use the present tense.*

Metaphorical thinking pervades mathematics to a remarkable degree. It is often reflected in the language mathematicians choose to define new concepts. In his construction of the system of real numbers, Dedekind could have referred to $A|B$ as a "type-2, order preserving equivalence class," or worse, whereas "cut" is the right metaphor. It corresponds closely to one's physical intuition about the real line. See Figure 3. In his book, *Where Mathematics Comes From*, George Lakoff gives a comprehensive view of metaphor in mathematics.

An analogy is a shallow form of metaphor. It just asserts that two things are similar. Although simple, analogies can be a great help in accepting abstract concepts. When you travel from home to school, at first you are closer to home, and then you are closer to school. Somewhere there is a halfway stage in your journey. You *know* this, long before you study mathematics. So when a curve connects two points in a metric space (Chapter 2), you should expect that as a point "travels along the curve," somewhere it will be equidistant between the curve's endpoints. Reasoning by analogy is also referred to as "intuitive reasoning."

Moral Try to translate what you know of the real world to guess what is true in mathematics.

Two Pieces of Advice

A colleague of mine regularly gives his students an excellent piece of advice. When you confront a general problem and do not see how to solve it, make some extra hypotheses, and try to solve it then. If the problem is posed in n dimensions, try it first in two dimensions. If the problem assumes that some function is continuous, does it get easier for a differentiable function? The idea is to reduce an abstract problem to its simplest concrete manifestation, rather like a metaphor in reverse. At the minimum, look for at least one instance in which you can solve the problem, and build from there.

Moral If you do not see how to solve a problem in complete generality, first solve it in some special cases.

Here is the second piece of advice. Buy a notebook. In it keep a diary of your own opinions about the mathematics you are learning. Draw a picture to illustrate every definition, concept, and theorem.

2 Cuts

We begin at the beginning and discuss $\mathbb{R} = $ the system of all real numbers from a somewhat theological point of view. The current mathematics teaching trend treats the real number system $\mathbb{R}$ as a given – it is defined axiomatically. Ten or so of its properties are listed, called axioms of a complete ordered field, and the game becomes to deduce its other properties from the axioms. This is something of a fraud, considering that the entire structure of analysis is built on the real number system. For what if a system satisfying the axioms failed to exist? Then one would be studying the empty set! However, you need not take the existence of the real numbers on faith alone – we will give a concise mathematical proof of it.

It is reasonable to accept all grammar school arithmetic facts about

The set $\mathbb{N}$ of natural numbers, $1, 2, 3, 4, \ldots$.
The set $\mathbb{Z}$ of integers, $0, 1, -1, -2, 2, \ldots$.
The set $\mathbb{Q}$ of rational numbers p/q where p, q are integers, $q \neq 0$.

For example, we will admit without question facts like $2 + 2 = 4$, and laws like $a + b = b + a$ for rational numbers a, b. All facts you know about arithmetic involving integers or rational numbers are fair to use in homework exercises too.[†] It is clear that $\mathbb{N} \subset \mathbb{Z} \subset \mathbb{Q}$. Now $\mathbb{Z}$ improves $\mathbb{N}$ because it contains negatives and $\mathbb{Q}$ improves $\mathbb{Z}$ because it contains reciprocals. $\mathbb{Z}$ legalizes subtraction and $\mathbb{Q}$ legalizes division. Still, $\mathbb{Q}$ needs further improvement. It doesn't admit irrational roots such as $\sqrt{2}$ or transcendental numbers such as π. We aim to go a step beyond $\mathbb{Q}$, completing it to form $\mathbb{R}$ so that

$$\mathbb{N} \subset \mathbb{Z} \subset \mathbb{Q} \subset \mathbb{R}.$$

As an example of the fact that $\mathbb{Q}$ is incomplete we have

1 Theorem *No number r in $\mathbb{Q}$ has square equal to 2; i.e., $\sqrt{2} \notin \mathbb{Q}$.*

Proof To prove that every $r = p/q$ has $r^2 \neq 2$ we show that $p^2 \neq 2q^2$. It is fair to assume that p and q have no common factors since we would have canceled them out beforehand.

<u>Case 1.</u> p is odd. Then p^2 is odd while $2q^2$ is not. Therefore $p^2 \neq 2q^2$.

[†]A subtler fact that you may find useful is the prime factorization theorem mentioned above. Any integer ≥ 2 can be factored into a product of prime numbers. For example, 120 is the product of primes $2 \cdot 2 \cdot 2 \cdot 3 \cdot 5$. Prime factorization is unique except for the order in which the factors appear. An easy consequence is that if a prime number p divides an integer k and if k is the product mn of integers then p divides m or it divides n. After all, by uniqueness, the prime factorization of k is just the product of the prime factorizations of m and n.

<u>Case 2.</u> p is even. Since p and q have no common factors, q is odd. Then p^2 is divisible by 4 while $2q^2$ is not. Therefore $p^2 \neq 2q^2$.

Since $p^2 \neq 2q^2$ for all integers p, there is no rational number $r = p/q$ whose square is 2. $\qquad\qquad\qquad\qquad\qquad\qquad\qquad\qquad\qquad\qquad\qquad\qquad\qquad\qquad\qquad\qquad$ □

The set $\mathbb{Q}$ of rational numbers is incomplete. It has "gaps," one of which occurs at $\sqrt{2}$. These gaps are really more like pinholes; they have zero width. Incompleteness is what is *wrong* with $\mathbb{Q}$. Our goal is to complete $\mathbb{Q}$ by filling in its gaps. An elegant method to arrive at this goal is **Dedekind cuts** in which one visualizes real numbers as places at which a line may be cut with scissors. See Figure 3.

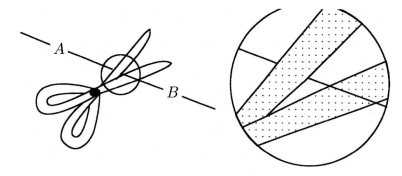

Figure 3 A Dedekind cut

Definition A **cut** in $\mathbb{Q}$ is a pair of subsets A, B of $\mathbb{Q}$ such that

(a) $A \cup B = \mathbb{Q}$, $A \neq \emptyset$, $B \neq \emptyset$, $A \cap B = \emptyset$.
(b) If $a \in A$ and $b \in B$ then $a < b$.
(c) A contains no largest element.

A is the left-hand part of the cut and B is the right-hand part. We denote the cut as $x = A|B$. Making a semantic leap, we now answer the question "what is a real number?"

Definition A **real number** is a cut in $\mathbb{Q}$.

$\mathbb{R}$ is the class[†] of all real numbers $x = A|B$. We will show that in a natural way $\mathbb{R}$ is a complete ordered field containing $\mathbb{Q}$. Before spelling out what this means, here are two examples of cuts.

[†]The word "class" is used instead of the word "set" to emphasize that for now the members of $\mathbb{R}$ are set-pairs $A|B$, and not the numbers that belong to A or B. The notation $A|B$ could be shortened to A since B is just the rest of $\mathbb{Q}$. We write $A|B$, however, as a mnemonic device. It *looks* like a cut.

(i) $A|B = \{r \in \mathbb{Q} : r < 1\} \,|\, \{r \in \mathbb{Q} : r \geq 1\}$.

(ii) $A|B = \{r \in \mathbb{Q} : r \leq 0 \text{ or } r^2 < 2\} \,|\, \{r \in \mathbb{Q} : r > 0 \text{ and } r^2 \geq 2\}$.

It is convenient to say that $A|B$ is a **rational cut** if it is like the cut in (i): For some fixed rational number c, A is the set of all rationals $< c$ while B is the rest of $\mathbb{Q}$. The B-set of a rational cut contains a smallest element c, and conversely, if $A|B$ is a cut in $\mathbb{Q}$ and B contains a smallest element c then $A|B$ is the rational cut at c. We write c^* for the rational cut at c. This lets us think of $\mathbb{Q} \subset \mathbb{R}$ by identifying c with c^*. It is like thinking of $\mathbb{Z}$ as a subset of $\mathbb{Q}$ since the integer n in $\mathbb{Z}$ can be thought of as the fraction $n/1$ in $\mathbb{Q}$. In the same way the rational number c in $\mathbb{Q}$ can be thought of as the cut at c. It is just a different way of looking at c. It is in this sense that we write

$$\mathbb{N} \subset \mathbb{Z} \subset \mathbb{Q} \subset \mathbb{R}.$$

There is an order relation $x \leq y$ on cuts that fairly cries out for attention.

Definition If $x = A|B$ and $y = C|D$ are cuts such that $A \subset C$ then x is **less than or equal** to y and we write $x \leq y$. If $A \subset C$ and $A \neq C$ then x is **less than** y and we write $x < y$.

The property distinguishing $\mathbb{R}$ from $\mathbb{Q}$ and which is at the bottom of every significant theorem about $\mathbb{R}$ involves upper bounds and least upper bounds or, equivalently, lower bounds and greatest lower bounds.

$M \in \mathbb{R}$ is an **upper bound** for a set $S \subset \mathbb{R}$ if each $s \in S$ satisfies

$$s \leq M.$$

We also say that the set S is **bounded above** by M. An upper bound for S that is less than all other upper bounds for S is a **least upper bound** for S. The least upper bound for S is denoted l.u.b.(S). For example,

3 is an upper bound for the set of negative integers.

-1 is the least upper bound for the set of negative integers.

1 is the least upper bound for the set of rational numbers $1 - 1/n$ with $n \in \mathbb{N}$.

-100 is an upper bound for the empty set.

A least upper bound for S may or may not belong to S. This is why you should say "least upper bound *for S*" rather than "least upper bound *of S*."

2 Theorem *The set $\mathbb{R}$, constructed by means of Dedekind cuts, is* **complete**[†] *in the sense that it satisfies the*

Least Upper Bound Property: *If S is a nonempty subset of $\mathbb{R}$ and is bounded above then in $\mathbb{R}$ there exists a least upper bound for S.*

Proof Easy! Let $\mathcal{C} \subset \mathbb{R}$ be any nonempty collection of cuts which is bounded above, say by the cut $X|Y$. Define

$$C = \{a \in \mathbb{Q} : \text{ for some cut} A|B \in \mathcal{C} \text{ we have } a \in A\} \text{ and } D = \text{the rest of } \mathbb{Q}.$$

It is easy to see that $z = C|D$ is a cut. Clearly, it is an upper bound for $\mathcal{C}$ since the A for every element of $\mathcal{C}$ is contained in C. Let $z' = C'|D'$ be any upper bound for $\mathcal{C}$. By the assumption that $A|B \leq C'|D'$ for all $A|B \in \mathcal{C}$, we see that the A for every member of $\mathcal{C}$ is contained in C'. Hence $C \subset C'$, so $z \leq z'$. That is, among all upper bounds for $\mathcal{C}$, z is least. $\qquad\square$

The simplicity of this proof is what makes cuts good. We go from $\mathbb{Q}$ to $\mathbb{R}$ by pure thought. To be more complete, as it were, we describe the natural arithmetic of cuts. Let cuts $x = A|B$ and $y = C|D$ be given. How do we add them? subtract them? ... Generally the answer is to do the corresponding operation to the elements comprising the two halves of the cuts, being careful about negative numbers. The sum of x and y is $x + y = E|F$ where

$$
\begin{aligned}
E &= \{r \in \mathbb{Q} : \text{ for some } a \in A \text{ and for some } c \in C \text{ we have } r = a + c\} \\
F &= \text{the rest of } \mathbb{Q}.
\end{aligned}
$$

It is easy to see that $E|F$ is a cut in $\mathbb{Q}$ and that it doesn't depend on the order in which x and y appear. That is, cut addition is well defined and $x + y = y + x$. The zero cut is 0^* and $0^* + x = x$ for all $x \in \mathbb{R}$. The additive inverse of $x = A|B$ is $-x = C|D$ where

$$
\begin{aligned}
C &= \{r \in \mathbb{Q} : \text{ for some } b \in B, \text{ not the smallest element of } B, \ r = -b\} \\
D &= \text{the rest of } \mathbb{Q}.
\end{aligned}
$$

Then $(-x) + x = 0^*$. Correspondingly, the difference of cuts is $x - y = x + (-y)$. Another property of cut addition is **associativity**:

$$(x + y) + z = x + (y + z).$$

[†] There is another, related, sense in which $\mathbb{R}$ is complete. See Theorem 5 below.

This follows from the corresponding property of $\mathbb{Q}$.

Multiplication is trickier to define. It helps to first say that the cut $x = A|B$ is **positive** if $0^* < x$ or **negative** if $x < 0^*$. Since 0 lies in A or B, a cut is either positive, negative, or zero. If $x = A|B$ and $y = C|D$ are positive cuts then their product is $x \cdot y = E|F$ where

$$E = \{r \in \mathbb{Q} : r \leq 0 \text{ or } \exists a \in A \text{ and } \exists c \in C \text{ such that } a > 0, c > 0, \text{ and } r = ac\}$$

and F is the rest of $\mathbb{Q}$. If x is positive and y is negative then we define the product to be $-(x \cdot (-y))$. Since x and $-y$ are both positive cuts this makes sense and is a negative cut. Similarly, if x is negative and y is positive then by definition their product is the negative cut $-((-x) \cdot y)$, while if x and y are both negative then their product is the positive cut $(-x) \cdot (-y)$. Finally, if x or y is the zero cut 0^* we define $x \cdot y$ to be 0^*. (This makes five cases in the definition.)

Verifying the arithmetic properties for multiplication is tedious, to say the least, and somehow nothing seems to be gained by writing out every detail. (To pursue cut arithmetic further you could read Landau's classically boring book, *Foundations of Analysis*.) To get the flavor of it, let's check the commutativity of multiplication: $x \cdot y = y \cdot x$ for cuts $x = A|B$, $y = C|D$. If x, y are positive then

$$\{ac : a \in A, c \in C, a > 0, c > 0\} = \{ca : c \in C, a \in A, c > 0, a > 0\}$$

implies that $x \cdot y = y \cdot x$. If x is positive and y is negative then

$$x \cdot y = -(x \cdot (-y)) = -((-y) \cdot x) = y \cdot x.$$

The second equality holds because we have already checked commutativity for positive cuts. The remaining three cases are checked similarly. There are twenty seven cases to check for associativity and twenty seven more for distributivity. All are simple and we omit their proofs. The real point is that cut arithmetic can be defined and it satisfies the same field properties that $\mathbb{Q}$ does:

The operation of cut addition is
well defined, natural, commutative, associative, and
has inverses with respect to the neutral element 0^*.

The operation of cut multiplication
is well defined, natural, commutative, associative,
distributive over cut addition, and has inverses of
nonzero elements with respect to the neutral element 1^*.

By definition, a **field** is a system consisting of a set of elements and two operations, addition and multiplication, that have the preceding algebraic properties – commutativity, associativity, etc. Besides just existing, cut arithmetic is consistent with $\mathbb{Q}$ arithmetic in the sense that if $c, r \in \mathbb{Q}$ then

$$c^* + r^* = (c + r)^* \quad \text{and} \quad c^* \cdot r^* = (cr)^*.$$

By definition, this is what we mean when we say that $\mathbb{Q}$ is a **subfield** of $\mathbb{R}$. The cut order enjoys the additional properties of

> **transitivity** $x < y < z$ implies $x < z$.
>
> **trichotomy** Either $x < y$, $y < x$, or $x = y$, but only one of the three things is true.
>
> **translation** $x < y$ implies $x + z < y + z$.

By definition, this is what we mean when we say that $\mathbb{R}$ is an **ordered field**. Besides, the product of positive cuts is positive and cut order is consistent with $\mathbb{Q}$ order: $c^* < r^*$ if and only if $c < r$ in $\mathbb{Q}$. By definition, this is what we mean when we say that $\mathbb{Q}$ is an ordered subfield of $\mathbb{R}$. To summarize

3 Theorem *The set $\mathbb{R}$ of all cuts in $\mathbb{Q}$ is a complete ordered field that contains $\mathbb{Q}$ as an ordered subfield.*

The **magnitude** or absolute value of $x \in \mathbb{R}$ is

$$|x| = \begin{cases} x & \text{if } x \geq 0 \\ -x & \text{if } x < 0. \end{cases}$$

Thus, $x \leq |x|$. A basic, constantly used fact about magnitude is the following.

4 Triangle Inequality *For all $x, y \in \mathbb{R}$ we have $|x + y| \leq |x| + |y|$.*

Proof The translation and transitivity properties of the order relation imply that adding y and $-y$ to the inequalities $x \leq |x|$ and $-x \leq |x|$ gives

$$\begin{aligned} x + y &\leq |x| + y \leq |x| + |y| \\ -x - y &\leq |x| - y \leq |x| + |y|. \end{aligned}$$

Since

$$|x + y| = \begin{cases} x + y & \text{if } x + y \geq 0 \\ -x - y & \text{if } x + y \leq 0 \end{cases}$$

and both $x + y$ and $-x - y$ are less than or equal to $|x| + |y|$, we infer that $|x + y| \leq |x| + |y|$ as asserted. $\square$

Next, suppose we try the same cut construction in $\mathbb{R}$ that we did in $\mathbb{Q}$. Are there gaps in $\mathbb{R}$ that can be detected by cutting $\mathbb{R}$ with scissors? The natural definition of a cut in $\mathbb{R}$ is a division $\mathcal{A}|\mathcal{B}$, where $\mathcal{A}$ and $\mathcal{B}$ are disjoint, nonempty subcollections of $\mathbb{R}$ with $\mathcal{A} \cup \mathcal{B} = \mathbb{R}$, and $a < b$ for all $a \in \mathcal{A}$ and $b \in \mathcal{B}$. Further, $\mathcal{A}$ contains no largest element. Each $b \in \mathcal{B}$ is an upper bound for $\mathcal{A}$. Therefore $y = \text{l.u.b.}(\mathcal{A})$ exists and $a \le y \le b$ for all $a \in \mathcal{A}$ and $b \in \mathcal{B}$. By trichotomy,

$$\mathcal{A}|\mathcal{B} = \{x \in \mathbb{R} : x < y\} \mid \{x \in \mathbb{R} : x \ge y\}.$$

In other words, $\mathbb{R}$ has no gaps. Every cut in $\mathbb{R}$ occurs exactly *at* a real number.

Allied to the existence of $\mathbb{R}$ is its uniqueness. Any complete ordered field $\mathbb{F}$ containing $\mathbb{Q}$ as an ordered subfield corresponds to $\mathbb{R}$ in a way preserving all the ordered field structure. To see this, take any $\varphi \in \mathbb{F}$ and associate to it the cut $A|B$ where

$$A = \{r \in \mathbb{Q} : r < \varphi \text{ in } \mathbb{F}\} \qquad B = \text{ the rest of } \mathbb{Q}.$$

This correspondence makes $\mathbb{F}$ equivalent to $\mathbb{R}$.

Upshot The real number system $\mathbb{R}$ exists and it satisfies the properties of a complete ordered field. The properties are not assumed as axioms, but are proved by logically analyzing the Dedekind construction of $\mathbb{R}$. Having gone through all this cut rigmarole, we must remark that it is a rare working mathematician who actually thinks of $\mathbb{R}$ as a complete ordered field or as the set of all cuts in $\mathbb{Q}$. Rather, he or she thinks of $\mathbb{R}$ as points on the x-axis, just as in calculus. You too should picture $\mathbb{R}$ this way, the only benefit of the cut derivation being that you should now unhesitatingly accept the least upper bound property of $\mathbb{R}$ as a true fact.

Note $\pm\infty$ are not real numbers, since $\mathbb{Q}|\emptyset$ and $\emptyset|\mathbb{Q}$ are not cuts. Although some mathematicians think of $\mathbb{R}$ together with $-\infty$ and $+\infty$ as an "extended real number system," it is simpler to leave well enough alone and just deal with $\mathbb{R}$ itself. Nevertheless, it is convenient to write expressions like "$x \to \infty$" to indicate that a real variable x grows larger and larger without bound.

If S is a nonempty subset of $\mathbb{R}$ then its **supremum** is its least upper bound when S is bounded above and is said to be $+\infty$ otherwise; its **infimum** is its greatest lower bound when S is bounded below and is said to be $-\infty$ otherwise. (In Exercise 19 you are asked to invent the notion of greatest lower bound.) By definition the supremum of the empty set is $-\infty$. This is reasonable, considering that every real number, no matter how negative, is an upper bound for $\emptyset$, and the least upper bound should be as far leftward as possible, namely $-\infty$. Similarly, the infimum of the empty set is $+\infty$. We write $\sup S$ and $\inf S$ for the supremum and infimum of S.

Cauchy sequences

As mentioned above there is a second sense in which $\mathbb{R}$ is complete. It involves the concept of convergent sequences. Let $a_1, a_2, a_3, a_4, \ldots = (a_n)$, $n \in \mathbb{N}$, be a sequence of real numbers. The sequence (a_n) **converges to the limit** $b \in \mathbb{R}$ as $n \to \infty$ provided that for each $\epsilon > 0$ there exists $N \in \mathbb{N}$ such that for all $n \geq N$ we have

$$|a_n - b| < \epsilon.$$

The statistician's language is evocative here. Think of $n = 1, 2, \ldots$ as a sequence of times and say that the sequence (a_n) converges to b provided that *eventually* all its terms nearly equal b. In symbols,

$$\forall \epsilon > 0 \ \exists N \in \mathbb{N} \text{ such that } n \geq N \Rightarrow |a_n - b| < \epsilon.$$

If the limit b exists it is not hard to see (Exercise 20) that it is unique, and we write

$$\lim_{n \to \infty} a_n = b \text{ or } a_n \to b.$$

Suppose that $\lim_{n \to \infty} a_n = b$. Since all the numbers a_n are eventually near b they are all near each other; i.e., every convergent sequence obeys a **Cauchy condition**:

$$\forall \epsilon > 0 \ \exists N \in \mathbb{N} \text{ such that if } n, k \geq N \text{ then } |a_n - a_k| < \epsilon.$$

The converse of this fact is a fundamental property of $\mathbb{R}$.

5 Theorem $\mathbb{R}$ *is* **complete** *with respect to Cauchy sequences in the sense that if* (a_n) *is a sequence of real numbers which obeys a Cauchy condition then it converges to a limit in* $\mathbb{R}$.

Proof First we show that (a_n) is bounded. Taking $\epsilon = 1$ in the Cauchy condition implies there is an N such that for all $n, k \geq N$ we have $|a_n - a_k| < 1$. Take K large enough that $-K \leq a_1, \ldots, a_N \leq K$. Set $M = K + 1$. Then for all n we have

$$-M < a_n < M,$$

which shows that the sequence is bounded.

Define a set X as

$$X = \{x \in \mathbb{R} : \exists \text{ infinitely many } n \text{ such that } a_n \geq x\}.$$

$-M \in X$ since for *all* n we have $a_n > -M$, while $M \notin X$ since *no* x_n is $\geq M$. Thus X is a nonempty subset of $\mathbb{R}$ which is bounded above by M. The least upper bound property applies to X and we have $b = \text{l.u.b.} X$ with $-M \leq b \leq M$.

We claim that a_n converges to b as $n \to \infty$. Given $\epsilon > 0$ we must show there is an N such that for all $n \geq N$ we have $|a_n - b| < \epsilon$. Since (a_n) is Cauchy and $\epsilon/2$ is positive there does exist an N such that if $n, k \geq N$ then

$$|a_n - a_k| < \frac{\epsilon}{2}.$$

Since $b - \epsilon/2$ is less than b it is not an upper bound for X, so there is $x \in X$ with $b - \epsilon/2 \leq x$. For infinitely many n we have $a_n \geq x$. Since $b + \epsilon/2 > b$ it does not belong to X, and therefore for only finitely many n do we have $a_n > b + \epsilon/2$. Thus, for infinitely many n we have

$$b - \frac{\epsilon}{2} \leq x \leq a_n \leq b + \frac{\epsilon}{2}.$$

Since there are infinitely many of these n there are infinitely many that are $\geq N$. Pick one, say a_{n_0} with $n_0 \geq N$ and $b - \epsilon/2 \leq a_{n_0} \leq b + \epsilon/2$. Then for *all* $n \geq N$ we have

$$|a_n - b| \ \leq \ |a_n - a_{n_0}| + |a_{n_0} - b| \ < \ \frac{\epsilon}{2} + \frac{\epsilon}{2} \ = \ \epsilon$$

which completes the verification that (a_n) converges. See Figure 4. □

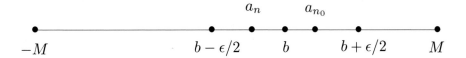

Figure 4 For all $n \geq N$ we have $|a_n - b| < \epsilon$.

Restating Theorem 5 gives the

6 Cauchy Convergence Criterion *A sequence (a_n) in $\mathbb{R}$ converges if and only if*

$$\forall \epsilon > 0 \ \exists N \in \mathbb{N} \ such \ that \ \ n, k \geq N \ \Rightarrow \ |a_n - a_k| < \epsilon.$$

Further description of $\mathbb{R}$

The elements of $\mathbb{R} \setminus \mathbb{Q}$ are irrational numbers. If x is irrational and r is rational then $y = x + r$ is irrational. For if y is rational then so is $y - r = x$, the difference of rationals being rational. Similarly, if $r \neq 0$ then rx is irrational. It follows that the reciprocal of an irrational number is irrational. From these observations we will show that the rational and irrational numbers are thoroughly mixed up with each other.

Let $a < b$ be given in $\mathbb{R}$. Define the intervals (a, b) and $[a, b]$ as

$$(a, b) \ = \ \{x \in \mathbb{R} : a < x < b\}$$
$$[a, b] \ = \ \{x \in \mathbb{R} : a \leq x \leq b\}.$$

7 Theorem *Every interval (a, b), no matter how small, contains both rational and irrational numbers. In fact it contains infinitely many rational numbers and infinitely many irrational numbers.*

Proof Think of a, b as cuts $a = A|A'$, $b = B|B'$. The fact that $a < b$ implies the set $B \setminus A$ is a nonempty set of rational numbers. Choose a rational $r \in B \setminus A$. Since B has no largest element, there is a rational s with $a < r < s < b$. Now consider the transformation

$$T : t \mapsto r + (s - r)t.$$

It sends the interval $[0, 1]$ to the interval $[r, s]$. Since r and $s - r$ are rational, T sends rationals to rationals and irrationals to irrationals. Clearly $[0, 1]$ contains infinitely many rationals, say $1/n$ with $n \in \mathbb{N}$, so $[r, s]$ contains infinitely many rationals. Also $[0, 1]$ contains infinitely many irrationals, say $1/n\sqrt{2}$ with $n \in \mathbb{N}$, so $[r, s]$ contains infinitely many irrationals. Since $[r, s]$ contains infinitely many rationals and infinitely many irrationals, the same is true of the larger interval (a, b). □

Theorem 7 expresses the fact that between any two rational numbers lies an irrational number, and between any two irrational numbers lies a rational number. This is a fact worth thinking about for it seems implausible at first. Spend some time trying to picture the situation, especially in light of the following related facts:

(a) There is no first (i.e., smallest) rational number in the interval $(0, 1)$.
(b) There is no first irrational number in the interval $(0, 1)$.
(c) There are strictly more irrational numbers in the interval $(0, 1)$ (in the cardinality sense explained in Section 4) than there are rational numbers.

The transformation in the proof of Theorem 7 shows that the real line is like rubber: stretch it out and it never breaks.

A somewhat obscure and trivial fact about $\mathbb{R}$ is its Archimedean property: for each $x \in \mathbb{R}$ there is an integer n that is greater than x. In other words, there exist arbitrarily large integers. The Archimedean property is true for $\mathbb{Q}$ since $p/q \leq |p|$. It follows that it is true for $\mathbb{R}$. Given $x = A|B$, just choose a rational number $r \in B$ and an integer $n > r$. Then $n > x$. An equivalent way to state the Archimedean property is that there exist arbitrarily small reciprocals of integers.

Mildly interesting is the existence of ordered fields for which the Archimedean property fails. One example is the field $\mathbb{R}(x)$ of rational functions with real coefficients. Each such function is of the form

$$R(x) = \frac{p(x)}{q(x)}$$

where p and q are polynomials with real coefficients and q is not the zero polynomial. (It does not matter that $q(x) = 0$ at a finite number of points.) Addition and multiplication are defined in the usual fashion of high school algebra, and it is easy to see that $\mathbb{R}(x)$ is a field. The order relation on $\mathbb{R}(x)$ is also easy to define. If $R(x) > 0$ for all sufficiently large x then we say that R is positive in $\mathbb{R}(x)$, and if $R - S$ is positive then we write $S < R$. Since a nonzero rational function vanishes (has value zero) at only finitely many $x \in \mathbb{R}$, we get trichotomy: either $R = S$, $R < S$, or $S < R$. (To be rigorous, we need to prove that the values of a rational function do not change sign for x large enough.) The other order properties are equally easy to check, and $\mathbb{R}(x)$ is an ordered field.

Is $\mathbb{R}(x)$ Archimedean? That is, given $R \in \mathbb{R}(x)$, does there exist a natural number $n \in \mathbb{R}(x)$ such that $R < n$? (A number n is the rational function whose numerator is the constant polynomial $p(x) = n$, a polynomial of degree zero, and whose denominator is the constant polynomial $q(x) = 1$.) The answer is "no." Take $R(x) = x/1$. The numerator is x and the denominator is 1. Clearly we have $n < x$, not the opposite, so $\mathbb{R}(x)$ fails to be Archimedean.

The same remarks hold for any positive rational function $R = p(x)/q(x)$ where the degree of p exceeds the degree of q. In $\mathbb{R}(x)$, R is never less than a natural number. (You might ask yourself: exactly which rational functions are less than n?)

The ϵ-principle

Finally let us note a nearly trivial principle that turns out to be invaluable in deriving inequalities and equalities in $\mathbb{R}$.

8 Theorem (ϵ-**principle**) *If a, b are real numbers and if for each $\epsilon > 0$ we have $a \leq b + \epsilon$ then $a \leq b$. If x, y are real numbers and for each $\epsilon > 0$ we have $|x - y| \leq \epsilon$ then $x = y$.*

Proof Trichotomy implies that either $a \leq b$ or $a > b$. In the latter case we can choose ϵ with $0 < \epsilon < a - b$ and get the absurdity

$$\epsilon < a - b \leq \epsilon.$$

Hence $a \leq b$. Similarly, if $x \neq y$ then choosing ϵ with $0 < \epsilon < |x - y|$ gives the contradiction $\epsilon < |x - y| \leq \epsilon$. Hence $x = y$. See also Exercise 12.   □

3 Euclidean Space

Given sets A and B, the **Cartesian product** of A and B is the set $A \times B$ of all ordered pairs (a, b) such that $a \in A$ and $b \in B$. (The name comes from Descartes who pioneered the idea of the xy-coordinate system in geometry.) See Figure 5.

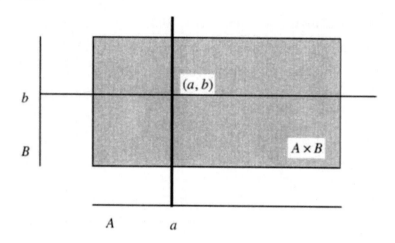

Figure 5 The Cartesian product $A \times B$

The Cartesian product of $\mathbb{R}$ with itself m times is denoted $\mathbb{R}^m$. Elements of $\mathbb{R}^m$ are vectors, ordered m-tuples of real numbers $(x_1, \ldots, x_m)$. In this terminology real numbers are called scalars and $\mathbb{R}$ is called the scalar field. When vectors are added, subtracted, and multiplied by scalars according to the rules

$$
\begin{aligned}
(x_1, \ldots, x_m) + (y_1, \ldots, y_m) &= (x_1 + y_1, \ldots, x_m + y_m) \\
(x_1, \ldots, x_m) - (y_1, \ldots, y_m) &= (x_1 - y_1, \ldots, x_m - y_m) \\
c(x_1, \ldots, x_m) &= (cx_1, \ldots, cx_m)
\end{aligned}
$$

then these operations obey the natural laws of linear algebra: commutativity, associativity, etc. There is another operation defined on $\mathbb{R}^m$, the **dot product** (also called the scalar product or inner product). The dot product of $x = (x_1, \ldots, x_m)$ and $y = (y_1, \ldots, y_m)$ is

$$
\langle x, y \rangle = x_1 y_1 + \ldots + x_m y_m.
$$

Remember: the dot product of two vectors is a scalar, not a vector. The dot product operation is bilinear, symmetric, and positive definite; i.e., for any vectors $x, y, z \in \mathbb{R}^m$

and any $c \in \mathbb{R}$ we have

$$
\begin{aligned}
\langle x, y + cz \rangle &= \langle x, y \rangle + c \langle x, z \rangle \\
\langle x, y \rangle &= \langle y, x \rangle \\
\langle x, x \rangle &\geq 0 \text{ and } \langle x, x \rangle = 0 \text{ if and only if } x \text{ is the zero vector.}
\end{aligned}
$$

The **length** or **magnitude** of a vector $x \in \mathbb{R}^m$ is defined to be

$$
|x| = \sqrt{\langle x, x \rangle} = \sqrt{x_1^2 + \ldots + x_m^2}.
$$

See Exercise 16 which legalizes taking roots. Expressed in coordinate-free language, the basic fact about the dot product is the

9 Cauchy-Schwarz Inequality *For all $x, y \in \mathbb{R}^m$ we have $\langle x, y \rangle \leq |x||y|$.*

Proof Tricky! For any vectors x, y consider the new vector $w = x + ty$, where $t \in \mathbb{R}$ is a varying scalar. Then

$$
Q(t) = \langle w, w \rangle = \langle x + ty, \ x + ty \rangle
$$

is a real-valued function of t. In fact, $Q(t) \geq 0$ since the dot product of any vector with itself is nonnegative. The bilinearity properties of the dot product imply that

$$
Q(t) = \langle x, x \rangle + 2t \langle x, y \rangle + t^2 \langle y, y \rangle = c + bt + at^2
$$

is a quadratic function of t. Nonnegative quadratic functions of $t \in \mathbb{R}$ have nonpositive discriminants, $b^2 - 4ac \leq 0$. For if $b^2 - 4ac > 0$ then $Q(t)$ has two real roots, between which $Q(t)$ is negative. See Figure 6.

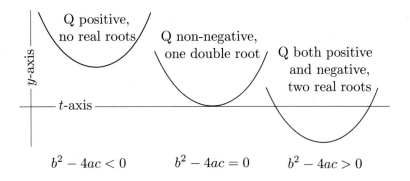

Figure 6 Quadratic graphs

But $b^2 - 4ac \leq 0$ means that $4\langle x, y \rangle^2 - 4\langle x, x \rangle \langle y, y \rangle \leq 0$, i.e.,

$$\langle x, y \rangle^2 \leq \langle x, x \rangle \langle y, y \rangle.$$

Taking the square root of both sides gives $\langle x, y \rangle \leq \sqrt{\langle x, x \rangle}\sqrt{\langle y, y \rangle} = |x||y|$. (We use Exercise 17 here and below without further mention.)   □

The Cauchy-Schwarz inequality implies easily the **Triangle Inequality for vectors**: For all $x, y \in \mathbb{R}^m$ we have

$$|x + y| \leq |x| + |y|.$$

For $|x + y|^2 = \langle x + y, \ x + y \rangle = \langle x, x \rangle + 2\langle x, y \rangle + \langle y, y \rangle$. By Cauchy-Schwarz, $2\langle x, y \rangle \leq 2|x||y|$. Thus,

$$|x + y|^2 \leq |x|^2 + 2|x||y| + |y|^2 = (|x| + |y|)^2.$$

Taking the square root of both sides gives the result.

The **Euclidean distance** between vectors $x, y \in \mathbb{R}^m$ is defined as the length of their difference,

$$|x - y| = \sqrt{\langle x - y, \ x - y \rangle} = \sqrt{(x_1 - y_1)^2 + \ldots + (x_m - y_m)^2}.$$

From the Triangle Inequality for vectors follows the **Triangle Inequality for distance**. For all $x, y, z \in \mathbb{R}^m$ we have

$$|x - z| \leq |x - y| + |y - z|.$$

To prove it, think of $x - z$ as the vector sum $(x - y) + (y - z)$ and apply the Triangle Inequality for vectors. See Figure 7.

Geometric intuition in Euclidean space can carry you a long way in real analysis, especially in being able to forecast whether a given statement is true or not. Your geometric intuition will grow with experience and contemplation. We begin with some vocabulary.

In real analysis, vectors in $\mathbb{R}^m$ are referred to as points in $\mathbb{R}^m$. The jth coordinate of the point $(x_1, \ldots, x_m)$ is the number x_j appearing in the jth position. The jth coordinate axis is the set of points $x \in \mathbb{R}^m$ whose k^{th} coordinates are zero for all $k \neq j$. The origin of $\mathbb{R}^m$ is the zero vector, $(0, \ldots, 0)$. The **first orthant** of $\mathbb{R}^m$ is the set of points $x \in \mathbb{R}^m$ all of whose coordinates are nonnegative. When $m = 2$, the first orthant is the first quadrant. The **integer lattice** is the set $\mathbb{Z}^m \subset \mathbb{R}^m$ of

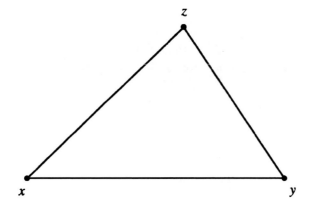

Figure 7 How the Triangle Inequality gets its name

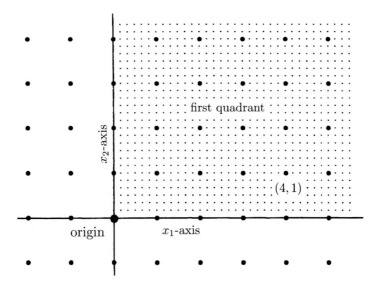

Figure 8 The integer lattice and first quadrant

ordered m-tuples of integers. The integer lattice is also called the **integer grid**. See Figure 8.

A **box** is a Cartesian product of intervals

$$[a_1, b_1] \times \cdots \times [a_m, b_m]$$

in $\mathbb{R}^m$. (A box is also called a **rectangular parallelepiped**.) The **unit cube** in $\mathbb{R}^m$ is the box $[0,1]^m = [0,1] \times \cdots \times [0,1]$. See Figure 9.

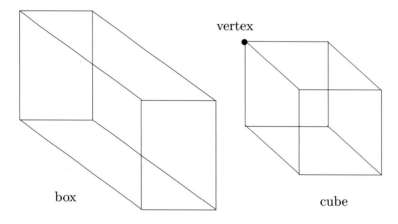

Figure 9 A box and a cube

The **unit ball** and **unit sphere** in $\mathbb{R}^m$ are the sets

$$
\begin{aligned}
B^m &= \{x \in \mathbb{R}^m : |x| \le 1\} \\
S^{m-1} &= \{x \in \mathbb{R}^m : |x| = 1\}.
\end{aligned}
$$

The reason for the exponent $m - 1$ is that the sphere is $(m - 1)$-dimensional as an object in its own right although it does *live* in m-space. In 3-space, the surface of a ball is a two-dimensional film, the 2-sphere S^2. See Figure 10.

A set $E \subset \mathbb{R}^m$ is **convex** if for each pair of points $x, y \in E$, the straight line segment between x and y is also contained in E. The unit ball is an example of a convex set. To see this, take any two points in B^m and draw the segment between them. It "obviously" lies in B^m. See Figure 11.

To give a mathematical proof, it is useful to describe the line segment between x and y with a formula. The straight line determined by distinct points $x, y \in \mathbb{R}^m$ is the set of all linear combinations $sx + ty$ where $s + t = 1$, and the line segment is the set of these linear combinations where s and t are ≤ 1. Such linear combinations

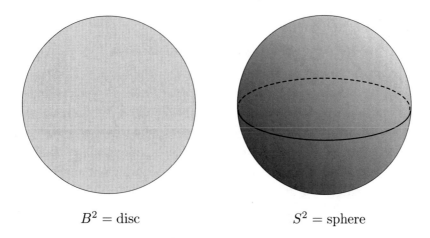

B^2 = disc S^2 = sphere

Figure 10 A 2-disc B^2 with its boundary circle, and a 2-sphere S^2 with its equator

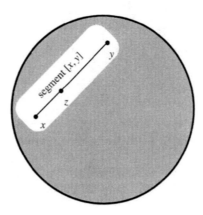

Figure 11 Convexity of the ball

$sx + ty$ with $s + t = 1$ and $0 \leq s, t \leq 1$ are called **convex combinations**. The line segment is denoted as $[x, y]$. (This notation is consistent with the interval notation $[a, b]$. See Exercise 27.) Now if $x, y \in B^m$ and $sx + ty = z$ is a convex combination of x and y then, using the Cauchy-Schwarz Inequality and the fact that $2st \geq 0$, we get

$$
\begin{aligned}
\langle z, z \rangle &= s^2 \langle x, x \rangle + 2st \langle x, y \rangle + t^2 \langle y, y \rangle \\
&\leq s^2 |x|^2 + 2st |x||y| + t^2 |y|^2 \\
&\leq s^2 + 2st + t^2 = (s + t)^2 = 1.
\end{aligned}
$$

Taking the square root of both sides gives $|z| \leq 1$, which proves convexity of the ball.

Inner product spaces

An **inner product** on a vector space V is an operation $\langle \, , \, \rangle$ on pairs of vectors in V that satisfies the same conditions that the dot product in Euclidean space does: Namely, bilinearity, symmetry, and positive definiteness. A vector space equipped with an inner product is an **inner product space**. The discriminant proof of the Cauchy-Schwarz Inequality is valid for any inner product defined on any real vector space, even if the space is infinite-dimensional and the standard coordinate proof would make no sense. For the discriminant proof uses only the inner product properties, and not the particular definition of the dot product in Euclidean space.

$\mathbb{R}^m$ has dimension m because it has a basis $e_1, \ldots, e_m$. Other vector spaces are more general. For example, let $C([a,b], \mathbb{R})$ denote the set of all of continuous real-valued functions defined on the interval $[a, b]$. (See Section 6 or your old calculus book for the definition of continuity.) It is a vector space in a natural way, the sum of continuous functions being continuous and the scalar multiple of a continuous function being continuous. The vector space $C([a,b], \mathbb{R})$, however, has no finite basis. It is infinite-dimensional. Even so, there is a natural inner product,

$$\langle f, g \rangle \;=\; \int_a^b f(x) g(x) \, dx.$$

Cauchy-Schwarz applies to this inner product, just as to any inner product, and we infer a general integral inequality valid for any two continuous functions,

$$\int_a^b f(x) g(x) \, dx \;\leq\; \sqrt{\int_a^b f(x)^2 \, dx} \, \sqrt{\int_a^b g(x)^2 \, dx}.$$

It would be challenging to prove such an inequality from scratch, would it not? See also the first paragraph of the next chapter.

A **norm** on a vector space V is any function $|\ \ | : V \to \mathbb{R}$ with the three properties of vector length: Namely, if $v, w \in V$ and $\lambda \in \mathbb{R}$ then

$$|v| \geq 0 \text{ and } |v| = 0 \text{ if and only if } v = 0,$$
$$|\lambda v| = |\lambda| \, |v| ,$$
$$|v + w| \leq |v| + |w| .$$

An inner product $\langle \, , \, \rangle$ defines a norm as $|v| = \sqrt{\langle v, v \rangle}$, but not all norms come from inner products. The unit sphere $\{v \in V : \langle v, v \rangle = 1\}$ for every inner product is smooth (has no corners) while for the norm

$$|v|_{\max} = \max\{|v_1|, |v_2|\}$$

defined on $v = (v_1, v_2) \in \mathbb{R}^2$, the unit sphere is the perimeter of the square $\{(v_1, v_2) \in \mathbb{R}^2 : |v_1| \leq 1 \text{ and } |v_2| \leq 1\}$. It has corners and so it does not arise from an inner product. See Exercises 46, 47, and the Manhattan metric on page 76.

The simplest Euclidean space beyond $\mathbb{R}$ is the plane $\mathbb{R}^2$. Its xy-coordinates can be used to define a multiplication,

$$(x, y) \bullet (x', y') = (xx' - yy', \ xy' + x'y).$$

The point $(1, 0)$ corresponds to the multiplicative unit element 1, while the point $(0, 1)$ corresponds to $i = \sqrt{-1}$, which converts the plane to the field $\mathbb{C}$ of complex numbers. Complex analysis is the study of functions of a complex variable, i.e., functions $f(z)$ where z and $f(z)$ lie in $\mathbb{C}$. Complex analysis is the good twin and real analysis the evil one: beautiful formulas and elegant theorems seem to blossom spontaneously in the complex domain, while toil and pathology rule the reals. Nevertheless, complex analysis relies more on real analysis than the other way around.

4 Cardinality

Let A and B be sets. A **function** $f : A \rightarrow B$ is a rule or mechanism which, when presented with any element $a \in A$, produces an element $b = f(a)$ of B. It need not be defined by a formula. Think of a function as a device into which you feed elements of A and out of which pour elements of B. See Figure 12. We also call f a **mapping**

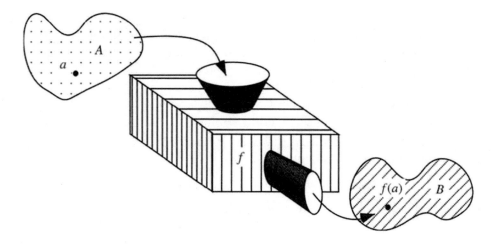

Figure 12 The function f as a machine

or a **map** or a **transformation**. The set A is the **domain** of the function and B is its **target**, also called its **codomain**. The **range** or **image** of f is the subset of the target

$$\{b \in B : \text{ there exists at least one element } a \in A \text{ with } f(a) = b\}.$$

See Figure 13.

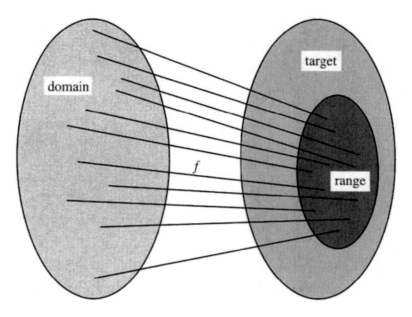

Figure 13 The domain, target, and range of a function

Try to write f instead of $f(x)$ to denote a function. The function is the device which when confronted with input x produces output $f(x)$. The function is the device, not the output.

Think also of a function dynamically. At time zero all the elements of A are sitting peacefully in A. Then the function applies itself to them and throws them into B. At time one all the elements that were formerly in A are now transferred into B. Each $a \in A$ gets sent to some element $f(a) \in B$.

A mapping $f : A \to B$ is an **injection** (or is **one-to-one**) if for each pair of distinct elements $a, a' \in A$, the elements $f(a), f(a')$ are distinct in B. That is,

$$a \neq a' \quad \Rightarrow \quad f(a) \neq f(a').$$

The mapping f is a **surjection** (or is **onto**) if for each $b \in B$ there is at least one $a \in A$ such that $f(a) = b$. That is, the range of f is B.

A mapping is a **bijection** if it is both injective and surjective. It is one-to-one and onto. If $f : A \to B$ is a bijection then the inverse map $f^{-1} : B \to A$ is a bijection where $f^{-1}(b)$ is by definition the unique element $a \in A$ such that $f(a) = b$.

The **identity map** of any set to itself is the bijection that takes each $a \in A$ and sends it to itself, $\mathrm{id}(a) = a$.

If $f : A \to B$ and $g : B \to C$ then the **composite** $g \circ f : A \to C$ is the function that sends $a \in A$ to $g(f(a)) \in C$. If f and g are injective then so is $g \circ f$, while if f and g are surjective then so is $g \circ f$,

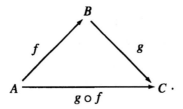

In particular the composite of bijections is a bijection. If there is a bijection from A onto B then A and B are said to have **equal cardinality**,[†] and we write $A \sim B$. The relation $\sim$ is an equivalence relation. That is,

(a) $A \sim A$.

(b) $A \sim B$ implies $B \sim A$.

(c) $A \sim B \sim C$ implies $A \sim C$.

(a) follows from the fact that the identity map bijects A to itself. (b) follows from the fact that the inverse of a bijection $A \to B$ is a bijection $B \to A$. (c) follows from the fact that the composite of bijections f and g is a bijection $g \circ f$.

A set S is

> **finite** if it is empty or for some $n \in \mathbb{N}$ we have $S \sim \{1, \ldots, n\}$.
>
> **infinite** if it is not finite.
>
> **denumerable** if $S \sim \mathbb{N}$.
>
> **countable** if it is finite or denumerable.
>
> **uncountable** if it is not countable.

[†]The word "cardinal" indicates the number of elements in the set. The cardinal numbers are $0, 1, 2, \ldots$ The first infinite cardinal number is **aleph null**, $\aleph_0$. One says the $\mathbb{N}$ has $\aleph_0$ elements. A mystery of math is the **Continuum Hypothesis** which states that $\mathbb{R}$ has cardinality $\aleph_1$, the second infinite cardinal. Equivalently, if $\mathbb{N} \subset S \subset \mathbb{R}$, the Continuum Hypothesis asserts that $S \sim \mathbb{N}$ or $S \sim \mathbb{R}$. No intermediate cardinalities exist. You can pursue this issue in Paul Cohen's book, *Set Theory and the Continuum Hypothesis*.

We also write $\operatorname{card} A = \operatorname{card} B$ and $\#A = \#B$ when A, B have equal cardinality.

If S is denumerable then there is a bijection $f : \mathbb{N} \to S$, and this gives a way to list the elements of S as $s_1 = f(1)$, $s_2 = f(2)$, $s_3 = f(3)$, etc. Conversely, if a set S is presented as an infinite list (without repetition) $S = \{s_1, s_2, s_3, \ldots\}$, then it is denumerable: Define $f(k) = s_k$ for all $k \in \mathbb{N}$. In brief, denumerable = listable.

Let's begin with a truly remarkable cardinality result, that although $\mathbb{N}$ and $\mathbb{R}$ are both infinite, $\mathbb{R}$ is more infinite than $\mathbb{N}$. Namely,

10 Theorem $\mathbb{R}$ *is uncountable.*

Proof There are other proofs of the uncountability of $\mathbb{R}$, but none so beautiful as this one. It is due to Cantor. I assume that you accept the fact that each real number x has a decimal expansion, $x = N.x_1x_2x_3\ldots$, and it is uniquely determined by x if one agrees never to terminate the expansion with an infinite string of 9s. (See also Exercise 18.) We want to prove that $\mathbb{R}$ is uncountable. Suppose it is not uncountable. Then it is countable and, being infinite, it must be denumerable. Accordingly let $f : \mathbb{N} \to \mathbb{R}$ be a bijection. Using f, we list the elements of $\mathbb{R}$ along with their decimal expansions as an array, and consider the digits x_{ii} that occur along the diagonal in this array. See Figure 14.

$$
\begin{array}{ccccccccc}
f(1) & = & N_1 & x_{11} & x_{12} & x_{13} & x_{14} & x_{15} & x_{16} & x_{17} \\
f(2) & = & N_2 & x_{21} & x_{22} & x_{23} & x_{24} & x_{25} & x_{26} & x_{27} \\
f(3) & = & N_3 & x_{31} & x_{32} & x_{33} & x_{34} & x_{35} & x_{36} & x_{37} \\
f(4) & = & N_4 & x_{41} & x_{42} & x_{43} & x_{44} & x_{45} & x_{46} & x_{47} \\
f(5) & = & N_5 & x_{51} & x_{52} & x_{53} & x_{54} & x_{55} & x_{56} & x_{57} \\
f(6) & = & N_6 & x_{61} & x_{62} & x_{63} & x_{64} & x_{65} & x_{66} & x_{67} \\
f(7) & = & N_7 & x_{71} & x_{72} & x_{73} & x_{74} & x_{75} & x_{76} & x_{77} \\
\vdots & & & & & & & & &
\end{array}
$$

Figure 14 Cantor's diagonal method

For each i, choose a digit y_i such that $y_i \neq x_{ii}$ and $y_i \neq 9$. Where is the number $y = 0.y_1y_2y_3\ldots$? Is it $f(1)$? No, because the first digit in the decimal expansion of

$f(1)$ is x_{11} and $y_1 \neq x_{11}$. Is it $f(2)$? No, because the second digit in the decimal expansion of $f(2)$ is x_{22} and $y_2 \neq x_{22}$. Is it $f(k)$? No, because the kth digit in the decimal expansion of $f(k)$ is x_{kk} and $y_k \neq x_{kk}$. Nowhere in the list do we find y. Nowhere! Therefore the list could not account for every real number, and $\mathbb{R}$ must have been uncountable. □

11 Corollary $[a, b]$ and (a, b) are uncountable.

Proof There are bijections from (a, b) onto $(-1, 1)$ onto the unit semicircle onto $\mathbb{R}$ shown in Figure 15. The composite f bijects (a, b) onto $\mathbb{R}$, so (a, b) is uncountable.

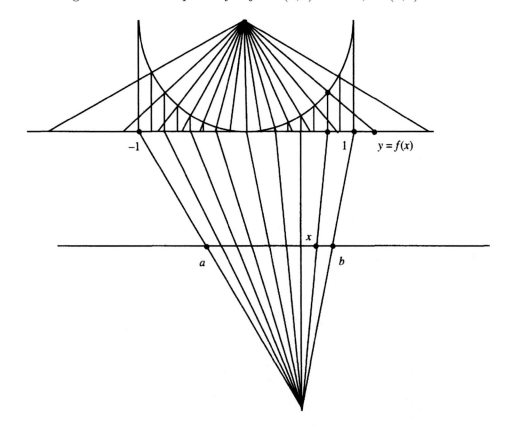

Figure 15 Equicardinality of (a, b), $(-1, 1)$, and $\mathbb{R}$

Since $[a, b]$ contains (a, b), it too is uncountable. □

The remaining results in this section are of a more positive flavor.

12 Theorem *Each infinite set S contains a denumerable subset.*

Proof Since S is infinite it is nonempty and contains an element s_1. Since S is infinite the set $S \setminus \{s_1\} = \{s \in S : s \neq s_1\}$ is nonempty and there exists $s_2 \in S \setminus \{s_1\}$. Since S is an infinite set, $S \setminus \{s_1, s_2\} = \{s \in S : s \neq s_1, s_2\}$ is nonempty and there exists $s_3 \in S \setminus \{s_1, s_2\}$. Continuing this way gives a list (s_n) of distinct elements of S. The set of these elements forms a denumerable subset of S. □

13 Theorem *An infinite subset A of a denumerable set B is denumerable.*

Proof There exists a bijection $f : \mathbb{N} \to B$. Each element of A appears exactly once in the list $f(1), f(2), f(3), \ldots$ of B. Define $g(k)$ to be the kth element of A appearing in the list. Since A is infinite, $g(k)$ is defined for all $k \in \mathbb{N}$. Thus $g : \mathbb{N} \to A$ is a bijection and A is denumerable. □

14 Corollary *The sets of even integers and of prime integers are denumerable.*

Proof They are infinite subsets of $\mathbb{N}$ which is denumerable. □

15 Theorem $\mathbb{N} \times \mathbb{N}$ *is denumerable.*

Proof Think of $\mathbb{N} \times \mathbb{N}$ as an $\infty \times \infty$ matrix and walk along the successive counter-diagonals. See Figure 16. This gives a list

$$(1, 1), (2, 1), (1, 2), (3, 1), (2, 2), (1, 3), (4, 1), (3, 2), (2, 3), (1, 4), (5, 1), \ldots$$

of $\mathbb{N} \times \mathbb{N}$ and proves that $\mathbb{N} \times \mathbb{N}$ is denumerable. □

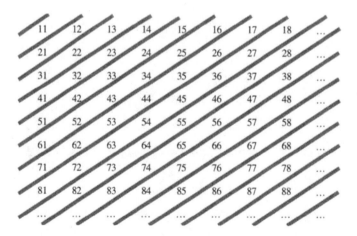

Figure 16 Counter-diagonals in an $\infty \times \infty$ matrix

16 Corollary *The Cartesian product of denumerable sets A and B is denumerable.*

Proof $\mathbb{N} \sim \mathbb{N} \times \mathbb{N} \sim A \times B.$ □

17 Theorem *If $f : \mathbb{N} \to B$ is a surjection and B is infinite then B is denumerable.*

Proof For each $b \in B$, the set $\{k \in \mathbb{N} : f(k) = b\}$ is nonempty and hence contains a smallest element; say $h(b) = k$ is the smallest integer that is sent to b by f. Clearly, if $b, b' \in B$ and $b \neq b'$ then $h(b) \neq h(b')$. That is, $h : B \to \mathbb{N}$ is an injection which bijects B to $hB \subset \mathbb{N}$. Since B is infinite, so is hB. By Theorem 13, hB is denumerable and therefore so is B.   □

18 Corollary *The denumerable union of denumerable sets is denumerable.*

Proof Suppose that $A_1, A_2, \ldots$ is a sequence of denumerable sets. List the elements of A_i as $a_{i1}, a_{i2}, \ldots$ and define

$$
\begin{aligned}
f : \mathbb{N} \times \mathbb{N} \quad &\to \quad A = \bigcup A_i \\
(i, j) \quad &\mapsto \quad a_{ij}
\end{aligned}
$$

Clearly f is a surjection. According to Theorem 15, there is a bijection $g : \mathbb{N} \to \mathbb{N} \times \mathbb{N}$. The composite $f \circ g$ is a surjection $\mathbb{N} \to A$. Since A is infinite, Theorem 17 implies it is denumerable.   □

19 Corollary $\mathbb{Q}$ *is denumerable.*

Proof $\mathbb{Q}$ is the denumerable union of the denumerable sets $A_q = \{p/q : p \in \mathbb{Z}\}$ as q ranges over $\mathbb{N}$.   □

20 Corollary *For each $m \in \mathbb{N}$ the set $\mathbb{Q}^m$ is denumerable.*

Proof Apply the induction principle. If $m = 1$ then the previous corollary states that $\mathbb{Q}^1$ is denumerable. Knowing inductively that $\mathbb{Q}^{m-1}$ is denumerable and $\mathbb{Q}^m = \mathbb{Q}^{m-1} \times \mathbb{Q}$, the result follows from Corollary 16.   □

Combination laws for countable sets are similar to those for denumerable sets. As is easily checked,

Every subset of a countable set is countable.

A countable set that contains a denumerable subset is denumerable.

The Cartesian product of finitely many countable sets is countable.

The countable union of countable sets is countable.

5* Comparing Cardinalities

The following result gives a way to conclude that two sets have the same cardinality. Roughly speaking the condition is that $\operatorname{card} A \leq \operatorname{card} B$ and $\operatorname{card} B \leq \operatorname{card} A$.

21 Schroeder-Bernstein Theorem *If A, B are sets and $f : A \to B$, $g : B \to A$ are injections then there exists a bijection $h : A \to B$.*

Proof-sketch Consider the dynamic Venn diagram, Figure 17. The disc labeled gfA

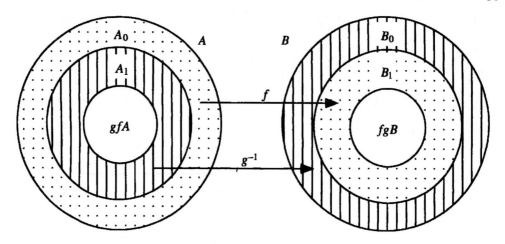

Figure 17 Pictorial proof of the Schroeder-Bernstein Theorem

is the image of A under the map $g \circ f$. It is a subset of A. The ring between A and gfA divides into two subrings. A_0 is the set of points in A that do not lie in the image of g, while A_1 is the set points in the image of g that do not lie in gfA. Similarly, B_0 is the set of points in B that do not lie in fA, while B_1 is the set of points in fA that do not lie in fgB. There is a natural bijection h from the pair of rings $A_0 \cup A_1 = A \setminus gfA$ to the pair of rings $B_0 \cup B_1 = B \setminus fgB$. It equals f on the outer ring $A_0 = A \setminus gB$ and it is g^{-1} on the inner ring $A_1 = gB \setminus gfA$. (The map g^{-1} is not defined on all of A, but it is defined on the set gB.) In this notation, h sends A_0 onto B_1 and sends A_1 onto B_0. It switches the indices. Repeat this on the next pair of rings for A and B. That is, look at gfA instead of A and fgB instead of B. The next two rings in A, B are

$$
\begin{aligned}
A_2 &= gfA \setminus gfgB & A_3 &= gfgB \setminus gfgfA \\
B_2 &= fgB \setminus fgfA & B_3 &= fgfA \setminus fgfgB.
\end{aligned}
$$

Send A_2 to B_3 by f and A_3 to B_2 by g^{-1}. The rings A_i are disjoint, and so are the rings B_i, so repetition gives a bijection

$$\phi : \bigsqcup A_i \quad \rightarrow \quad \bigsqcup B_i,$$

($\bigsqcup$ indicates disjoint union) defined by

$$\phi(x) = \begin{cases} f(x) & \text{if } x \in A_i \text{ and } i \text{ is even} \\ g^{-1}(x) & \text{if } x \in A_i \text{ and } i \text{ is odd.} \end{cases}$$

Let $A_* = A \setminus (\bigsqcup A_i)$ and $B_* = B \setminus (\bigsqcup B_i)$ be the rest of A and B. Then f bijects A_* to B_* and ϕ extends to a bijection $h : A \rightarrow B$ defined by

$$h(x) = \begin{cases} \phi(x) & \text{if } x \in \bigsqcup A_i \\ f(x) & \text{if } x \in A_*. \end{cases} \qquad \square$$

A supplementary aid in understanding the Schroeder Bernstein proof is the following crossed ladder diagram, Figure 18.

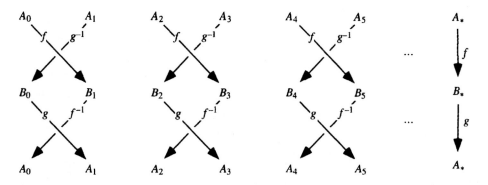

Figure 18 Diagramatic proof of the Schroeder-Bernstein Theorem

Exercise 36 asks you to show directly that $(a, b) \sim [a, b]$. This makes sense since $(a, b) \subset [a, b] \subset \mathbb{R}$ and $(a, b) \sim \mathbb{R}$ should certainly imply $(a, b) \sim [a, b] \sim \mathbb{R}$. The Schroeder-Bernstein theorem gives a quick indirect solution to the exercise. The inclusion map $i : (a, b) \hookrightarrow [a, b]$ sending x to x injects (a, b) into $[a, b]$, while the function $j(x) = x/2 + (a + b)/4$ injects $[a, b]$ into (a, b). The existence of the two injections implies by the Schroeder-Bernstein Theorem that there is a bijection $(a, b) \sim [a, b]$.

6* The Skeleton of Calculus

The behavior of a continuous function defined on an interval $[a, b]$ is at the root of all calculus theory. Using solely the Least Upper Bound Property of the real numbers we rigorously derive the basic properties of such functions. The function $f : [a, b] \to \mathbb{R}$ is **continuous** if for each $\epsilon > 0$ and each $x \in [a, b]$ there is a $\delta > 0$ such that

$$t \in [a, b] \text{ and } |t - x| < \delta \quad \Rightarrow \quad |f(t) - f(x)| < \epsilon.$$

See Figure 19.

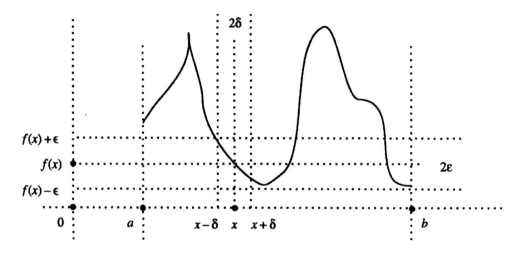

Figure 19 The graph of a continuous function of a real variable

Continuous functions are found everywhere in analysis and topology. Theorems 22, 23, and 24 present their simplest properties. Later we generalize these results to functions that are neither real valued nor dependent on a real variable. Although it is possible to give a combined proof of Theorems 22 and 23 I prefer to highlight the Least Upper Bound Property and keep them separate.

22 Theorem *The values of a continuous function defined on an interval $[a, b]$ form a bounded subset of $\mathbb{R}$. That is, there exist $m, M \in \mathbb{R}$ such that for all $x \in [a, b]$ we have $m \leq f(x) \leq M$.*

Proof For $x \in [a, b]$, let V_x be the value set of $f(t)$ as t varies from a to x,

$$V_x = \{y \in \mathbb{R} : \text{ for some } t \in [a, x] \text{ we have } y = f(t)\}.$$

Set

$$X = \{x \in [a, b] : V_x \text{ is a bounded subset of } \mathbb{R}\}.$$

We must prove that $b \in X$. Clearly $a \in X$ and b is an upper bound for X. Since X is nonempty and bounded above, there exists in $\mathbb{R}$ a least upper bound $c \leq b$ for X. Take $\epsilon = 1$ in the definition of continuity at c. There exists a $\delta > 0$ such that $|x - c| < \delta$ implies $|f(x) - f(c)| < 1$. Since c is the least upper bound for X, there exists $x \in X$ in the interval $[c - \delta, c]$. (Otherwise $c - \delta$ is a smaller upper bound for X.) Now as t varies from a to c, the value $f(t)$ varies first in the bounded set V_x and then in the bounded set $J = (f(c) - 1, f(c) + 1)$. See Figure 20.

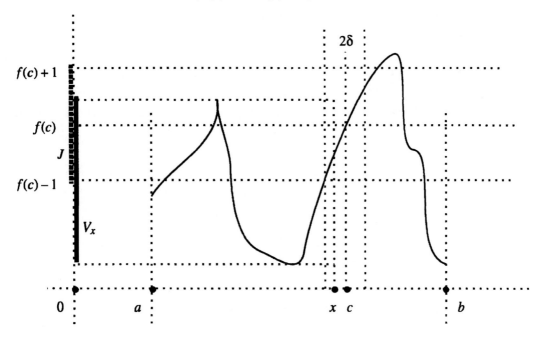

Figure 20 The value set V_x and the interval J

The union of two bounded sets is a bounded set and it follows that V_c is bounded, so $c \in X$. Besides, if $c < b$ then $f(t)$ continues to vary in the bounded set J for $t > c$, contrary to the fact that c is an upper bound for X. Thus, $c = b$, $b \in X$, and the values of f form a bounded subset of $\mathbb{R}$.   □

23 Theorem *A continuous function f defined on an interval $[a, b]$ takes on absolute minimum and absolute maximum values: For some $x_0, x_1 \in [a, b]$ and for all $x \in [a, b]$ we have*

$$f(x_0) \leq f(x) \leq f(x_1).$$

Proof Let $M = \mathrm{l.\,u.\,b.}\ f(t)$ as t varies in $[a, b]$. By Theorem 22 M exists. Consider the set $X = \{x \in [a, b] : \mathrm{l.u.b.} V_x < M\}$ where, as above, V_x is the set of values of $f(t)$ as t varies on $[a, x]$.

Case 1. $f(a) = M$. Then f takes on a maximum at a and the theorem is proved.

Case 2. $f(a) < M$. Then $X \neq \emptyset$ and we can consider the least upper bound of X, say c. If $f(c) < M$, we choose $\epsilon > 0$ with $\epsilon < M - f(c)$. By continuity at c, there exists a $\delta > 0$ such that $|t - c| < \delta$ implies $|f(t) - f(c)| < \epsilon$. Thus, l.u.b.$V_c < M$. If $c < b$ this implies there exist points t to the right of c at which l.u.b.$V_t < M$, contrary to the fact that c is an upper bound of such points. Therefore, $c = b$, which implies that $M < M$, a contradiction. Having arrived at a contradiction from the supposition that $f(c) < M$, we duly conclude that $f(c) = M$, so f assumes a maximum at c. The situation with minima is similar. $\qquad\square$

24 Intermediate Value Theorem *A continuous function defined on an interval* $[a, b]$ *takes on (or "achieves," "assumes," or "attains") all intermediate values: That is, if $f(a) = \alpha$, $f(b) = \beta$, and γ is given, $\alpha \leq \gamma \leq \beta$, then there is some $c \in [a, b]$ such that $f(c) = \gamma$. The same conclusion holds if $\beta \leq \gamma \leq \alpha$.*

The theorem is pictorially obvious. A continuous function has a graph that is a curve without break points. Such a graph can not jump from one height to another. It must pass through all intermediate heights.

Proof Set $X = \{x \in [a, b] : \text{l.u.b.}V_x \leq \gamma\}$ and $c = \text{l.u.b.}X$. Now c exists because X is nonempty (it contains a) and it is bounded above (by b). We claim that $f(c) = \gamma$, as shown in Figure 21.

To prove it we just eliminate the other two possibilities which are $f(c) < \gamma$ and $f(c) > \gamma$, by showing that each leads to a contradiction. Suppose that $f(c) < \gamma$ and take $\epsilon = \gamma - f(c)$. Continuity at c gives $\delta > 0$ such that $|t - c| < \delta$ implies $|f(t) - f(c)| < \epsilon$. That is,

$$t \in (c - \delta,\ c + \delta) \quad \Rightarrow \quad f(t) < \gamma,$$

so $c + \delta/2 \in X$, contrary to c being an upper bound of X.

Suppose that $f(c) > \gamma$ and take $\epsilon = f(c) - \gamma$. Continuity at c gives $\delta > 0$ such that $|t - c| < \delta$ implies $|f(t) - f(c)| < \epsilon$. That is,

$$t \in (c - \delta,\ c + \delta) \quad \Rightarrow \quad f(t) > \gamma,$$

so $c - \delta/2$ is an upper bound for X, contrary to c being the least upper bound for X. Since $f(c)$ is neither $< \gamma$ nor $> \gamma$ we get $f(c) = \gamma$. $\qquad\square$

A combination of Theorems 22, 23, 24, and Exercise 43 could well be called the

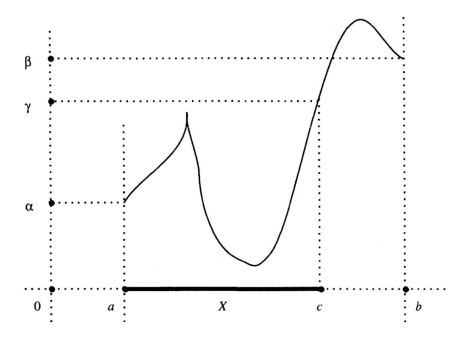

Figure 21 $x \in X$ implies that $f(x) \le \gamma$.

25 Fundamental Theorem of Continuous Functions *Every continuous real valued function of a real variable $x \in [a, b]$ is bounded, achieves minimum, intermediate, and maximum values, and is uniformly continuous.*

7* Visualizing the Fourth Dimension

A lot of real analysis takes place in $\mathbb{R}^m$ but the full m-dimensionality is rarely important. Rather, most analysis facts which are true when $m = 1, 2, 3$ remain true for $m \ge 4$. Still, I suspect you would be happier if you could visualize $\mathbb{R}^4$, $\mathbb{R}^5$, etc. Here is how to do it.

It is often said that time is *the* fourth dimension and that $\mathbb{R}^4$ should be thought of as $xyzt$-space where a point has position (x, y, z) in 3-space at time t. This is only one possible way to think of *a* fourth dimension. Instead, you can think of color as a fourth dimension. Imagine our usual 3-space with its xyz-coordinates in which points are colorless. Then imagine that you can give color to points ("paint" them) with shades of red indicating positive fourth coordinate and blue indicating negative fourth coordinate. This gives $xyzc$-coordinates. Points with equal xyz-coordinates but different colors are different points.

How is this useful? We have not used time as a coordinate, reserving it to describe motion in 4-space. Figure 22 shows two circles – the unit circle C in the horizontal xy-plane and the circle V with radius 1 and center $(1, 0, 0)$ in the vertical xz-plane. They are linked. No continuous motion can unlink them in 3-space without one

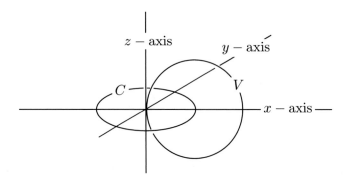

Figure 22 C and V are linked circles.

crossing the other. However, in Figure 23 you can *watch* them unlink in 4-space as follows.

Just gradually give redness to C while dragging it leftward parallel to the x-axis, until it is to the left of V. (Leave V always fixed.) Then diminish the redness of C until it becomes colorless. It ends up to the left of V and no longer links it. In formulas we can let

$$C(t) = \{(x, y, z, c) \in \mathbb{R}^4 : (x + 2t)^2 + y^2 = 1, \; z = 0, \; \text{and } c(t) = t(t - 1)\}$$

for $0 \leq t \leq 1$. See Figure 23.

The moving circle $C(t)$ never touches the stationary circle V. In particular, at time $t = 1/2$ we have $C(t) \cap V = \emptyset$. For $(-1, 0, 0, 1/4) \neq (-1, 0, 0, 0)$.

Other parameters can be used for higher dimensions. For example we could use pressure, temperature, chemical concentration, monetary value, etc. In theoretical mechanics one uses six parameters for a moving particle – three coordinates of position and three more for momentum.

Moral Choosing a new parameter as the fourth dimension (color instead of time) lets one visualize 4-space and observe motion there.

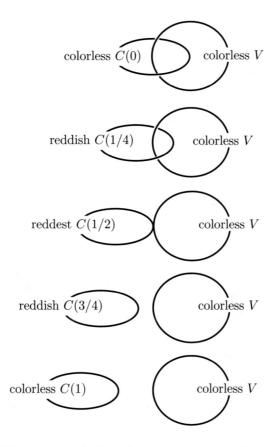

Figure 23 How to unlink linked circles using the fourth dimension

Exercises

0. Prove that for all sets A, B, C the formula

$$A \cap (B \cup C) = (A \cap B) \cup (A \cap C)$$

is true. Here is the solution written out in gory detail. *Imitate this style in writing out proofs in this course.* See also the guidelines for writing a rigorous proof on page 5. Follow them!

Hypothesis. A, B, C are sets.

Conclusion. $A \cap (B \cup C) = (A \cap B) \cup (A \cap C)$.

Proof. To prove two sets are equal we must show that every element of the first set is an element of the second set and vice versa. Referring to Figure 24, let x denote an element of the set $A \cap (B \cup C)$. It belongs to A and it belongs to B or to C. Therefore x belongs to $A \cap B$ or it belongs to $A \cap C$. Thus x belongs to the set $(A \cap B) \cup (A \cap C)$ and we have shown that every element of the first set $A \cap (B \cup C)$ belongs to the second set $(A \cap B) \cup (A \cap C)$.

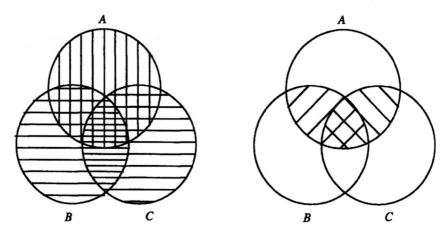

Figure 24 A is ruled vertically, B and C are ruled horizontally, $A \cap B$ is ruled diagonally, and $A \cap C$ is ruled counter-diagonally.

On the other hand let y denote an element of the set $(A \cap B) \cup (A \cap C)$. It belongs to $A \cap B$ or it belongs to $A \cap C$. Therefore it belongs to A and it belongs to B or to C. Thus y belongs to $A \cap (B \cup C)$ and we have shown that every element of the second set $(A \cap B) \cup (A \cap C)$ belongs to the first set $A \cap (B \cup C)$. Since each element of the first set belongs to the second set and each element of the second belongs to the first, the two sets are equal, $A \cap (B \cup C) = (A \cap B) \cup (A \cap C)$. QED

1. Prove that for all sets A, B, C the formula

$$A \cup (B \cap C) = (A \cup B) \cap (A \cup C)$$

is true.

2. If several sets $A, B, C, \ldots$ all are subsets of the same set X then the differences $X \setminus A$, $X \setminus B$, $X \setminus C, \ldots$ are the **complements** of $A, B, C, \ldots$ in X and are denoted $A^c, B^c, C^c, \ldots$. The symbol A^c is read "A complement."

   (a) Prove that $(A^c)^c = A$.

   (b) Prove **De Morgan's Law**: $(A \cap B)^c = A^c \cup B^c$ and derive from it the law $(A \cup B)^c = A^c \cap B^c$.

   (c) Draw Venn diagrams to illustrate the two laws.

   (d) Generalize these laws to more than two sets.

3. Recast the following English sentences in mathematics, using correct mathematical grammar. Preserve their meaning.

   (a) 2 is the smallest prime number.

   (b) The area of any bounded plane region is bisected by some line parallel to the x-axis.

   *(c) "All that glitters is not gold."

*4. What makes the following sentence ambiguous? "A death row prisoner can't have too much hope."

5. Negate the following sentences in English using correct mathematical grammar.

   (a) If roses are red, violets are blue.

   *(b) He will sink unless he swims.

6. Why is the square of an odd integer odd and the square of an even integer even? What is the situation for higher powers? [Hint: Prime factorization.]

7. (a) Why does 4 divide every even integer square?

   (b) Why does 8 divide every even integer cube?

   (c) Why can 8 never divide twice an odd cube?

   (d) Prove that the cube root of 2 is irrational.

8. Suppose that the natural number k is not a perfect n^{th} power.

   a Prove that its nth root is irrational.

   b Infer that the nth root of a natural number is either a natural number or it is irrational. It is never a fraction.

9. Let $x = A|B$, $x' = A'|B'$ be cuts in $\mathbb{Q}$. We defined

$$x + x' = (A + A') \mid \text{ rest of } \mathbb{Q}.$$

   (a) Show that although $B + B'$ is disjoint from $A + A'$, it may happen in degenerate cases that $\mathbb{Q}$ is not the union of $A + A'$ and $B + B'$.

   (b) Infer that the definition of $x + x'$ as $(A + A') \mid (B + B')$ would be incorrect.

   (c) Why did we not define $x \cdot x' = (A \cdot A') \mid \text{ rest of } \mathbb{Q}$?

10. Prove that for each cut x we have $x + (-x) = 0^*$. [This is not entirely trivial.]

11. A multiplicative inverse of a nonzero cut $x = A|B$ is a cut $y = C|D$ such that $x \cdot y = 1^*$.

(a) If $x > 0^*$, what are C and D?

(b) If $x < 0^*$, what are they?

(c) Prove that x uniquely determines y.

12. Prove that there exists no smallest positive real number. Does there exist a smallest positive rational number? Given a real number x, does there exist a smallest real number $y > x$?

13. Let $b = $ l.u.b. S, where S is a bounded nonempty subset of $\mathbb{R}$.

(a) Given $\epsilon > 0$ show that there exists an $s \in S$ with

$$b - \epsilon \le s \le b.$$

(b) Can $s \in S$ always be found so that $b - \epsilon < s < b$?

(c) If $x = A|B$ is a cut in $\mathbb{Q}$, show that $x = $ l.u.b. A.

14. Prove that $\sqrt{2} \in \mathbb{R}$ by showing that $x \cdot x = 2$ where $x = A|B$ is the cut in $\mathbb{Q}$ with $A = \{r = \mathbb{Q} : r \le 0 \text{ or } r^2 < 2\}$. [Hint: Use Exercise 13. See also Exercise 16, below.]

15. Given $y \in \mathbb{R}$, $n \in \mathbb{N}$, and $\epsilon > 0$, show that for some $\delta > 0$, if $u \in \mathbb{R}$ and $|u - y| < \delta$ then $|u^n - y^n| < \epsilon$. [Hint: Prove the inequality when $n = 1$, $n = 2$, and then do induction on n using the identity

$$u^n - y^n = (u - y)(u^{n-1} + u^{n-2}y + \ldots + y^{n-1}).]$$

16. Given $x > 0$ and $n \in \mathbb{N}$, prove that there is a unique $y > 0$ such that $y^n = x$. That is, the nth root of x exists and is unique. [Hint: Consider

$$y = \text{l. u. b.}\{s \in \mathbb{R} : s^n \le x\}.$$

Then use Exercise 15 to show that y^n can be neither $< x$ nor $> x$.]

17. Let $x, y \in \mathbb{R}$ with $x, y > 0$, and $n \in \mathbb{N}$ be given.

(a) Prove that $x < y$ if and only if $x^n < y^n$.

(b) Infer from Exercise 16 that $x < y$ if and only if the nth root of x is less than the nth root of y.

18. Prove that real numbers correspond bijectively to decimal expansions not terminating in an infinite strings of nines, as follows. The decimal expansion of $x \in \mathbb{R}$ is $N.x_1x_2 \ldots$, where N is the largest integer $\le x$, x_1 is the largest integer $\le 10(x - N)$, x_2 is the largest integer $\le 100(x - (N + x_1/10))$, and so on.

(a) Show that each x_k is a digit between 0 and 9.

(b) Show that for each k there is an $\ell \ge k$ such that $x_\ell \ne 9$.

(c) Conversely, show that for each such expansion $N.x_1x_2 \ldots$ not terminating in an infinite string of nines, the set

$$\{N, \ N + \frac{x_1}{10}, \ N + \frac{x_1}{10} + \frac{x_2}{100}, \ \ldots\}$$

        is bounded and its least upper bound is a real number x with decimal expansion $N.x_1x_2\ldots$.

    (d) Repeat the exercise with a general base in place of 10.

19. Formulate the definition of the **greatest lower bound** (g.l.b.) of a set of real numbers. State a g.l.b. property of $\mathbb{R}$ and show it is equivalent to the l.u.b. property of $\mathbb{R}$.

20. Prove that limits are unique, i.e., if (a_n) is a sequence of real numbers that converges to a real number b and also converges to a real number b', then $b = b'$.

21. Let $f : A \to B$ be a function. That is, f is some rule or device which, when presented with any element $a \in A$, produces an element $b = f(a)$ of B. The **graph** of f is the set S of all pairs $(a, b) \in A \times B$ such that $b = f(a)$.

    (a) If you are given a subset $S \subset A \times B$, how can you tell if it is the graph of some function? (That is, what are the set theoretic properties of a graph?)

    (b) Let $g : B \to C$ be a second function and consider the composed function $g \circ f : A \to C$. Assume that $A = B = C = [0, 1]$, draw $A \times B \times C$ as the unit cube in 3-space, and try to relate the graphs of f, g, and $g \circ f$ in the cube.

22. A **fixed-point** of a function $f : A \to A$ is a point $a \in A$ such that $f(a) = a$. The **diagonal** of $A \times A$ is the set of all pairs (a, a) in $A \times A$.

    (a) Show that $f : A \to A$ has a fixed-point if and only if the graph of f intersects the diagonal.

    (b) Prove that every continuous function $f : [0, 1] \to [0, 1]$ has at least one fixed-point.

    (c) Is the same true for continuous functions $f : (0, 1) \to (0, 1)$?[†]

    (d) Is the same true for discontinuous functions?

23. A rational number p/q is **dyadic** if q is a power of 2, $q = 2^k$ for some nonnegative integer k. For example, $0, 3/8, 3/1, -3/256$, are dyadic rationals, but $1/3, 5/12$ are not. A dyadic interval is $[a, b]$ where $a = p/2^k$ and $b = (p + 1)/2^k$. For example, $[.75, 1]$ is a dyadic interval but $[1, \pi]$, $[0, 2]$, and $[.25, .75]$ are not. A dyadic cube is the product of dyadic intervals having equal length. The set of dyadic rationals may be denoted as $\mathbb{Q}_2$ and the dyadic lattice as $\mathbb{Q}_2^m$.

    (a) Prove that any two dyadic squares (i.e., planar dyadic cubes) of the same size are either identical, intersect along a common edge, intersect at a common vertex, or do not intersect at all.

    (b) Show that the corresponding intersection property is true for dyadic cubes in $\mathbb{R}^m$.

[†]A question posed in this manner means that, as well as answering the question with a "yes" or a "no," you should give a proof if your answer is "yes" or a specific counterexample if your answer is "no." Also, to do this exercise you should read Theorems 22, 23, 24.

24. Given a cube in $\mathbb{R}^m$, what is the largest ball it contains? Given a ball in $\mathbb{R}^m$, what is the largest cube it contains? What are the largest ball and cube contained in a given box in $\mathbb{R}^m$?

25. (a) Given $\epsilon > 0$, show that the unit disc contains finitely many dyadic squares whose total area exceeds $\pi - \epsilon$, and which intersect each other only along their boundaries.

   **(b) Show that the assertion remains true if we demand that the dyadic squares are disjoint.

   (c) Formulate (a) in dimension $m = 3$ and $m \geq 4$.

   **(d) Do the analysis with squares and discs interchanged. That is, given $\epsilon > 0$ prove that finitely many disjoint closed discs can be drawn inside the unit square so that the total area of the discs exceeds $1 - \epsilon$. [Hint: The Pile of Sand Principle. On the first day of work, take away $1/16$ of a pile of sand. On the second day take away $1/16$ of the remaining pile of sand. Continue. What happens to the pile of sand after n days when $n \to \infty$? Instead of sand, think of your obligation to place finitely many disjoint dyadic squares (or discs) that occupy at least $1/16$ of the area of the unit disc (or unit square).]

*26. Let $b(R)$ and $s(R)$ be the number of integer unit cubes in $\mathbb{R}^m$ that intersect the ball and sphere of radius R, centered at the origin.

   (a) Let $m = 2$ and calculate the limits

   $$\lim_{R \to \infty} \frac{s(R)}{b(R)} \quad \text{and} \quad \lim_{R \to \infty} \frac{s(R)^2}{b(R)}.$$

   (b) Take $m \geq 3$. What exponent k makes the limit

   $$\lim_{R \to \infty} \frac{s(R)^k}{b(R)}$$

   interesting?

   (c) Let $c(R)$ be the number of integer unit cubes that are contained in the ball of radius R, centered at the origin. Calculate

   $$\lim_{R \to \infty} \frac{c(R)}{b(R)}$$

   (d) Shift the ball to a new, arbitrary center (not on the integer lattice) and re-calculate the limits.

27. Prove that the interval $[a, b]$ in $\mathbb{R}$ is the same as the segment $[a, b]$ in $\mathbb{R}^1$. That is,

$$\{x \in \mathbb{R} : a \leq x \leq b\}$$
$$= \{y \in \mathbb{R} : \exists s, t \in [0, 1] \text{ with } s + t = 1 \text{ and } y = sa + tb\}.$$

[Hint: How do you prove that two sets are equal?]

28. A **convex combination** of $w_1, \ldots, w_k \in \mathbb{R}^m$ is a vector sum

$$w = s_1 w_1 + \cdots + s_k w_k$$

such that $s_1 + \cdots + s_k = 1$ and $0 \le s_1, \ldots, s_k \le 1$.
   (a) Prove that if a set E is convex then E contains the convex combination of any finite number of points in E.
   (b) Why is the converse obvious?

29. (a) Prove that the ellipsoid

$$E = \{(x, y, z) \in \mathbb{R}^3 : \frac{x^2}{a^2} + \frac{y^2}{b^2} + \frac{z^2}{c^2} \le 1\}$$

is convex. [Hint: E is the unit ball for a different dot product. What is it? Does the Cauchy-Schwarz inequality not apply to all dot products?]
   (b) Prove that all boxes in $\mathbb{R}^m$ are convex.

30. A function $f : (a, b) \to \mathbb{R}$ is a **convex function** if for all $x, y \in (a, b)$ and all $s, t \in [0, 1]$ with $s + t = 1$ we have

$$f(sx + ty) \le sf(x) + tf(y).$$

   (a) Prove that f is convex if and only if the set S of points above its graph is convex in $\mathbb{R}^2$. The set S is $\{(x, y) : f(x) \le y\}$.
  *(b) Prove that every convex function is continuous.
   (c) Suppose that f is convex and $a < x < u < b$. The slope σ of the line through $(x, f(x))$ and $(u, f(u))$ depends on x and u, say $\sigma = \sigma(x, u)$. Prove that σ increases when x increases, and σ increases when u increases.
   (d) Suppose that f is second-order differentiable. That is, f is differentiable and its derivative f' is also differentiable. As is standard, we write $(f')' = f''$. Prove that f is convex if and only if $f''(x) \ge 0$ for all $x \in (a, b)$.
   (e) Formulate a definition of convexity for a function $f : M \to \mathbb{R}$ where $M \subset \mathbb{R}^m$ is a convex set. [Hint: Start with $m = 2$.]

*31. Suppose that a function $f : [a, b] \to \mathbb{R}$ is monotone nondecreasing. That is, $x_1 \le x_2$ implies $f(x_1) \le f(x_2)$.
   (a) Prove that f is continuous except at a countable set of points. [Hint: Show that at each $x \in (a, b)$, f has **right limit** $f(x+)$ and a **left limit** $f(x-)$, which are limits of $f(x + h)$ as h tends to 0 through positive and negative values respectively. The **jump** of f at x is $f(x+) - f(x-)$. Show that f is continuous at x if and only if it has zero jump at x. At how many points can f have jump ≥ 1? At how many points can the jump be between $1/2$ and 1? Between $1/3$ and $1/2$?]
   (b) Is the same assertion true for a monotone function defined on all of $\mathbb{R}$?

*32. Suppose that E is a convex region in the plane bounded by a curve C.

(a) Show that C has a tangent line except at a countable number of points. [For example, the circle has a tangent line at all its points. The triangle has a tangent line except at three points, and so on.]

(b) Similarly, show that a convex function has a derivative except at a countable set of points.

*33. Let $f(k, m)$ be the number of k-dimensional faces of the m-cube. See Table 1.

	$m = 1$	$m = 2$	$m = 3$	$m = 4$	$m = 5$	$\cdots$	m	$m + 1$
$k = 0$	2	4	8	$f(0, 4)$	$f(0, 5)$	$\cdots$	$f(0, m)$	$f(0, m + 1)$
$k = 1$	1	4	12	$f(1, 4)$	$f(1, 5)$	$\cdots$	$f(1, m)$	$f(1, m + 1)$
$k = 2$	0	1	6	$f(2, 4)$	$f(2, 5)$	$\cdots$	$f(2, m)$	$f(2, m + 1)$
$k = 3$	0	0	1	$f(3, 4)$	$f(3, 5)$	$\cdots$	$f(3, m)$	$f(3, m + 1)$
$k = 4$	0	0	0	$f(4, 4)$	$f(4, 5)$	$\cdots$	$f(4, m)$	$f(4, m + 1)$
$\cdots$	$\cdots$	$\cdots$	$\cdots$	$\cdots$	$\cdots$	$\cdots$	$\cdots$	$\cdots$

Table 1 $f(k, m)$ is the number of k-dimensional faces of the m-cube.

(a) Verify the numbers in the first three columns.

(b) Calculate the columns $m = 4$, $m = 5$, and give the formula for passing from the mth column to the $(m + 1)^{\text{st}}$.

(c) What would an $m = 0$ column mean?

(d) Prove that the alternating sum of the entries in any column is 1. That is, $2 - 1 = 1$, $4 - 4 + 1 = 1$, $8 - 12 + 6 - 1 = 1$, and in general $\sum (-1)^k f(k, m) = 1$. This alternating sum is called the **Euler characteristic**.

34. Find an exact formula for a bijection $f : \mathbb{N} \times \mathbb{N} \to \mathbb{N}$. Is one

$$f(i, j) = j + (1 + 2 + \cdots + (i + j - 2)) = \frac{i^2 + j^2 + i(2j - 3) - j + 2}{2}?$$

35. Prove that the union of denumerably many sets B_k, each of which is countable, is countable. How could it happen that the union is finite?

*36. Without using the Schroeder-Bernstein Theorem,

(a) Prove that $[a, b] \sim (a, b] \sim (a, b)$.

(b) More generally, prove that if C is countable then

$$\mathbb{R} \setminus C \sim \mathbb{R} \sim \mathbb{R} \cup C.$$

(c) Infer that the set of irrational numbers has the same cardinality as $\mathbb{R}$, i.e., $\mathbb{R} \setminus \mathbb{Q} \sim \mathbb{R}$. [Hint: Imagine that you are the owner of denumerably many hotels, $H_1, H_2, \ldots$, all fully occupied, and that a traveler arrives and asks you for accommodation. How could you re-arrange your current guests to make room for the traveler?]

*37. Prove that $\mathbb{R}^2 \sim \mathbb{R}$. [Hint: Think of shuffling two digit strings

$$(a_1 a_2 a_3 \ldots)\&(b_1 b_2 b_3 \ldots) \to (a_1 b_1 a_2 b_2 a_3 b_3 \ldots).$$

In this way you could transform a pair of reals into a single real. Be sure to face the nines-termination issue.]

38. Let S be a set and let $\mathcal{P} = \mathcal{P}(S)$ be the collection of all subsets of S. [$\mathcal{P}(S)$ is called the **power set** of S.] Let $\mathcal{F}$ be the set of functions $f : S \to \{0, 1\}$.

  (a) Prove that there is a natural bijection from $\mathcal{F}$ onto $\mathcal{P}$ defined by

$$f \mapsto \{s \in S : f(s) = 1\}.$$

  *(b) Prove that the cardinality of $\mathcal{P}$ is greater than the cardinality of S, even when S is empty or finite.

  [Hints: The notation Y^X is sometimes used for the set of all functions $X \to Y$. In this notation $\mathcal{F} = \{0, 1\}^S$ and assertion (b) becomes $\#(S) < \#(\{0, 1\}^S)$. The empty set has one subset, itself, whereas it has no elements, so $\#(\emptyset) = 0$, while $\#(\{0, 1\}^\emptyset) = 1$, which proves (b) for the empty set. Assume there is a bijection from S onto $\mathcal{P}$. Then there is a bijection $\beta : S \to \mathcal{F}$, and for each $s \in S$, $\beta(s)$ is a function, say $f_s : S \to \{0, 1\}$. Think like Cantor and try to find a function which corresponds to no s. Infer that β could not have been onto.]

39. A real number is **algebraic** if it is a root of a nonconstant polynomial with integer coefficients.

  (a) Prove that the set A of algebraic numbers is denumerable. [Hint: Each polynomial has how many roots? How many linear polynomials are there? How many quadratics? ...]

  (b) Repeat the exercise for roots of polynomials whose coefficients belong to some fixed, arbitrary denumerable set $S \subset \mathbb{R}$.

  *(c) Repeat the exercise for roots of trigonometric polynomials with integer coefficients.

  (d) Real numbers that are not algebraic are said to be **transcendental**. Trying to find transcendental numbers is said to be like looking for hay in a haystack. Why?

40. A **finite word** is a finite string of letters, say from the roman alphabet.

  (a) What is the cardinality of the set of all finite words, and thus of the set of all possible poems and mathematical proofs?

  (b) What if the alphabet had only two letters?

  (c) What if it had countably many letters?

  (d) Prove that the cardinality of the set Σ_n of all infinite words formed using a finite alphabet of n letters, $n \geq 2$, is equal to the cardinality of $\mathbb{R}$.

(e) Give a solution to Exercise 37 by justifying the equivalence chain

$$\mathbb{R}^2 = \mathbb{R} \times \mathbb{R} \quad \sim \quad \Sigma_2 \times \Sigma_2 \quad \sim \quad \Sigma_4 \times \Sigma_4 \quad \sim \quad \mathbb{R}.$$

(f) How many decimal expansions terminate in an infinite string of 9's? How many don't?

41. If v is a value of a continuous function $f : [a, b] \to \mathbb{R}$ use the Least Upper Bound Property to prove that there are smallest and largest $x \in [a, b]$ such that $f(x) = v$.

42. A function defined on an interval $[a, b]$ or (a, b) is **uniformly continuous** if for each $\epsilon > 0$ there exists a $\delta > 0$ such that $|x - t| < \delta$ implies that $|f(x) - f(t)| < \epsilon$. (Note that this δ cannot depend on x, it can only depend on ϵ. With ordinary continuity, the δ can depend on both x and ϵ.)

   (a) Show that a uniformly continuous function is continuous but continuity does not imply uniform continuity. (For example, prove that $\sin(1/x)$ is continuous on the interval $(0, 1)$ but is not uniformly continuous there. Graph it.)

   (b) Is the function $2x$ uniformly continuous on the unbounded interval $(-\infty, \infty)$?

   (c) What about x^2?

*43. Prove that a continuous function defined on an interval $[a, b]$ is uniformly continuous. [Hint: Let $\epsilon > 0$ be given. Think of ϵ as fixed and consider the sets

$$A(\delta) = \{u \in [a, b] : \text{if } x, t \in [a, u] \text{ and } |x - t| < \delta$$
$$\text{then } |f(x) - f(t)| < \epsilon\}$$

$$A = \bigcup_{\delta > 0} A(\delta).$$

Using the Least Upper Bound Property, prove that $b \in A$. Infer that f is uniformly continuous. The fact that continuity on $[a, b]$ implies uniform continuity is one of the important, fundamental principles of continuous functions.]

*44. Define injections $f : \mathbb{N} \to \mathbb{N}$ and $g : \mathbb{N} \to \mathbb{N}$ by $f(n) = 2n$ and $g(n) = 2n$. From f and g, the Schroeder-Bernstein Theorem produces a bijection $\mathbb{N} \to \mathbb{N}$. What is it?

*45. Let (a_n) be a sequence of real numbers. It is **bounded** if the set $A = \{a_1, a_2, \ldots\}$ is bounded. The **limit supremum**, or lim sup, of a bounded sequence (a_n) as $n \to \infty$ is

$$\limsup_{n \to \infty} a_n = \lim_{n \to \infty} \left(\sup_{k \geq n} a_k \right)$$

   (a) Why does the lim sup exist?

   (b) If $\sup\{a_n\} = \infty$, how should we define $\limsup_{n \to \infty} a_n$?

   (c) If $\lim_{n \to \infty} a_n = -\infty$, how should we define $\limsup_{n \to \infty} a_n$?

(d) When is it true that

$$\limsup_{n\to\infty}(a_n + b_n) \;\leq\; \limsup_{n\to\infty} a_n + \limsup_{n\to\infty} b_n$$
$$\limsup_{n\to\infty} c a_n \;=\; c\limsup_{n\to\infty} a_n?$$

When is it true they are unequal? Draw pictures that illustrate these relations.

(e) Define the **limit infimum**, or lim inf, of a sequence of real numbers, and find a formula relating it to the limit supremum.

(f) Prove that $\lim\limits_{n\to\infty} a_n$ exists if and only if the sequence (a_n) is bounded and $\liminf\limits_{n\to\infty} a_n = \limsup\limits_{n\to\infty} a_n$.

**46. The unit ball with respect to a norm $\|\ \ \|$ on $\mathbb{R}^2$ is

$$\{v \in \mathbb{R}^2 : \|v\| \leq 1\}.$$

(a) Find necessary and sufficient geometric conditions on a subset of $\mathbb{R}^2$ that it is the unit ball for some norm.

(b) Give necessary and sufficient geometric conditions that a subset be the unit ball for a norm arising from an inner product.

(c) Generalize to $\mathbb{R}^m$. [You may find it useful to read about closed sets in the next chapter, and to consult Exercise 41 there.]

47. Assume that V is an inner product space whose inner product induces a norm as $|x| = \sqrt{\langle x, x \rangle}$.

(a) Show that $|\ \ |$ obeys the **parallelogram law**

$$|x + y|^2 + |x - y|^2 = 2|x|^2 + 2|y|^2$$

for all $x, y \in V$.

*(b) Show that any norm obeying the parallelogram law arises from a unique inner product. [Hints: Define the prospective inner product as

$$\langle x, y \rangle = \left|\frac{x + y}{2}\right|^2 - \left|\frac{x - y}{2}\right|^2$$

Checking that $\langle\ ,\ \rangle$ satisfies the inner product properties of symmetry and positive definiteness is easy. Also, it is immediate that $|x|^2 = \langle x, x \rangle$, so $\langle\ ,\ \rangle$ induces the given norm. Checking bilinearity is another story.

(i) Let $x, y, z \in V$ be arbitrary. Show that the parallelogram law implies

$$\langle x + y, z \rangle + \langle x - y, z \rangle = 2\langle x, y \rangle,$$

and infer that $\langle 2x, z\rangle = 2\langle x, z\rangle$. For arbitrary $u, v \in V$ set $x = \frac{1}{2}(u+v)$ and $y = \frac{1}{2}(u - v)$, plug in to the previous equation, and deduce

$$\langle u, z\rangle + \langle v, z\rangle = \langle u + v, z\rangle,$$

which is additive bilinearity in the first variable. Why does it now follow at once that $\langle\ ,\ \rangle$ is also additively bilinear in the second variable?

(ii) To check multiplicative bilinearity, prove by induction that if $m \in \mathbb{Z}$ then $m\langle x, y\rangle = \langle mx, y\rangle$, and if $n \in \mathbb{N}$ then $\frac{1}{n}\langle x, y\rangle = \langle \frac{1}{n}x, y\rangle$. Infer that $r\langle x, y\rangle = \langle rx, y\rangle$ when r is rational. Is $\lambda \mapsto \langle \lambda x, y\rangle - \lambda\langle x, y\rangle$ a continuous function of $\lambda \in \mathbb{R}$, and does this give multiplicative bilinearity?]

48. Consider a knot in 3-space as shown in Figure 25. In 3-space it cannot be

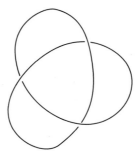

Figure 25 An overhand knot in 3-space

unknotted. How can you unknot it in 4-space?

*49. Prove that there exists no continuous three dimensional motion de-linking the two circles shown in Figure 22 which keeps both circles flat at all times.

50. The Klein bottle is a surface that has an oval of self intersection when it is shown in 3-space. See Figure 26. It can live in 4-space with no self-intersection.

oval of self-intersection

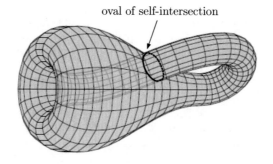

Figure 26 The Klein Bottle in 3-space has an oval of self-intersection.

How?

51. Read *Flatland* by Edwin Abbott. Try to imagine a Flatlander using color to visualize 3-space.

52. Can you visualize a 4-dimensional cube – its vertices, edges, and faces? [Hint: It may be easier (and equivalent) to picture a 4-dimensional parallelepiped whose eight red vertices have xyz-coordinates that differ from the xyz-coordinates of its eight colorless vertices. It is a 4-dimensional version of a rectangle or parallelogram whose edges are not parallel to the coordinate axes.]

2

A Taste of Topology

1 Metric Spaces

It may seem paradoxical at first, but a specific math problem can be harder to solve than some abstract generalization of it. For instance if you want to know how many roots the equation

$$t^5 - 4t^4 + t^3 - t + 1 = 0$$

can have then you could use calculus and figure it out. It would take a while. But thinking more abstractly, and with less work, you could show that every n^{th}-degree polynomial has at most n roots. In the same way many general results about functions of a real variable are more easily grasped at an abstract level – the level of metric spaces.

Metric space theory can be seen as a special case of general topology, and many books present it that way, explaining compactness primarily in terms of open coverings. In my opinion, however, the sequence/subsequence approach provides the easiest and simplest route to mastering the subject. Accordingly it gets top billing throughout this chapter.

A **metric space** is a set M, the elements of which are referred to as points of M, together with a **metric** d having the three properties that distance has in Euclidean space. The metric $d = d(x, y)$ is a real number defined for all points $x, y \in M$ and $d(x, y)$ is called the **distance** from the point x to the point y. The three distance properties are as follows: For all $x, y, z \in M$ we have

© Springer International Publishing Switzerland 2015
C.C. Pugh, *Real Mathematical Analysis*, Undergraduate Texts
in Mathematics, DOI 10.1007/978-3-319-17771-7_2

(a) **positive definiteness**: $d(x, y) \geq 0$, and $d(x, y) = 0$ if and only if $x = y$.

(b) **symmetry**: $d(x, y) = d(y, x)$.

(c) **triangle inequality**: $d(x, z) \leq d(x, y) + d(y, z)$.

The function d is also called the **distance function**. Strictly speaking, it is the pair (M, d) which is a metric space, but we will follow the common practice of referring to "the metric space M," and leave to you the job of inferring the correct metric.

The main examples of metric spaces are $\mathbb{R}$, $\mathbb{R}^m$, and their subsets. The metric on $\mathbb{R}$ is $d(x, y) = |x - y|$ where $x, y \in \mathbb{R}$ and $|x - y|$ is the magnitude of $x - y$. The metric on $\mathbb{R}^m$ is the Euclidean length of $x - y$ where x, y are vectors in $\mathbb{R}^m$. Namely,

$$d(x, y) = \sqrt{(x_1 - y_1)^2 + \ldots + (x_m - y_m)^2}$$

for $x = (x_1, \ldots, x_m)$ and $y = (y_1, \ldots, y_m)$.

Since Euclidean length satisfies the three distance properties, d is a bona fide metric and it makes $\mathbb{R}^m$ into a metric space. A subset $M \subset \mathbb{R}^m$ becomes a metric space when we declare the distance between points of M to be their Euclidean distance apart as points in $\mathbb{R}^m$. We say that M **inherits** its metric from $\mathbb{R}^m$ and is a **metric subspace** of $\mathbb{R}^m$. Figure 27 shows a few subsets of $\mathbb{R}^2$ to suggest some interesting metric spaces.

There is also one metric that is hard to picture but valuable as a source for counterexamples, the **discrete metric**. Given any set M, define the distance between distinct points of M to be 1 and the distance between every point and itself to be 0. This is a metric. See Exercise 4. If M consists of three points, say $M = \{a, b, c\}$, you can think of the vertices of the unit equilateral triangle as a model for M. See Figure 28. They have mutual distance 1 from each other. If M consists of one, two, or four points can you think of a model for the discrete metric on M? More challenging is to imagine the discrete metric on $\mathbb{R}$. All points, by definition of the discrete metric, lie at unit distance from each other.

Convergent Sequences and Subsequences

A sequence of points in a metric space M is a list $p_1, p_2, \ldots$ where the points p_n belong to M. Repetition is allowed, and not all the points of M need to appear in the list. Good notation for a sequence is (p_n), or $(p_n)_{n \in \mathbb{N}}$. The notation $\{p_n\}$ is also used but it is too easily confused with the set of points making up the sequence. The difference between $(p_n)_{n \in \mathbb{N}}$ and $\{p_n : n \in \mathbb{N}\}$ is that in the former case

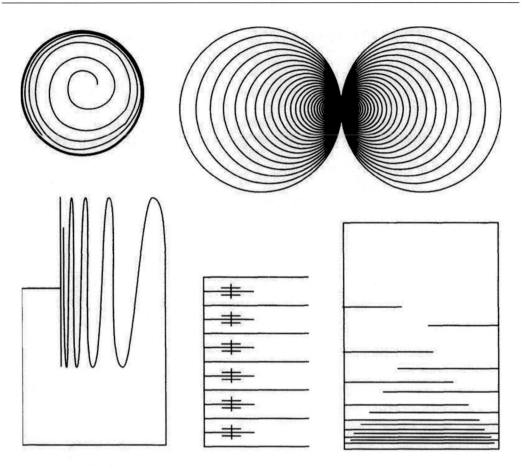

Figure 27 Five metric spaces – a closed outward spiral, a Hawaiian earring, a topologist's sine circle, an infinite television antenna, and Zeno's maze

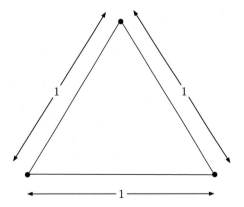

Figure 28 The vertices of the unit equilateral triangle form a discrete metric space.

the sequence prescribes an ordering of the points, while in the latter the points get jumbled together. For example, the sequences $1, 2, 3, \ldots$ and $1, 2, 1, 3, 2, 1, 4, 3, 2, 1, \ldots$ are different sequences but give the same set of points, namely $\mathbb{N}$.

Formally, a sequence in M is a function $f : \mathbb{N} \to M$. The nth term in the sequence is $f(n) = p_n$. Clearly, every sequence defines a function $f : \mathbb{N} \to M$ and conversely, every function $f : \mathbb{N} \to M$ defines a sequence in M. The sequence (p_n) **converges to the limit** p in M if

$$\forall \epsilon > 0 \; \exists N \in \mathbb{N} \text{ such that}$$
$$n \in \mathbb{N} \text{ and } n \geq N \quad \Rightarrow \quad d(p_n, p) < \epsilon.$$

Limits are unique in the sense that if (p_n) converges to p and if (p_n) also converges to p' then $p = p'$. The reason is this. Given any $\epsilon > 0$, there are integers N and N' such that if $n \geq N$ then $d(p_n, p) < \epsilon$, while if $n \geq N'$ then $d(p_n, p') < \epsilon$. Then for all $n \geq \max\{N, N'\}$ we have

$$d(p, p') \; \leq \; d(p, p_n) + d(p_n, p') \; < \; \epsilon + \epsilon \; = \; 2\epsilon.$$

But ϵ is arbitrary and so $d(p, p') = 0$ and $p = p'$. (This is the ϵ-principle again.)

We write $p_n \to p$, or $p_n \to p$ as $n \to \infty$, or

$$\lim_{n \to \infty} p_n \; = \; p$$

to indicate convergence. For example, in $\mathbb{R}$ the sequence $p_n = 1/n$ converges to 0 as $n \to \infty$. In $\mathbb{R}^2$ the sequence $(1/n, \; \sin n)$ does not converge as $n \to \infty$. In the metric space $\mathbb{Q}$ (with the metric it inherits from $\mathbb{R}$) the sequence $1, 1.4, 1.414, 1.4142, \ldots$ does not converge.

Just as a set can have a subset, a sequence can have a subsequence. For example, the sequence $2, 4, 6, 8, \ldots$ is a subsequence of $1, 2, 3, 4, \ldots$. The sequence $3, 5, 7, 11, 13, 17, \ldots$ is a subsequence of $1, 3, 5, 7, 9, \ldots$, which in turn is a subsequence of $1, 2, 3, 4, \ldots$. In general, if $(p_n)_{n \in \mathbb{N}}$ and $(q_k)_{k \in \mathbb{N}}$ are sequences and if there is a sequence $n_1 < n_2 < n_3 < \ldots$ of positive integers such that for each $k \in \mathbb{N}$ we have $q_k = p_{n_k}$ then (q_k) is a **subsequence** of (p_n). Note that the terms in the subsequence occur in the same order as in the mother sequence.

1 Theorem *Every subsequence of a convergent sequence in M converges and it converges to the same limit as does the mother sequence.*

Proof Let (q_k) be a subsequence of (p_n), $q_k = p_{n_k}$, where $n_1 < n_2 < \ldots$. Assume that (p_n) converges to p in M. Given $\epsilon > 0$, there is an N such that for all $n \geq N$ we have $d(p_n, p) < \epsilon$. Since $n_1, n_2, \ldots$ are positive integers we have $k \leq n_k$ for all k. Thus, if $k \geq N$ then $n_k \geq N$ and $d(q_k, p) < \epsilon$. Hence (q_k) converges to p. $\square$

A common way to state Theorem 1 is that limits are unaffected when we pass to a subsequence.

2 Continuity

In linear algebra the objects of interest are linear transformations. In real analysis the objects of interest are functions, especially continuous functions. A function f from the metric space M to the metric space N is just that; $f : M \to N$ and f sends points $p \in M$ to points $fp \in N$. The function maps M to N. The way you should think of functions – as devices, not formulas – is discussed in Section 4 of Chapter 1. The most common type of function maps M to $\mathbb{R}$. It is a real-valued function of the variable $p \in M$.

Definition A function $f : M \to N$ is **continuous** if it **preserves sequential convergence**: f sends convergent sequences in M to convergent sequences in N, limits being sent to limits. That is, for each sequence (p_n) in M which converges to a limit p in M, the image sequence $(f(p_n))$ converges to fp in N.

Here and in what follows, the notation fp is often used as convenient shorthand for $f(p)$. The metrics on M and N are d_M and d_N. We will often refer to either metric as just d.

2 Theorem *The composite of continuous functions is continuous.*

Proof Let $f : M \to N$ and $g : N \to P$ be continuous and assume that

$$\lim_{n \to \infty} p_n = p$$

in M. Since f is continuous, $\lim_{n \to \infty} f(p_n) = fp$. Since g is continuous, $\lim_{n \to \infty} g(f(p_n)) = g(fp)$ and therefore $g \circ f : M \to P$ is continuous. See Figure 29 on the next page. $\square$

Moral The sequence condition is the easy way to tell at a glance whether a function is continuous.

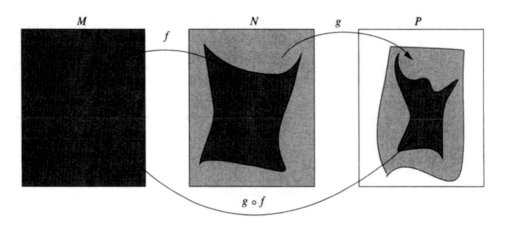

Figure 29 The composite function $g \circ f$

There are two "obviously" continuous functions.

3 Proposition *For every metric space M the identity map* $\mathrm{id} : M \to M$ *is continuous, and so is every constant function* $f : M \to N$.

Proof Let $p_n \to p$ in M. Then $\mathrm{id}(p_n) = p_n \to p = \mathrm{id}(p)$ as $n \to \infty$ which gives continuity of the identity map. Likewise, if $f(x) = q \in N$ for all $x \in M$ and if $p_n \to p$ in M then $fp = q$ and $f(p_n) = q$ for all n. Thus $f(p_n) \to fp$ as $n \to \infty$ which gives continuity of the constant function f.   □

Homeomorphism

Vector spaces are isomorphic if there is a linear bijection from one to the other. When are metric spaces isomorphic? They should "look the same." The letters Y and T look the same; and they look different from the letter O. If $f : M \to N$ is a bijection and f is continuous and the inverse bijection $f^{-1} : N \to M$ is also continuous then f is a **homeomorphism**[†](or a "homeo" for short) and M, N are **homeomorphic**. We write $M \cong N$ to indicate that M and N are homeomorphic. $\cong$ is an equivalence relation: $M \cong M$ since the identity map is a homeomorphism $M \to M$; $M \cong N$ clearly implies that $N \cong M$; and the previous theorem shows that the composite of homeomorphisms is a homeomorphism.

Geometrically speaking, a homeomorphism is a bijection that can bend, twist, stretch, and wrinkle the space M to make it coincide with N, but it cannot rip,

---

[†]This is a rare case in mathematics in which spelling is important. Homeomorphism $\neq$ homomorphism.

puncture, shred, or pulverize M in the process. The basic questions to ask about metric spaces are:

(a) Given M and N, are they homeomorphic?

(b) What are the continuous functions from M to N?

A major goal of this chapter is to show you how to answer these questions in many cases. For example, is the circle homeomorphic to the interval? To the sphere? etc. Figure 30 indicates that the circle and the (perimeter of the) triangle are homeomorphic, while Figure 15 shows that (a, b), the semicircle, and $\mathbb{R}$ are homeomorphic.

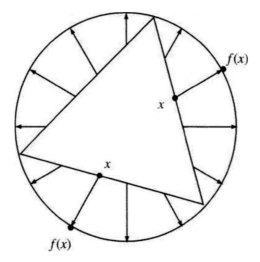

Figure 30 The circle and triangle are homeomorphic.

A natural question that should occur to you is whether continuity of f^{-1} is actually implied by continuity of a bijection f. It is not. Here is an instructive example.

Consider the interval $[0, 2\pi) = \{x \in \mathbb{R} : 0 \leq x < 2\pi\}$ and define $f : [0, 2\pi) \to S^1$ to be the mapping $f(x) = (\cos x, \sin x)$ where S^1 is the unit circle in the plane. The mapping f is a continuous bijection, but the inverse bijection is not continuous. For there is a sequence of points (z_n) on S^1 in the fourth quadrant that converges to $p = (1, 0)$ from below, and $f^{-1}(z_n)$ does not converge to $f^{-1}(p) = 0$. Rather it converges to 2π. Thus, f is a continuous bijection whose inverse bijection fails to be continuous. See Figure 31. In Exercises 49 and 50 you are asked to find worse examples of continuous bijections that are not homeomorphisms.

To build your intuition about continuous mappings and homeomorphisms, consider the following examples shown in Figure 32 – the unit circle in the plane, a trefoil knot in $\mathbb{R}^3$, the perimeter of a square, the surface of a donut (the 2-torus), the surface

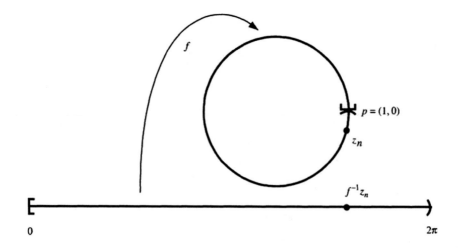

Figure 31 f wraps $[0, 2\pi)$ bijectively onto the circle.

of a ceramic coffee cup, the unit interval $[0,1]$, the unit disc including its boundary. Equip all with the inherited metric. Which should be homeomorphic to which?

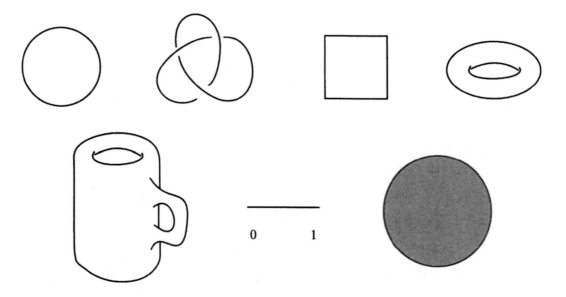

Figure 32 Seven metric spaces

The (ϵ, δ)-Condition

The following theorem presents the more familiar (but equivalent!) definition of continuity using ϵ and δ. It corresponds to the definition given in Chapter 1 for real-valued functions of a real variable.

4 Theorem $f : M \to N$ *is continuous if and only if it satisfies the* (ϵ, δ)*-condition: For each* $\epsilon > 0$ *and each* $p \in M$ *there exists* $\delta > 0$ *such that if* $x \in M$ *and* $d_M(x, p) < \delta$ *then* $d_N(fx, fp) < \epsilon$.

Proof Suppose that f is continuous. It preserves sequential convergence. From the supposition that f fails to satisfy the (ϵ, δ)-condition at some $p \in M$ we will derive a contradiction. If the (ϵ, δ)-condition fails at p then there exists $\epsilon > 0$ such that for each $\delta > 0$ there is a point x with $d(x, p) < \delta$ and $d(fx, fp) \geq \epsilon$. Taking $\delta = 1/n$ we get a sequence (x_n) with $d(x_n, p) < 1/n$ and $d(f(x_n), fp) \geq \epsilon$, which contradicts preservation of sequential convergence. For $x_n \to p$ but $f(x_n)$ does not converge to fp, which contradicts the fact that f preserves sequential convergence. Since the supposition that f fails to satisfy the (ϵ, δ)-condition leads to a contradiction, f actually does satisfy the (ϵ, δ)-condition.

To check the converse, suppose that f satisfies the (ϵ, δ)-condition at p. For each sequence (x_n) in M that converges to p we must show $f(x_n) \to fp$ in N as $n \to \infty$. Let $\epsilon > 0$ be given. There is $\delta > 0$ such that $d_M(x, p) < \delta \Rightarrow d_N(fx, fp) < \epsilon$. Convergence of x_n to p implies there is an integer K such that for all $n \geq K$ we have $d_M(x_n, p) < \delta$, and therefore $d_N(f(x_n), fp) < \epsilon$. That is, $f(x_n) \to fp$ as $n \to \infty$. See also Exercise 13.   □

3 The Topology of a Metric Space

Now we come to two basic concepts in a metric space theory – closedness and openness. Let M be a metric space and let S be a subset of M. A point $p \in M$ is a **limit** of S if there exists a sequence (p_n) in S that converges to it.[†]

[†] A limit of S is also sometimes called a **limit point** of S. Take care though: Some mathematicians require that a limit point of S be the limit of a sequence of *distinct* points of S. They would say that a finite set has no limit points. We will *not* adopt their point of view. Another word used in this context, especially by the French, is "adherence." A point p **adheres** to the set S if and only if p is a limit of S. In more general circumstances, limits are defined using "nets" instead of sequences. They are like "uncountable sequences." You can read more about nets in graduate-level topology books such as *Topology* by James Munkres.

Definition S is a **closed set** if it contains all its limits.[†]

Definition S is an **open set** if for each $p \in S$ there exists an $r > 0$ such that

$$d(p, q) < r \quad \Rightarrow \quad q \in S.$$

5 Theorem *Openness is dual to closedness: The complement of an open set is a closed set and the complement of a closed set is an open set.*

Proof Suppose that $S \subset M$ is an open set. We claim that S^c is a closed set. If $p_n \to p$ and $p_n \in S^c$ we must show that $p \in S^c$. Well, if $p \notin S^c$ then $p \in S$ and, since S is open, there is an $r > 0$ such that

$$d(p, q) < r \quad \Rightarrow \quad q \in S.$$

Since $p_n \to p$, we have $d(p, p_n) < r$ for all large n, which implies that $p_n \in S$, contrary to the sequence being in S^c. Since the supposition that p lies in S leads to a contradiction, p actually does lie in S^c, proving that S^c is a closed set.

Suppose that S is a closed set. We claim that S^c is open. Take any $p \in S^c$. If there fails to exist an $r > 0$ such that

$$d(p, q) < r \quad \Rightarrow \quad q \in S^c$$

then for each $r = 1/n$ with $n = 1, 2, \ldots$ there exists a point $p_n \in S$ such that $d(p, p_n) < 1/n$. This sequence in S converges to $p \in S^c$, contrary to closedness of S. Therefore there actually does exist an $r > 0$ such that

$$d(p, q) < r \quad \Rightarrow \quad q \in S^c$$

which proves that S^c is an open set.   □

Most sets, like doors, are neither open nor closed, but ajar. Keep this in mind. For example neither $(a, b]$ nor its complement is closed in $\mathbb{R}$; $(a, b]$ is neither closed nor open. Unlike doors, however, sets can be both open and closed at the same time. For example, the empty set $\emptyset$ is a subset of every metric space and it is always closed. There are no sequences and no limits to even worry about. Similarly the full metric space M is a closed subset of itself: For it certainly contains the limit of

---

[†]Note how similarly algebraists use the word "closed." A field (or group or ring, etc.) is closed under its arithmetic operations: Sums, differences, products, and quotients of elements in the field still lie in the field. In our case it is limits. Limits of sequences in S must lie in S.

every sequence that converges in M. Thus, $\emptyset$ and M are closed subsets of M. Their complements, M and $\emptyset$, are therefore open: $\emptyset$ and M are both closed and open.

Subsets of M that are both closed and open are **clopen**. See also Exercise 125. It turns out that in $\mathbb{R}$ the only clopen sets are $\emptyset$ and $\mathbb{R}$. In $\mathbb{Q}$, however, things are quite different, sets such as $\{r \in \mathbb{Q} : -\sqrt{2} < r < \sqrt{2}\}$ being clopen in $\mathbb{Q}$. To summarize,

A subset of a metric space can be
closed, open, both, or neither.

You should expect the "typical" subset of a metric space to be neither closed nor open.

The **topology** of M is the collection $\mathcal{T}$ of all open subsets of M.

6 Theorem *$\mathcal{T}$ has three properties:[†] as a system it is closed under union, finite intersection, and it contains $\emptyset$, M. That is,*

 (a) *Every union of open sets is an open set.*
 (b) *The intersection of finitely many open sets is an open set.*
 (c) *$\emptyset$ and M are open sets.*

Proof (a) If $\{U_\alpha\}$ is any collection[‡] of open subsets of M and $V = \bigcup U_\alpha$ then V is open. For if $p \in V$ then p belongs to at least one U_α and there is an $r > 0$ such that

$$d(p,q) < r \quad \Rightarrow \quad q \in U_\alpha.$$

Since $U_\alpha \subset V$, this implies that all such q lie in V, proving that V is open.

(b) If $U_1, \ldots, U_n$ are open sets and $W = \bigcap U_k$ then W is open. For if $p \in W$ then for each k, $1 \leq k \leq n$, then there is an $r_k > 0$ such that

$$d(p,q) < r_k \quad \Rightarrow \quad q \in U_k.$$

Take $r = \min\{r_1, \ldots, r_n\}$. Then $r > 0$ and

$$d(p,q) < r \quad \Rightarrow \quad q \in U_k,$$

[†]Any collection $\mathcal{T}$ of subsets of a set X that satisfies these three properties is called a topology on X, and X is called a **topological space**. Topological spaces are more general than metric spaces: There exist topologies that do not arise from a metric. Think of them as pathological. The question of which topologies can be generated by a metric and which cannot is discussed in *Topology* by Munkres. See also Exercise 30.

[‡]The α in the notation U_α "indexes" the sets. If $\alpha = 1, 2, \ldots$ then the collection is countable, but we are just as happy to let α range through uncountable index sets.

for each k; i.e., $q \in W = \bigcap U_k$, proving that W is open.

(c) It is clear that $\emptyset$ and M are open sets. $\square$

7 Corollary *The intersection of any number of closed sets is a closed set; the finite union of closed sets is a closed set; $\emptyset$ and M are closed sets.*

Proof Take complements and use De Morgan's laws. If $\{K_\alpha\}$ is a collection of closed sets then $U_\alpha = (K_\alpha)^c$ is open and

$$K = \bigcap K_\alpha = \left(\bigcup U_\alpha \right)^c.$$

Since $\bigcup U_\alpha$ is open, its complement K is closed. Similarly, a finite union of closed sets is the complement of the finite intersection of their complements, and is a closed set. $\square$

What about an infinite union of closed sets? Generally, it is not closed. For example, the interval $[1/n, 1]$ is closed in $\mathbb{R}$, but the union of these intervals as n ranges over $\mathbb{N}$ is the interval $(0, 1]$ which is not closed in $\mathbb{R}$. Neither is the infinite intersection of open sets open in general.

Two sets whose closedness/openness properties are basic are:

$$\begin{aligned} \lim S &= \{p \in M : p \text{ is a limit of } S\} \\ M_r p &= \{q \in M : d(p, q) < r\}. \end{aligned}$$

The former is the **limit set** of S; the latter is the **r-neighborhood** of p.

8 Theorem $\lim S$ *is a closed set and $M_r p$ is an open set.*

Proof Simple but not immediate! See Figure 33.

Suppose that $p_n \to p$ and each p_n lies in $\lim S$. We claim that $p \in \lim S$. Since p_n is a limit of S there is a sequence $(p_{n,k})_{k \in \mathbb{N}}$ in S that converges to p_n as $k \to \infty$. Thus there exists $q_n = p_{n,k(n)} \in S$ such that

$$d(p_n, q_n) < \frac{1}{n}.$$

Then, as $n \to \infty$ we have

$$d(p, q_n) \leq d(p, p_n) + d(p_n, q_n) \to 0$$

which implies that $q_n \to p$, so $p \in \lim S$, which completes the proof that $\lim S$ is a closed set.

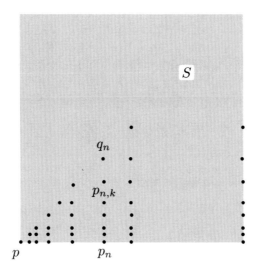

Figure 33 $S = (0,1) \times (0,1)$ and $p_n = (1/n, 0)$ converges to $p = (0,0)$ as $n \to \infty$. Each p_n is the limit of the sequence $p_{n,k} = (1/n, 1/k)$ as $k \to \infty$. The sequence $q_n = (1/n, 1/n)$ lies in S and converges to $(0,0)$. Hence: *The limits of limits are limits.*

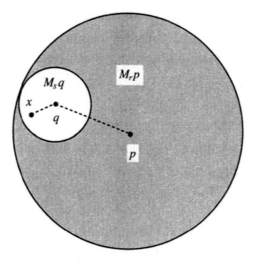

Figure 34 Why the r-neighborhood of p is an open set

To check that $M_r p$ is an open set, take any $q \in M_r p$ and observe that

$$s = r - d(p, q) > 0.$$

By the triangle inequality, if $d(q, x) < s$ then

$$d(p, x) \leq d(p, q) + d(q, x) < r,$$

and hence $M_s q \subset M_r p$. See Figure 34. Since each $q \in M_r p$ has some $M_s q$ that is contained in $M_r p$, $M_r p$ is an open set. $\qquad\square$

9 Corollary *The interval (a, b) is open in $\mathbb{R}$ and so are $(-\infty, b)$, (a, ∞), and $(-\infty, \infty)$. The interval $[a, b]$ is closed in $\mathbb{R}$.*

Proof (a, b) is the r-neighborhood of its midpoint $m = (a+b)/2$ where $r = (b-a)/2$. Therefore (a, b) is open in $\mathbb{R}$. Since the union of open sets is open we see that

$$\bigcup_{n \in \mathbb{N}} (b - n,\ b - 1/n) = (-\infty, b)$$

is open. The same applies to (a, ∞). The whole metric space $\mathbb{R} = (-\infty, \infty)$ is always open in itself.

Since the complement of $[a, b]$ is the open set $(-\infty, a) \cup (b, \infty)$, the interval $[a, b]$ is closed. $\qquad\square$

10 Corollary $\lim S$ *is the "smallest" closed set that contains S in the sense that if $K \supset S$ and K is closed then $K \supset \lim S$.*

Proof Obvious. K must contain the limit of each sequence in K that converges in M and therefore it must contain the limit of each sequence in $S \subset K$ that converges in M. These limits are exactly $\lim S$. $\qquad\square$

We refer to $\lim S$ as the **closure** of S and denote it also as $\overline{S}$. You start with S and make it closed by adding all its limits. You don't need to add any more points because limits of limits are limits. That is, $\lim(\lim S) = \lim S$. An operation like this is called **idempotent**. Doing the operation twice produces the same outcome as doing it once.

A **neighborhood** of a point p in M is any open set V that contains p. Theorem 8 implies that $V = M_r p$ is a neighborhood of p. Eventually, you will run across the phrase "closed neighborhood" of p, which refers to a closed set that contains an open set that contains p. However, until further notice all neighborhoods are open.

Usually, sets defined by strict inequalities are open while those defined by equalities or nonstrict inequalities are closed. Examples of closed sets in $\mathbb{R}$ are finite sets, $[a, b]$, $\mathbb{N}$, and the set $\{0\} \cup \{1/n : n \in N\}$. Each contains all its limits. Examples of open sets in $\mathbb{R}$ are open intervals, bounded or not.

Topological Description of Continuity

A property of a metric space or of a mapping between metric spaces that can be described solely in terms of open sets (or equivalently, in terms of closed sets) is called a **topological property**. The next result describes continuity topologically.

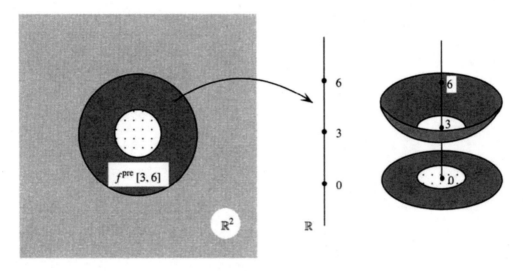

Figure 35 The function $f : (x, y) \mapsto x^2 + y^2 + 2$ and its graph over the preimage of $[3, 6]$

Let $f : M \to N$ be given. The **preimage**[†] of a set $V \subset N$ is

$$f^{\mathrm{pre}}(V) = \{p \in M : f(p) \in V\}.$$

For example, if $f : \mathbb{R}^2 \to \mathbb{R}$ is the function defined by the formula

$$f(x, y) = x^2 + y^2 + 2$$

then the preimage of the interval $[3, 6]$ in $\mathbb{R}$ is the annulus in the plane with inner radius 1 and outer radius 2. Figure 35 shows the domain of f as $\mathbb{R}^2$ and the target

---

[†]The preimage of V is also called the **inverse image** of V and is denoted by $f^{-1}(V)$. Unless f is a bijection, this notation leads to confusion. There may be no map f^{-1} and yet expressions like $V \supset f(f^{-1}(V))$ are written that mix maps and nonmaps. By the way, if f sends no point of M into V then $f^{\mathrm{pre}}(V)$ is the empty set.

as $\mathbb{R}$. The range is the set of real numbers ≥ 2. The graph of f is a paraboloid with lowest point $(0,0,2)$. The second part of the figure shows the portion of the graph lying above the annulus. You will find it useful to keep in mind the distinctions among the concepts: function, range, and graph.

11 Theorem *The following are equivalent for continuity of* $f : M \to N$.

(i) *The* (ϵ, δ)*-condition.*

(ii) *The sequential convergence preservation condition.*

(iii) *The* **closed set condition**: *The preimage of each closed set in* N *is closed in* M.

(iv) *The* **open set condition**: *The preimage of each open set in* N *is open in* M.

Proof Totally natural! By Theorem 4, (i) implies (ii).

(ii) implies (iii). If $K \subset N$ is closed in N and $p_n \in f^{\mathrm{pre}}(K)$ converges to $p \in M$ then we claim that $p \in f^{\mathrm{pre}}(K)$. By (ii), f preserves sequential convergence so

$$\lim_{n\to\infty} f(p_n) = fp.$$

Since K is closed in N, $fp \in K$, so $p \in f^{\mathrm{pre}}(K)$, and we see that $f^{\mathrm{pre}}(K)$ is closed in M. Thus (ii) implies (iii).

(iii) implies (iv). This follows by taking complements: $(f^{\mathrm{pre}}(U))^c = f^{\mathrm{pre}}(U^c)$.

(iv) implies (i). Let $\epsilon > 0$ and $p \in M$ be given. $N_\epsilon(fp)$ is open in N, so its preimage $U = f^{\mathrm{pre}}(N_\epsilon(fp))$ is open in M. The point p belongs to the preimage so openness of U implies there is a $\delta > 0$ such that $M_\delta(p) \subset U$. Then

$$f(M_\delta(p)) \subset fU \subset N_\epsilon(fp)$$

gives the ϵ, δ condition, $d_M(p,x) < \delta \implies d_N(fp, fx) < \epsilon$. See Figure 36. $\square$

I hope you find the closed and open set characterizations of continuity elegant. Note that no explicit mention is made of the metric. The open set condition is purely topological. It would be perfectly valid to take as a *definition* of continuity that the preimage of each open set is open. In fact this is exactly what's done in general topology.

12 Corollary *A homeomorphism* $f : M \to N$ *bijects the collection of open sets in* M *to the collection of open sets in* N. *It bijects the topologies.*

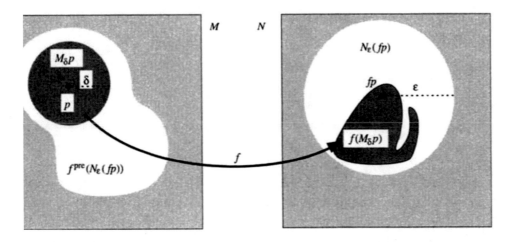

Figure 36 The ϵ, δ - condition for a continuous function $f : M \to N$

Proof Let V be an open set in N. By Theorem 11, since f is continuous, the preimage of V is open in M. Since f is a bijection, this preimage $U = \{p \in M : fp \in V\}$ is exactly the image of V by the inverse bijection, $U = f^{-1}(V)$. The same thing can be said about f^{-1} since f^{-1} is also a homeomorphism. That is, $V = fU$. Thus, sending U to fU bijects the topology of M to the topology of N. $\qquad \square$

Because of this corollary, a homeomorphism is also called a **topological equivalence**.

In general, continuous maps do not need to send open sets to open sets. For example, the squaring map $x \mapsto x^2$ from $\mathbb{R}$ to $\mathbb{R}$ is continuous but it sends the open interval $(-1, 1)$ to the nonopen interval $[0, 1)$. See also Exercise 28.

Inheritance

If a set S is contained in a metric subspace $N \subset M$ you need to be careful when you say that S is open or closed. For example,

$$S = \{x \in \mathbb{Q} : -\pi < x < \pi\}$$

is a subset of the metric subspace $\mathbb{Q} \subset \mathbb{R}$. It is both open and closed with respect to $\mathbb{Q}$ but is neither open nor closed with respect to $\mathbb{R}$. To avoid this kind of ambiguity it is best to say that S is clopen "with respect to $\mathbb{Q}$ but not with respect to $\mathbb{R}$," or more briefly that S is clopen "in $\mathbb{Q}$ but not in $\mathbb{R}$." Nevertheless there is a simple relation between the topologies of M and N when N is a metric subspace of M.

13 Inheritance Principle *Every metric subspace N of M **inherits its topology** from M in the sense that each subset $V \subset N$ which is open in N is actually the intersection $V = N \cap U$ for some $U \subset M$ that is open in M, and vice versa.*

Proof It all boils down to the fact that for each $p \in N$ we have

$$N_r p = N \cap M_r p.$$

After all, $N_r p$ is the set of $x \in N$ such that $d_N(x, p) < r$ and this is exactly the same as the set of those $x \in M_r p$ that belong to N. Therefore N inherits its r-neighborhoods from M. Since its open sets are unions of its r-neighborhoods, N also inherits its open sets from M.

In more detail, if V is open in N then it is the union of those $N_r p$ with $N_r p \subset V$. Each such $N_r p$ is $N \cap M_r p$ and the union of these $M_r p$ is U, an open subset of M. The intersection $N \cap U$ equals V. Conversely, if U is any open subset of M and $p \in V = N \cap U$ then openness of U implies there is an $M_r p \subset U$. Thus $N_r p = N \cap M_r p \subset V$, which shows that V is open in N. $\square$

14 Corollary *Every metric subspace of M inherits its closed sets from M.*

Proof By "inheriting its closed sets" we mean that each closed subset $L \subset N$ is the intersection $N \cap K$ for some closed subset $K \subset M$ and vice versa. The proof consists of two words: "Take complements." See also Exercise 34. $\square$

Let's return to the example with $\mathbb{Q} \subset \mathbb{R}$ and $S = \{x \in \mathbb{Q} : -\pi < x < \pi\}$. The set S is clopen in $\mathbb{Q}$, so we should have $S = \mathbb{Q} \cap U$ for some open set $U \subset \mathbb{R}$ and $S = \mathbb{Q} \cap K$ for some closed set $K \subset \mathbb{R}$. In fact U and K are

$$U = (-\pi, \pi) \quad \text{and} \quad K = [-\pi, \pi].$$

15 Corollary *Assume that N is a metric subspace of M and also is a closed subset of M. A set $L \subset N$ is closed in N if and only if it is closed in M. Similarly, if N is a metric subspace of M and also is an open subset of M then $U \subset N$ is open in N if and only if it is open in M.*

Proof The proof is left to the reader as Exercise 34. $\square$

Product Metrics

We next define a metric on the Cartesian product $M = X \times Y$ of two metric spaces. There are three natural ways to do so:

$$
\begin{aligned}
d_E(p, p') &= \sqrt{d_X(x, x')^2 + d_Y(y, y')^2} \\
d_{\max}(p, p') &= \max\{d_X(x, x'),\, d_Y(y, y')\} \\
d_{\text{sum}}(p, p') &= d_X(x, x') + d_Y(y, y')
\end{aligned}
$$

where $p = (x, y)$ and $p' = (x', y')$ belong to M. (d_E is the **Euclidean product metric**.) The proof that these expressions actually define metrics on M is left as Exercise 38.

16 Proposition $d_{\max} \leq d_E \leq d_{\text{sum}} \leq 2d_{\max}$.

Proof Dropping the smaller term inside the square root shows that $d_{\max} \leq d_E$; comparing the square of d_E and the square of d_{sum} shows that the latter has the terms of the former and the cross term besides, so $d_E \leq d_{\text{sum}}$; and clearly d_{sum} is no larger than twice its greater term, so $d_{\text{sum}} \leq 2d_{\max}$.   □

17 Convergence in a Product Space *The following are equivalent for a sequence $p_n = (p_{1n}, p_{2n})$ in $M = M_1 \times M_2$:*

(a) (p_n) converges with respect to the metric d_{max}.
(b) (p_n) converges with respect to the metric d_E.
(c) (p_n) converges with respect to the metric d_{sum}.
(d) (p_{1n}) and (p_{2n}) converge in M_1 and M_2 respectively.

Proof This is immediate from Proposition 16.   □

18 Corollary *If $f : M \to N$ and $g : X \to Y$ are continuous then so is their Cartesian product $f \times g$ which sends $(p, x) \in M \times X$ to $(fp, gx) \in N \times Y$.*

Proof If $(p_n, x_n) \to (p, x)$ in $M \times X$ then Theorem 17 implies $p_n \to p$ in M and $x_n \to x$ in X. By continuity, $f(p_n) \to fp$ and $g(x_n) \to gx$. By Theorem 17, $(f(p_n), g(x_n)) \to (fp, gx)$ which gives continuity of $f \times g$.   □

The three metrics make sense in the obvious way for a Cartesian product of $m \geq 3$ metric spaces. The inequality

$$
d_{\max} \leq d_E \leq d_{\text{sum}} \leq m d_{\max}
$$

is proved in the same way, and we see that Theorem 17 holds also for the product of m metric spaces. This gives

19 Corollary (**Convergence in** $\mathbb{R}^m$) *A sequence of vectors* (v_n) *in* $\mathbb{R}^m$ *converges in* $\mathbb{R}^m$ *if and only if each of its component sequences* (v_{in}) *converges,* $1 \leq i \leq m$. *The limit of the vector sequence is the vector*

$$v = \lim_{n\to\infty} v_n = \left(\lim_{n\to\infty} v_{1n}, \ \lim_{n\to\infty} v_{2n}, \ldots, \ \lim_{n\to\infty} v_{mn} \right).$$

The distance function $d : M \times M \to \mathbb{R}$ is a function from the metric space $M \times M$ to the metric space $\mathbb{R}$, so the following assertion makes sense.

20 Theorem *d is continuous.*

Proof We check (ϵ, δ)-continuity with respect to the metric d_{sum} on $M \times M$. Given $\epsilon > 0$ we take $\delta = \epsilon$. If $d_{\text{sum}}((p, q), (p', q')) < \delta$ then the triangle inequality gives

$$
\begin{aligned}
d(p, q) &\leq d(p, p') + d(p', q') + d(q', q) &< d(p', q') + \epsilon \\
d(p', q') &\leq d(p', p) + d(p, q) + d(q, q') &< d(p, q) + \epsilon
\end{aligned}
$$

which gives

$$d(p, q) - \epsilon \ < \ d(p', q') \ < \ d(p, q) + \epsilon.$$

Thus $|d(p', q') - d(p, q)| < \epsilon$ and we get continuity with respect to the metric d_{sum}. By Theorem 17 it does not matter which metric we use on $M \times M$. $\qquad\square$

As you can see, the use of d_{sum} simplifies the proof by avoiding square root manipulations. The sum metric is also called the **Manhattan metric** or the **taxicab metric**. Figure 37 shows the "unit discs" with respect to these metrics in $\mathbb{R}^2$. See also Exercise 2.

21 Corollary *The metrics* $d_{\max}$, d_E, *and* d_{sum} *are continuous.*

Proof Theorem 20 asserts that all metrics are continuous. $\qquad\square$

22 Corollary *The absolute value is a continuous mapping* $\mathbb{R} \to \mathbb{R}$. *In fact the norm is a continuous mapping from any normed space to* $\mathbb{R}$.

Proof $\|v\| = d(v, 0)$. $\qquad\square$

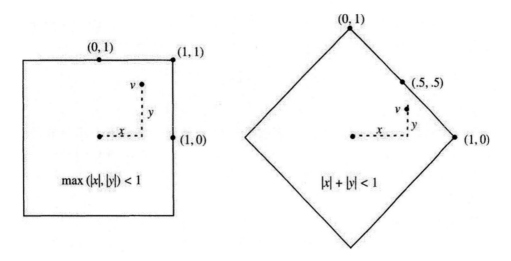

Figure 37 The unit disc in the max metric is a square, and in the sum
metric it is a rhombus.

Completeness

In Chapter 1 we discussed the Cauchy criterion for convergence of a sequence of
real numbers. There is a natural way to carry these ideas over to a metric space M.
The sequence (p_n) in M satisfies a **Cauchy condition** provided that for each $\epsilon > 0$
there is an integer N such that for all $k, n \geq N$ we have $d(p_k, p_n) < \epsilon$, and (p_n) is
said to be a **Cauchy sequence**. In symbols,

$$\forall \epsilon > 0 \ \exists N \text{ such that } k, n \geq N \ \Rightarrow \ d(p_k, p_n) < \epsilon.$$

The terms of a Cauchy sequence "bunch together" as $n \to \infty$. Each convergent
sequence (p_n) is Cauchy. For if (p_n) converges to p as $n \to \infty$ then, given $\epsilon > 0$, there
is an N such that for all $n \geq N$ we have

$$d(p_n, p) < \frac{\epsilon}{2}.$$

By the triangle inequality, if $k, n \geq N$ then

$$d(p_k, p_n) \ \leq \ d(p_k, p) + d(p, p_n) < \epsilon,$$

so convergence $\Rightarrow$ Cauchy.

Theorem 1.5 states that the converse is true in the metric space $\mathbb{R}$. Every Cauchy
sequence in $\mathbb{R}$ converges to a limit in $\mathbb{R}$. In the general metric space, however, this

need not be true. For example, consider the metric space $\mathbb{Q}$ of rational numbers, equipped with the inherited metric $d(x, y) = |x - y|$, and consider the sequence

$$(r_n) = (1.4,\ 1.41,\ 1.414,\ 1.4142,\ \ldots).$$

It is Cauchy. Given $\epsilon > 0$, choose $N > -\log_{10} \epsilon$. If $k, n \geq N$ then $|r_k - r_n| \leq 10^{-N} < \epsilon$. Nevertheless, (r_n) refuses to converge in $\mathbb{Q}$. After all, as a sequence in $\mathbb{R}$ it converges to $\sqrt{2}$, and if it also converges to some $r \in \mathbb{Q}$, then by uniqueness of limits in $\mathbb{R}$ we have $r = \sqrt{2}$, something we know is false. In brief, convergence $\Rightarrow$ Cauchy but not conversely.

A metric space M is **complete** if each Cauchy sequence in M converges to a limit in M. Theorem 1.5 states that $\mathbb{R}$ is complete.

23 Theorem $\mathbb{R}^m$ *is complete.*

Proof Let (p_n) be a Cauchy sequence in $\mathbb{R}^m$. Express p_n in components as

$$p_n = (p_{1n}, \ldots,\ p_{mn}).$$

Because (p_n) is Cauchy, each component sequence $(p_{in})_{n \in \mathbb{N}}$ is Cauchy. Completeness of $\mathbb{R}$ implies that the component sequences converge, and therefore the vector sequence converges. □

24 Theorem *Every closed subset of a complete metric space is a complete metric subspace.*

Proof Let A be a closed subset of the complete metric space M and let (p_n) be a Cauchy sequence in A with respect to the inherited metric. It is of course also a Cauchy sequence in M and therefore it converges to a limit p in M. Since A is closed we have $p \in A$. □

25 Corollary *Every closed subset of Euclidean space is a complete metric space.*

Proof Obvious from the previous theorem and completeness of $\mathbb{R}^m$. □

Remark Completeness is *not* a topological property. For example, consider $\mathbb{R}$ with its usual metric and $(-1, 1)$ with the metric it inherits from $\mathbb{R}$. Although they are homeomorphic metric spaces, $\mathbb{R}$ is complete but $(-1, 1)$ is not.

In Section 10 we explain how every metric space can be completed.

4 Compactness

Compactness is the single most important concept in real analysis. It is what reduces the infinite to the finite.

Definition A subset A of a metric space M is (sequentially) **compact** if every sequence (a_n) in A has a subsequence (a_{n_k}) that converges to a limit in A.

The empty set and finite sets are trivial examples of compact sets. For a sequence (a_n) contained in a finite set repeats a term infinitely often, and the corresponding constant subsequence converges.

Compactness is a *good* feature of a set. We will develop criteria to decide whether a set is compact. The first is the most often used, but beware! – its converse is generally false.

26 Theorem *Every compact set is closed and bounded.*

Proof Suppose that A is a compact subset of the metric space M and that p is a limit of A. Does p belong to A? There is a sequence (a_n) in A converging to p. By compactness, some subsequence (a_{n_k}) converges to some $q \in A$, but every subsequence of a convergent sequence converges to the same limit as does the mother sequence, so $q = p$ and $p \in A$. Thus A is closed.

To see that A is bounded, choose and fix any point $p \in M$. Either A is bounded or else for each $n \in \mathbb{N}$ there is a point $a_n \in A$ such that $d(p, a_n) \geq n$. Compactness implies that some subsequence (a_{n_k}) converges. Convergent sequences are bounded, which contradicts the fact that $d(p, a_{n_k}) \to \infty$ as $k \to \infty$. Therefore (a_n) cannot exist and for some large r we have $A \subset M_r p$, which is what it means that A is bounded. □

27 Theorem *The closed interval $[a, b] \subset \mathbb{R}$ is compact.*

Proof Let (x_n) be a sequence in $[a, b]$ and set

$$C = \{x \in [a, b] : x_n < x \text{ only finitely often}\}.$$

Equivalently, for all but finitely many n, $x_n \geq x$. Since $a \in C$ we know that $C \neq \emptyset$. Clearly b is an upper bound for C. By the least upper bound property of $\mathbb{R}$ there exists $c = \mathrm{l.u.b.}\, C$ with $c \in [a, b]$. We claim that a subsequence of (x_n) converges to c. Suppose not, i.e., no subsequence of (x_n) converges to c. Then for some $r > 0$, x_n lies in $(c - r,\ c + r)$ only finitely often, which implies that $c + r \in C$, contrary to c being an upper bound for C. Hence some subsequence of (x_n) does converge to c and $[a, b]$ is compact. □

To pass from $\mathbb{R}$ to $\mathbb{R}^m$ we think about compactness for Cartesian products.

28 Theorem *The Cartesian product of two compact sets is compact.*

Proof Let $(a_n, b_n) \in A \times B$ be given where $A \subset M$ and $B \subset N$ are compact. There exists a subsequence (a_{n_k}) that converges to some point $a \in A$ as $k \to \infty$. The subsequence (b_{n_k}) has a sub-subsequence $(b_{n_{k(\ell)}})$ that converges to some $b \in B$ as $\ell \to \infty$. The sub-subsequence $(a_{n_{k(\ell)}})$ continues to converge to the point a. Thus

$$(a_{n_{k(\ell)}}, b_{n_{k(\ell)}}) \to (a, b)$$

as $\ell \to \infty$. This implies that $A \times B$ is compact. $\qquad\square$

29 Corollary *The Cartesian product of m compact sets is compact.*

Proof Write $A_1 \times A_2 \times \cdots \times A_m = A_1 \times (A_2 \times \cdots \times A_m)$ and perform induction on m. (Theorem 28 handles the bottom case $m = 2$.) $\qquad\square$

30 Corollary *Every box $[a_1, b_1] \times \cdots \times [a_m, b_m]$ in $\mathbb{R}^m$ is compact.*

Proof Obvious from Theorem 27 and the previous corollary. $\qquad\square$

An equivalent formulation of these results is the

31 Bolzano-Weierstrass Theorem *Every bounded sequence in $\mathbb{R}^m$ has a convergent subsequence.*

Proof A bounded sequence is contained in a box, which is compact, and therefore the sequence has a subsequence that converges to a limit in the box. See also Exercise 11. $\qquad\square$

Here is a simple fact about compacts.

32 Theorem *Every closed subset of a compact is compact.*

Proof If A is a closed subset of the compact set K and if (a_n) is a sequence of points in A then clearly (a_n) is also a sequence of points in K, so by compactness of K there is a subsequence (a_{n_k}) converging to a limit $p \in K$. Since A is closed, p lies in A which proves that A is compact. $\qquad\square$

Now we come to the first partial converse to Theorem 26.

33 Heine-Borel Theorem *Every closed and bounded subset of $\mathbb{R}^m$ is compact.*

Proof Let $A \subset \mathbb{R}^m$ be closed and bounded. Boundedness implies that A is contained in some box, which is compact. Since A is closed, Theorem 32 implies that A is compact. See also Exercise 11. $\qquad\square$

The Heine-Borel Theorem states that closed and bounded subsets of Euclidean space are compact, but it is *vital*[†] to remember that a closed and bounded subset of a general metric space may fail to be compact. For example, the set $\mathbb{N}$ of natural numbers equipped with the discrete metric is a complete metric space, it is closed in itself (as is every metric space), and it is bounded. But it is not compact. After all, what subsequence of $1, 2, 3, \ldots$ converges?

A more striking example occurs in the metric space $C([0,1], \mathbb{R})$ whose metric is $d(f, g) = \max\{|f(x) - g(x)|\}$. The space is complete but its closed unit ball is not compact. For example, the sequence of functions $f_n = x^n$ has no subsequence that converges with respect to the metric d. In fact every closed ball is noncompact.

Ten Examples of Compact Sets

1. Any finite subset of a metric space, for instance the empty set.
2. Any closed subset of a compact set.
3. The union of finitely many compact sets.
4. The Cartesian product of finitely many compact sets.
5. The intersection of arbitrarily many compact sets.
6. The closed unit ball in $\mathbb{R}^3$.
7. The boundary of a compact set, for instance the unit 2-sphere in $\mathbb{R}^3$.
8. The set $\{x \in \mathbb{R} : \exists n \in \mathbb{N} \text{ and } x = 1/n\} \cup \{0\}$.
9. The Hawaiian earring. See page 58.
10. The Cantor set. See Section 8.

Nests of Compacts

If $A_1 \supset A_2 \supset \cdots \supset A_n \supset A_{n+1} \supset \ldots$ then (A_n) is a **nested sequence** of sets. Its intersection is

$$\bigcap_{n=1}^{\infty} A_n = \{p : \text{for each } n \text{ we have } p \in A_n\}.$$

[†] I have asked variants of the following True or False question on every analysis exam I've given: "Every closed and bounded subset of a complete metric space is compact." You would be surprised at how many students answer "True."

See Figure 38.

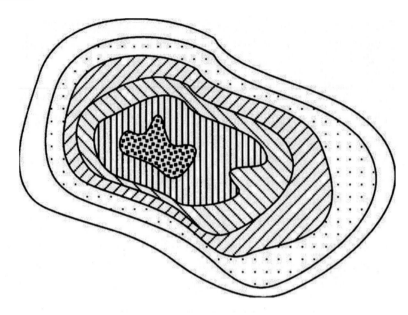

Figure 38 A nested sequence of sets

For example, we could take A_n to be the disc $\{z \in \mathbb{R}^2 : |z| \leq 1/n\}$. The intersection of all the sets A_n is then the singleton $\{0\}$. On the other hand, if A_n is the ball $\{z \in \mathbb{R}^3 : |z| \leq 1 + 1/n\}$ then $\bigcap A_n$ is the closed unit ball B^3.

34 Theorem *The intersection of a nested sequence of compact nonempty sets is compact and nonempty.*

Proof Let (A_n) be such a sequence. By Theorem 26, A_n is closed. The intersection of closed sets is always closed. Thus, $\bigcap A_n$ is a closed subset of the compact set A_1 and is therefore compact. It remains to show that the intersection is nonempty.

A_n is nonempty, so for each $n \in \mathbb{N}$ we can choose $a_n \in A_n$. The sequence (a_n) lies in A_1 since the sets are nested. Compactness of A_1 implies that (a_n) has a subsequence (a_{n_k}) converging to some point $p \in A_1$. The limit p also lies in the set A_2 since except possibly for the first term, the subsequence (a_{n_k}) lies in A_2 and A_2 is a closed set. The same is true for A_3 and for all the sets in the nested sequence. Thus, $p \in \bigcap A_n$ and $\bigcap A_n$ is shown to be nonempty. $\qquad\square$

The **diameter** of a nonempty set $S \subset M$ is the supremum of the distances $d(x, y)$ between points of S.

35 Corollary *If in addition to being nested, nonempty, and compact, the sets A_n have diameter that tends to 0 as $n \to \infty$ then $A = \bigcap A_n$ is a single point.*

Proof For each $n \in \mathbb{N}$, A is a subset of A_n, which implies that A has diameter zero. Since distinct points lie at positive distance from each other, A consists of at most one point, while by Theorem 34 it consists of at least one point. See also Exercise 52. $\square$

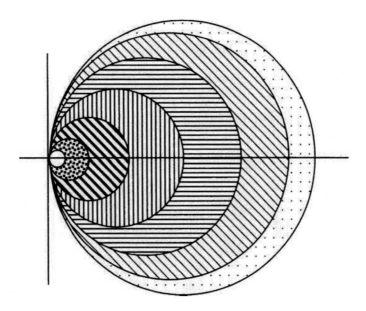

Figure 39 This nested sequence has empty intersection.

Figure 39 shows a nested sequence of nonempty *noncompact* sets with empty intersection. They are the open discs with center $(1/n, 0)$ on the x-axis and radius $1/n$. They contain no common point. (Their closures do intersect at a common point, the origin.)

Continuity and Compactness

Next we discuss how compact sets behave under continuous transformations.

36 Theorem *If $f : M \to N$ is continuous and A is a compact subset of M then fA is a compact subset of N. That is, the continuous image of a compact is compact.*

Proof Suppose that (b_n) is a sequence in fA. For each $n \in \mathbb{N}$ choose a point $a_n \in A$ such that $f(a_n) = b_n$. By compactness of A there exists a subsequence (a_{n_k}) that converges to some point $p \in A$. By continuity of f it follows that

$$b_{n_k} = f(a_{n_k}) \to fp \in fA$$

as $k \to \infty$. Thus, every sequence (b_n) in fA has a subsequence converging to a limit in fA, which shows that fA is compact. □

From Theorem 36 follows the natural generalization of the min/max theorem in Chapter 1 which concerns continuous real-valued functions defined on an interval $[a, b]$. See Theorem 1.23.

37 Corollary *A continuous real-valued function defined on a compact set is bounded; it assumes maximum and minimum values.*

Proof Let $f : M \to \mathbb{R}$ be continuous and let A be a compact subset of M. Theorem 36 implies that fA is a compact subset of $\mathbb{R}$, so by Theorem 26 it is closed and bounded. Thus, the greatest lower bound, v, and least upper bound, V, of fA exist and belong to fA; i.e., there exist points $p, P \in A$ such that for all $a \in A$ we have $v = fp \leq fa \leq fP = V$. □

Homeomorphisms and Compactness

A homeomorphism is a bicontinuous bijection. Originally, compactness was called bicompactness. This is reflected in the next theorem.

38 Theorem *If M is compact and M is homeomorphic to N then N is compact. Compactness is a topological property.*

Proof N is the continuous image of M, so by Theorem 36 it is compact. □

39 Corollary $[0, 1]$ *and* $\mathbb{R}$ *are not homeomorphic.*

Proof One is compact and the other isn't. □

40 Theorem *If M is compact then a continuous bijection $f : M \to N$ is a homeomorphism – its inverse bijection $f^{-1} : N \to M$ is automatically continuous.*

Proof Suppose that $q_n \to q$ in N. Since f is a bijection, $p_n = f^{-1}(q_n)$ and $p = f^{-1}(q)$ are well defined points in M. To check continuity of f^{-1} we must show that $p_n \to p$.

If (p_n) refuses to converge to p then there is a subsequence (p_{n_k}) and a $\delta > 0$ such that for all k we have $d(p_{n_k}, p) \geq \delta$. Compactness of M gives a sub-subsequence $(p_{n_{k(\ell)}})$ that converges to a point $p^* \in M$ as $\ell \to \infty$.

Necessarily, $d(p, p^*) \geq \delta$, which implies that $p \neq p^*$. Since f is continuous we have

$$f(p_{n_{k(\ell)}}) \to f(p^*)$$

as $\ell \to \infty$. The limit of a convergent sequence is unchanged by passing to a subsequence, and so $f(p_{n_{k(\ell)}}) = q_{n_{k(\ell)}} \to q$ as $\ell \to \infty$. Thus, $f(p^*) = q = f(p)$, contrary to f being a bijection. It follows that $p_n \to p$ and therefore that f^{-1} is continuous. $\qquad \square$

If M is not compact then Theorem 40 becomes false. For example, the function $f : [0, 2\pi) \to \mathbb{R}^2$ defined by $f(x) = (\cos x, \sin x)$ is a continuous bijection onto the unit circle in the plane, but it is not a homeomorphism. This useful example was discussed on page 65. Not only does this f fail to be a homeomorphism, but there is no homeomorphism at all from $[0, 2\pi)$ to S^1. The circle is compact while $[0, 2\pi)$ is not. Therefore they are not homeomorphic. See also Exercises 49 and 50.

Embedding a Compact

Not only is a compact space M closed in itself, as is every metric space, but it is also a closed subset of each metric space in which it is embedded. More precisely we say that $h : M \to N$ **embeds** M into N if h is a homeomorphism from M onto hM. (The metric on hM is the one it inherits from N.) Topologically M and hM are equivalent. A property of M that holds also for every embedded copy of M is an **absolute** or **intrinsic** property of M.

41 Theorem *A compact is absolutely closed and absolutely bounded.*

Proof Obvious from Theorems 26 and 36. $\qquad\qquad\qquad\qquad\qquad\qquad \square$

For example, no matter how the circle is embedded in $\mathbb{R}^3$, its image is always closed and bounded. See also Exercises 48 and 120.

Uniform Continuity and Compactness

In Chapter 1 we defined the concept of uniform continuity for real-valued functions of a real variable. The definition in metric spaces is analogous. A function $f : M \to N$ is **uniformly continuous** if for each $\epsilon > 0$ there exists a $\delta > 0$ such that

$$p, q \in M \text{ and } d_M(p, q) < \delta \quad \Rightarrow \quad d_N(fp, fq) < \epsilon.$$

42 Theorem *Every continuous function defined on a compact is uniformly continuous.*

Proof Suppose not, and $f : M \to N$ is continuous, M is compact, but f fails to be uniformly continuous. Then there is some $\epsilon > 0$ such that no matter how small

δ is, there exist points $p, q \in M$ with $d(p, q) < \delta$ but $d(fp, fq) \geq \epsilon$. Take $\delta = 1/n$ and let (p_n) and (q_n) be sequences of points in M such that $d(p_n, q_n) < 1/n$ while $d(f(p_n), f(q_n)) \geq \epsilon$. Compactness of M implies that there is a subsequence p_{n_k} which converges to some $p \in M$ as $k \to \infty$. Since $d(p_n, q_n) < 1/n \to 0$ as $n \to \infty$, (q_{n_k}) converges to the same limit as does (p_{n_k}) as $k \to \infty$; namely $q_{n_k} \to p$. Continuity at p implies that $f(p_{n_k}) \to fp$ and $f(q_{n_k}) \to fp$. If k is large then

$$d(f(p_{n_k}), f(q_{n_k})) \ \leq \ d(f(p_{n_k}), fp) + d(fp, f(q_{n_k})) \ < \ \epsilon,$$

contrary to the supposition that $d(f(p_n), f(q_n)) \geq \epsilon$ for all n. $\qquad\square$

Theorem 42 gives a second proof that continuity implies uniform continuity on an interval $[a, b]$. For $[a, b]$ is compact.

5 Connectedness

As another application of these ideas, we consider the general notion of connectedness. Let A be a subset of a metric space M. If A is neither the empty set nor M then A is a **proper** subset of M. Recall that if A is both closed and open in M it is said to be clopen. The complement of a clopen set is clopen. The complement of a proper subset is proper.

If M has a proper clopen subset A then M is **disconnected**. For there is a **separation** of M into proper, disjoint clopen subsets,

$$M \ = \ A \sqcup A^c.$$

(The notation $\sqcup$ indicates disjoint union.) M is **connected** if it is not disconnected, i.e., it contains no proper clopen subset. Connectedness of M does not mean that M is connected *to* something, but rather that M is one unbroken thing. See Figure 40.

Figure 40 M and N illustrate the difference between being connected and being disconnected.

43 Theorem *If M is connected, $f : M \to N$ is continuous, and f is onto then N is connected. The continuous image of a connected is connected.*

Proof Simple! If A is a clopen proper subset of N then, according to the open and closed set conditions for continuity, $f^{\mathrm{pre}}(A)$ is a clopen subset of M. Since f is onto and $A \neq \emptyset$, we have $f^{\mathrm{pre}}(A) \neq \emptyset$. Similarly, $f^{\mathrm{pre}}(A^c) \neq \emptyset$. Therefore $f^{\mathrm{pre}}(A)$ is a proper clopen subset of M, contrary to M being connected. It follows that A cannot exist and that N is connected. $\qquad\square$

44 Corollary *If M is connected and M is homeomorphic to N then N is connected. Connectedness is a topological property.*

Proof N is the continuous image of M, so Theorem 43 implies it is connected. $\quad\square$

45 Corollary (Generalized Intermediate Value Theorem) *Every continuous real-valued function defined on a connected domain has the intermediate value property.*

Proof Assume that $f : M \to \mathbb{R}$ is continuous and M is connected. If f assumes values $\alpha < \beta$ in $\mathbb{R}$ and if it fails to assume some value γ with $\alpha < \gamma < \beta$, then

$$M = \{x \in M : f(x) < \gamma\} \sqcup \{x \in M : f(x) > \gamma\}$$

is a separation of M, contrary to connectedness. $\qquad\square$

46 Theorem $\mathbb{R}$ *is connected.*

Proof If $U \subset \mathbb{R}$ is nonempty and clopen we claim that $U = \mathbb{R}$. Choose some $p \in U$ and consider the set

$$X = \{x \in U : \text{the open interval } (p, x) \text{ is contained in } U\}.$$

X is nonempty since U is open. Let s be the supremum of X. If s is finite (i.e., X is bounded above) then $s = \mathrm{l.\,u.\,b.}\, X$ and s is a limit of X. Since $X \subset U$ and U is closed we have $s \in U$. Since U is open there is an interval $(s - r, s + r) \subset U$, which gives an interval $(p, s + r) \subset U$, contrary to s being an upper bound for X. Hence $s = \infty$ and $U \supset (p, \infty)$. The same reasoning gives $U \supset (-\infty, p)$, so $U = \mathbb{R}$ as claimed. Thus there are no proper clopen subsets of $\mathbb{R}$ and $\mathbb{R}$ is connected. $\qquad\square$

47 Corollary (Intermediate Value Theorem for $\mathbb{R}$) *Every continuous function $f : \mathbb{R} \to \mathbb{R}$ has the intermediate value property.*

Proof Immediate from the Generalized Intermediate Value Theorem and connectedness of $\mathbb{R}$.   □

48 Corollary *The following metric spaces are connected: The intervals (a, b), $[a, b]$, the circle, and all capital letters of the Roman alphabet.*

Proof The interval (a, b) is homeomorphic to $\mathbb{R}$, while $[a, b]$ is the continuous image of $\mathbb{R}$ under the map whose graph is shown in Figure 41. The circle is the continuous image of $\mathbb{R}$ under the map $t \mapsto (\cos t, \sin t)$. Connectedness of the letters $A, \ldots, Z$ is equally clear.   □

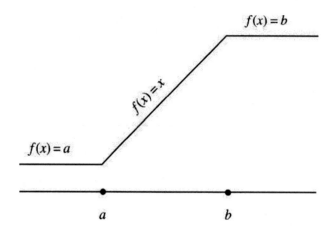

Figure 41 The function f surjects $\mathbb{R}$ continuously to $[a, b]$.

Connectedness properties give a good way to distinguish nonhomeomorphic sets.

Example The union of two disjoint closed intervals is not homeomorphic to a single interval. One set is disconnected and the other is connected.

Example The closed interval $[a, b]$ is not homeomorphic to the circle S^1. For removal of a point $x \in (a, b)$ disconnects $[a, b]$ while the circle remains connected upon removal of any point. More precisely, suppose that $h : [a, b] \to S^1$ is a homeomorphism. Choose a point $x \in (a, b)$, and consider $X = [a, b] \setminus \{x\}$. The restriction of h to X is a homeomorphism from X onto Y, where Y is the circle with the point hx removed. But X is disconnected while Y is connected. Hence h cannot exist and the segment is not homeomorphic to the circle.

Example The circle is not homeomorphic to the figure eight. Removing any two points of the circle disconnects it, but this is not true of the figure eight. Or, removing the crossing point disconnects the figure eight but removing any point of the circle leaves it connected.

Example The circle is not homeomorphic to the disc. For removing two points disconnects the circle but does not disconnect the disc.

As you can see, it is useful to be able to recognize disconnected subsets S of a metric space M. By definition, S is a disconnected subset of M if it is disconnected when considered in its own right as a metric space with the metric it inherits from M; i.e., it has a separation $S = A \sqcup B$ such that A and B are proper clopen subsets of S. The sets A, B are separated in S but they need not be separated in M. Their closures in M may intersect.

Example The punctured interval $X = [a, b] \setminus \{c\}$ is disconnected if $a < c < b$. For $X = [a, c) \sqcup (c, b]$ is a separation of X. The closures of the two sets with respect to the metric space X do not intersect, even though their closures with respect to $\mathbb{R}$ do intersect. Pay attention to this phenomenon which is related to the Inheritance Principle.

Example Any subset Y of the punctured interval is disconnected if it meets both $[a, c)$ and $(c, b]$. For $Y = ([a, c) \cap Y) \sqcup ((c, b] \cap Y)$ is a separation of Y.

49 Theorem *The closure of a connected set is connected. More generally, if $S \subset M$ is connected and $S \subset T \subset \overline{S}$ then T is connected.*

Proof It is equivalent to show that if T is disconnected then S is disconnected. Disconnectedness of T implies that

$$T = A \sqcup B$$

where A, B are clopen and proper in T. It is natural to expect that

$$S = K \sqcup L$$

is a separation of S where $K = A \cap S$ and $L = B \cap S$. The sets K and L are disjoint, their union is S, and by the Inheritance Principle they are clopen. But are they proper?

If $K = \emptyset$ then $A \subset S^c$. Since A is proper there exists $p \in A$. Since A is open in T, there exists a neighborhood $M_r p$ such that

$$T \cap M_r p \ \subset \ A \ \subset \ S^c.$$

The neighborhood $M_r p$ contains no points of S, which is contrary to p belonging to $\overline{S}$. Thus, $K \neq \emptyset$. Similarly, $L = B \cap S \neq \emptyset$, so $S = K \sqcup L$ is a separation of S, proving that S is disconnected. $\qquad \square$

Example The outward spiral expressed in polar coordinates as

$$S = \{(r, \theta) : (1 - r)\theta = 1 \text{ and } \theta \geq \pi/2\}$$

has $\overline{S} = S \cup S^1$, where S^1 is the unit circle. Since S is connected, so is $\overline{S}$. (Recall that $\overline{S}$ is the closure of S.) See Figure 27.

50 Theorem *The union of connected sets sharing a common point p is connected.*

Proof Let $S = \bigcup S_\alpha$, where each S_α is connected and $p \in \bigcap S_\alpha$. If S is disconnected then it has a separation $S = A \sqcup A^c$ where A, A^c are proper and clopen. One of them contains p; say it is A. Then $A \cap S_\alpha$ is a nonempty clopen subset of S_α. Since S_α is connected, $A \cap S_\alpha = S_\alpha$ for each α, and $A = S$. This implies that $A^c = \emptyset$, a contradiction. Therefore S is connected. $\qquad \square$

Example The 2-sphere S^2 is connected. For S^2 is the union of great circles, each passing through the poles.

Example Every convex set C in $\mathbb{R}^m$ (or in any metric space with a compatible linear structure) is connected. If we choose a point $p \in C$ then each $q \in C$ lies on a line segment $[p, q] \subset C$. Thus, C is the union of connected sets sharing the common point p. It is connected.

Definition A **path** joining p to q in a metric space M is a continuous function $f : [a, b] \to M$ such that $fa = p$ and $fb = q$. If each pair of points in M can be joined by a path in M then M is **path-connected**. See Figure 42.

51 Theorem *Path-connected implies connected.*

Proof Assume that M is path-connected but not connected. Then $M = A \sqcup A^c$ for some proper clopen $A \subset M$. Choose $p \in A$ and $q \in A^c$. There is a path $f : [a, b] \to M$ from p to q. The separation $f^{\text{pre}}(A) \sqcup f^{\text{pre}}(A^c)$ contradicts connectedness of $[a, b]$. Therefore M is connected. $\qquad \square$

Example All connected subsets of $\mathbb{R}$ are path-connected. See Exercise 67.

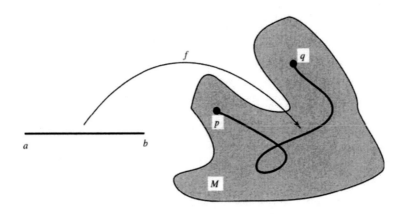

Figure 42 A path f in M that joins p to q

Example Every open connected subset of $\mathbb{R}^m$ is path-connected. See Exercises 61 and 66.

Example The **topologist's sine curve** is a compact connected set that is not path-connected. It is $M = G \cup Y$ where

$$G = \{(x,y) \in \mathbb{R}^2 : y = \sin 1/x \text{ and } 0 < x \le 1/\pi\}$$
$$Y = \{(0,y) \in \mathbb{R}^2 : -1 \le y \le 1\}.$$

See Figure 43. The metric on M is just Euclidean distance. Is M connected? Yes!

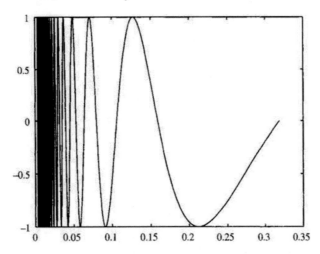

Figure 43 The topologist's sine curve M is a connected set. It includes the vertical segment Y at $x = 0$.

The graph G is connected and $M = \overline{G}$. By Theorem 49 M is connected.

6 Other Metric Space Concepts

Here are a few standard metric space topics related to what appears above. If $S \subset M$ then its **closure** is the smallest closed subset of M that contains S, its **interior** is the largest open subset of M contained in S, and its **boundary** is the difference between its closure and its interior. Their notations are

$$\overline{S} = \operatorname{cl} S = \text{ closure of } S \quad \operatorname{int} S = \text{ interior of } S \quad \partial S = \text{ boundary of } S.$$

To avoid inheritance ambiguity it would be better (but too cumbersome) to write $\operatorname{cl}_M S$, $\operatorname{int}_M S$, and $\partial_M S$ to indicate the ambient space M. In Exercise 95 you are asked to check various simple facts about them, such as $\overline{S} = \lim S =$ the intersection of all closed sets that contain S.

Clustering and Condensing

Two concepts similar to limits are clustering and condensing. The set S "clusters" at p (and p is a **cluster point**[†] of S) if each $M_r p$ contains infinitely many points of S. The set S **condenses** at p (and p is a **condensation point** of S) if each $M_r p$ contains uncountably many points of S. Thus, S limits at p, clusters at p, or condenses at p according to whether each $M_r p$ contains some, infinitely many, or uncountably many points of S. See Figure 44.

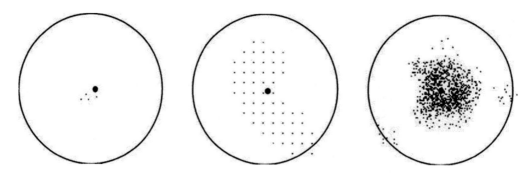

Figure 44 Limiting, clustering, and condensing behavior

[†]Cluster points are also called **accumulation points**. As mentioned above, they are also sometimes called limit points, a usage that conflicts with the limit idea. A finite set S has no cluster points, but of course, each of its points p is a limit of S since the constant sequence $(p, p, p, \ldots)$ converges to p.

52 Theorem *The following are equivalent conditions to S clustering at p.*

(i) *There is a sequence of distinct points in S that converges to p.*

(ii) *Each neighborhood of p contains infinitely many points of S.*

(iii) *Each neighborhood of p contains at least two points of S.*

(iv) *Each neighborhood of p contains at least one point of S other than p.*

Proof Clearly (i) $\Rightarrow$ (ii) $\Rightarrow$ (iii) $\Rightarrow$ (iv), and (ii) is the definition of clustering. It remains to check (iv) $\Rightarrow$ (i).

Assume (iv) is true: Each neighborhood of p contains a point of S other than p. In $M_1 p$ choose a point $p_1 \in (S \setminus \{p\})$. Set $r_2 = \min(1/2, d(p_1, p))$, and in the smaller neighborhood $M_{r_2} p$, choose $p_2 \in (S \setminus \{p\})$. Proceed inductively: Set $r_n = \min(1/n, d(p_{n-1}, p))$ and in $M_{r_n} p$, choose $p_n \in (S \setminus \{p\})$. Since $r_n \to 0$ the sequence (p_n) converges to p. The points p_n are distinct since they have different distances to p,

$$d(p_1, p) \geq r_2 > d(p_2, p) \geq r_3 > d(p_3, p) \geq \ldots.$$

Thus (iv) $\Rightarrow$ (i) and the four conditions are equivalent. $\qquad\square$

Condition (iv) is the form of the definition of clustering most frequently used, although it is the hardest to grasp. It is customary to denote by S' the set of cluster points of S.

53 Proposition $S \cup S' = \overline{S}$.

Proof A cluster point is a type of limit of S, so $S' \subset \lim S = \overline{S}$ and

$$S \cup S' \subset \overline{S}$$

On the other hand, if $p \in \overline{S}$ then either $p \in S$ or else $p \notin S$ and each neighborhood of p contains points of S other than p. This implies that $p \in S \cup S'$, so $\overline{S} \subset S \cup S'$, and the two sets are equal. $\qquad\square$

54 Corollary S *is closed if and only if $S' \subset S$.*

Proof S is closed if and only if $S = \overline{S}$. Since $\overline{S} = S \cup S'$, equivalent to $S' \subset S$ is $\overline{S} = S$. $\qquad\square$

55 Corollary *The least upper bound and greatest lower bound of a nonempty bounded set $S \subset \mathbb{R}$ belong to the closure of S. Thus, if S is closed then they belong to S.*

Proof If $b = \mathrm{l.\,u.\,b.}\, S$ then each interval $(b - r, b]$ contains points of S. The same is true for intervals $[a, a + r)$ where $a = \mathrm{g.\,l.\,b.}\, S$ $\qquad\square$

Perfect Metric Spaces

A metric space M is **perfect** if $M' = M$, i.e., each $p \in M$ is a cluster point of M. Recall that M clusters at p if each $M_r p$ is an infinite set. For example $[a, b]$ is perfect and $\mathbb{Q}$ is perfect. $\mathbb{N}$ is not perfect since none of its points are cluster points.

56 Theorem *Every nonempty, perfect, complete metric space is uncountable.*

Proof Suppose not: Assume M is nonempty, perfect, complete, and countable. Since M consists of cluster points it must be denumerable and not finite. Say

$$M = \{x_1, x_2, \ldots\}$$

is a list of all the elements of M. We will derive a contradiction by finding a point of M not in the list. Define

$$\widehat{M_r} p = \{q \in M \ : \ d(p, q) \leq r\}.$$

It is the **closed neighborhood** of radius r at p. Choose any $y_1 \in M$ with $y_1 \neq x_1$ and choose $r_1 > 0$ so that $Y_1 = \widehat{M_{r_1}}(y_1)$ "excludes" x_1 in the sense that $x_1 \notin Y_1$. We can take r_1 as small as we want, say $r_1 < 1$.

Since M clusters at y_1 we can choose $y_2 \in M_{r_1}(y_1)$ with $y_2 \neq x_2$ and choose $r_2 > 0$ so that $Y_2 = \widehat{M_{r_2}}(y_2)$ excludes x_2. Taking r_2 small ensures $Y_2 \subset Y_1$. (Here we are using openness of $M_{r_1}(y_1)$.) Also we take $r_2 < 1/2$. Since $Y_2 \subset Y_1$, it excludes x_1 as well as x_2. See Figure 45.

Nothing stops us from continuing inductively, and we get a nested sequence of closed neighborhoods $Y_1 \supset Y_2 \supset Y_3 \ldots$ such that Y_n excludes $x_1, \ldots, x_n$, and has radius $r_n \leq 1/n$. Thus the center points y_n form a Cauchy sequence. Completeness of M implies that

$$\lim_{n \to \infty} y_n = y \in M$$

exists. Since the sets Y_n are closed and nested, $y \in Y_n$ for each n. Does y equal x_1? No, for Y_1 excludes x_1. Does it equal x_2? No, for Y_2 excludes x_2. In fact, for each n we have $y \neq x_n$. The point y is nowhere in the supposedly complete list of elements of M, a contradiction. Hence M is uncountable. $\qquad \square$

57 Corollary $\mathbb{R}$ *and* $[a, b]$ *are uncountable.*

Proof $\mathbb{R}$ is complete and perfect, while $[a, b]$ is compact, therefore complete, and perfect. Neither is empty. $\qquad \square$

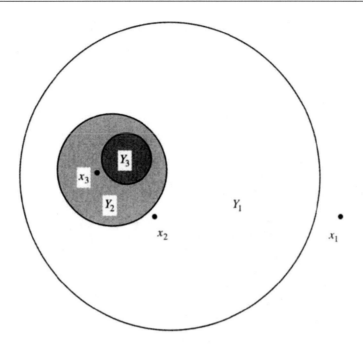

Figure 45 The exclusion of successively more points of the sequence (x_n)
that supposedly lists all the elements of M

58 Corollary *Every nonempty perfect complete metric space is everywhere uncountable in the sense that each r-neighborhood is uncountable.*

Proof The $r/2$-neighborhood $M_{r/2}(p)$ is perfect: It clusters at each of its points.
The closure of a perfect set is perfect. Thus, $\overline{M_{r/2}(p)}$ is perfect. Being a closed
subset of a complete metric space, it is complete. According to Theorem 56, $\overline{M_{r/2}(p)}$
is uncountable. Since $\overline{M_{r/2}(p)} \subset M_r p$, $M_r p$ is uncountable. $\qquad\square$

Continuity of Arithmetic in $\mathbb{R}$

Addition is a mapping $\mathrm{Sum} : \mathbb{R} \times \mathbb{R} \to \mathbb{R}$ that assigns to (x, y) the real number
$x + y$. Subtraction and multiplication are also such mappings. Division is a mapping
$\mathbb{R} \times (\mathbb{R} \setminus \{0\}) \to \mathbb{R}$ that assigns to (x, y) the number x/y.

59 Theorem *The arithmetic operations of $\mathbb{R}$ are continuous.*

60 Lemma *For each real number c the function $\mathrm{Mult}_c : \mathbb{R} \to \mathbb{R}$ that sends x to cx
is continuous.*

Proof If $c = 0$ the function is constantly equal to 0 and is therefore continuous. If $c \neq 0$ and $\epsilon > 0$ is given, choose $\delta = \epsilon / |c|$. If $|x - y| < \delta$ then

$$|\mathrm{Mult}_c(x) - \mathrm{Mult}_c(y)| = |c|\,|x - y| < |c|\,\delta = \epsilon$$

which shows that Mult_c is continuous. $\square$

Proof of Theorem 59 We use the preservation of sequential convergence criterion for continuity. It's simplest. Let $(x_n, y_n) \to (x, y)$ as $n \to \infty$.

By the triangle inequality we have

$$|\mathrm{Sum}(x_n, y_n) - \mathrm{Sum}(x, y)| \leq |x_n - x| + |y_n - y| = d_{\mathrm{sum}}((x_n, y_n), (x, y)).$$

By Corollary 21 d_{sum} is continuous, so $d_{\mathrm{sum}}((x_n, y_n), (x, y)) \to 0$ as $n \to \infty$, which completes the proof that Sum is continuous. (By Theorem 17 it does not matter which metric we use on $\mathbb{R} \times \mathbb{R}$.)

Subtraction is the composition of continuous functions

$$\mathrm{Sub}(x, y) = \mathrm{Sum} \circ (\mathrm{id} \times \mathrm{Mult}_{-1})(x, y)$$

and is therefore continuous. (Proposition 3 implies id is continuous, Lemma 60 implies Mult_{-1} is continuous, and Corollary 18 implies $\mathrm{id} \times \mathrm{Mult}_{-1}$ is continuous.)

Multiplication is continuous since

$$
\begin{aligned}
|\mathrm{Mult}(x_n, y_n) - \mathrm{Mult}(x, y)| &= |x_n y_n - xy| \\
&\leq |x_n - x|\,|y_n| + |x|\,|y_n - y| \\
&\leq B(|x - x_n| + |y - y_n|) \\
&= \mathrm{Mult}_B(d_{\mathrm{sum}}((x_n, y_n), (x, y))) \to 0
\end{aligned}
$$

as $n \to \infty$, where we use the fact that convergent sequences are bounded to write $|y_n| + |x| \leq B$ for all n.

Reciprocation is the function $\mathrm{Rec} : \mathbb{R} \setminus \{0\} \to \mathbb{R} \setminus \{0\}$ that sends x to $1/x$. If $x_n \to x \neq 0$ then there is a constant $b > 0$ such that for all large n we have $|1/x_n| \leq b$ and $|1/x| \leq b$. Since

$$|\mathrm{Rec}(x_n) - \mathrm{Rec}(x)| = \left| \frac{1}{x_n} - \frac{1}{x} \right| = \frac{|x_n - x|}{|x x_n|} \leq \mathrm{Mult}_{b^2}(|x_n - x|) \to 0$$

as $n \to \infty$ we see that Rec is continuous.

Division is continuous on $\mathbb{R} \times (\mathbb{R} \setminus \{0\})$ since it is the composite of continuous mappings $\mathrm{Mult} \circ (\mathrm{id} \times \mathrm{Rec}) : (x, y) \mapsto (x, 1/y) \mapsto x \cdot 1/y$. $\square$

The absolute value is a mapping $\mathrm{Abs} : \mathbb{R} \to \mathbb{R}$ that sends x to $|x|$. It is continuous since it is $d(x, 0)$ and the distance function is continuous. The maximum and minimum are functions $\mathbb{R} \times \mathbb{R} \to \mathbb{R}$ given by the formulas

$$\max(x, y) = \frac{x + y}{2} + \frac{|x - y|}{2} \qquad \min(x, y) = \frac{x + y}{2} - \frac{|x - y|}{2},$$

so they are also continuous.

61 Corollary *The sums, differences, products, and quotients, absolute values, maxima, and minima of real-valued continuous functions are continuous. (The denominator functions should not equal zero.)*

Proof Take, for example, the sum $f + g$ where $f, g : M \to \mathbb{R}$ are continuous. It is the composite of continuous functions

$$M \xrightarrow{\ f \times g\ } \mathbb{R} \times \mathbb{R} \xrightarrow{\ \mathrm{Sum}\ } \mathbb{R}$$
$$x \ \mapsto \ (fx, gx) \ \mapsto \ \mathrm{Sum}(fx, gx),$$

and is therefore continuous. The same applies to the other operations. $\qquad \square$

62 Corollary *Polynomials are continuous functions.*

Proof Proposition 3 states that constant functions and the identity function are continuous. Thus Corollary 61 and induction imply that the polynomial $a_0 + a_1 x + \cdots + a_n x^n$ is continuous. $\qquad \square$

The same reasoning shows that polynomials of m variables are continuous functions $\mathbb{R}^m \to \mathbb{R}$.

Boundedness

A subset S of a metric space M is **bounded** if for some $p \in M$ and some $r > 0$,

$$S \subset M_r p.$$

A set which is not bounded is **unbounded**. For example, the elliptical region $4x^2 + y^2 < 4$ is a bounded subset of $\mathbb{R}^2$, while the hyperbola $xy = 1$ is unbounded. It is easy to see that if S is bounded then for each $q \in M$ there is an s such that $M_s q$ contains S.

Distinguish the word "bounded" from the word "finite." The first refers to physical size, the second to the number of elements. The concepts are totally different.

Also, boundedness has little connection to the existence of a boundary – a clopen subset of a metric space has empty boundary, but some clopen sets are bounded, others not.

Exercise 39 asks you to show that every convergent sequence is bounded, and to decide whether it is also true that every Cauchy sequence is bounded, even when the metric space is not complete.

Boundedness is not a topological property. For example, $(-1, 1)$ and $\mathbb{R}$ are homeomorphic although $(-1, 1)$ is bounded and $\mathbb{R}$ is unbounded. The same example shows that completeness is not a topological property.

A function from M to another metric space N is a **bounded function** if its range is a bounded subset of N. That is, there exist $q \in N$ and $r > 0$ such that

$$fM \subset N_r q.$$

Note that a function can be bounded even though its graph is not. For example, $x \mapsto \sin x$ is a bounded function $\mathbb{R} \to \mathbb{R}$ although its graph, $\{(x, y) \in \mathbb{R}^2 : y = \sin x\}$, is an unbounded subset of $\mathbb{R}^2$.

7 Coverings

For the sake of simplicity we have postponed discussing compactness in terms of open coverings until this point. Typically, students find coverings a challenging concept. It is central, however, to much of analysis – for example, measure theory.

Definition A collection $\mathcal{U}$ of subsets of M **covers** $A \subset M$ if A is contained in the union of the sets belonging to $\mathcal{U}$. The collection $\mathcal{U}$ is a **covering** of A. If $\mathcal{U}$ and $\mathcal{V}$ both cover A and if $\mathcal{V} \subset \mathcal{U}$ in the sense that each set $V \in \mathcal{V}$ belongs also to $\mathcal{U}$ then we say that $\mathcal{U}$ **reduces to** $\mathcal{V}$, and that $\mathcal{V}$ is a **subcovering** of A.

Definition If all the sets in a covering $\mathcal{U}$ of A are open then $\mathcal{U}$ is an **open covering** of A. If every open covering of A reduces to a finite subcovering of A then we say that A is **covering compact**[†].

The idea is that if A is covering compact and $\mathcal{U}$ is an open covering of A then just a finite number of the open sets are actually doing the work of covering A. The rest are redundant.

[†]You will frequently find it said that an open covering of A *has* a finite subcovering. "Has" means "reduces to."

A covering $\mathcal{U}$ of A is also called a **cover** of A. The members of $\mathcal{U}$ are *not* called covers. Instead, you could call them **scraps** or **patches**. Imagine the covering as a patchwork quilt that covers a bed, the quilt being sewn together from overlapping scraps of cloth. See Figure 46.

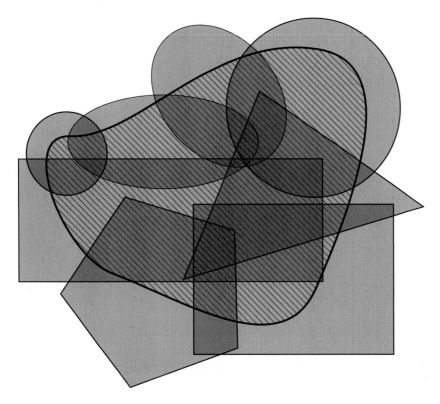

Figure 46 A covering of A by eight scraps. The set A is cross-hatched. The scraps are two discs, two rectangles, two ellipses, a pentagon, and a triangle. Each point of A belongs to at least one scrap.

The mere existence of a finite open covering of A is trivial and utterly worthless. *Every* set A has such a covering, namely the single open set M. Rather, for A to be covering compact, each and every open covering of A must reduce to a finite subcovering of A. Deciding directly whether this is so is daunting. How could you hope to verify the finite reducibility of all open coverings of A? There are so many of them. For this reason we concentrated on sequential compactness; it is relatively easy to check by inspection whether every sequence in a set has a convergent subsequence.

To check that a set is not covering compact it suffices to find an open covering which fails to reduce to a finite subcovering. Occasionally this is simple. For example, the set $(0, 1]$ is not covering compact in $\mathbb{R}$ because its covering

$$\mathcal{U} = \{(1/n, 2) : n \in \mathbb{N}\}$$

fails to reduce to a finite subcovering.

63 Theorem *For a subset A of a metric space M the following are equivalent:*

(a) *A is covering compact.*
(b) *A is sequentially compact.*

Proof that (a) implies (b) We assume A is covering compact and prove it is sequentially compact. Suppose not. Then there is a sequence (p_n) in A, no subsequence of which converges in A. Each point $a \in A$ therefore has some neighborhood $M_r a$ such that $p_n \in M_r a$ only finitely often. (The radius r may depend on the point a.) The collection $\{M_r a : a \in A\}$ is an open covering of A and by covering compactness it reduces to a finite subcovering

$$\{M_{r_1}(a_1), \, M_{r_2}(a_2), \, \ldots, \, M_{r_k}(a_k)\}$$

of A. Since p_n appears in each of these finitely many neighborhoods $M_{r_i}(a_i)$ only finitely often, it follows from the pigeonhole principle that (p_n) has only finitely many terms, a contradiction. Thus (p_n) cannot exist, and A is sequentially compact. $\qquad\square$

The following presentation of the proof that (b) implies (a) appears in Royden's book, *Real Analysis*. A **Lebesgue number** for a covering $\mathcal{U}$ of A is a positive real number λ such that for each $a \in A$ there is some $U \in \mathcal{U}$ with $M_\lambda a \subset U$. Of course, the choice of this U depends on a. It is crucial, however, that the Lebesgue number λ is independent of $a \in A$.

The idea of a Lebesgue number is that we know each point $a \in A$ is contained in some $U \in \mathcal{U}$, and if λ is extremely small then $M_\lambda a$ is just a slightly swollen point – so the same should be true for it too. No matter where in A the neighborhood $M_\lambda a$ is placed, it should lie wholly in some member of the covering. See Figure 47.

If A is noncompact then it may have open coverings with no positive Lebesgue number. For example, let $A = (0, 1) \subset \mathbb{R} = M$. The singleton collection $\{A\}$ is an open covering of A, but there is no $\lambda > 0$ such that for every $a \in A$ we have $(a - \lambda, a + \lambda) \subset A$. See Exercise 86.

64 Lebesgue Number Lemma *Every open covering of a sequentially compact set has a Lebesgue number $\lambda > 0$.*

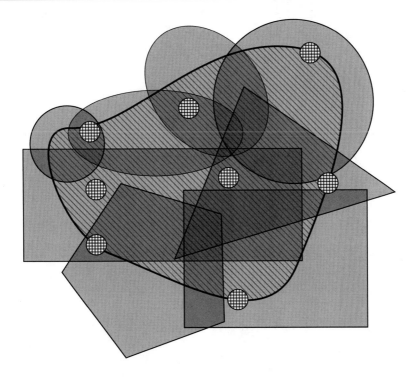

Figure 47 Small neighborhoods are like swollen points. $\mathcal{U}$ has a positive
Lebesgue number λ. The λ-neighborhood of each point in the cross-hatched
set A is wholly contained in at least one member of the covering.

Proof Suppose not: $\mathcal{U}$ is an open covering of a sequentially compact set A, and yet
for each $\lambda > 0$ there exists an $a \in A$ such that no $U \in \mathcal{U}$ contains $M_\lambda a$. Take $\lambda = 1/n$
and let $a_n \in A$ be a point such that no $U \in \mathcal{U}$ contains $M_{1/n}(a_n)$. By sequential
compactness, there is a subsequence (a_{n_k}) converging to some point $p \in A$. Since $\mathcal{U}$
is an open covering of A, there exist $r > 0$ and $U \in \mathcal{U}$ with $M_r p \subset U$. If k is large
then $d(a_{n_k}, p) < r/2$ and $1/n_k < r/2$, which implies by the triangle inequality that

$$M_{1/n_k}(a_{n_k}) \subset M_r p \subset U,$$

contrary to the supposition that no $U \in \mathcal{U}$ contains $M_{1/n}(a_n)$. We conclude that,
after all, $\mathcal{U}$ does have a Lebesgue number $\lambda > 0$. See Figure 48. $\square$

Proof that (b) implies (a) in Theorem 63 Let $\mathcal{U}$ be an open covering of the
sequentially compact set A. We want to reduce $\mathcal{U}$ to a finite subcovering. By the
Lebesgue Number Lemma, $\mathcal{U}$ has a Lebesgue number $\lambda > 0$. Choose any $a_1 \in A$ and
some $U_1 \in \mathcal{U}$ such that

$$M_\lambda(a_1) \subset U_1.$$

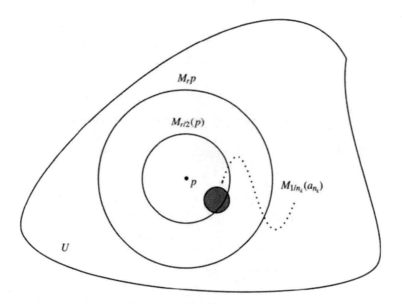

Figure 48 The neighborhood $M_r p$ engulfs the smaller neighborhood $M_{1/n_k}(a_{n_k})$.

If $U_1 \supset A$ then $\mathcal{U}$ reduces to the finite subcovering $\{U_1\}$ consisting of a single set, and the implication (b) $\Rightarrow$ (a) is proved. On the other hand, as is more likely, if U_1 does not contain A then we choose a point $a_2 \in A \setminus U_1$ and $U_2 \in \mathcal{U}$ such that

$$M_\lambda(a_2) \subset U_2.$$

Either $\mathcal{U}$ reduces to the finite subcovering $\{U_1, U_2\}$ (and the proof is finished) or else we can continue, eventually producing a sequence (a_n) in A and a sequence (U_n) in $\mathcal{U}$ such that

$$M_\lambda(a_n) \subset U_n \text{ and } a_{n+1} \in (A \setminus (U_1 \cup \cdots \cup U_n)).$$

We will show that such sequences (a_n), (U_n) lead to a contradiction. By sequential compactness, there is a subsequence (a_{n_k}) that converges to some $p \in A$. For a large k we have $d(a_{n_k}, p) < \lambda$ and

$$p \in M_\lambda(a_{n_k}) \subset U_{n_k}.$$

See Figure 49.

All a_{n_ℓ} with $\ell > k$ lie outside U_{n_k}, which contradicts their convergence to p. Thus, at some finite stage the process of choosing points a_n and sets U_n terminates, and $\mathcal{U}$

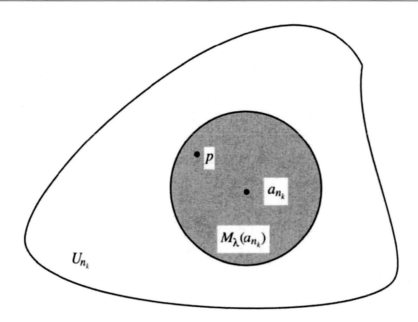

Figure 49 The point a_{n_k} is so near p that the neighborhood $M_\lambda(a_{n_k})$
engulfs p.

reduces to a finite subcovering $\{U_1, \ldots, U_n\}$ of A, which implies that A is covering
compact. See also the remark on page 421. □

Upshot In light of Theorem 63, the term "compact" may now be applied equally to
any set obeying (a) or (b).

Total Boundedness

The Heine-Borel Theorem states that a subset of $\mathbb{R}^m$ is compact if and only if
it is closed and bounded. In more general metric spaces, such as $\mathbb{Q}$, the assertion is
false. But what if the metric space is complete? As remarked on page 81 it is still
false.

But mathematicians do not quit easily. The Heine-Borel Theorem ought to generalize beyond $\mathbb{R}^m$ somehow. Here is the concept we need: A set $A \subset M$ is **totally
bounded** if for each $\epsilon > 0$ there exists a finite covering of A by ϵ-neighborhoods. No
mention is made of a covering reducing to a subcovering. How close total boundedness
is to the worthless fact that every metric space has a finite open covering!

65 Generalized Heine-Borel Theorem *A subset of a complete metric space is
compact if and only if it is closed and totally bounded.*

Proof Let A be a compact subset of M. Therefore it is closed. To see that it is totally bounded, let $\epsilon > 0$ be given and consider the covering of A by ϵ-neighborhoods,

$$\{M_\epsilon\, x\,:\, x \in A\}.$$

Compactness of A implies that this covering reduces to a finite subcovering and therefore A is totally bounded.

Conversely, assume that A is a closed and totally bounded subset of the complete metric space M. We claim that A is sequentially compact. That is, every sequence (a_n) in A has a subsequence that converges in A. Set $\epsilon_k = 1/k$, $k = 1, 2, \ldots$. Since A is totally bounded we can cover it by finitely many ϵ_1-neighborhoods

$$M_{\epsilon_1}(q_1), \ \ldots, \ M_{\epsilon_1}(q_m).$$

By the pigeonhole principle, terms of the sequence a_n lie in at least one of these neighborhoods infinitely often, say it is $M_{\epsilon_1}(p_1)$. Choose

$$a_{n_1} \in A_1 = A \cap M_{\epsilon_1}(p_1).$$

Every subset of a totally bounded set is totally bounded, so we can cover A_1 by finitely many ϵ_2-neighborhoods. For one of them, say $M_{\epsilon_2}(p_2)$, a_n lies in $A_2 = A_1 \cap M_{\epsilon_2}(p_2)$ infinitely often. Choose $a_{n_2} \in A_2$ with $n_2 > n_1$.

Proceeding inductively, cover A_{k-1} by finitely many ϵ_k-neighborhoods, one of which, say $M_{\epsilon_k}(p_k)$, contains terms of the sequence (a_n) infinitely often. Then choose $a_{n_k} \in A_k = A_{k-1} \cap M_{\epsilon_k}(p_k)$ with $n_k > n_{k-1}$. Then (a_{n_k}) is a subsequence of (a_n). It is Cauchy. For if $\epsilon > 0$ is given we choose N such that $2/N < \epsilon$. If $k, \ell \geq N$ then

$$a_{n_k}, a_{n_\ell} \in A_N \quad \text{and} \quad \operatorname{diam} A_N \leq 2\epsilon_N = \frac{2}{N} < \epsilon,$$

which shows that (a_{n_k}) is Cauchy. Completeness of M implies that (a_{n_k}) converges to some $p \in M$ and since A is closed we have $p \in A$. Hence A is compact. $\qquad\square$

66 Corollary *A metric space is compact if and only if it is complete and totally bounded.*

Proof Every compact metric space M is complete. This is because, given a Cauchy sequence (p_n) in M, compactness implies that some subsequence converges in M, and if a Cauchy sequence has a convergent subsequence then the mother sequence converges too. As observed above, compactness immediately gives total boundedness.

Conversely, assume that M is complete and totally bounded. Every metric space is closed in itself. By Theorem 65, M is compact. $\qquad\square$

8 Cantor Sets

Cantor sets are fascinating examples of compact sets that are maximally disconnected. (To emphasize the disconnectedness, one sometimes refers to a Cantor set as "Cantor dust.") Here is how to construct the standard **Cantor set**. Start with the unit interval $[0, 1]$ and remove its open middle third, $(1/3, 2/3)$. Then remove the open middle third from the remaining two intervals, and so on. This gives a nested sequence $C^0 \supset C^1 \supset C^2 \supset \ldots$ where $C^0 = [0, 1]$, C^1 is the union of the two intervals $[0, 1/3]$ and $[2/3, 1]$, C^2 is the union of four intervals $[0, 1/9]$, $[2/9, 1/3]$, $[2/3, 7/9]$, and $[8/9, 1]$, C^3 is the union of eight intervals, and so on. See Figure 50.

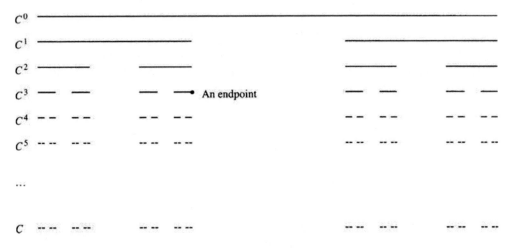

Figure 50 The construction of the standard middle-thirds Cantor set C

In general C^n is the union of 2^n closed intervals, each of length $1/3^n$. Each C^n is compact. The **standard middle thirds Cantor set** is the nested intersection

$$C = \bigcap_{n=0}^{\infty} C^n.$$

We refer to C as "the" Cantor set. Clearly it contains the endpoints of each of the intervals comprising C^n. Actually, it contains uncountably many more points than these endpoints! There are other Cantor sets defined by removing, say, middle fourths, pairs of middle tenths, etc. All Cantor sets turn out to be homeomorphic to the standard Cantor set. See Section 9.

A metric space M is **totally disconnected** if each point $p \in M$ has arbitrarily small clopen neighborhoods. That is, given $\epsilon > 0$ and $p \in M$, there exists a clopen set U such that

$$p \in U \subset M_\epsilon p.$$

For example, every discrete space is totally disconnected. So is $\mathbb{Q}$.

67 Theorem *The Cantor set is a compact, nonempty, perfect, and totally disconnected metric space.*

Proof The metric on C is the one it inherits from $\mathbb{R}$, the usual distance $|x - y|$. Let E be the set of endpoints of all the C^n-intervals,

$$E = \{0,\ 1,\ 1/3,\ 2/3,\ 1/9,\ 2/9,\ 7/9,\ 8/9,\ \ldots\}.$$

Clearly E is denumerable and contained in C, so C is nonempty and infinite. It is compact because it is the intersection of compacts.

To show C is perfect and totally disconnected, take any $x \in C$ and any $\epsilon > 0$. Fix n so large that $1/3^n < \epsilon$. The point x lies in one of the 2^n intervals I of length $1/3^n$ that comprise C^n. Fix this I. The set $E \cap I$ is infinite and contained in the interval $(x - \epsilon,\ x + \epsilon)$. Thus C clusters at x and C is perfect. See Figure 51.

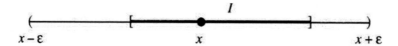

Figure 51 The endpoints of C cluster at x.

The interval I is closed in $\mathbb{R}$ and therefore in C^n. The complement $J = C^n \setminus I$ consists of finitely many closed intervals and is therefore closed too. Thus, I and J are clopen in C^n. By the Inheritance Principle their intersections with C are clopen in C, so $C \cap I$ is a clopen neighborhood of x in C which is contained in the ϵ-neighborhood of x, completing the proof that C is totally disconnected. $\square$

68 Corollary *The Cantor set is uncountable.*

Proof Being compact, C is complete, and by Theorem 56, every complete, perfect, nonempty metric space is uncountable. $\square$

A more direct way to see that the Cantor set is uncountable involves a geometric coding scheme. Take the code $0 = $ left and $2 = $ right. Then

$$C_0 = \text{ left interval } = [0, 1/3] \qquad C_2 = \text{ right interval } = [2/3, 1],$$

and $C^1 = C_0 \cup C_2$. Similarly, the left and right subintervals of C_0 are coded C_{00} and C_{02}, while the left and right subintervals of C_2 are C_{20} and C_{22}. This gives

$$C^2 = C_{00} \sqcup C_{02} \sqcup C_{20} \sqcup C_{22}.$$

The intervals that comprise C^3 are specified by strings of length 3. For instance, C_{220} is the left subinterval of C_{22}. In general an interval of C^n is coded by an **address string** of n symbols, each a 0 or a 2. Read it like a zip code. The first symbol gives the interval's gross location (left or right), the second symbol refines the location, the third refines it more, and so on.

Imagine now an **infinite address string** $\omega = \omega_1\omega_2\omega_3\ldots$ of zeros and twos. Corresponding to ω, we form a nested sequence of intervals

$$C_{\omega_1} \supset C_{\omega_1\omega_2} \supset C_{\omega_1\omega_2\omega_3} \supset \cdots \supset C_{\omega_1\ldots\omega_n} \supset \cdots,$$

the intersection of which is a point $p = p(\omega) \in C$. Specifically,

$$p(\omega) = \bigcap_{n \in \mathbb{N}} C_{\omega|n}$$

where $\omega|n = \omega_1\ldots\omega_n$ **truncates** ω to an address of length n. See Theorem 34.

As we have observed, each infinite address string defines a point in the Cantor set. Conversely, each point $p \in C$ has an address $\omega = \omega(p)$: its first n symbols $\alpha = \omega|n$ are specified by the interval C_α of C^n in which p lies. A second point q has a different address, since there is some n for which p and q lie in distinct intervals C_α and C_β of C^n.

In sum, the Cantor set is in one-to-one correspondence with the set Ω of addresses. Each address $\omega \in \Omega$ defines a point $p(\omega) \in C$ and each point $p \in C$ has a unique address $\omega(p)$. The set Ω is uncountable. In fact it corresponds bijectively to $\mathbb{R}$. See Exercise 112.

If $S \subset M$ and $\overline{S} = M$ then S is **dense** in M. For example, $\mathbb{Q}$ is dense in $\mathbb{R}$. The set S is **somewhere dense** if there exists an open nonempty set $U \subset M$ such that $\overline{S \cap U} \supset U$. If S is not somewhere dense then it is **nowhere dense**.

69 Theorem *The Cantor set contains no interval and is nowhere dense in $\mathbb{R}$.*

Proof Suppose not and C contains (a, b). Then $(a, b) \subset C^n$ for all $n \in \mathbb{N}$. Take n with $1/3^n < b - a$. Since (a, b) is connected it lies wholly in a single C^n-interval, say I. But I has smaller length than (a, b), which is absurd, so C contains no interval.

Next, suppose C is dense in some nonempty open set $U \subset \mathbb{R}$, i.e., the closure of $C \cap U$ contains U. Thus

$$C = \overline{C} \supset \overline{C \cap U} \supset U \supset (a, b),$$

contrary to the fact that C contains no interval. $\qquad\qquad\square$

The existence of an uncountable nowhere dense set is astonishing. Even more is true: The Cantor set is a **zero set** – it has "outer measure zero." By this we mean that, given any $\epsilon > 0$, there is a countable covering of C by open intervals (a_k, b_k), and the **total length** of the covering is

$$\sum_{k=1}^{\infty} b_k - a_k < \epsilon.$$

(Outer measure is one of the central concepts of Lebesgue Theory. See Chapter 6.) After all, C is a subset of C^n, which consists of 2^n closed intervals, each of length $1/3^n$. If n is large enough then $2^n/3^n < \epsilon$. Enlarging each of these closed intervals to an open interval keeps the sum of the lengths $< \epsilon$, and it follows that C is a zero set.

If we discard subintervals of $[0,1]$ in a different way, we can make a **fat Cantor set** – one that has positive outer measure. Instead of discarding the middle-thirds of intervals at the nth stage in the construction, we discard only the middle $1/n!$ portion. The discards are grossly smaller than the remaining intervals. See Figure 52. The total amount discarded from $[0,1]$ is < 1, and the total amount remaining, the outer measure of the fat Cantor set, is positive. See Exercise 3.31.

Figure 52 In forming a fat Cantor set, the gap intervals occupy a progressively smaller proportion of the Cantor set intervals.

9* Cantor Set Lore

In this section, we explore some arcane features of Cantor sets.

Although the continuous image of a connected set is connected, the continuous image of a disconnected set may well be connected. Just crush the disconnected set to a single point. Nevertheless, I hope you find the following result striking, for it means that the Cantor set C is the **universal compact metric space**, of which all others are merely shadows.

70 Cantor Surjection Theorem *Given a compact nonempty metric space M, there is a continuous surjection of C onto M.*

See Figure 53. Exercise 114 suggests a direct construction of a continuous surjection $C \rightarrow [0,1]$, which is already an interesting fact. The proof of Theorem 70

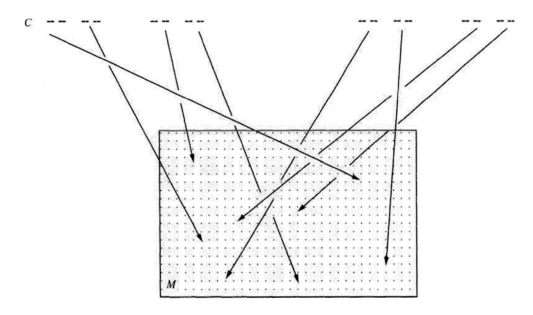

Figure 53 σ surjects C onto M.

involves a careful use of the address notation from Section 8 and the following simple lemma about dividing a compact metric space M into small pieces. A **piece** of M is any compact nonempty subset of M.

71 Lemma *If M is a nonempty compact metric space and $\epsilon > 0$ is given then M can be expressed as the finite union of pieces, each of diameter $\leq \epsilon$.*

Proof Reduce the covering $\{M_{\epsilon/2}(x) : x \in M\}$ of M to a finite subcovering and take the closure of each member of the subcovering. $\qquad\square$

We say that M **divides into** these small pieces. The metaphor is imperfect because the pieces may overlap. The strategy of the proof of Theorem 70 is to divide M into large pieces, divide the large pieces into small pieces, divide the small pieces into smaller pieces and continue indefinitely. Labeling the pieces coherently with words in two letters leads to the Cantor surjection.

Let $W(n)$ be the set of words in two letters, say a and b, having length n. Then $\#W(n) = 2^n$. For example $W(2)$ consists of the four words aa, bb, ab, and ba.

Using Lemma 71 we divide M into a finite number of pieces of diameter ≤ 1 and we denote by $\mathcal{M}_1$ the collection of these pieces. We choose n_1 with $2^{n_1} \geq \#\mathcal{M}_1$ and choose any surjection $w_1 : W(n_1) \to \mathcal{M}_1$. Since there are enough words in $W(n_1)$, w_1 exists. We say w_1 **labels** $\mathcal{M}_1$ and if $w_1(\alpha) = L$ then α is a **label** of L.

Then we divide each $L \in \mathcal{M}_1$ into finitely many smaller pieces. Let $\mathcal{M}_2(L)$ be the collection of these smaller pieces and let

$$\mathcal{M}_2 = \bigcup_{L \in \mathcal{M}_1} \mathcal{M}_2(L).$$

Choose n_2 such that $2^{n_2} \geq \max\{\#\mathcal{M}_2(L) : L \in \mathcal{M}_1\}$ and label $\mathcal{M}_2$ with words $\alpha\beta \in W(n_1 + n_2)$ such that

> If $L = w_1(\alpha)$ then $\alpha\beta$ labels the pieces $S \in \mathcal{M}_2(L)$
>
> as β varies in $W(n_2)$.

This labeling amounts to a surjection $w_2 : W(n_1 + n_2) \to \mathcal{M}_2$ that is **coherent** with w_1 in the sense that $\beta \mapsto w_2(\alpha\beta)$ labels the pieces $S \in w_1(\alpha)$. Since there are enough words in $W(n_2)$, w_2 exists. If there are other labels α' of $L \in \mathcal{M}_1$ then we get other labels $\alpha'\beta'$ for the pieces $S \in \mathcal{M}_2(L)$. We make no effort to correlate them.

Proceeding by induction we get finer and finer divisions of M coherently labeled with longer and longer words. More precisely there is a sequence of divisions $(\mathcal{M}_k)$ and surjections $w_k : W_k = W(n_1 + \cdots + n_k) \to \mathcal{M}_k$ such that

(a) The maximum diameter of the pieces $L \in \mathcal{M}_k$ tends to zero as $k \to \infty$.

(b) $\mathcal{M}_{k+1}$ **refines** $\mathcal{M}_k$ in the sense that each $S \in \mathcal{M}_{k+1}$ is contained in some $L \in \mathcal{M}_k$. ("The small pieces S are contained in the large pieces L.")

(c) If $L \in \mathcal{M}_k$ and $\mathcal{M}_{k+1}(L)$ denotes $\{S \in \mathcal{M}_{k+1} : S \subset L\}$ then

$$L = \bigcup_{S \in \mathcal{M}_{k+1}(L)} S.$$

(d) The labelings w_k are **coherent** in the sense that if $w_k(\alpha) = L \in \mathcal{M}_k$ then $\beta \mapsto w_{k+1}(\alpha\beta)$ labels $\mathcal{M}_{k+1}(L)$ as β varies in $W(n_{k+1})$.

See Figure 54.

Proof of the Cantor Surjection Theorem We are given a nonempty compact metric space M and we seek a continuous surjection $\sigma : C \to M$ where C is the standard Cantor set.

$C = \bigcap C^n$ where C^n is the disjoint union of 2^n closed intervals of length $1/3^n$. In Section 8 we labeled these C^n-intervals with words in the letters 0 and 2 having length n. (For instance C_{220} is the left C^3-interval of $C_{22} = [8/9, 1]$, namely $C_{220} = [8/9, 25/27]$.) We showed there is a natural bijection between C and the set of all infinite words in the letters 0 and 2 defined by

$$p = \bigcap_{n \in \mathbb{N}} C_{\omega|n}.$$

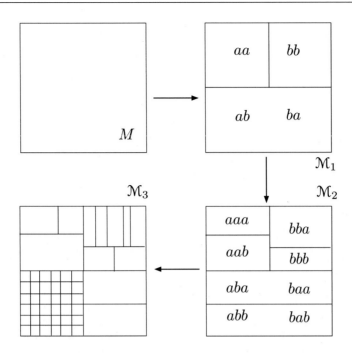

Figure 54 Coherently labeled successive divisions of M. They have
$n_1 = 2$, $n_2 = 1$, and $n_3 = 6$. Note that overlabeling is necessary.

We referred to $\omega = \omega(p)$ as the address of p. ($\omega|n$ is the truncation of ω to its first n letters.) See page 107.

For $k = 1, 2, \ldots$ let $\mathcal{M}_k$ be the fine divisions of M constructed above, coherently labeled by w_k. They obey (a)-(d). Given $p \in C$ we look at the nested sequence of pieces $L_k(p) \in \mathcal{M}_k$ such that $L_k(p) = w_k(\omega|(n_1 + \cdots + n_k))$ where $\omega = \omega(p)$. That is, we truncate $\omega(p)$ to its first $n_1 + \cdots + n_k$ letters and look at the piece in $\mathcal{M}_k$ with this label. (We replace the letters 0 and 2 with a and b.) Then $(L_k(p))$ is a nested decreasing sequence of nonempty compact sets whose diameters tend to 0 as $k \to \infty$. Thus $\bigcap L_k(p)$ is a well defined point in M and we set

$$\sigma(p) = \bigcap_{k \in \mathbb{N}} L_k(p).$$

We must show that σ is a continuous surjection $C \to M$. Continuity is simple. If $p, p' \in C$ are close together then for large n the first n entries of their addresses are equal. This implies that $\sigma(p)$ and $\sigma(p')$ belong to a common L_k and k is large. Since the diameter of L_k tends to 0 as $k \to \infty$ we get continuity.

Surjectivity is also simple. Each $q \in M$ is the intersection of at least one nested sequence of pieces $L_k \in \mathcal{M}_k$. For q belongs to some piece $L_1 \in \mathcal{M}_1$, and it also belongs to some subpiece $L_2 \in \mathcal{M}_2(L_1)$, etc. Coherence of the labeling of the $\mathcal{M}_k$ implies that for each nested sequence (L_k) there is an infinite word $\alpha = \alpha_1\alpha_2\alpha_3 \ldots$ such that $\alpha_i \in W(n_i)$ and $L_k = w_k(\alpha_1 \ldots \alpha_m)$ with $m = n_1 + \cdots + n_k$. The point $p \in C$ with address α is sent by σ to q.   □

Peano Curves

72 Theorem *There exists a **Peano curve**, a continuous path in the plane which is **space-filling** in the sense that its image has nonempty interior. In fact there is a Peano curve whose image is the closed unit disc B^2.*

Proof Let $\sigma : C \to B^2$ be a continuous surjection supplied by Theorem 70. Extend σ to a map $\tau : [0,1] \to B^2$ by setting

$$\tau(x) = \begin{cases} \sigma(x) & \text{if } x \in C \\ (1-t)\sigma(a) + t\sigma(b) & \text{if } x = (1-t)a + tb \in (a,b) \\ & \text{and } (a,b) \text{ is a gap interval.} \end{cases}$$

A **gap interval** is an interval $(a,b) \subset C^c$ such that $a, b \in C$. Because σ is continuous, $|\sigma(a) - \sigma(b)| \to 0$ as $|a - b| \to 0$. Hence τ is continuous. Its image includes the disc B^2 and thus has nonempty interior. In fact the image of τ is exactly B^2, since the disc is convex and τ just extends σ via linear interpolation. See Figure 55.   □

This Peano curve cannot be one-to-one since C is not homeomorphic to B^2. (C is disconnected while B^2 is connected.) In fact no Peano curve τ can be one-to-one. See Exercise 102.

Cantor Spaces

We say that M is a **Cantor space** if, like the standard Cantor set C, it is compact, nonempty, perfect, and totally disconnected.

73 Moore-Kline Theorem *Every Cantor space M is homeomorphic to the standard middle-thirds Cantor set C.*

A **Cantor piece** is a nonempty clopen subset S of a Cantor space M. It is easy to see that S is also a Cantor space. See Exercise 100. Since a Cantor space is totally disconnected, each point has a small clopen neighborhood N. Thus, a Cantor space can always be divided into two disjoint Cantor pieces, $M = U \sqcup U^c$.

gap interval

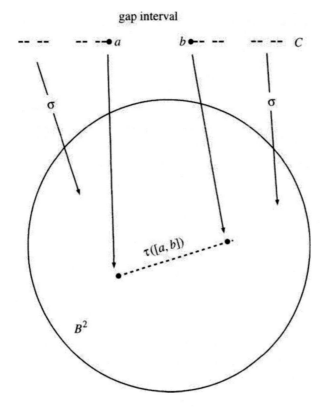

Figure 55 Filling in the Cantor surjection σ to make a Peano space-filling curve τ

74 Cantor Partition Lemma *Given a Cantor space M and $\epsilon > 0$, there is a number N such that for each $d \geq N$ there is a partition of M into d Cantor pieces of diameter $\leq \epsilon$. (We care most about dyadic d.)*

Proof A **partition** of a set is a division of it into disjoint subsets. In this case the small Cantor pieces form a partition of the Cantor space M. Since M is totally disconnected and compact, we can cover it with finitely many clopen neighborhoods $U_1, \ldots, U_m$ having diameter $\leq \epsilon$. To make the sets U_i disjoint, define

$$
\begin{aligned}
V_1 &= U_1 \\
V_2 &= U_1 \setminus U_2 \\
&\ldots \\
V_m &= U_m \setminus (U_1 \cup \cdots \cup U_{m-1}).
\end{aligned}
$$

If any V_i is empty, discard it. This gives a partition $M = X_1 \sqcup \cdots \sqcup X_N$ into $N \leq m$ Cantor pieces of diameter $\leq \epsilon$.

If $d = N$ this finishes the proof. If $d > N$ then we inductively divide X_N into two, and then three, and eventually $d - N + 1$ disjoint Cantor pieces; say

$$X_N = Y_1 \sqcup \cdots \sqcup Y_{d-N+1}.$$

The partition $M = X_1 \sqcup \cdots \sqcup X_{N-1} \sqcup Y_1 \sqcup \cdots \sqcup Y_{d-N+1}$ finishes the proof. $\qquad\square$

Proof of the Moore-Kline Theorem We are given a Cantor space M and we seek a homeomorphism from the standard Cantor set C onto M.

By Lemma 74 there is a partition $\mathcal{M}_1$ of M into d_1 nonempty Cantor pieces where $d_1 = 2^{n_1}$ is dyadic and the pieces have diameter ≤ 1. Thus there is a bijection $w_1 : W_1 \to \mathcal{M}_1$ where $W_1 = W(n_1)$.

According to the same lemma, each $L \in \mathcal{M}_1$ can be partitioned into $N(L)$ Cantor pieces of diameter $\leq 1/2$. Choose a dyadic number

$$d_2 = 2^{n_2} \geq \max\{N(L) : L \in \mathcal{M}_1\}$$

and use the lemma again to partition each L into d_2 smaller Cantor pieces. These pieces constitute $\mathcal{M}_2(L)$, and we set $\mathcal{M}_2 = \bigcup_L \mathcal{M}_2(L)$. It is a partition of M having cardinality $d_1 d_2$ and in the natural way described in the proof of Theorem 70 it is coherently labeled by $W_2 = W(n_1 + n_2)$. Specifically, for each $L \in \mathcal{M}_1$ there is a bijection $w_L : W(n_2) \to \mathcal{M}_2(L)$ and we define $w_2 : W_2 \to \mathcal{M}_2$ by $w_2(\alpha\beta) = S \in \mathcal{M}_2$ if and only if $w_1(\alpha) = L$ and $w_L(\beta) = S$. This w_2 is a bijection.

Proceeding in exactly the same way, we pass from 2 to 3, from 3 to 4, and eventually from k to $k + 1$, successively refining the partitions and extending the bijective labelings.

The Cantor surjection constructed in the proof of Theorem 70 is

$$\sigma(p) = \bigcap_k L_k(p)$$

where $L_k(p) \in \mathcal{M}_k$ has label $\omega(p)|m$ with $m = n_1 + \cdots + n_k$. Distinct points $p, p' \in C$ have distinct addresses ω, ω'. Because the labelings w_k are bijections and the divisions $\mathcal{M}_k$ are partitions, $\omega \neq \omega'$ implies that for some k, $L_k(p) \neq L_k(p')$, and thus $\sigma(p) \neq \sigma(p')$. That is, σ is a continuous bijection $C \to M$. A continuous bijection from one compact to another is a homeomorphism. $\qquad\square$

75 Corollary *Every two Cantor spaces are homeomorphic.*

Proof Immediate from the Moore-Kline Theorem: Each is homeomorphic to C. $\qquad\square$

76 Corollary *The fat Cantor set is homeomorphic to the standard Cantor set.*

Proof Immediate from the Moore-Kline Theorem.   □

77 Corollary *A Cantor set is homeomorphic to its own Cartesian square; that is,* $C \cong C \times C$.

Proof It is enough to check that $C \times C$ is a Cantor space. It is. See Exercise 99. □

The fact that a nontrivial space is homeomorphic to its own Cartesian square is disturbing, is it not?

Ambient Topological Equivalence

Although all Cantor spaces are homeomorphic to each other when considered as abstract metric spaces, they can present themselves in very different ways as subsets of Euclidean space. Two sets A, B in $\mathbb{R}^m$ are **ambiently homeomorphic** if there is a homeomorphism of $\mathbb{R}^m$ to itself that sends A onto B. For example, the sets

$$A = \{0\} \cup [1,2] \cup \{3\} \quad \text{and} \quad B = \{0\} \cup \{1\} \cup [2,3]$$

are homeomorphic when considered as metric spaces, but there is no ambient homeomorphism of $\mathbb{R}$ that carries A to B. Similarly, the trefoil knot in $\mathbb{R}^3$ is homeomorphic but not ambiently homeomorphic in $\mathbb{R}^3$ to a planar circle. See also Exercise 105.

78 Theorem *Every two Cantor spaces in $\mathbb{R}$ are ambiently homeomorphic.*

Let M be a Cantor space contained in $\mathbb{R}$. According to Theorem 73, M is homeomorphic to the standard Cantor set C. We want to find a homeomorphism of $\mathbb{R}$ to itself that carries C to M.

The **convex hull** of $S \subset \mathbb{R}^m$ is the smallest convex set H that contains S. When $m = 1$, H is the smallest interval that contains S.

79 Lemma *A Cantor space $M \subset \mathbb{R}$ splits into two Cantor pieces with disjoint convex hulls, each having diameter $< 2\,\mathrm{diam}(M)/3$.*

Proof Obvious from one-dimensionality of $\mathbb{R}$: Let $[a,b]$ be the convex hull of M. Its diameter is $d = b - a$ and its midpoint is $m = (a+b)/2$. Total disconnectedness of M implies there exists $c \in (a,b) \setminus M$ with $m \le c$ and $c - a < 2d/3$. Then

$$M = M \cap [a,c] \;\sqcup\; [c,b] \cap M$$

divides M into disjoint Cantor pieces as required.   □

Proof of Theorem 78 Let $M \subset \mathbb{R}$ be a Cantor space. We will find a homeomorphism $\tau : \mathbb{R} \to \mathbb{R}$ sending C to M. Lemma 79 leads to Cantor divisions $\mathfrak{M}_k$ whose pieces have disjoint convex hulls with diameters tending to zero as $k \to \infty$. With respect to the left/right order of $\mathbb{R}$, label these pieces in the same way that the Cantor middle third intervals are labeled: L_0 and L_2 in $\mathfrak{M}_1$ are the left and right pieces of M, L_{00} and L_{02} are the left and right pieces of L_0, and so on. Then the homeomorphism $\sigma : C \to M$ constructed in Theorems 70 and 73 is automatically monotone increasing. Extend σ across the gap intervals affinely as was done in the proof of Theorem 72, and extend it to $\mathbb{R} \setminus [0,1]$ in any affine increasing fashion such that $\tau(0) = \sigma(0)$ and $\tau(1) = \sigma(1)$. Then $\tau : \mathbb{R} \to \mathbb{R}$ extends σ to $\mathbb{R}$. The monotonicity of σ implies that τ is one-to-one, while the continuity of σ implies that τ is continuous. $\tau : \mathbb{R} \to \mathbb{R}$ is a homeomorphism that carries C onto M.

If $M' \subset \mathbb{R}$ is a second Cantor space and $\tau' : \mathbb{R} \to \mathbb{R}$ is a homeomorphism that sends C onto M' then $\tau' \circ \tau^{-1}$ is a homeomorphism of $\mathbb{R}$ that sends M onto M'. $\square$

As an example, one may construct a Cantor set in $\mathbb{R}$ by removing from $[0,1]$ its middle third, then removing from each of the remaining intervals nine symmetrically placed subintervals; then removing from each of the remaining twenty intervals, four asymmetrically placed subintervals; and so forth. In the limit (if the lengths of the remaining intervals tend to zero) we get a nonstandard Cantor set M. According to Theorem 78, there is a homeomorphism of $\mathbb{R}$ to itself sending the standard Cantor set C onto M.

Another example is the fat Cantor set mentioned on page 108. It too is ambiently homeomorphic to C.

Theorem *Every two Cantor spaces in $\mathbb{R}^2$ are ambiently homeomorphic.*

We do not prove this theorem here. The key step is to show M has a dyadic disc partition. That is, M can be divided into a dyadic number of Cantor pieces, each piece contained in the interior of a small topological disc D_i, the D_i being mutually disjoint. (A topological disc is any homeomorph of the closed unit disc B^2. Smallness refers to diam D_i.) The proofs I know of the existence of such dyadic partitions are tricky cut-and-paste arguments and are beyond the scope of this book. See Moise's book, *Geometric Topology in Dimensions 2 and 3* and also Exercise 138.

Antoine's Necklace

A Cantor space $M \subset \mathbb{R}^m$ is **tame** if there is an ambient homeomorphism $h : \mathbb{R}^m \to \mathbb{R}^m$ that carries the standard Cantor set C (imagined to lie on the x_1-axis in $\mathbb{R}^m$) onto M. If M is not tame it is **wild**. Cantor spaces contained in the line or plane are tame. In 3-space, however, there are wild ones, Cantor sets A so badly embedded in $\mathbb{R}^3$ that they act like curves. It is the lack of a "ball dyadic partition lemma" that causes the problem.

The first wild Cantor set was discovered by Louis Antoine, and is known as **Antoine's Necklace**. The construction involves the solid torus or anchor ring, which is homeomorphic to the Cartesian product $B^2 \times S^1$. It is easy to imagine a necklace of solid tori: Take an ordinary steel chain and modify it so its first and last links are also linked. See Figure 56.

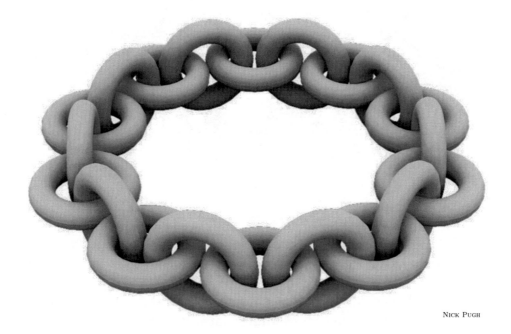

NICK PUGH

Figure 56 A necklace of twenty solid tori

Antoine's construction then goes like this. Draw a solid torus A^0. Interior to A^0, draw a necklace A^1 of several small solid tori, and make the necklace encircle the hole of A^0. Repeat the construction on each solid torus T comprising A^1. That is, interior to each T, draw a necklace of very small solid tori so that it encircles the hole of T. The result is a set $A^2 \subset A^1$ which is a necklace of necklaces. In Figure 56, A^2 would consist of 400 solid tori. Continue indefinitely, producing a nested decreasing sequence $A^0 \supset A^1 \supset A^2 \supset \dots$. The set A^n is compact and consists of a large number (20^n) of extremely small solid tori arranged in a hierarchy of necklaces. It is an nth order necklace. The intersection $A = \bigcap A^n$ is a Cantor space, since it is compact, perfect, nonempty, and totally disconnected. It is homeomorphic to C. See Exercise 139.

Certainly A is bizarre, but is it wild? Is there no ambient homeomorphism h of $\mathbb{R}^3$ that sends the standard Cantor set C onto A? The reason that h cannot exist is explained next.

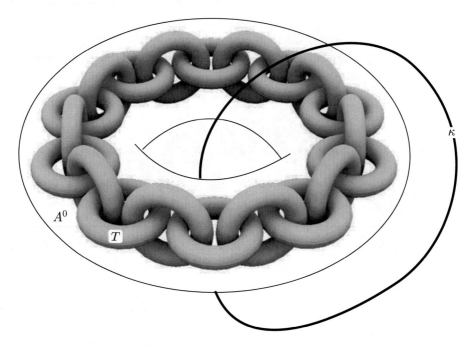

Figure 57 κ loops through A^0, which contains the necklace of solid tori.

Referring to Figure 57, the loop κ passing through the hole of A^0 cannot be continuously shrunk to a point in $\mathbb{R}^3$ without hitting A. For if such a motion of κ avoids A then, by compactness, it also avoids one of the high-order necklaces A^n. In $\mathbb{R}^3$ it is impossible to continuously de-link two linked loops, and it is also impossible to continuously de-link a loop from a necklace of loops. (These facts are intuitively believable but hard to prove. See Dale Rolfsen's book, *Knots and Links*.)

On the other hand, each loop λ in $\mathbb{R}^3 \setminus C$ can be continuously shrunk to a point without hitting C. For there is no obstruction to pushing λ through the gap intervals of C.

Now suppose that there is an ambient homeomorphism h of $\mathbb{R}^3$ that sends C to A. Then $\lambda = h^{-1}(\kappa)$ is a loop in $\mathbb{R}^3 \setminus C$, and it can be shrunk to a point in $\mathbb{R}^3 \setminus C$, avoiding C. Applying h to this motion of λ continuously shrinks κ to a point, avoiding A, which we have indicated is impossible. Hence h cannot exist, and A is wild.

10* Completion

Many metric spaces are complete (for example, every closed subset of Euclidean space is complete), and completeness is a reasonable property to require of a metric space, especially in light of the following theorem.

80 Completion Theorem *Every metric space can be completed.*

This means that just as $\mathbb{R}$ completes $\mathbb{Q}$, we can take any metric space M and find a complete metric space $\widehat{M}$ containing M whose metric extends the metric of M. To put it another way, M is always a metric subspace of a complete metric space. In a natural sense the completion is uniquely determined by M.

81 Lemma *Given four points $p, q, x, y \in M$, we have*

$$|d(p,q) - d(x,y)| \leq d(p,x) + d(q,y).$$

Proof The triangle inequality implies that

$$d(x,y) \leq d(x,p) + d(p,q) + d(q,y)$$
$$d(p,q) \leq d(p,x) + d(x,y) + d(y,q),$$

and hence

$$-(d(p,x) + d(q,y)) \leq d(p,q) - d(x,y) \leq (d(p,x) + d(q,y)).$$

A number sandwiched between $-k$ and k has magnitude $\leq k$, which completes the proof. ☐

Proof of the Completion Theorem 80 We consider the collection $\mathcal{C}$ of all Cauchy sequences in M, convergent or not, and convert *it* into the completion of M. (This is a bold idea, is it not?) Cauchy sequences (p_n) and (q_n), are **co-Cauchy** if $d(p_n, q_n) \to 0$ as $n \to \infty$. Co-Cauchyness is an equivalence relation on $\mathcal{C}$. (This is easy to check.)

Define $\widehat{M}$ to be $\mathcal{C}$ modulo the equivalence relation of being co-Cauchy. Points of $\widehat{M}$ are equivalence classes $P = [(p_n)]$ such that (p_n) is a Cauchy sequence in M. The metric on $\widehat{M}$ is

$$D(P,Q) = \lim_{n \to \infty} d(p_n, q_n),$$

where $P = [(p_n)]$ and $Q = [(q_n)]$. It only remains to verify three things:

(a) D is a well defined metric on $\widehat{M}$.
(b) $M \subset \widehat{M}$.
(c) $\widehat{M}$ is complete.

None of these assertions is really hard to prove, although the details are somewhat messy because of possible equivalence class/representative ambiguity.

(a) By Lemma 81

$$|d(p_m, q_m) - d(p_n, q_n)| \leq d(p_m, p_n) + d(q_m, q_n).$$

Thus $(d(p_n, q_n))$ is a Cauchy sequence in $\mathbb{R}$, and because $\mathbb{R}$ is complete,

$$L = \lim_{n \to \infty} d(p_n, q_n)$$

exists. Let (p'_n) and (q'_n) be sequences that are co-Cauchy with (p_n) and (q_n), and let

$$L' = \lim_{n \to \infty} d(p'_n, q'_n).$$

Then

$$|L - L'| \leq |L - d(p_n, q_n)| + |d(p_n, q_n) - d(p'_n, q'_n)| + |d(p'_n, q'_n) - L'|.$$

As $n \to \infty$, the first and third terms tend to 0. By Lemma 81, the middle term is

$$|d(p_n, q_n) - d(p'_n, q'_n)| \leq d(p_n, p'_n) + d(q_n, q'_n),$$

which also tends to 0 as $n \to \infty$. Hence $L = L'$ and D is well defined on $\widehat{M}$. The d-distance on M is symmetric and satisfies the triangle inequality. Taking limits, these properties carry over to D on $\widehat{M}$, while positive definiteness follows directly from the co-Cauchy definition.

(b) Think of each $p \in M$ as a constant sequence, $\bar{p} = (p, p, p, p, \ldots)$. Clearly it is Cauchy and clearly the D-distance between two constant sequences $\bar{p}$ and $\bar{q}$ is the same as the d-distance between the points p and q. In this way M is naturally a metric subspace of $\widehat{M}$.

(c) Let $(P_k)_{k \in \mathbb{N}}$ be a Cauchy sequence in $\widehat{M}$. We must find $Q \in \widehat{M}$ to which P_k converges as $k \to \infty$. (Note that (P_k) is a sequence of equivalence classes, not a sequence of points in M, and convergence refers to D not d.) Because D is well defined we can use a trick to shorten the proof. Observe that every subsequence of a Cauchy sequence is Cauchy, and it and the mother sequence are co-Cauchy. For all the terms far along in the subsequence are also far along in the mother sequence. This lets us take a representative of P_k *all* of whose terms are at distance $< 1/k$ from each other. Call this sequence $(p_{k,n})_{n \in \mathbb{N}}$. We have $[(p_{k,n})] = P_k$.

Set $q_n = p_{n,n}$. We claim that (q_n) is Cauchy and $D(P_k, Q) \to 0$ as $k \to \infty$, where $Q = [(q_n)]$. That is, $\widehat{M}$ is complete.

Let $\epsilon > 0$ be given. There exists $N \geq 3/\epsilon$ such that if $k, \ell \geq N$ then

$$D(P_k, P_\ell) \leq \frac{\epsilon}{3}$$

and

$$
\begin{aligned}
d(q_k, q_\ell) &= d(p_{k,k}, p_{\ell,\ell}) \\
&\leq d(p_{k,k}, p_{k,n}) + d(p_{k,n}, p_{\ell,n}) + d(p_{\ell,n}, p_{\ell,\ell}) \\
&\leq \frac{1}{k} + d(p_{k,n}, p_{\ell,n}) + \frac{1}{\ell} \\
&\leq \frac{2\epsilon}{3} + d(p_{k,n}, p_{\ell,n}).
\end{aligned}
$$

The inequality is valid for all n and the left-hand side, $d(q_k, q_\ell)$, does not depend on n. The limit of $d(p_{k,n}, p_{\ell,n})$ as $n \to \infty$ is $D(P_k, P_\ell)$, which we know to be $< \epsilon/3$. Thus, if $k, \ell \geq N$ then $d(q_k, q_\ell) < \epsilon$ and (q_n) is Cauchy. Similarly we see that $P_k \to Q$ as $k \to \infty$. For, given $\epsilon > 0$, we choose $N \geq 2/\epsilon$ such that if $k, n \geq N$ then $d(q_k, q_n) < \epsilon/2$, from which it follows that

$$
\begin{aligned}
d(p_{k,n}, q_n) &\leq d(p_{k,n}, p_{k,k}) + d(p_{k,k}, q_n) \\
&= d(p_{k,n}, p_{k,k}) + d(q_k, q_n) \\
&\leq \frac{1}{k} + \frac{\epsilon}{2} < \epsilon.
\end{aligned}
$$

The limit of the left-hand side of this inequality, as $n \to \infty$, is $D(P_k, Q)$. Thus

$$\lim_{k \to \infty} P_k = Q$$

and $\widehat{M}$ is complete. $\square$

Uniqueness of the completion is not surprising, and is left as Exercise 106. A different proof of the Completion Theorem is sketched in Exercise 4.39.

A Second Construction of $\mathbb{R}$ from $\mathbb{Q}$

In the particular case that the metric space M is $\mathbb{Q}$, the Completion Theorem leads to a construction of $\mathbb{R}$ from $\mathbb{Q}$ via Cauchy sequences. Note, however, that applying the theorem as it stands involves circular reasoning, for its proof uses completeness of $\mathbb{R}$ to define the metric D. Instead, we use only the Cauchy sequence *strategy*.

Convergence and Cauchyness for sequences of rational numbers are concepts that make perfect sense without a priori knowledge of $\mathbb{R}$. Just take all epsilons and deltas in the definitions to be rational. The **Cauchy completion** $\widehat{\mathbb{Q}}$ of $\mathbb{Q}$ is the collection $\mathcal{C}$ of Cauchy sequences in $\mathbb{Q}$ modulo the equivalence relation of being co-Cauchy.

We claim that $\widehat{\mathbb{Q}}$ is a complete ordered field. That is, $\widehat{\mathbb{Q}}$ is just another version of $\mathbb{R}$. The arithmetic on $\widehat{\mathbb{Q}}$ is defined by

$$
\begin{aligned}
P + Q &= [(p_n + q_n)] & P - Q &= [(p_n - q_n)] \\
PQ &= [(p_n q_n)] & P/Q &= [(p_n/q_n)]
\end{aligned}
$$

where $P = [(p_n)]$ and $Q = [(q_n)]$. Of course $Q \neq [(0,0,\ldots)]$ in the fraction P/Q. Exercise 134 asks you to check that these natural definitions make $\widehat{\mathbb{Q}}$ a field. Although there are many things to check – well definedness, commutativity, and so forth – all are effortless. There are no sixteen case proofs as with cuts. Also, just as with metric spaces, $\mathbb{Q}$ is naturally a subfield of $\widehat{\mathbb{Q}}$ when we think of $r \in \mathbb{Q}$ as the equivalence class $\overline{r} = [(r,r,\ldots)]$. Here, $\overline{r}$ is an equivalence class, not the constant sequence itself.

That's the easy part – now the rest.

To define the order relation on $\widehat{\mathbb{Q}}$ we rework some of the cut ideas. If $P \in \widehat{\mathbb{Q}}$ has a representative $[(p_n)]$, such that for some $\epsilon > 0$, we have $p_n \geq \epsilon$ for all n then P is positive. If $-P$ is positive then P is negative.

Then we define $P \prec Q$ if $Q - P$ is positive. Exercise 135 asks you to check that this defines an order on $\widehat{\mathbb{Q}}$, consistent with the standard order $<$ on $\mathbb{Q}$ in the sense that for all $p, q \in \mathbb{Q}$ we have $p < q \iff \overline{p} \prec \overline{q}$. In particular, you are asked to prove the trichotomy property: Each $P \in \widehat{\mathbb{Q}}$ is either positive, negative, or zero, and these possibilities are mutually exclusive.

Combining Cauchyness with the definition of $\prec$ gives

(1)
$$
\begin{aligned}
P = [(p_n)] \prec Q = [(q_n)] \quad &\iff \quad \text{there exist } \epsilon > 0 \text{ and } N \in \mathbb{N} \\
&\qquad \text{such that for all } m, n \geq N, \\
&\qquad \text{we have } p_m + \epsilon < q_n.
\end{aligned}
$$

It remains to check the least upper bound property. Let $\mathcal{P}$ be a nonempty subset of $\widehat{\mathbb{Q}}$ that is bounded above. We must find a least upper bound for $\mathcal{P}$.

First of all, since $\mathcal{P}$ is bounded there is a $B = [(b_n)] \in \widehat{\mathbb{Q}}$ such that $P \preceq B$ for all $P \in \mathcal{P}$. We can choose (b_n) so its terms lie at distance ≤ 1 from each other. Set $b = b_1 + 1$. Then $\overline{b}$ is an upper bound for $\mathcal{P}$. Since $\mathbb{Q}$ is Archimedean there is an integer $m \geq b$, and $\overline{m}$ is also an upper bound for $\mathcal{P}$. By the same reasoning $\mathcal{P}$ has upper bounds $\overline{r}$ such that r is a dyadic fraction with arbitrarily large denominator 2^n, e.g., $r = m2^n/2^n$.

Since $\mathcal{P}$ is nonempty, the same reasoning shows that there are dyadic fractions s with large denominators such that $\bar{s}$ is not an upper bound for $\mathcal{P}$.

We assert that the least upper bound for $\mathcal{P}$ is the equivalence class Q of the following Cauchy sequence $(q_0, q_1, q_2, \ldots)$.

(a) q_0 is the smallest integer such that $\overline{q_0}$ is an upper bound for $\mathcal{P}$.

(b) q_1 is the smallest fraction with denominator 2 such that $\overline{q_1}$ is an upper bound for $\mathcal{P}$.

(c) q_2 is the smallest fraction with denominator 4 such that $\overline{q_2}$ is an upper bound for $\mathcal{P}$.

(d) $\ldots$

(e) q_n is the smallest fraction with denominator 2^n such that $\overline{q_n}$ is an upper bound for $\mathcal{P}$.

The sequence (q_n) is well defined because for some but not all dyadic fractions r with denominator 2^n, $\bar{r}$ is an upper bound for $\mathcal{P}$. By construction (q_n) is monotone decreasing and $q_{n-1} - q_n \leq 1/2^n$. Thus, if $m \leq n$ then

$$0 \leq q_m - q_n = q_m - q_{m+1} + q_{m+1} - q_{m+2} + \cdots + q_{n-1} - q_n$$
$$\leq \frac{1}{2^{m+1}} + \cdots + \frac{1}{2^n} < \frac{1}{2^m}.$$

It follows that (q_n) is Cauchy and $Q = [(q_n)] \in \widehat{\mathbb{Q}}$.

Suppose that Q is *not* an upper bound for $\mathcal{P}$. Then there is some $P = [(p_n)] \in \mathcal{P}$ with $Q \prec P$. By (1), there is an $\epsilon > 0$ and an N such that for all $n \geq N$,

$$q_N + \epsilon < p_n.$$

It follows that $\overline{q_N} \prec P$, a contradiction to $\overline{q_N}$ being an upper bound for $\mathcal{P}$.

On the other hand suppose there is a smaller upper bound for $\mathcal{P}$, say $R = [(r_n)] \prec Q$. By (1) there are $\epsilon > 0$ and N such that for all $m, n \geq N$,

$$r_m + \epsilon < q_n.$$

Fix a $k \geq N$ with $1/2^k < \epsilon$. Then for all $m \geq N$,

$$r_m < q_k - \epsilon < q_k - \frac{1}{2^k}.$$

By (1), $R \prec \overline{q_k - 1/2^k}$. Since R is an upper bound for $\mathcal{P}$, so is $\overline{q_k - 1/2^k}$, a contradiction to q_k being the *smallest* fraction with denominator 2^k such that $\overline{q_k}$ is an upper bound for $\mathcal{P}$. Therefore, Q is indeed a least upper bound for $\mathcal{P}$.

This completes the verification that the Cauchy completion of $\mathbb{Q}$ is a complete ordered field. Uniqueness implies that it is isomorphic to the complete ordered field $\mathbb{R}$ constructed by means of Dedekind cuts in Section 2 of Chapter 1. Decide for yourself which of the two constructions of the real number system you like better – cuts or Cauchy sequences. Cuts make least upper bounds straightforward and algebra awkward, while with Cauchy sequences it is the reverse.

Exercises

1. An ant walks on the floor, ceiling, and walls of a cubical room. What metric is natural for the ant's view of its world? What metric would a spider consider natural? If the ant wants to walk from a point p to a point q, how could it determine the shortest path?

2. Why is the sum metric on $\mathbb{R}^2$ called the Manhattan metric and the taxicab metric?

3. What is the set of points in $\mathbb{R}^3$ at distance exactly $1/2$ from the unit circle S^1 in the plane,

$$T = \{p \in \mathbb{R}^3 : \exists q \in S^1 \text{ and } d(p,q) = 1/2$$
$$\text{and for all } q' \in S^1 \text{ we have } d(p,q) \leq d(p,q')\}?$$

4. Write out a proof that the discrete metric on a set M is actually a metric.

5. For $p, q \in S^1$, the unit circle in the plane, let

$$d_a(p,q) = \min\{|\angle(p) - \angle(q)|, \ 2\pi - |\angle(p) - \angle(q)|\}$$

   where $\angle(z) \in [0, 2\pi)$ refers to the angle that z makes with the positive x-axis. Use your geometric talent to prove that d_a is a metric on S^1.

6. For $p, q \in [0, \pi/2)$ let

$$d_s(p,q) = \sin|p - q|.$$

   Use your calculus talent to decide whether d_s is a metric.

7. Prove that every convergent sequence (p_n) in a metric space M is bounded, i.e., that for some $r > 0$, some $q \in M$, and all $n \in \mathbb{N}$, we have $p_n \in M_r q$.

8. Consider a sequence (x_n) in the metric space $\mathbb{R}$.
   (a) If (x_n) converges in $\mathbb{R}$ prove that the sequence of absolute values $(|x_n|)$ converges in $\mathbb{R}$.
   (b) State the converse.
   (c) Prove or disprove it.

9. A sequence (x_n) in $\mathbb{R}$ **increases** if $n < m$ implies $x_n \leq x_m$. It **strictly increases** if $n < m$ implies $x_n < x_m$. It **decreases** or **strictly decreases** if $n < m$ always implies $x_n \geq x_m$ or always implies $x_n > x_m$. A sequence is **monotone** if it increases or it decreases. Prove that every sequence in $\mathbb{R}$ which is monotone and bounded converges in $\mathbb{R}$.[†]

10. Prove that the least upper bound property is equivalent to the "monotone sequence property" that every bounded monotone sequence converges.

[†]This is nicely is expressed by Pierre Teilhard de Chardin, *"Tout ce qui monte converge,"* in a different context.

11. Let (x_n) be a sequence in $\mathbb{R}$.

    *(a) Prove that (x_n) has a monotone subsequence.

    (b) How can you deduce that every bounded sequence in $\mathbb{R}$ has a convergent subsequence?

    (c) Infer that you have a second proof of the Bolzano-Weierstrass Theorem in $\mathbb{R}$.

    (d) What about the Heine-Borel Theorem?

12. Let (p_n) be a sequence and $f : \mathbb{N} \to \mathbb{N}$ be a bijection. The sequence $(q_k)_{k \in \mathbb{N}}$ with $q_k = p_{f(k)}$ is a **rearrangement** of (p_n).

    (a) Are limits of a sequence unaffected by rearrangement?

    (b) What if f is an injection?

    (c) A surjection?

13. Assume that $f : M \to N$ is a function from one metric space to another which satisfies the following condition: If a sequence (p_n) in M converges then the sequence $(f(p_n))$ in N converges. Prove that f is continuous. [This result improves Theorem 4.]

14. The simplest type of mapping from one metric space to another is an **isometry**. It is a bijection $f : M \to N$ that preserves distance in the sense that for all $p, q \in M$ we have

$$d_N(fp, fq) = d_M(p, q).$$

If there exists an isometry from M to N then M and N are said to be **isometric**, $M \equiv N$. You might have two copies of a unit equilateral triangle, one centered at the origin and one centered elsewhere. They are isometric. Isometric metric spaces are indistinguishable as metric spaces.

    (a) Prove that every isometry is continuous.

    (b) Prove that every isometry is a homeomorphism.

    (c) Prove that $[0, 1]$ is not isometric to $[0, 2]$.

15. Prove that isometry is an equivalence relation: If M is isometric to N, show that N is isometric to M; show that each M is isometric to itself (what mapping of M to M is an isometry?); if M is isometric to N and N is isometric to P, show that M is isometric to P.

16. Is the perimeter of a square isometric to the circle? Homeomorphic? Explain.

17. Which capital letters of the Roman alphabet are homeomorphic? Are any isometric? Explain.

18. Is $\mathbb{R}$ homeomorphic to $\mathbb{Q}$? Explain.

19. Is $\mathbb{Q}$ homeomorphic to $\mathbb{N}$? Explain.

20. What function (given by a formula) is a homeomorphism from $(-1, 1)$ to $\mathbb{R}$? Is every open interval homeomorphic to $(0, 1)$? Why or why not?

21. Is the plane minus four points on the x-axis homeomorphic to the plane minus four points in an arbitrary configuration?

22. If every closed and bounded subset of a metric space M is compact, does it follow that M is complete? (Proof or counterexample.)

23. $(0,1)$ is an open subset of $\mathbb{R}$ but not of $\mathbb{R}^2$, when we think of $\mathbb{R}$ as the x-axis in $\mathbb{R}^2$. Prove this.

24. For which intervals $[a,b]$ in $\mathbb{R}$ is the intersection $[a,b] \cap \mathbb{Q}$ a clopen subset of the metric space $\mathbb{Q}$?

25. Prove directly from the definition of closed set that every singleton subset of a metric space M is a closed subset of M. Why does this imply that every finite set of points is also a closed set?

26. Prove that a set $U \subset M$ is open if and only if none of its points are limits of its complement.

27. If $S, T \subset M$, a metric space, and $S \subset T$, prove that
    (a) $\overline{S} \subset \overline{T}$.
    (b) $\text{int}(S) \subset \text{int}(T)$.

28. A map $f : M \to N$ is **open** if for each open set $U \subset M$, the image set $f(U)$ is open in N.
    (a) If f is open, is it continuous?
    (b) If f is a homeomorphism, is it open?
    (c) If f is an open, continuous bijection, is it a homeomorphism?
    (d) If $f : \mathbb{R} \to \mathbb{R}$ is a continuous surjection, must it be open?
    (e) If $f : \mathbb{R} \to \mathbb{R}$ is a continuous, open surjection, must it be a homeomorphism?
    (f) What happens in (e) if $\mathbb{R}$ is replaced by the unit circle S^1?

29. Let $\mathcal{T}$ be the collection of open subsets of a metric space M, and $\mathcal{K}$ the collection of closed subsets. Show that there is a bijection from $\mathcal{T}$ onto $\mathcal{K}$.

30. Consider a two-point set $M = \{a, b\}$ whose topology consists of the two sets, M and the empty set. Why does this topology not arise from a metric on M?

31. Prove the following.
    (a) If U is an open subset of $\mathbb{R}$ then it consists of countably many disjoint intervals $U = \bigsqcup U_i$. (Unbounded intervals $(-\infty, b)$, (a, ∞), and $(-\infty, \infty)$ are permitted.)
    (b) Prove that these intervals U_i are uniquely determined by U. In other words, there is only one way to express U as a disjoint union of open intervals.
    (c) If $U, V \subset \mathbb{R}$ are both open, so $U = \bigsqcup U_i$ and $V = \bigsqcup V_j$ where U_i and V_j are open intervals, show that U and V are homeomorphic if and only if there are equally many U_i and V_j.

32. Show that every subset of $\mathbb{N}$ is clopen. What does this tell you about every function $f : \mathbb{N} \to M$, where M is a metric space?

33. (a) Find a metric space in which the boundary of $M_r p$ is not equal to the sphere of radius r at p, $\partial(M_r p) \neq \{x \in M : d(x,p) = r\}$.

    (b) Need the boundary be contained in the sphere?

34. Use the Inheritance Principle to prove Corollary 15.

35. Prove that S clusters at p if and only if for each $r > 0$ there is a point $q \in M_r p \cap S$, such that $q \neq p$.

36. Construct a set with exactly three cluster points.

37. Construct a function $f : \mathbb{R} \to \mathbb{R}$ that is continuous only at points of $\mathbb{Z}$.

38. Let X, Y be metric spaces with metrics d_X, d_Y, and let $M = X \times Y$ be their Cartesian product. Prove that the three natural metrics d_E, $d_{\max}$, and d_{sum} on M are actually metrics. [Hint: Cauchy-Schwarz.]

39. (a) Prove that every convergent sequence is bounded. That is, if (p_n) converges in the metric space M, prove that there is some neighborhood $M_r q$ containing the set $\{p_n : n \in \mathbb{N}\}$.

    (b) Is the same true for a Cauchy sequence in an incomplete metric space?

40. Let M be a metric space with metric d. Prove that the following are equivalent.

    (a) M is homeomorphic to M equipped with the discrete metric.

    (b) Every function $f : M \to M$ is continuous.

    (c) Every bijection $g : M \to M$ is a homeomorphism.

    (d) M has no cluster points.

    (e) Every subset of M is clopen.

    (f) Every compact subset of M is finite.

41. Let $\| \ \|$ be any norm on $\mathbb{R}^m$ and let $B = \{x \in \mathbb{R}^m : \|x\| \leq 1\}$. Prove that B is compact. [Hint: It suffices to show that B is closed and bounded with respect to the Euclidean metric.]

42. What is wrong with the following "proof" of Theorem 28? "Let $((a_n, b_n))$ be any sequence in $A \times B$ where A and B are compact. Compactness implies the existence of subsequences (a_{n_k}) and (b_{n_k}) converging to $a \in A$ and $b \in B$ as $k \to \infty$. Therefore $((a_{n_k}, b_{n_k}))$ is a subsequence of $((a_n, b_n))$ that converges to a limit in $A \times B$, proving that $A \times B$ is compact."

43. Assume that the Cartesian product of two nonempty sets $A \subset M$ and $B \subset N$ is compact in $M \times N$. Prove that A and B are compact.

44. Consider a function $f : M \to \mathbb{R}$. Its graph is the set

$$\{(p, y) \in M \times \mathbb{R} : y = fp\}.$$

    (a) Prove that if f is continuous then its graph is closed (as a subset of $M \times \mathbb{R}$).

    (b) Prove that if f is continuous and M is compact then its graph is compact.

    (c) Prove that if the graph of f is compact then f is continuous.

    (d) What if the graph is merely closed? Give an example of a discontinuous function $f : \mathbb{R} \to \mathbb{R}$ whose graph is closed.

45. Draw a Cantor set C on the circle and consider the set A of all chords between points of C.

    (a) Prove that A is compact.

    *(b) Is A convex?

46. Assume that A, B are compact, disjoint, nonempty subsets of M. Prove that there are $a_0 \in A$ and $b_0 \in B$ such that for all $a \in A$ and $b \in B$ we have

$$d(a_0, b_0) \leq d(a, b).$$

    [The points a_0, b_0 are closest together.]

47. Suppose that $A, B \subset \mathbb{R}^2$.

    (a) If A and B are homeomorphic, are their complements homeomorphic?

    *(b) What if, in addition, A and B are compact?

    ***(c) What if A and B are compact and connected?

48. Prove that there is an embedding of the line as a closed subset of the plane, and there is an embedding of the line as a bounded subset of the plane, but there is no embedding of the line as a closed and bounded subset of the plane.

*49. Construct a subset $A \subset \mathbb{R}$ and a continuous bijection $f : A \to A$ that is not a homeomorphism. [Hint: By Theorem 36 A must be noncompact.]

**50. Construct nonhomeomorphic connected, closed subsets $A, B \subset \mathbb{R}^2$ for which there exist continuous bijections $f : A \to B$ and $g : B \to A$. [Hint: By Theorem 36 A and B must be noncompact.]

***51. Do there exist nonhomeomorphic closed sets $A, B \subset \mathbb{R}$ for which there exist continuous bijections $f : A \to B$ and $g : B \to A$?

52. Let (A_n) be a nested decreasing sequence of nonempty closed sets in the metric space M.

    (a) If M is complete and diam $A_n \to 0$ as $n \to \infty$, show that $\bigcap A_n$ is exactly one point.

    (b) To what assertions do the sets $[n, \infty)$ provide counterexamples?

53. Suppose that (K_n) is a nested sequence of compact nonempty sets, $K_1 \supset K_2 \supset \ldots$, and $K = \bigcap K_n$. If for some $\mu > 0$, diam $K_n \geq \mu$ for all n, is it true that diam $K \geq \mu$?

54. If $f : A \to B$ and $g : C \to B$ such that $A \subset C$ and for each $a \in A$ we have $f(a) = g(a)$ then g **extends** f. We also say that f **extends to** g. Assume that $f : S \to \mathbb{R}$ is a uniformly continuous function defined on a subset S of a metric space M.

    (a) Prove that f extends to a uniformly continuous function $\overline{f} : \overline{S} \to \mathbb{R}$.

    (b) Prove that $\overline{f}$ is the unique continuous extension of f to a function defined on $\overline{S}$.

    (c) Prove the same things when $\mathbb{R}$ is replaced with a complete metric space N.

55. The **distance** from a point p in a metric space M to a nonempty subset $S \subset M$ is defined to be $\text{dist}(p, S) = \inf\{d(p, s) : s \in S\}$.
    (a) Show that p is a limit of S if and only if $\text{dist}(p, S) = 0$.
    (b) Show that $p \mapsto \text{dist}(p, S)$ is a uniformly continuous function of $p \in M$.

56. Prove that the 2-sphere is not homeomorphic to the plane.

57. If S is connected, is the interior of S connected? Prove this or give a counterexample.

58. Theorem 49 states that the closure of a connected set is connected.
    (a) Is the closure of a disconnected set disconnected?
    (b) What about the interior of a disconnected set?

*59. Prove that every countable metric space (not empty and not a singleton) is disconnected. [Astonishingly, there exists a countable topological space which is connected. Its topology does not arise from a metric.]

60. (a) Prove that a continuous function $f : M \to \mathbb{R}$, all of whose values are integers, is constant provided that M is connected.
    (b) What if all the values are irrational?

61. Prove that the (double) cone $\{(x, y, z) \in \mathbb{R}^3 : x^2 + y^2 = z^2\}$ is path-connected.

62. Prove that the annulus $A = \{z \in \mathbb{R}^2 : r \le |z| \le R\}$ is connected.

63. A subset E of $\mathbb{R}^m$ is **starlike** if it contains a point p_0 (called a **center** for E) such that for each $q \in E$, the segment between p_0 and q lies in E.
    (a) If E is convex and nonempty prove that it is starlike.
    (b) Why is the converse false?
    (c) Is every starlike set connected?
    (d) Is every connected set starlike? Why or why not?

*64. Suppose that $E \subset \mathbb{R}^m$ is open, bounded, and starlike, and p_0 is a center for E.
    (a) Is it true or false that all points p_1 in a small enough neighborhood of p_0 are also centers for E?
    (b) Is the set of centers convex?
    (c) Is it closed as a subset of E?
    (d) Can it consist of a single point?

65. Suppose that $A, B \subset \mathbb{R}^2$ are convex, closed, and have nonempty interiors.
    (a) Prove that A, B are the closure of their interiors.
    (b) If A, B are compact, prove that they are homeomorphic.
    [Hint: Draw a picture.]

66. (a) Prove that every connected open subset of $\mathbb{R}^m$ is path-connected.
    (b) Is the same true for open connected subsets of the circle?
    (c) What about connected nonopen subsets of the circle?

67. List the convex subsets of $\mathbb{R}$ up to homeomorphism. How many are there and how many are compact?

68. List the closed convex sets in $\mathbb{R}^2$ up to homeomorphism. There are nine. How many are compact?

*69. Generalize Exercises 65 and 68 to $\mathbb{R}^3$; to $\mathbb{R}^m$.

70. Prove that (a,b) and $[a,b)$ are not homeomorphic metric spaces.

71. Let M and N be nonempty metric spaces.
    (a) If M and N are connected prove that $M \times N$ is connected.
    (b) What about the converse?
    (c) Answer the questions again for path-connectedness.

72. Let H be the hyperbola $\{(x,y) \in \mathbb{R}^2 : xy = 1 \text{ and } x,y > 0\}$ and let X be the x-axis.
    (a) Is the set $S = X \cup H$ connected?
    (b) What if we replace H with the graph G of any continuous positive function $f : \mathbb{R} \to (0,\infty)$; is $X \cup G$ connected?
    (c) What if f is everywhere positive but discontinuous at just one point.

73. Is the disc minus a countable set of points connected? Path-connected? What about the sphere or the torus instead of the disc?

74. Let $S = \mathbb{R}^2 \setminus \mathbb{Q}^2$. (Points $(x,y) \in S$ have at least one irrational coordinate.) Is S connected? Path-connected? Prove or disprove.

*75. An **arc** is a path with no self-intersection. Define the concept of arc-connectedness and prove that a metric space is path-connected if and only if it is arc-connected.

76. (a) The intersection of connected sets need not be connected. Give an example.
    (b) Suppose that $S_1, S_2, S_3, \ldots$ is a sequence of connected, closed subsets of the plane and $S_1 \supset S_2 \supset \ldots$. Is $S = \bigcap S_n$ connected? Give a proof or counterexample.
    *(c) Does the answer change if the sets are compact?
    (d) What is the situation for a nested decreasing sequence of compact path-connected sets?

77. If a metric space M is the union of path-connected sets S_α, all of which have the nonempty path-connected set K in common, is M path-connected?

78. $(p_1, \ldots, p_n)$ is an **ϵ-chain** in a metric space M if for each i we have $p_i \in M$ and $d(p_i, p_{i+1}) < \epsilon$. The metric space is **chain-connected** if for each $\epsilon > 0$ and each pair of points $p, q \in M$ there is an ϵ-chain from p to q.
    (a) Show that every connected metric space is chain-connected.
    (b) Show that if M is compact and chain-connected then it is connected.
    (c) Is $\mathbb{R} \setminus \mathbb{Z}$ chain-connected?
    (d) If M is complete and chain-connected, is it connected?

79. Prove that if M is nonempty, compact, locally path-connected, and connected then it is path-connected. (See Exercise 143, below.)

80. The **Hawaiian earring** is the union of circles of radius $1/n$ and center $x = \pm 1/n$ on the x-axis, for $n \in \mathbb{N}$. See Figure 27 on page 58.

    (a) Is it connected?
    (b) Path-connected?
    (c) Is it homeomorphic to the one-sided Hawaiian earring?

*81. The topologist's sine curve is the set

$$\{(x, y) : x = 0 \text{ and } |y| \leq 1 \text{ or } 0 < x \leq 1 \text{ and } y = \sin 1/x\}.$$

See Figure 43. The **topologist's sine circle** is shown in Figure 58. (It is the union of a circular arc and the topologist's sine curve.) Prove that it is path-connected but not locally path-connected. (M is **locally path-connected** if for each $p \in M$ and each neighborhood U of p there is a path-connected subneighborhood V of p.)

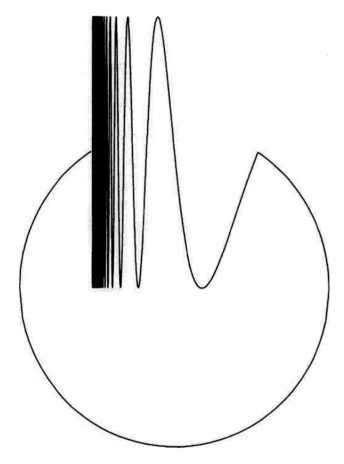

Figure 58 The topologist's sine circle

82. The graph of $f : M \to \mathbb{R}$ is the set $\{(x, y) \in M \times \mathbb{R} : y = fx\}$.
    (a) If M is connected and f is continuous, prove that the graph of f is connected.
    (b) Give an example to show that the converse is false.
    (c) If M is path-connected and f is continuous, show that the graph is path-connected.
    (d) What about the converse?

83. The open cylinder is $(0, 1) \times S^1$. The punctured plane is $\mathbb{R}^2 \setminus \{0\}$.
    (a) Prove that the open cylinder is homeomorphic to the punctured plane.
    (b) Prove that the open cylinder, the double cone, and the plane are not homeomorphic.

84. Is the closed strip $\{(x, y) \in \mathbb{R}^2 : 0 \le x \le 1\}$ homeomorphic to the closed half-plane $\{(x, y) \in \mathbb{R}^2 : x \ge 0\}$? Prove or disprove.

85. Suppose that M is compact and that $\mathcal{U}$ is an open covering of M which is "redundant" in the sense that each $p \in M$ is contained in at least two members of $\mathcal{U}$. Show that $\mathcal{U}$ reduces to a finite subcovering with the same property.

86. Suppose that every open covering of M has a positive Lebesgue number. Give an example of such an M that is not compact.

Exercises 87–94 treat the basic theorems in the chapter, avoiding the use of sequences. The proofs will remain valid in general topological spaces.

87. Give a direct proof that $[a, b]$ is covering compact. [Hint: Let $\mathcal{U}$ be an open covering of $[a, b]$ and consider the set

$$C = \{x \in [a, b] : \text{finitely many members of } \mathcal{U} \text{ cover } [a, x]\}.$$

Use the least upper bound principle to show that $b \in C$.]

88. Give a direct proof that a closed subset A of a covering compact set K is covering compact. [Hint: If $\mathcal{U}$ is an open covering of A, adjoin the set $W = M \setminus A$ to $\mathcal{U}$. Is $\mathcal{W} = \mathcal{U} \cup \{W\}$ an open covering of K? If so, so what?]

89. Give a proof of Theorem 36 using open coverings. That is, assume A is a covering compact subset of M and $f : M \to N$ is continuous. Prove directly that fA is covering compact. [Hint: What is the criterion for continuity in terms of preimages?]

90. Suppose that $f : M \to N$ is a continuous bijection and M is covering compact. Prove directly that f is a homeomorphism.

91. Suppose that M is covering compact and that $f : M \to N$ is continuous. Use the Lebesgue number lemma to prove that f is uniformly continuous. [Hint: Consider the covering of N by $\epsilon/2$-neighborhoods $\{N_{\epsilon/2}(q) : q \in N\}$ and its preimage in M, $\{f^{\mathrm{pre}}(N_{\epsilon/2}(q)) : q \in N\}$.]

92. Give a direct proof that the nested decreasing intersection of nonempty covering compact sets is nonempty. [Hint: If $A_1 \supset A_2 \supset \ldots$ are covering compact, consider the open sets $U_n = A_n^c$. If $\bigcap A_n = \emptyset$, what does $\{U_n\}$ cover?]

93. Generalize Exercise 92 as follows. Suppose that M is covering compact and $\mathcal{C}$ is a collection of closed subsets of M such that every intersection of finitely many members of $\mathcal{C}$ is nonempty. (Such a collection $\mathcal{C}$ is said to have the **finite intersection property**.) Prove that the **grand intersection** $\bigcap_{C \in \mathcal{C}} C$ is nonempty. [Hint: Consider the collection of open sets $\mathcal{U} = \{C^c : C \in \mathcal{C}.\}$]

94. If every collection of closed subsets of M which has the finite intersection property also has a nonempty grand intersection, prove that M is covering compact. [Hint: Given an open covering $\mathcal{U} = \{U_\alpha\}$, consider the collection of closed sets $\mathcal{C} = \{U_\alpha^c\}$.]

95. Let S be a subset of a metric space M. With respect to the definitions on page 92 prove the following.
    (a) The closure of S is the intersection of all closed subsets of M that contain S.
    (b) The interior of S is the union of all open subsets of M that are contained in S.
    (c) The boundary of S is a closed set.
    (d) Why does (a) imply the closure of S equals $\lim S$?
    (e) If S is clopen, what is ∂S?
    (f) Give an example of $S \subset \mathbb{R}$ such that $\partial(\partial S) \neq \emptyset$, and infer that "the boundary of the boundary $\partial \circ \partial$ is not always zero."

96. If $A \subset B \subset C$, A is dense in B, and B is dense in C prove that A is dense in C.

97. Is the set of dyadic rationals (the denominators are powers of 2) dense in $\mathbb{Q}$? In $\mathbb{R}$? Does one answer imply the other? (Recall that A is dense in B if $A \subset B$ and $\overline{A} \supset B$.)

98. Show that $S \subset M$ is somewhere dense in M if and only if $\text{int}(\overline{S}) \neq \emptyset$. Equivalently, S is nowhere dense in M if and only if its closure has empty interior.

99. Let M, N be nonempty metric spaces and $P = M \times N$.
    (a) If M, N are perfect prove that P is perfect.
    (b) If M, N are totally disconnected prove that P is totally disconnected.
    (c) What about the converses?
    (d) Infer that the Cartesian product of Cantor spaces is a Cantor space. (We already know that the Cartesian product of compacts is compact.)
    (e) Why does this imply that $C \times C = \{(x, y) \in \mathbb{R}^2 : x \in C \text{ and } y \in C\}$ is homeomorphic to C, C being the standard Cantor set?

100. Prove that every Cantor piece is a Cantor space. (Recall that M is a Cantor space if it is compact, nonempty, totally disconnected and perfect, and that $A \subset M$ is a Cantor piece if it is nonempty and clopen.)

*101. Let Σ be the set of all infinite sequences of zeroes and ones. For example, $(100111000011111\ldots) \in \Sigma$. Define the metric

$$d(a, b) = \sum \frac{|a_n - b_n|}{2^n}$$

where $a = (a_n)$ and $b = (b_n)$ are points in Σ.
  (a) Prove that Σ is compact.
  (b) Prove that Σ is homeomorphic to the Cantor set.

102. Prove that no Peano curve is one-to-one. (Recall that a Peano curve is a continuous map $f : [0, 1] \to \mathbb{R}^2$ whose image has a nonempty interior.)

103. Prove that there is a continuous surjection $\mathbb{R} \to \mathbb{R}^2$. What about $\mathbb{R}^m$?

104. Find two nonhomeomorphic compact subsets of $\mathbb{R}$ whose complements are homeomorphic.

105. As on page 115, consider the subsets of $\mathbb{R}$,

$$A = \{0\} \cup [1, 2] \cup \{3\} \quad \text{and} \quad B = \{0\} \cup \{1\} \cup [2, 3].$$

  (a) Why is there no ambient homeomorphism of $\mathbb{R}$ to itself that carries A onto B?
  (b) Thinking of $\mathbb{R}$ as the x-axis, is there an ambient homeomorphism of $\mathbb{R}^2$ to itself that carries A onto B?

106. Prove that the completion of a metric space is unique in the following natural sense: A completion of a metric space M is a complete metric X space containing M as a metric subspace such that M is dense in X. That is, every point of X is a limit of M.

  (a) Prove that M is dense in the completion $\widehat{M}$ constructed in the proof of Theorem 80.
  (b) If X and X' are two completions of M prove that there is an isometry $i : X \to X'$ such that $i(p) = p$ for all $p \in M$.
  (c) Prove that i is the unique such isometry.
  (d) Infer that $\widehat{M}$ is unique.

107. If M is a metric subspace of a complete metric space S prove that $\overline{M}$ is a completion of M.

*108. Consider the identity map $\mathrm{id} : C_{\max} \to C_{\mathrm{int}}$ where $C_{\max}$ is the metric space $C([0, 1], \mathbb{R})$ of continuous real-valued functions defined on $[0, 1]$, equipped with the max-metric $d_{\max}(f, g) = \max |f(x) - g(x)|$, and C_{int} is $C([0, 1], \mathbb{R})$ equipped with the integral metric,

$$d_{\mathrm{int}}(f, g) = \int_0^1 |f(x) - g(x)| \, dx.$$

Show that id is a continuous linear bijection (an isomorphism) but its inverse is not continuous.

*109. A metric on M is an **ultrametric** if for all $x, y, z \in M$ we have

$$d(x, z) \leq \max\{d(x, y), d(y, z)\}.$$

(Intuitively this means that the trip from x to z cannot be broken into shorter legs by making a stopover at some y.)

(a) Show that the ultrametric property implies the triangle inequality.
(b) In an ultrametric space show that "all triangles are isosceles."
(c) Show that a metric space with an ultrametric is totally disconnected.
(d) Define a metric on the set Σ of strings of zeroes and ones in Exercise 101 as

$$d_*(a, b) = \begin{cases} \dfrac{1}{2^n} & \text{if } n \text{ is the smallest index for which } a_n \neq b_n \\ 0 & \text{if } a = b. \end{cases}$$

Show that d_* is an ultrametric and prove that the identity map is a homeomorphism $(\Sigma, d) \to (\Sigma, d_*)$.

*110. $\mathbb{Q}$ inherits the Euclidean metric from $\mathbb{R}$ but it also carries a very different metric, the **p-adic** metric. Given a prime number p and an integer n, the p-adic norm of n is

$$|n|_p = \frac{1}{p^k}$$

where p^k is the largest power of p that divides n. (The norm of 0 is by definition 0.) The more factors of p, the smaller the p-norm. Similarly, if $x = a/b$ is a fraction, we factor x as

$$x = p^k \cdot \frac{r}{s}$$

where p divides neither r nor s, and we set

$$|x|_p = \frac{1}{p^k}.$$

The p-adic metric on $\mathbb{Q}$ is

$$d_p(x, y) = |x - y|_p.$$

(a) Prove that d_p is a metric with respect to which $\mathbb{Q}$ is perfect – every point is a cluster point.
(b) Prove that d_p is an ultrametric.
(c) Let $\mathbb{Q}_p$ be the metric space completion of $\mathbb{Q}$ with respect to the metric d_p, and observe that the extension of d_p to $\mathbb{Q}_p$ remains an ultrametric. Infer from Exercise 109 that $\mathbb{Q}_p$ is totally disconnected.

(d) Prove that $\mathbb{Q}_p$ is locally compact, in the sense that every point has small compact neighborhoods.

(e) Infer that $\mathbb{Q}_p$ is covered by neighborhoods homeomorphic to the Cantor set. See Gouvêa's book, *p-adic Numbers*.

111. Let $M = [0, 1]$ and let $\mathcal{M}_1$ be its division into two intervals $[0, 1/2]$ and $[1/2, 1]$. Let $\mathcal{M}_2$ be its division into four intervals $[0, 1/4]$, $[1/4, 1/2]$, $[1/2, 3/4]$, and $[3/4, 1]$. Continuing these bisections generates natural divisions of $[0, 1]$. The pieces are intervals. We label them with words using the letters 0 and 1 as follows: 0 means "left" and 1 means "right," so the four intervals in $\mathcal{M}_2$ are labeled as 00, 01, 10, and 11 respectively.

(a) Verify that all endpoints of the intervals (except 0 and 1) have two addresses. For instance,

$$\bigcap_k \left[\frac{2^{k-1} - 1}{2^k}, \frac{1}{2} \right] = \left\{ \frac{1}{2} \right\} = \bigcap_k \left[\frac{1}{2}, \frac{2^{k-1} + 1}{2^k} \right].$$

(b) Verify that the points $0, 1$, and all nonendpoints have unique addresses.

*112. Prove that $\#C = \#\mathbb{R}$. [Hint: According to the Schroeder-Bernstein Theorem from Chapter 1 it suffices to find injections $C \to \mathbb{R}$ and $\mathbb{R} \to C$. The inclusion $C \subset \mathbb{R}$ is an injection $C \to \mathbb{R}$. Each $t \in [0, 1)$ has a unique base-2 expansion $\tau(t)$ that does not terminate in an infinite string of ones. Replacing each 1 by 2 converts $\tau(t)$ to $\omega(t)$, an infinite address in the symbols 0 and 2. It does not terminate in an infinite string of twos. Set $h(t) = \sum_{i=1}^{\infty} \omega_i/3^i$ and verify that $h : [0, 1) \to C$ is an injection. Since there is an injection $\mathbb{R} \to [0, 1)$, conclude that there is an injection $\mathbb{R} \to C$, and hence that $\#C = \#\mathbb{R}$.]

Remark The Continuum Hypothesis states that if S is any uncountable subset of $\mathbb{R}$ then S and $\mathbb{R}$ have equal cardinality. The preceding coding shows that C is not only uncountable (as is implied by Theorem 56) but actually has the same cardinality as $\mathbb{R}$. That is, C is not a counterexample to the Continuum Hypothesis. The same is true of all uncountable closed subsets of $\mathbb{R}$. See Exercise 151.

113. Let M be the standard Cantor set C. In the notation of Section 8, C^n is the collection of 2^n Cantor intervals of length $1/3^n$ that nest down to C as $n \to \infty$. Verify that setting $\mathcal{C}_k = C \cap C^k$ gives divisions of C into disjoint clopen pieces.

*114. (a) Prove directly that there is a continuous surjection of the middle-thirds Cantor set C onto the closed interval $[0, 1]$. [Hint: Each $x \in C$ has a base 3 expansion (x_n), all of whose entries are zeroes and twos. (For example, $2/3 = (2\overline{0})_{\text{base } 3}$ and $1/3 = (0\overline{2})_{\text{base } 3}$. Write $y = (y_n)$ by replacing the twos in (x_n) by ones and interpreting the answer base 2. Show that the map $x \mapsto y$ works.]

(b) Compare this surjection to the one constructed from the bisection divisions in Exercise 113.

115. Rotate the unit circle S^1 by a fixed angle α, say $R : S^1 \to S^1$. (In polar coordinates, the transformation R sends $(1, \theta)$ to $(1, \theta + \alpha)$.)

   (a) If α/π is rational, show that each orbit of R is a finite set.

  *(b) If α/π is irrational, show that each orbit is infinite and has closure equal to S^1.

116. A metric space M with metric d can always be remetrized so the metric becomes bounded. Simply define the **bounded metric**

$$\rho(p, q) = \frac{d(p, q)}{1 + d(p, q)}.$$

   (a) Prove that ρ is a metric. Why is it obviously bounded?

   (b) Prove that the identity map $M \to M$ is a homeomorphism from M with the d-metric to M with the ρ-metric.

   (c) Infer that boundedness of M is not a topological property.

   (d) Find homeomorphic metric spaces, one bounded and the other not.

117. Fold a piece of paper in half.

   (a) Is this a continuous transformation of one rectangle into another?

   (b) Is it injective?

   (c) Draw an open set in the target rectangle, and find its preimage in the original rectangle. Is it open?

   (d) What if the open set meets the crease?

The **baker's transformation** is a similar mapping. A rectangle of dough is stretched to twice its length and then folded back on itself. Is the transformation continuous? A formula for the baker's transformation in one variable is $f(x) = 1 - |1 - 2x|$. The nth **iterate** of f is $f^n = f \circ f \circ \ldots \circ f$, n times. The **orbit** of a point x is

$$\{x, \ f(x), \ f^2(x), \ \ldots, \ f^n(x), \ \ldots\}.$$

[For clearer but more awkward notation one can write $f^{\circ n}$ instead of f^n. This distinguishes composition $f \circ f$ from multiplication $f \cdot f$.]

   (e) If x is rational prove that the orbit of x is a finite set.

   (f) If x is irrational what is the orbit?

*118. The implications of compactness are frequently equivalent to it. Prove

   (a) If every continuous function $f : M \to \mathbb{R}$ is bounded then M is compact.

   (b) If every continuous bounded function $f : M \to \mathbb{R}$ achieves a maximum or minimum then M is compact.

   (c) If every continuous function $f : M \to \mathbb{R}$ has compact range fM then M is compact.

(d) If every nested decreasing sequence of nonempty closed subsets of M has nonempty intersection then M is compact.

Together with Theorems 63 and 65, (a)–(d) give seven equivalent definitions of compactness. [Hint: Reason contrapositively. If M is not compact then it contains a sequence (p_n) that has no convergent subsequence. It is fair to assume that the points p_n are distinct. Find radii $r_n > 0$ such that the neighborhoods $M_{r_n}(p_n)$ are disjoint and no sequence $q_n \in M_{r_n}(p_n)$ has a convergent subsequence. Using the metric define a function $f_n : M_{r_n}(p_n) \to \mathbb{R}$ with a spike at p_n, such as

$$f_n(x) = \frac{r_n - d(x, p_n)}{a_n + d(x, p_n)}$$

where $a_n > 0$. Set $f(x) = f_n(x)$ if $x \in M_{r_n}(p_n)$, and $f(x) = 0$ if x belongs to no $M_{r_n}(p_n)$. Show that f is continuous. With the right choice of a_n show that f is unbounded. With a different choice of a_n, it is bounded but achieves no maximum, and so on.]

119. Let M be a metric space of diameter ≤ 2. The **cone** for M is the set

$$C = C(M) = \{p_0\} \cup M \times (0, 1]$$

with the **cone metric**

$$\begin{aligned} \rho((p, s), (q, t)) &= |s - t| + \min\{s, t\} d(p, q) \\ \rho((p, s), p_0) &= s \\ \rho(p_0, p_0) &= 0. \end{aligned}$$

The point p_0 is the vertex of the cone. Prove that ρ is a metric on C. [If M is the unit circle, think of it in the plane $z = 1$ in $\mathbb{R}^3$ centered at the point $(0, 0, 1)$. Its cone is the 45-degree cone with vertex the origin.]

120. Recall that if for each embedding of M, $h : M \to N$, hM is closed in N then M is said to be absolutely closed. If each hM is bounded then M is absolutely bounded. Theorem 41 implies that compact sets are absolutely closed and absolutely bounded. Prove:

(a) If M is absolutely bounded then M is compact.

*(b) If M is absolutely closed then M is compact.

Thus these are two more conditions equivalent to compactness. [Hint: From Exercise 118(a), if M is noncompact there is a continuous function $f : M \to \mathbb{R}$ that is unbounded. For Exercise 120(a), show that $F(x) = (x, f(x))$ embeds M onto a nonbounded subset of $M \times \mathbb{R}$. For 120(b), justify the additional assumption that the metric on M is bounded by 2. Then use Exercise 118(b) to show that if M is noncompact then there is a continuous function $g : M \to (0, 1]$ such that for some nonclustering sequence (p_n), we have $g(p_n) \to 0$ as $n \to \infty$. Finally, show that $G(x) = (x, gx)$ embeds M onto a nonclosed subset S of the cone $C(M)$ discussed in Exercise 119. S will be nonclosed because it limits at p_0 but does not contain it.]

121. (a) Prove that every function defined on a discrete metric space is uniformly continuous.

    (b) Infer that it is false to assert that if every continuous function $f : M \to \mathbb{R}$ is uniformly continuous then M is compact.

    (c) Prove, however, that if M is a metric subspace of a compact metric space K and every continuous function $f : M \to \mathbb{R}$ is uniformly continuous then M is compact.

122. Recall that p is a cluster point of S if each $M_r p$ contains infinitely many points of S. The set of cluster points of S is denoted as S'. Prove:

    (a) If $S \subset T$ then $S' \subset T'$.

    (b) $(S \cup T)' = S' \cup T'$.

    (c) $S' = (\overline{S})'$.

    (d) S' is closed in M; that is, $S'' \subset S'$ where $S'' = (S')'$.

    (e) Calculate $\mathbb{N}'$, $\mathbb{Q}'$, $\mathbb{R}'$, $(\mathbb{R} \setminus \mathbb{Q})'$, and $\mathbb{Q}''$.

    (f) Let T be the set of points $\{1/n : n \in \mathbb{N}\}$. Calculate T' and T''.

    (g) Give an example showing that S'' can be a proper subset of S'.

123. Recall that p is a condensation point of S if each $M_r p$ contains uncountably many points of S. The set of condensation points of S is denoted as S^*. Prove:

    (a) If $S \subset T$ then $S^* \subset T^*$.

    (b) $(S \cup T)^* = S^* \cup T^*$.

    (c) $S^* \subset \overline{S}^*$ where $\overline{S}^* = (\overline{S})^*$.

    (d) S^* is closed in M; that is, $S^{*\prime} \subset S^*$ where $S^{*\prime} = (S^*)'$.

    (e) $S^{**} \subset S^*$ where $S^{**} = (S^*)^*$.

    (f) Calculate $\mathbb{N}^*$, $\mathbb{Q}^*$, $\mathbb{R}^*$, and $(\mathbb{R} \setminus \mathbb{Q})^*$.

    (g) Give an example showing that S^* can be a proper subset of $(\overline{S})^*$. Thus, (c) is not in general an equality.

  **(h) Give an example that S^{**} can be a proper subset of S^*. Thus, (e) is not in general an equality. [Hint: Consider the set M of all functions $f : [a, b] \to [0, 1]$, continuous or not, and let the metric on M be the sup metric, $d(f, g) = \sup\{|f(x) - g(x)| : x \in [a, b]\}$. Consider the set S of all "δ-functions with rational values."]

  **(i) Give examples that show in general that S^* neither contains nor is contained in S'^* where $S'^* = (S')^*$. [Hint: δ-functions with values $1/n$, $n \in \mathbb{N}$.]

124. Recall that p is an interior point of $S \subset M$ if some $M_r p$ is contained in S. The set of interior points of S is the interior of S and is denoted int S. For all subsets S, T of the metric space M prove:

    (a) int $S = S \setminus \partial S$.

    (b) int $S = (\overline{S^c})^c$.

    (c) $\text{int}(\text{int } S) = \text{int } S$.

    (d) $\text{int}(S \cap T) = \text{int}(S \cap \text{int } T$.

    (e) What are the dual equations for the closure?

    (f) Prove that $\text{int}(S \cup T) \supset \text{int } S \cup \text{int } T$. Show by example that the inclusion can be strict, i.e., not an equality.

125. A point p is a boundary point of a set $S \subset M$ if every neighborhood $M_r p$ contains points of both S and S^c. The boundary of S is denoted ∂S. For all subsets S, T of a metric space M prove:

    (a) S is clopen if and only if $\partial S = \emptyset$.

    (b) $\partial S = \partial S^c$.

    (c) $\partial \partial S \subset \partial S$.

    (d) $\partial \partial \partial S = \partial \partial S$.

    (e) $\partial(S \cup T) \subset \partial S \cup \partial T$.

    (f) Give an example in which (c) is a strict inclusion, $\partial \partial S \neq \partial S$.

    (g) What about (e)?

*126. Suppose that E is an uncountable subset of $\mathbb{R}$. Prove that there exists a point $p \in \mathbb{R}$ at which E condenses. [Hint: Use decimal expansions. Why must there be an interval $[n, n+1)$ containing uncountably many points of E? Why must it contain a decimal subinterval with the same property? (A decimal subinterval $[a, b)$ has endpoints $a = n+k/10$, $b = n+(k+1)/10$ for some digit k, $0 \leq k \leq 9$.) Do you see lurking the decimal expansion of a condensation point?] Generalize to $\mathbb{R}^2$ and to $\mathbb{R}^m$.

127. The metric space M is **separable** if it contains a countable dense subset. [Note the confusion of language: "Separable" has nothing to do with "separation."]

    (a) Prove that $\mathbb{R}^m$ is separable.

    (b) Prove that every compact metric space is separable.

128. *(a) Prove that every metric subspace of a separable metric space is separable, and deduce that every metric subspace of $\mathbb{R}^m$ or of a compact metric space is separable.

    (b) Is the property of being separable topological?

    (c) Is the continuous image of a separable metric space separable?

129. Think up a nonseparable metric space.

130. Let $\mathcal{B}$ denote the collection of all ϵ-neighborhoods in $\mathbb{R}^m$ whose radius ϵ is rational and whose center has all coordinates rational.

    (a) Prove that $\mathcal{B}$ is countable.

    (b) Prove that every open subset of $\mathbb{R}^m$ can be expressed as the countable union of members of $\mathcal{B}$.

(The union need not be disjoint, but it is at most a countable union because there are only countably many members of $\mathcal{B}$. A collection such as $\mathcal{B}$ is called a **countable base** for the topology of $\mathbb{R}^m$.)

131. (a) Prove that every separable metric space has a countable base for its topology, and conversely that every metric space with a countable base for its topology is separable.

    (b) Infer that every compact metric space has a countable base for its topology.

*132. Referring to Exercise 123, assume now that M is separable, $S \subset M$, and, as before S' is the set of cluster points of S while S^* is the set of condensation points of S. Prove:

    (a) $S^* \subset (S')^* = (\overline{S})^*$.

    (b) $S^{**} = S^{*\prime} = S^*$.

    (c) Why is (a) not in general an equality?

    [Hints: For (a) write $S \subset (S \setminus S') \cup S'$ and $\overline{S} = (S \setminus S') \cup S'$, show that $(S \setminus S')^* = \emptyset$, and use Exercise 123(a). For (b), Exercise 123(d) implies that $S^{**} \subset S^{*\prime} \subset S^*$. To prove that $S^* \subset S^{**}$, write $S \subset (S \setminus S^*) \cup S^*$ and show that $(S \setminus S^*)^* = \emptyset$.]

*133. Prove that

    (a) An uncountable subset of $\mathbb{R}$ clusters at some point of $\mathbb{R}$.

    (b) An uncountable subset of $\mathbb{R}$ clusters at some point of itself.

    (c) An uncountable subset of $\mathbb{R}$ condenses at uncountably many points of itself.

    (d) What about $\mathbb{R}^m$ instead of $\mathbb{R}$?

    (e) What about any compact metric space?

    (f) What about any separable metric space?

    [Hint: Review Exercise 126.]

*134. Prove that $\widehat{\mathbb{Q}}$, the Cauchy sequences in $\mathbb{Q}$ modulo the equivalence relation of being co-Cauchy, is a field with respect to the natural arithmetic operations defined on page 122, and that $\mathbb{Q}$ is naturally a subfield of $\widehat{\mathbb{Q}}$.

135. Prove that the order on $\widehat{\mathbb{Q}}$ defined on page 122 is a bona fide order which agrees with the standard order on $\mathbb{Q}$.

*136. Let M be the square $[0,1]^2$, and let aa, ba, bb, ab label its four quadrants – upper right, upper left, lower left, and lower right.

    (a) Define nested bisections of the square using this pattern repeatedly, and let τ_k be a curve composed of line segments that visit the kth-order quadrants systematically. Let $\tau = \lim_k \tau_k$ be the resulting Peano curve à la the Cantor Surjection Theorem.

    (b) Compare τ to the Peano curve $f : I \to I^2$ directly constructed on pages 271- 274 of the second edition of Munkres' book *Topology*.

*137. Let P be a closed perfect subset of a separable complete metric space M. Prove that each point of P is a condensation point of P. In symbols, $P = P' \Rightarrow P = P^*$.

**138. Given a Cantor space $M \subset \mathbb{R}^2$, given a line segment $[p,q] \subset \mathbb{R}^2$ with $p, q \notin M$, and given an $\epsilon > 0$, prove that there exists a path A in the ϵ-neighborhood of $[p, q]$ that joins p to q and is disjoint from M. [Hint: Think of A as a bisector of M. From this bisection fact a dyadic disc partition of M can be constructed, which leads to the proof that M is tame.]

139. To prove that Antoine's Necklace A is a Cantor set, you need to show that A is compact, perfect, nonempty, and totally disconnected.

   (a) Do so. [Hint: What is the diameter of any connected component of A^n, and what does that imply about A?]

   **(b) If, in the Antoine construction two linked solid tori are placed *very cleverly* inside each larger solid torus, show that the intersection $A = \bigcap A^n$ is a Cantor set.

*140. Consider the **Hilbert cube**

$$H = \{(x_1, x_2, \ldots) \in [0, 1]^\infty : \text{ for each } n \in \mathbb{N} \text{ we have } |x_n| \leq 1/2^n\}.$$

Prove that H is compact with respect to the metric

$$d(x, y)) = \sup_n |x_n - y_n|$$

where $x = (x_n)$, $y = (y_n)$. [Hint: Sequences of sequences.]

Remark Although compact, H is infinite-dimensional and is homeomorphic to no subset of $\mathbb{R}^m$.

141. Prove that the Hilbert cube is perfect and homeomorphic to its Cartesian square, $H \cong H \times H$.

***142. Assume that M is compact, nonempty, perfect, and homeomorphic to its Cartesian square, $M \cong M \times M$. Must M be homeomorphic to the Cantor set, the Hilbert cube, or some combination of them?

143. A **Peano space** is a metric space M that is the continuous image of the unit interval: There is a continuous surjection $\tau : [0, 1] \to M$. Theorem 72 states the amazing fact that the 2-disc is a Peano space. Prove that every Peano space is

   (a) compact,

   (b) nonempty,

   (c) path-connected,

   *(d) and **locally path-connected**, in the sense that for each $p \in M$ and each neighborhood U of p there is a smaller neighborhood V of p such that any two points of V can be joined by a path in U.

*144. The converse to Exercise 143 is the **Hahn-Mazurkiewicz Theorem**. Assume that a metric space M is a compact, nonempty, path-connected, and locally path-connected. Use the Cantor Surjection Theorem 70 to show that M is a Peano space. [The key is to make uniformly short paths to fill in the gaps of $[0, 1] \setminus C$.]

145. One of the famous theorems in plane topology is the **Jordan Curve Theorem**. A **Jordan curve** J is a homeomorph of the unit circle in the plane. (Equivalently it is $f([a,b])$ where $f : [a,b] \to \mathbb{R}^2$ is continuous, $f(a) = f(b)$, and for no other pair of distinct $s, t \in [a, b]$ does $f(s)$ equal $f(t)$. It is also called a **simple closed curve**.) The Jordan Curve Theorem asserts that $\mathbb{R}^2 \setminus J$ consists of two disjoint, connected open sets, its inside and its outside, and every path between them must meet J. Prove the Jordan Curve Theorem for the circle, the square, the triangle, and – if you have courage – every simple closed polygon.

146. The **utility problem** gives three houses 1, 2, 3 in the plane and the three utilities, Gas, Water, and Electricity. You are supposed to connect each house to the three utilities without crossing utility lines. (The houses and utilities are disjoint.)

    (a) Use the Jordan curve theorem to show that there is no solution to the utility problem in the plane.

    *(b) Show also that the utility problem cannot be solved on the 2-sphere S^2.

    *(c) Show that the utility problem can be solved on the surface of the torus.

    *(d) What about the surface of the Klein bottle?

    ***(e) Given utilities $U_1, \ldots, U_m$ and houses $H_1, \ldots, H_n$ located on a surface with g handles, find necessary and sufficient conditions on m, n, g so that the utility problem can be solved.

147. Let M be a metric space and let $\mathcal{K}$ denote the class of nonempty compact subsets of M. The r-neighborhood of $A \in \mathcal{K}$ is

$$M_r A = \{x \in M : \exists a \in A \text{ and } d(x, a) < r\} = \bigcup_{a \in A} M_r a.$$

For $A, B \in \mathcal{K}$ define

$$D(A, B) = \inf\{r > 0 : A \subset M_r B \text{ and } B \subset M_r A\}.$$

    (a) Show that D is a metric on $\mathcal{K}$. (It is called the **Hausdorff metric** and $\mathcal{K}$ is called the **hyperspace** of M.)

    (b) Denote by $\mathcal{F}$ the collection of finite nonempty subsets of M and prove that $\mathcal{F}$ is dense in $\mathcal{K}$. That is, given $A \in \mathcal{K}$ and given $\epsilon > 0$ show there exists $F \in \mathcal{F}$ such that $D(A, F) < \epsilon$.

    *(c) If M is compact prove that $\mathcal{K}$ is compact.

    (d) If M is connected prove that $\mathcal{K}$ is connected.

    **(e) If M is path-connected is $\mathcal{K}$ path-connected?

    (f) Do homeomorphic metric spaces have homeomorphic hyperspaces?

    Remark The converse to (f), $\mathcal{K}(M) \cong \mathcal{K}(N) \Rightarrow M \cong N$ is false. The hyperspace of every Peano space is the Hilbert cube. This is a difficult result but a good place to begin reading about hyperspaces is Sam Nadler's book *Continuum Theory*.

**148. Start with a set $S \subset \mathbb{R}$ and successively take its closure, the complement of its closure, the closure of that, and so on: S, $\mathrm{cl}(S)$, $(\mathrm{cl}(S))^c, \ldots$. Do the same to S^c. In total, how many distinct subsets of $\mathbb{R}$ can be produced this way? In particular decide whether each chain S, $\mathrm{cl}(S)$, $\ldots$ consists of only finitely many sets. For example, if $S = \mathbb{Q}$ then we get $\mathbb{Q}$, $\mathbb{R}$, $\emptyset$, $\emptyset$, $\mathbb{R}$, $\mathbb{R}, \ldots$ and $\mathbb{Q}^c$, $\mathbb{R}$, $\emptyset$, $\emptyset$, $\mathbb{R}$, $\mathbb{R}, \ldots$ for a total of four sets.

**149. Consider the letter T.

(a) Prove that there is no way to place uncountably many copies of the letter T disjointly in the plane. [Hint: First prove this when the unit square replaces the plane.]

(b) Prove that there is no way to place uncountably many homeomorphic copies of the letter T disjointly in the plane.

(c) For which other letters of the alphabet is this true?

(d) Let U be a set in $\mathbb{R}^3$ formed like an umbrella: It is a disc with a perpendicular segment attached to its center. Prove that uncountably many copies of U cannot be placed disjointly in $\mathbb{R}^3$.

(e) What if the perpendicular segment is attached to the boundary of the disc?

150. Let M be a complete, separable metric space such as $\mathbb{R}^m$. Prove the **Cupcake Theorem: Each closed set $K \subset M$ can be expressed uniquely as the disjoint union of a countable set and a perfect closed set,

$$C \sqcup P = K.$$

**151. Let M be an uncountable compact metric space.

(a) Prove that M contains a homeomorphic copy of the Cantor set. [Hint: Imitate the construction of the standard Cantor set C.]

(b) Infer that Cantor sets are ubiquitous. There is a continuous surjection $\sigma : C \to M$ and there is a continuous injection $i : C \to M$.

(c) Infer that every uncountable closed set $S \subset \mathbb{R}$ has $\#S = \#\mathbb{R}$, and hence that the Continuum Hypothesis is valid for closed sets in $\mathbb{R}$. [Hint: Cupcake and Exercise 112.]

(d) Is the same true if M is separable, uncountable, and complete?

**152. Write jingles at least as good as the following. Pay attention to the meter as well as the rhyme.

> When a set in the plane
> is closed and bounded,
> you can always draw
> a curve around it.
>
>   Peter Přibik

If a clopen set can be detected,

Your metric space is disconnected.

David Owens

A coffee cup feeling quite dazed,

said to a donut, amazed,

an open surjective continuous injection,

You'd be plastic and I'd be glazed.

Norah Esty

'Tis a most indisputable fact

If you want to make something compact

Make it bounded and closed

For you're totally hosed

If either condition you lack.

Lest the reader infer an untruth

(Which I think would be highly uncouth)

I must hasten to add

There are sets to be had

Where the converse is false, fo'sooth.

Karla Westfahl

For ev'ry a and b in S

if there exists a path that's straight

from a to b and it's inside

then "S must be convex," we state.

Alex Wang

Prelim Problems[†]

1. Suppose that $f : \mathbb{R}^m \to \mathbb{R}$ satisfies two conditions:
   (i) For each compact set K, $f(K)$ is compact.
   (ii) For every nested decreasing sequence of compacts (K_n),
   $$f\left(\bigcap K_n\right) = \bigcap f(K_n).$$
   Prove that f is continuous.

2. Let $X \subset \mathbb{R}^m$ be compact and $f : X \to \mathbb{R}$ be continuous. Given $\epsilon > 0$, show that there is a constant M such that for all $x, y \in X$ we have $|f(x) - f(y)| \leq M|x - y| + \epsilon$.

3. Consider $f : \mathbb{R}^2 \to \mathbb{R}$. Assume that for each fixed x_0, $y \mapsto f(x_0, y)$ is continuous and for each fixed y_0, $x \mapsto f(x, y_0)$ is continuous. Find such an f that is not continuous.

4. Let $f : \mathbb{R}^2 \to \mathbb{R}$ satisfy the following properties. For each fixed $x_0 \in \mathbb{R}$ the function $y \mapsto f(x_0, y)$ is continuous and for each fixed $y_0 \in \mathbb{R}$ the function $x \mapsto f(x, y_0)$ is continuous. Also assume that if K is any compact subset of $\mathbb{R}^2$ then $f(K)$ is compact. Prove that f is continuous.

5. Let $f(x, y)$ be a continuous real-valued function defined on the unit square $[0, 1] \times [0, 1]$. Prove that
   $$g(x) = \max\{f(x, y) : y \in [0, 1]\}$$
   is continuous.

6. Let $\{U_k\}$ be a cover of $\mathbb{R}^m$ by open sets. Prove that there is a cover $\{V_k\}$ of $\mathbb{R}^m$ by open sets V_k such that $V_k \subset U_k$ and each compact subset of $\mathbb{R}^m$ is disjoint from all but finitely many of the V_k.

7. A function $f : [0, 1] \to \mathbb{R}$ is said to be **upper semicontinuous** if given $x \in [0, 1]$ and $\epsilon > 0$ there exists a $\delta > 0$ such that $|y - x| < \delta$ implies that $f(y) < f(x) + \epsilon$. Prove that an upper semicontinuous function on $[0, 1]$ is bounded above and attains its maximum value at some point $p \in [0, 1]$.

8. Prove that a continuous function $f : \mathbb{R} \to \mathbb{R}$ which sends open sets to open sets must be monotonic.

9. Show that $[0, 1]$ cannot be written as a countably infinite union of disjoint closed subintervals.

10. A **connected component** of a metric space M is a maximal connected subset of M. Give an example of $M \subset \mathbb{R}$ having uncountably many connected components. Can such a subset be open? Closed? Does your answer change if $\mathbb{R}^2$ replaces $\mathbb{R}$?

[†]These are questions taken from the exam given to first-year math graduate students at U.C. Berkeley.

11. Let $U \subset \mathbb{R}^m$ be an open set. Suppose that the map $h : U \to \mathbb{R}^m$ is a homeomorphism from U onto $\mathbb{R}^m$ which is uniformly continuous. Prove that $U = \mathbb{R}^m$.

12. Let X be a nonempty connected set of real numbers. If every element of X is rational prove that X has only one element.

13. Let $A \subset \mathbb{R}^m$ be compact, $x \in A$. Let (x_n) be a sequence in A such that every convergent subsequence of (x_n) converges to x.

    (a) Prove that the sequence (x_n) converges.

    (b) Give an example to show if A is not compact, the result in (a) is not necessarily true.

14. Assume that $f : \mathbb{R} \to \mathbb{R}$ is uniformly continuous. Prove that there are constants A, B such that $|f(x)| \leq A + B|x|$ for all $x \in \mathbb{R}$.

15. Let $h : [0, 1) \to \mathbb{R}$ be a uniformly continuous function where $[0, 1)$ is the half-open interval. Prove that there is a unique continuous map $g : [0, 1] \to \mathbb{R}$ such that $g(x) = h(x)$ for all $x \in [0, 1)$.

3

Functions of a Real Variable

1 Differentiation

The function $f : (a,b) \to \mathbb{R}$ is **differentiable at** x if

$$(1) \qquad \lim_{t \to x} \frac{f(t) - f(x)}{t - x} = L$$

exists. This means L is a real number and for each $\epsilon > 0$ there exists a $\delta > 0$ such that if $0 < |t - x| < \delta$ then the **differential quotient** above differs from L by $< \epsilon$. The limit L is the **derivative** of f at x, $L = f'(x)$. In calculus language, $\Delta x = t - x$ is the change in the independent variable x while $\Delta f = f(t) - f(x)$ is the resulting change in the dependent variable $y = f(x)$. Differentiability at x means that

$$f'(x) = \lim_{\Delta x \to 0} \frac{\Delta f}{\Delta x}.$$

We begin by reviewing the proofs of some standard calculus facts.

1 The Rules of Differentiation

(a) *Differentiability implies continuity.*

(b) *If f and g are differentiable at x then so is $f + g$, the derivative being*

$$(f + g)'(x) = f'(x) + g'(x).$$

(c) *If f and g are differentiable at x then so is their product $f \cdot g$, the derivative being given by the **Leibniz Formula***

$$(f \cdot g)'(x) = f'(x) \cdot g(x) + f(x) \cdot g'(x).$$

© Springer International Publishing Switzerland 2015
C.C. Pugh, *Real Mathematical Analysis*, Undergraduate Texts
in Mathematics, DOI 10.1007/978-3-319-17771-7_3

(d) *The derivative of a constant is zero, $c' = 0$.*

(e) *If f and g are differentiable at x and $g(x) \neq 0$ then their ratio f/g is differentiable at x, the derivative being*

$$\left(\frac{f}{g}\right)'(x) = \frac{f'(x) \cdot g(x) - f(x) \cdot g'(x)}{g(x)^2}.$$

(f) *If f is differentiable at x and g is differentiable at $y = f(x)$ then their composite $g \circ f$ is differentiable at x, the derivative being given as the **Chain Rule***

$$(g \circ f)'(x) = g'(y) \cdot f'(x).$$

Proof (a) Continuity in the calculus notation amounts to the assertion that $\Delta f \to 0$ as $\Delta x \to 0$. This is obvious: If the fraction $\Delta f/\Delta x$ tends to a finite limit while its denominator tends to zero, then its numerator must also tend to zero.

(b) Since $\Delta(f + g) = \Delta f + \Delta g$, we have

$$\frac{\Delta(f + g)}{\Delta x} = \frac{\Delta f}{\Delta x} + \frac{\Delta g}{\Delta x} \to f'(x) + g'(x)$$

as $\Delta x \to 0$.

(c) Since $\Delta(f \cdot g) = \Delta f \cdot g(x + \Delta x) + f(x) \cdot \Delta g$, continuity of g at x implies that

$$\frac{\Delta(f \cdot g)}{\Delta x} = \frac{\Delta f}{\Delta x} g(x + \Delta x) + f(x)\frac{\Delta g}{\Delta x} \to f'(x)g(x) + f(x)g'(x),$$

as $\Delta x \to 0$.

(d) If c is a constant then $\Delta c = 0$ and $c' = 0$.

(e) Since

$$\Delta(f/g) = \frac{g(x)\Delta f - f(x)\Delta g}{g(x + \Delta x)g(x)},$$

the formula follows when we divide by Δx and take the limit.

(f) The shortest proof of the chain rule for $y = f(x)$ is by cancellation:

$$\frac{\Delta g}{\Delta x} = \frac{\Delta g}{\Delta y}\frac{\Delta y}{\Delta x} \to g'(y)f'(x).$$

A slight flaw is present: Δy may be zero when Δx is not. This is not a big problem. Differentiability of g at y implies that

$$\frac{\Delta g}{\Delta y} = g'(y) + \sigma$$

where $\sigma = \sigma(\Delta y) \to 0$ as $\Delta y \to 0$. *Define* $\sigma(0) = 0$. The formula

$$\Delta g = (g'(y) + \sigma)\Delta y$$

holds for all small Δy, including $\Delta y = 0$. Continuity of f at x (which is true by (a)) implies that $\Delta f \to 0$ as $\Delta x \to 0$. Thus

$$\frac{\Delta g}{\Delta x} = (g'(y) + \sigma(\Delta f))\frac{\Delta y}{\Delta x} \to g'(y)f'(x)$$

as $\Delta x \to 0$. □

2 Corollary *The derivative of a polynomial $a_0 + a_1 x + \cdots + a_n x^n$ exists at every $x \in \mathbb{R}$ and equals $a_1 + 2a_2 x + \cdots + n a_n x^{n-1}$.*

Proof Immediate from the differentiation rules. □

A function $f : (a, b) \to \mathbb{R}$ that is differentiable at each $x \in (a, b)$ is **differentiable**.

3 Mean Value Theorem *A continuous function $f : [a, b] \to \mathbb{R}$ that is differentiable on the interval (a, b) has the **mean value property**: There exists a point $\theta \in (a, b)$ such that*

$$f(b) - f(a) = f'(\theta)(b - a).$$

4 Lemma *If $f : (a, b) \to \mathbb{R}$ is differentiable and achieves a minimum or maximum at some $\theta \in (a, b)$ then $f'(\theta) = 0$.*

Proof Assume that f has a minimum at θ. The derivative $f'(\theta)$ is the limit of the differential quotient $(f(t) - f(\theta))/(t - \theta)$ as $t \to \theta$. Since $f(t) \geq f(\theta)$ for all $t \in (a, b)$, the differential quotient is nonnegative for $t > \theta$ and nonpositive for $t < \theta$. Thus $f'(\theta)$ is a limit of both nonnegative and nonpositive quantities, so $f'(\theta) = 0$. Similarly $f'(\theta) = 0$ when f has a maximum at θ. □

Proof of the Mean Value Theorem See Figure 59, where

$$S = \frac{f(b) - f(a)}{b - a}$$

is the slope of the secant of the graph of f. The function $\phi(x) = f(x) - S(x - a)$ is continuous on $[a, b]$ and differentiable on (a, b). It has the same value, namely $f(a)$, at $x = a$ and $x = b$. Since $[a, b]$ is compact ϕ takes on maximum and minimum values, and since it has the same value at both endpoints, ϕ has a maximum or a minimum that occurs at an interior point $\theta \in (a, b)$. See Figure 60. By Lemma 4 we have $\phi'(\theta) = 0$ and $f(b) - f(a) = f'(\theta)(b - a)$. □

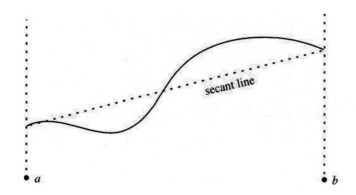

Figure 59 The secant line for the graph of f

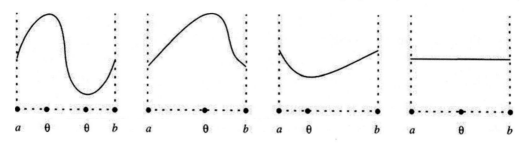

Figure 60 $\phi'(\theta) = 0$.

5 Corollary *If f is differentiable and $|f'(x)| \leq M$ for all $x \in (a, b)$ then f satisfies the global Lipschitz condition – for all $t, x \in (a, b)$ we have*

$$|f(t) - f(x)| \leq M|t - x|.$$

In particular, if $f'(x) = 0$ for all $x \in (a, b)$ then $f(x)$ is constant.

Proof $|f(t) - f(x)| = |f'(\theta)(t - x)|$ for some θ between x and t.   □

Remark The Mean Value Theorem and this corollary are the most important tools in calculus for making estimates.

It is often convenient to deal with two functions simultaneously, and for that we have the following result.

6 Ratio Mean Value Theorem *Suppose that the functions f and g are continuous on an interval $[a, b]$ and differentiable on the interval (a, b). Then there is a $\theta \in (a, b)$ such that*

$$\Delta f \cdot g'(\theta) \; = \; \Delta g \cdot f'(\theta)$$

where $\Delta f = f(b) - f(a)$ and $\Delta g = g(b) - g(a)$. (If $g(x) \equiv x$, the Ratio Mean Value Theorem becomes the ordinary Mean Value Theorem.)

Proof If $\Delta g \neq 0$ then the theorem states that for some $\theta \in (a, b)$ we have

$$\frac{\Delta f}{\Delta g} = \frac{f'(\theta)}{g'(\theta)}.$$

This ratio expression is how to remember the theorem. The whole point here is that f' and g' are evaluated at the same θ. The function

$$\Phi(x) = \Delta f \cdot (g(x) - g(a)) - \Delta g \cdot (f(x) - f(a))$$

is differentiable and its value at both endpoints a, b is 0. Since Φ is continuous it takes on a maximum and a minimum somewhere in the interval $[a, b]$. Since Φ has equal values at the endpoints of the interval, it must take on a maximum or minimum at some point $\theta \in (a, b)$; i.e., $\theta \neq a, b$. Then $\Phi'(\theta) = 0$ and $\Delta f \cdot g'(\theta) = \Delta g \cdot f'(\theta)$ as claimed. $\qquad\square$

7 L'Hôpital's Rule *If f and g are differentiable functions defined on an interval (a, b), both of which tend to 0 at b, and if the ratio of their derivatives $f'(x)/g'(x)$ tends to a finite limit L at b then $f(x)/g(x)$ also tends to L at b. (We assume that $g(x), g'(x) \neq 0$.)*

Rough Proof Let $x \in (a, b)$ tend to b. Imagine a point $t \in (a, b)$ tending to b much faster than x does. It is a kind of "advance guard" for x. Then $f(t)/f(x)$ and $g(t)/g(x)$ are as small as we wish, and by the Ratio Mean Value Theorem there is a $\theta \in (x, t)$ such that

$$\frac{f(x)}{g(x)} = \frac{f(x) - 0}{g(x) - 0} \overset{\cdot}{=} \frac{f(x) - f(t)}{g(x) - g(t)} = \frac{f'(\theta)}{g'(\theta)}.$$

The latter tends to L because θ is sandwiched between x and t as they tend to b. The symbol $\overset{\cdot}{=}$ means approximately equal. See Figure 61. $\qquad\square$

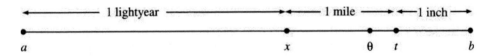

Figure 61 x and t escort θ toward b.

Complete Proof Given $\epsilon > 0$ we must find $\delta > 0$ such that if $|x - b| < \delta$ then $|f(x)/g(x) - L| < \epsilon$. Since $f'(x)/g'(x)$ tends to L as x tends to b there does exist $\delta > 0$ such that if $x \in (b - \delta, b)$ then

$$\left| \frac{f'(x)}{g'(x)} - L \right| < \frac{\epsilon}{2}.$$

For each $x \in (b - \delta, b)$ determine a point $t \in (b - \delta, b)$ which is so near to b that

$$|f(t)| + |g(t)| \;\; < \;\; \frac{g(x)^2 \epsilon}{4(|f(x)| + |g(x)|)}$$

$$|g(t)| \;\; < \;\; \frac{|g(x)|}{2}.$$

Since $f(t)$ and $g(t)$ tend to 0 as t tends to b, and since $g(x) \neq 0$ such a t exists. It depends on x, of course. By this choice of t and the Ratio Mean Value Theorem we have

$$\left| \frac{f(x)}{g(x)} - L \right| \;=\; \left| \frac{f(x)}{g(x)} - \frac{f(x) - f(t)}{g(x) - g(t)} + \frac{f(x) - f(t)}{g(x) - g(t)} - L \right|$$

$$\leq \;\; \left| \frac{g(x)f(t) - f(x)g(t)}{g(x)(g(x) - g(t))} \right| + \left| \frac{f'(\theta)}{g'(\theta)} - L \right| < \epsilon,$$

which completes the proof that $f(x)/g(x) \to L$ as $x \to b$.   □

It is clear that L'Hôpital's Rule holds equally well as x tends to b or to a. It is also true that it holds when x tends to $\pm\infty$ or when f and g tend to $\pm\infty$. See Exercises 6 and 7.

From now on feel free to use L'Hôpital's Rule!

8 Theorem *If f is differentiable on (a, b) then its derivative function $f'(x)$ has the intermediate value property.*

Differentiability of f implies continuity of f, and so the Intermediate Value Theorem from Chapter 2 applies to f and states that f takes on all intermediate values, but this is not what Theorem 8 is about. Not at all. Theorem 8 concerns f' not f. The function f' can well be discontinuous, but nevertheless it too takes on all intermediate values. In a clear abuse of language, functions like f' possessing the intermediate value property are called **Darboux continuous**, even when they are discontinuous! Darboux was the first to realize how badly discontinuous a derivative function can be. Despite the fact that f' has the intermediate value property, it can be discontinuous at almost every point of $[a, b]$. Strangely enough, however, f' cannot be discontinuous at every point. If f is differentiable, f' must be continuous at a dense, thick set of points. See Exercise 25 and the next section for the definitions.

Proof of Theorem 8 Suppose that $a < x_1 < x_2 < b$ and

$$\alpha = f'(x_1) < \gamma < f'(x_2) = \beta.$$

We must find $\theta \in (x_1, x_2)$ such that $f'(\theta) = \gamma$.

Choose a small h, $0 < h < x_2 - x_1$, and draw the secant segment $\sigma(x)$ between the points $(x, f(x))$ and $(x+h, f(x+h))$ on the graph of f. Slide x from x_1 to $x_2 - h$ continuously. This is the **sliding secant method**. See Figure 62.

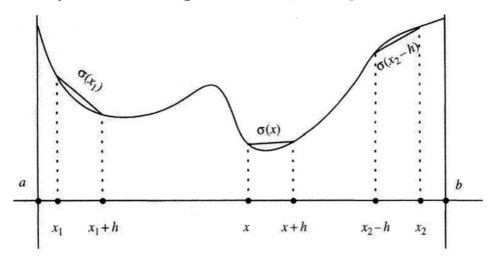

Figure 62 The sliding secant

When h is small enough, slope $\sigma(x_1) \doteq f'(x_1)$ and slope $\sigma(x_2 - h) \doteq f'(x_2)$. Thus

$$\text{slope } \sigma(x_1) < \gamma < \text{slope } \sigma(x_2 - h).$$

Continuity of f implies that the slope of $\sigma(x)$ depends continuously on x, so by the Intermediate Value Theorem for continuous functions there is an $x \in (x_1, x_2 - h)$ with slope $\sigma(x) = \gamma$. The Mean Value Theorem then gives a $\theta \in (x, x+h)$ such that $f'(\theta) = \gamma$. ☐

9 Corollary *The derivative of a differentiable function never has a jump discontinuity.*

Proof Near a jump, a function omits intermediate values. ☐

Pathological Examples

Nonjump discontinuities of f' may very well occur. The function

$$f(x) = \begin{cases} x^2 \sin \dfrac{1}{x} & \text{if } x > 0 \\ 0 & \text{if } x \leq 0 \end{cases}$$

is differentiable everywhere, even at $x = 0$, where $f'(0) = 0$. Its derivative function for $x > 0$ is

$$f'(x) = 2x \sin \frac{1}{x} - \cos \frac{1}{x},$$

which oscillates more and more rapidly with amplitude approximately 1 as $x \to 0$. Since $f'(x) \nrightarrow 0$ as $x \to 0$, f' is discontinuous at $x = 0$. Figure 63 shows why f is differentiable at $x = 0$ and has $f'(0) = 0$. Although the graph oscillates wildly at 0, it does so between the envelopes $y = \pm x^2$, and any curve between these envelopes is tangent to the x-axis at the origin. *Study* this example, Figure 63.

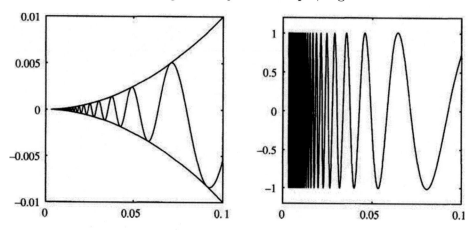

Figure 63 The graphs of the function $y = x^2 \sin(1/x)$ and its envelopes $y = \pm x^2$; and the graph of its derivative

A similar but worse example is illustrated in Figure 64, where

$$g(x) = \begin{cases} x^{3/2} \sin \dfrac{1}{x} & \text{if } x > 0 \\ 0 & \text{if } x \leq 0 \end{cases}$$

Its derivative at $x = 0$ is $g'(0) = 0$, while at $x \neq 0$ its derivative is

$$g'(x) = \frac{3}{2}\sqrt{x} \sin \frac{1}{x} - \frac{1}{\sqrt{x}} \cos \frac{1}{x},$$

which oscillates with increasing frequency and unbounded amplitude as $x \to 0$ because $1/\sqrt{x}$ blows up at $x = 0$.

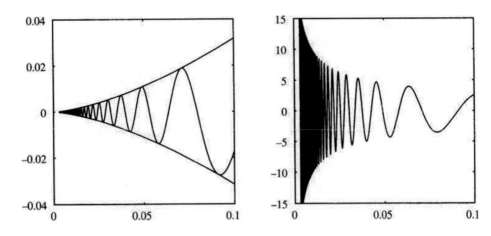

Figure 64 The function $y = x^{3/2} \sin(1/x)$, its envelopes $y = \pm x^{3/2}$, and its derivative.

Higher Derivatives

The derivative of f', if it exists, is the second derivative of f,

$$(f')'(x) = f''(x) = \lim_{t \to x} \frac{f'(t) - f'(x)}{t - x}.$$

Higher derivatives are defined inductively and written $f^{(r)} = (f^{(r-1)})'$. If $f^{(r)}(x)$ exists then f is r^{th}-**order differentiable at** x. If $f^{(r)}(x)$ exists for each $x \in (a, b)$ then f is r^{th}-**order differentiable**. If $f^{(r)}(x)$ exists for all r and all x then f is **infinitely differentiable** or **smooth**. The **zeroth derivative** of f is f itself, $f^{(0)}(x) = f(x)$.

10 Theorem *If f is r^{th}-order differentiable and $r \geq 1$ then $f^{(r-1)}(x)$ is a continuous function of $x \in (a, b)$.*

Proof Differentiability implies continuity and $f^{(r-1)}(x)$ is differentiable. $\square$

11 Corollary *A smooth function is continuous. Each derivative of a smooth function is smooth and hence continuous.*

Proof Obvious from the definition of smoothness and Theorem 10. $\square$

Smoothness Classes

If f is differentiable and its derivative function $f'(x)$ is a continuous function of x then f is **continuously differentiable** and we say that f is of **class C^1**. If f is r^{th}-order differentiable and $f^{(r)}(x)$ is a continuous function of x then f is **continuously**

r^{th}**-order differentiable** and we say that f is of **class C^r**. If f is smooth then by the preceding corollary it is of class C^r for all finite r and we say that f is of **class C^∞**. To round out the notation we say that a continuous function is of class C^0.

Thinking of C^r as the set of functions of class C^r, we have the **regularity hierarchy**

$$C^0 \supset C^1 \supset \cdots \supset \bigcap_{r \in \mathbb{N}} C^r = C^\infty.$$

Each inclusion $C^r \supset C^{r+1}$ is proper. There exist continuous functions that are not of class C^1, C^1 functions that are not of class C^2, and so on. For example,

$$
\begin{aligned}
f(x) &= |x| &&\text{is of class } C^0 \text{ but not of class } C^1, \\
f(x) &= x|x| &&\text{is of class } C^1 \text{ but not of class } C^2, \\
f(x) &= |x|^3 &&\text{is of class } C^2 \text{ but not of class } C^3,
\end{aligned}
$$

$$\cdots$$

Analytic Functions

A function that can be expressed locally as a convergent power series is **analytic**. More precisely, the function $f : (a, b) \to \mathbb{R}$ is analytic if for each $x \in (a, b)$, there exist a power series

$$\sum a_r h^r$$

and a $\delta > 0$ such that if $|h| < \delta$ then the series converges and

$$f(x + h) = \sum_{r=0}^{\infty} a_r h^r.$$

The concept of series convergence will be discussed further in Section 3 and Chapter 4. Among other things we show in Section 2 of Chapter 4 that analytic functions are smooth, and if $f(x + h) = \sum a_r h^r$ then

$$f^{(r)}(x) = r! a_r.$$

This gives uniqueness of the power series expression of a function: if two power series express the same function f at x then they have identical coefficients, namely $f^{(r)}(x)/r!$. See Exercise 4.38 for a stronger type of uniqueness, namely the identity theorem for analytic functions.

We write C^ω for the class of analytic functions.

A Nonanalytic Smooth Function

The fact that smooth functions need not be analytic is somewhat surprising; i.e., C^ω is a proper subset of C^∞. A standard example is

$$e(x) = \begin{cases} e^{-1/x} & \text{if } x > 0 \\ 0 & \text{if } x \le 0. \end{cases}$$

Its smoothness is left as an exercise in the use of L'Hôpital's Rule and induction, Exercise 17. At $x = 0$ the graph of $e(x)$ is infinitely tangent to the x-axis. Every derivative $e^{(r)}(0) = 0$. See Figure 65.

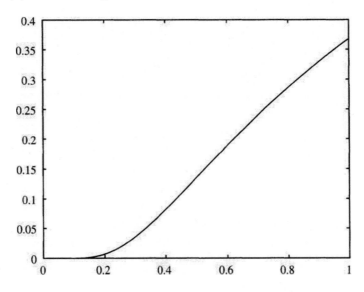

Figure 65 The graph of $e(x) = e^{-1/x}$

It follows that $e(x)$ is not analytic. For if it were then it could be expressed near $x = 0$ as a convergent series $e(h) = \sum a_r h^r$, and $a_r = e^{(r)}(0)/r!$. Thus $a_r = 0$ for each r, and the series converges to zero, whereas $e(h)$ is different from zero when $h > 0$. Although not analytic at $x = 0$, $e(x)$ is analytic elsewhere. See also Exercise 4.37.

Taylor Approximation

The rth-order **Taylor polynomial** of an rth-order differentiable function f at x is

$$P(h) = f(x) + f'(x)h + \frac{f''(x)}{2!}h^2 + \ldots + \frac{f^{(r)}(x)}{r!}h^r = \sum_{k=0}^{r} \frac{f^{(k)}(x)}{k!}h^k.$$

The coefficients $f^{(k)}(x)/k!$ are constants, the variable is h. Differentiation of P with respect to h at $h = 0$ gives

$$
\begin{aligned}
P(0) &= f(x) \\
P'(0) &= f'(x) \\
&\cdots \\
P^{(r)}(0) &= f^{(r)}(x).
\end{aligned}
$$

12 Taylor Approximation Theorem *Assume that $f : (a, b) \to \mathbb{R}$ is r^{th} order differentiable at x. Then*

(a) *P approximates f to order r at x in the sense that the Taylor remainder*

$$
R(h) = f(x + h) - P(h)
$$

is r^{th} order flat at $h = 0$; i.e., $R(h)/h^r \to 0$ as $h \to 0$.

(b) *The Taylor polynomial is the only polynomial of degree $\leq r$ with this approximation property.*

(c) *If, in addition, f is $(r+1)^{\text{st}}$-order differentiable on (a, b) then for some θ between x and $x + h$ we have*

$$
R(h) = \frac{f^{(r+1)}(\theta)}{(r+1)!} h^{r+1}.
$$

Remark (c) is the **Lagrange form** of the remainder. If $\left| f^{(r+1)}(\theta) \right| \leq M$ for all $\theta \in (a, b)$ then

$$
R(h) \leq \frac{Mh^{r+1}}{(r+1)!},
$$

an estimate that is valid uniformly with respect to x and $x+h$ in (a, b), whereas (a) is only an infinitesimal pointwise estimate. Of course (c) requires stronger hypotheses than (a).

Proof (a) The first r derivatives of $R(h)$ exist and equal 0 at $h = 0$. If $h > 0$ then repeated applications of the Mean Value Theorem give

$$
\begin{aligned}
R(h) &= R(h) - 0 = R'(\theta_1)h = (R'(\theta_1) - 0)h = R''(\theta_2)\theta_1 h \\
&= \cdots = R^{(r-1)}(\theta_{r-1})\theta_{r-2}\ldots\theta_1 h
\end{aligned}
$$

where $0 < \theta_{r-1} < \cdots < \theta_1 < h$. Thus

$$
\left| \frac{R(h)}{h^r} \right| = \left| \frac{R^{(r-1)}(\theta_{r-1})\theta_{r-2}\ldots\theta_1 h}{h^r} \right| \leq \left| \frac{R^{(r-1)}(\theta_{r-1}) - 0}{\theta_{r-1}} \right| \to 0
$$

as $h \to 0$. On the other hand, if $h < 0$ the same is true with $h < \theta_1 < \cdots < \theta_{r-1} < 0$.

(b) If $Q(h)$ is a polynomial of degree $\leq r$, $Q \neq P$, then $Q - P$ is not r^{th}-order flat at $h = 0$, so $f(x + h) - Q(h)$ cannot be r^{th}-order flat either.

(c) Fix $h > 0$ and define

$$g(t) = f(x + t) - P(t) - \frac{R(h)}{h^{r+1}} t^{r+1} = R(t) - R(h)\frac{t^{r+1}}{h^{r+1}}$$

for $0 \leq t \leq h$. Note that since $P(t)$ is a polynomial of degree r, $P^{(r+1)}(t) = 0$ for all t, and

$$g^{(r+1)}(t) = f^{(r+1)}(x + t) - (r + 1)!\frac{R(h)}{h^{r+1}}.$$

Also, $g(0) = g'(0) = \cdots = g^{(r)}(0) = 0$, and $g(h) = R(h) - R(h) = 0$. Since $g = 0$ at 0 and h, the Mean Value Theorem gives a $t_1 \in (0, h)$ such that $g'(t_1) = 0$. Since $g'(0)$ and $g'(t_1) = 0$, the Mean Value Theorem gives a $t_2 \in (0, t_1)$ such that $g'(t_2) = 0$. Continuing, we get a sequence $t_1 > t_2 > \cdots > t_{r+1} > 0$ such that $g^{(k)}(t_k) = 0$. The $(r + 1)^{\text{st}}$ equation, $g^{(r+1)}(t_{r+1}) = 0$, implies that

$$0 = f^{(r+1)}(x + t_{r+1}) - (r + 1)!\frac{R(h)}{h^{r+1}}.$$

Thus, $\theta = x + t_{r+1}$ makes the equation in (c) true. If $h < 0$ the argument is symmetric. $\square$

13 Corollary *For each $r \in \mathbb{N}$ the smooth nonanalytic function $e(x)$ satisfies $\lim\limits_{h \to 0} e(h)/h^r$* $= 0$.

Proof Obvious from the theorem and the fact that $e^{(r)}(0) = 0$ for all r. $\square$

The **Taylor series** at x of a smooth function f is the infinite Taylor polynomial

$$T(h) = \sum_{r=0}^{\infty} \frac{f^{(r)}(x)}{r!} h^r.$$

In calculus, you compute the Taylor series of functions such as $\sin x$, $\arctan x$, e^x, etc. These functions are analytic: Their Taylor series converge and express them as power series. In general, however, the Taylor series of a smooth function need not converge to the function, and in fact it may fail to converge at all. The function $e(x)$ is an example of the first phenomenon. Its Taylor series at $x = 0$ converges, but gives the wrong answer. Examples of divergent and totally divergent Taylor series are indicated in Exercise 4.37.

The convergence of a Taylor series is related to how quickly the rth derivative grows (in magnitude) as $r \to \infty$. In Section 6 of Chapter 4 you will find necessary and sufficient conditions on the growth rate that determine whether a smooth function is analytic.

Inverse Functions

A strictly monotone continuous function $f : (a, b) \to \mathbb{R}$ bijects (a, b) onto some interval (c, d) where $c = \lim_{t \to a} f(t)$ and $d = \lim_{t \to b}$ in the increasing case. (We permit c or d to be infinite.) It is a homeomorphism $(a, b) \to (c, d)$ and its inverse function $f^{-1} : (c, d) \to (a, b)$ is also a homeomorphism. These facts were proved in Chapter 2.

Does differentiability of f imply differentiability of f^{-1}? If $f' \neq 0$ the answer is "yes." Keep in mind, however, the function $f : x \mapsto x^3$. It shows that differentiability of f^{-1} fails when $f'(x) = 0$. For the inverse function is $y \mapsto y^{1/3}$, which is not differentiable at $y = 0$.

14 Inverse Function Theorem in dimension 1 *If $f : (a, b) \to (c, d)$ is a differentiable surjection and $f'(x)$ is never zero then f is a homeomorphism. Its inverse is differentiable and its derivative at $y \in (c, d)$ is*

$$(f^{-1})'(y) = \frac{1}{f' \circ f^{-1}(y)}$$

Proof If f' is never zero then by the intermediate value property of derivatives, it is either always positive or always negative. We assume for all x that $f'(x) > 0$. If $a < s < t < b$ then by the Mean Value Theorem there exists $\theta \in (s, t)$ such that $f(t) - f(s) = f'(\theta)(t - s) > 0$. Thus f is strictly monotone. Differentiability implies continuity, so f is a homeomorphism $(a, b) \to (c, d)$. To check differentiability of f^{-1} at $y \in (c, d)$, define

$$x = f^{-1}(y) \quad \text{and} \quad \Delta x = f^{-1}(y + \Delta y) - x.$$

Then $y = f(x)$ and $\Delta y = f(x + \Delta x) - fx = \Delta f$. Thus

$$\frac{\Delta f^{-1}}{\Delta y} = \frac{f^{-1}(y + \Delta y) - f^{-1}(y)}{\Delta y} = \frac{\Delta x}{\Delta y} = \frac{1}{\Delta y / \Delta x} = \frac{1}{\Delta f / \Delta x}.$$

Since f is a homeomorphism, $\Delta x \to 0$ if and only if $\Delta y \to 0$, so the limit of $\Delta f^{-1}/\Delta y$ exists as $\Delta y \to 0$ and equals $1/f'(x) = 1/f' \circ f^{-1}(y)$. $\qquad\square$

A longer but more geometric proof of the one-dimensional inverse function theorem can be done in two steps.

(i) A function is differentiable if and only if its graph is differentiable.

(ii) The graph of f^{-1} is the reflection of the graph of f across the diagonal, and is thus differentiable.

See Figure 66.

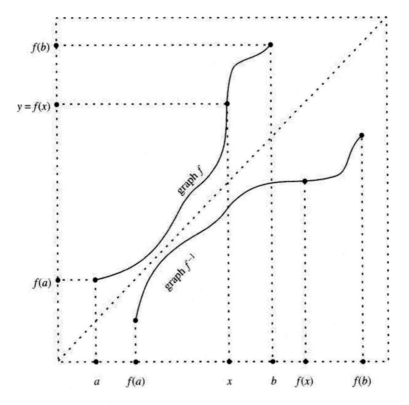

Figure 66 A picture proof of the inverse function theorem in $\mathbb{R}$

If a homeomorphism f and its inverse are both of class C^r, $r \geq 1$, then f is a $\boldsymbol{C^r}$ **diffeomorphism**.

15 Corollary *If $f : (a, b) \to (c, d)$ is a homeomorphism of class C^r, $1 \leq r \leq \infty$, and for all $x \in (a, b)$ we have $f'(x) \neq 0$ then f is a C^r diffeomorphism.*

Proof If $r = 1$, the formula $(f^{-1})'(y) = 1/f' \circ f^{-1}(y)$ implies that $(f^{-1})'(y)$ is continuous, so f is a C^1 diffeomorphism. The Rules of Differentiation and induction on $r \geq 2$ complete the proof. □

The corollary remains true for analytic functions – the inverse of an analytic function with nonvanishing derivative is analytic. The generalization of the inverse function theorem to higher dimensions is a principal goal of Chapter 5.

2 Riemann Integration

Let $f : [a, b] \to \mathbb{R}$ be given. Intuitively, the integral of f is the area under its graph; i.e., for $f \geq 0$ we have

$$\int_a^b f(x)\,dx = \text{area}\,\mathcal{U}$$

where $\mathcal{U}$ is the **undergraph** of f,

$$\mathcal{U} = \{(x, y) : a \leq x \leq b \text{ and } 0 \leq y < f(x)\}.$$

The precise definition involves approximation. A **partition pair** consists of two finite sets of points $P, T \subset [a, b]$ where $P = \{x_0, \ldots, x_n\}$ and $T = \{t_1, \ldots, t_n\}$ are interlaced as

$$a = x_0 \leq t_1 \leq x_1 \leq t_2 \leq x_2 \leq \cdots \leq t_n \leq x_n = b.$$

We assume the points $x_0, \ldots, x_n$ are distinct. The **Riemann sum** corresponding to f, P, T is

$$R(f, P, T) = \sum_{i=1}^n f(t_i)\Delta x_i = f(t_1)\Delta x_1 + f(t_2)\Delta x_2 + \ldots + f(t_n)\Delta x_n$$

where $\Delta x_i = x_i - x_{i-1}$. The Riemann sum R is the area of rectangles which approximate the area under the graph of f. See Figure 67. Think of the points t_i as **sample points**. We sample the value of the function f at t_i.

The **mesh** of the partition P is the length of the largest subinterval $[x_{i-1}, x_i]$. A partition with large mesh is coarse; one with small mesh is fine. In general, the finer the better. A real number I is the **Riemann integral** of f over $[a, b]$ if it satisfies the following approximation condition:

$$\forall \epsilon > 0 \ \exists \delta > 0 \text{ such that if } P, T \text{ is any partition pair then}$$
$$\text{mesh}\, P < \delta \quad \Rightarrow \quad |R - I| < \epsilon$$

where $R = R(f, P, T)$. If such an I exists it is unique, we denote it as

$$\int_a^b f(x)\,dx = I = \lim_{\text{mesh}\, P \to 0} R(f, P, T),$$

and we say that f is **Riemann integrable** with Riemann integral I.

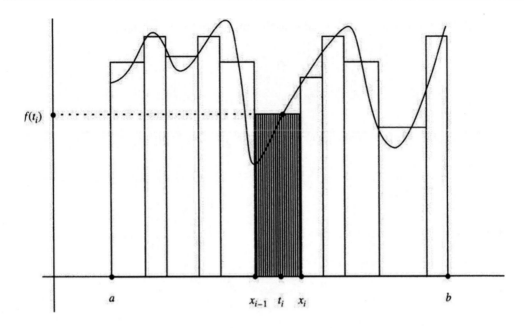

Figure 67 The area of the strip is $f(t_i)\Delta x_i$.

16 Theorem *If f is Riemann integrable then it is bounded.*

Proof Suppose not. Let $I = \int_a^b f(x)\,dx$. There is some $\delta > 0$ such that for all partition pairs with mesh $P < \delta$, we have $|R - I| < 1$. Fix such a partition pair $P = \{x_0, \ldots, x_n\}$, $T = \{t_1, \ldots, t_n\}$. If f is unbounded on $[a, b]$ then there is also a subinterval $[x_{i_0-1}, x_{i_0}]$ on which it is unbounded. Choose a new set $T' = \{t'_1, \ldots, t'_n\}$ with $t'_i = t_i$ for all $i \neq i_0$ and choose t'_{i_0} such that

$$|f(t'_{i_0}) - f(t_{i_0})|\Delta x_{i_0} > 2.$$

This is possible since the supremum of $\{|f(t)| : x_{i_0-1} \leq t \leq x_{i_0}\}$ is ∞. Let $R' = R(f, P, T')$. Then $|R - R'| > 2$, contrary to the fact that both R and R' differ from I by < 1. $\square$

Let $\mathcal{R}$ denote the set of all functions that are Riemann integrable over $[a, b]$.

17 Theorem (Linearity of the Integral)

(a) *$\mathcal{R}$ is a vector space and $f \mapsto \int_a^b f(x)\,dx$ is a linear map $\mathcal{R} \to \mathbb{R}$.*
(b) *The constant function $h(x) = k$ is integrable and its integral is $k(b - a)$.*

Proof (a) Riemann sums behave naturally under linear combination:

$$R(f + cg, P, T) = R(f, P, T) + cR(g, P, T),$$

and it follows that their limits, as mesh $P \to 0$, give the expected formula

$$\int_a^b f(x) + cg(x)\, dx = \int_a^b f(x)\, dx + c\int_a^b g(x)\, dx.$$

(b) Every Riemann sum for the constant function $h(x) = k$ is $k(b - a)$, so its integral equals this number too.   □

18 Theorem (Monotonicity of the Integral) *If $f, g \in \mathcal{R}$ and $f \le g$ then*

$$\int_a^b f(x)\, dx \;\le\; \int_a^b g(x)\, dx.$$

Proof For each partition pair P, T, we have $R(f, P, T) \le R(g, P, T)$.   □

19 Corollary *If $f \in \mathcal{R}$ and $|f| \le M$ then $|\int_a^b f(x)\, dx| \le M(b - a)$.*

Proof By Theorem 17, the constant functions $\pm M$ are integrable. By Theorem 18, $-M \le f(x) \le M$ implies that

$$-M(b - a) \;\le\; \int_a^b f(x)\, dx \;\le\; M(b - a).$$

     □

Darboux Integrability

The **lower sum** and **upper sum** of a function $f : [a, b] \to [-M, M]$ with respect to a partition P of $[a, b]$ are

$$L(f, P) \;=\; \sum_{i=1}^{n} m_i \Delta x_i \quad \text{and} \quad U(f, P) \;=\; \sum_{i=1}^{n} M_i \Delta x_i$$

where

$$m_i \;=\; \inf\{f(t) : x_{i-1} \le t \le x_i\} \qquad M_i \;=\; \sup\{f(t) : x_{i-1} \le t \le x_i\}.$$

We assume f is bounded in order to be sure that m_i and M_i are real numbers. Clearly

$$L(f, P) \;\le\; R(f, P, T) \;\le\; U(f, P)$$

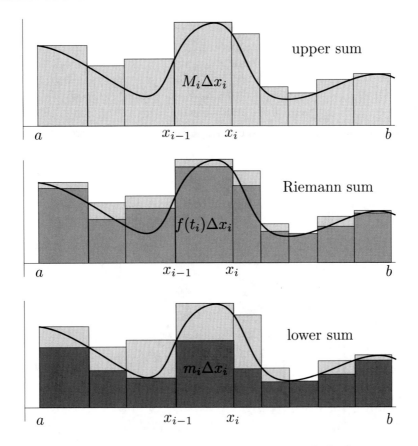

Figure 68 The upper sum, a Riemann sum, and the lower sum

for all partition pairs P, T. See Figure 68.

The **lower integral** and **upper integral** of f over $[a, b]$ are

$$\underline{I} \; = \; \sup_P L(f, P) \quad \text{and} \quad \overline{I} \; = \; \inf_P U(f, P).$$

P ranges over all partitions of $[a, b]$ when we take the supremum and infimum. If the lower and upper integrals of f are equal, $\underline{I} = \overline{I}$, then f is **Darboux integrable** and their common value is its **Darboux integral**.

20 Theorem *Riemann integrability is equivalent to Darboux integrability, and when a function is integrable, its three integrals – lower, upper, and Riemann – are equal.*

To prove Theorem 20 it is convenient to refine a partition P by adding more partition points. The partition P' **refines** P if $P' \supset P$.

Suppose first that $P' = P \cup \{w\}$ where $w \in (x_{i_0-1}, x_{i_0})$. The lower sums for P and P' are the same except that $m_{i_0}\Delta x_{i_0}$ in $L(f, P)$ splits into two terms in $L(f, P')$. The sum of the two terms is at least as large as $m_{i_0}\Delta x_{i_0}$. For the infimum of f over the intervals $[x_{i_0-1}, w]$ and $[w, x_{i_0}]$ is at least as large as m_{i0}. Similarly, $U(f, P') \leq U(f, P)$. See Figure 69.

Repetition continues the pattern and we formalize it as the

Refinement Principle *Refining a partition causes the lower sum to increase and the upper sum to decrease.*

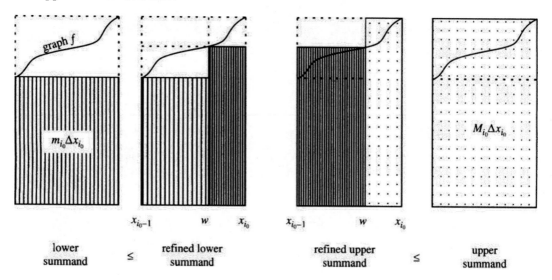

Figure 69 Refinement increases L and decreases U.

The **common refinement** P^* of two partitions P, P' of $[a, b]$ is

$$P^* = P \cup P'.$$

According to the Refinement Principle we have

$$L(f, P) \ \leq \ L(f, P^*) \ \leq \ U(f, P^*) \ \leq \ U(f, P').$$

We conclude that each lower sum is less than or equal to each upper sum, the lower integral is less than or equal to the upper, and thus

(2)   A bounded function $f : [a, b] \to \mathbb{R}$ is Darboux integrable

              if and only if $\forall \epsilon > 0 \ \exists P$ such that $U(f, P) - L(f, P) < \epsilon$.

Proof of Theorem 20 Let $f : [a, b] \to \mathbb{R}$. We assert that f is Riemann integrable if and only if it is Darboux integrable. One direction is easy: Riemann $\Rightarrow$ Darboux. Riemann integrability implies that f is bounded and that for each $\epsilon > 0$ there exists a $\delta > 0$ such that if P is any partition with mesh $P < \delta$ then

$$|R - I| < \frac{\epsilon}{4}$$

where $R = R(f, P, T)$ and I is the Riemann integral of f. Fix such a partition P and choose a set of sample points $T = \{t_i\}$ such that $f(t_i)$ is so near m_i that

$$R(f, P, T) - L(f, P) < \frac{\epsilon}{4}.$$

(It is enough to choose $t_i \in [x_{i-1}, x_i]$ such that $f(t_i) - m_i < \epsilon/4(b - a)$.) Choose a second set of sample points $T' = \{t'_i\}$ so that

$$U(f, P) - R(f, P, T') < \frac{\epsilon}{4}.$$

Both $R = R(f, P, T)$ and $R' = R(f, P, T')$ differ from I by $< \epsilon/4$. Thus

$$U - L \; = \; (U - R') + (R' - I) + (I - R) + (R - L) \; < \; \epsilon,$$

from which (2) gives Darboux integrability. Since $\underline{I}$, I, $\overline{I}$ are fixed numbers that belong to the interval $[L, U]$ of length ϵ, and ϵ is arbitrary, the ϵ-principle implies that

$$\underline{I} = I = \overline{I},$$

which completes the proof that f is Darboux integrable and its Darboux integral equals its Riemann integral.

Next, we assume Darboux integrability and prove Riemann integrability. (The proof is messier because checking Riemann integrability requires that we look at all fine partitions P, not just at those for which $U - L$ is small.) Darboux integrability implies that f is bounded, say $f : [a, b] \to [-M, M]$ for $M > 0$. By (2) we know that for each $\epsilon > 0$ there is a partition P_1 such that

$$U_1 - L_1 < \frac{\epsilon}{3}$$

where $L_1 = L(f, P_1)$ and $U_1 = U(f, P_1)$. Fix P_1 and choose

$$\delta = \frac{\epsilon}{6n_1 M}$$

where n_1 is the number of P_1-intervals. Consider any partition P with mesh $P < \delta$, and fix a set T of sample points for P. We claim that the Riemann sum $R(f, P, T)$

for every such partition pair P, T differs from the Darboux integral I by less than ϵ. Then, by the ϵ-principle, f is Riemann integrable and its Riemann integral is I.

According to the Refinement Principle, the common refinement $P^* = P_1 \cup P$ has

$$L_1 \leq L^* \leq U^* \leq U_1$$

where $L^* = L(f, P^*)$ and $U^* = U(f, P^*)$. Hence $U^* - L^* < \epsilon/3$.

We claim that $0 \leq U - U^* \leq \epsilon/3$ where $U = U(f, P)$. Since P^* refines P, each P^*-interval $I_j^* = [x_{j-1}^*, x_j^*]$ is contained in some P-interval $I_i = [x_{i-1}, x_i]$. Except for n_1 exceptional P^*-intervals I_j^*, we have $I_j^* = I_i$ and $M_j^* = M_i$. Thus $U - U^*$ reduces to a sum over the exceptional intervals. See Figure 70.

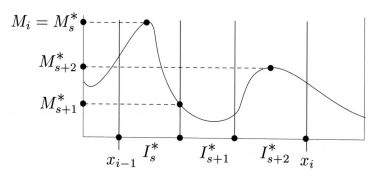

Figure 70 I_i contains three exceptional subintervals I_j^* for $j = s, s+1, s+2$.

Setting

$$\mathcal{I} = \{i : I_i \text{ contains exceptional subintervals}\}$$
$$\mathcal{I}(i) = \{j : I_j^* \text{ is an exceptional subinterval of } I_i\}$$

therefore gives

$$0 \leq U - U^* = \sum_{i \in \mathcal{I}} M_i \Delta x_i - \sum_{i \in \mathcal{I}} \sum_{j \in \mathcal{I}(i)} M_j^* \Delta x_j^* = \sum_{i \in \mathcal{I}} \sum_{j \in \mathcal{I}(i)} (M_i - M_j^*) \Delta x_j^*.$$

For Δx_i telescopes to $\sum_{j \in \mathcal{I}(i)} \Delta x_j^*$. Since $0 \leq M_i - M_j^* \leq 2M$, and since there are at most n_1 exceptional intervals, each of length $\leq \delta$, this implies

$$0 \leq U - U^* \leq 2Mn_1\delta < \epsilon/3.$$

Similarly, $L^* - L < \epsilon/3$. Thus

$$U - L = (U - U^*) + (U^* - L^*) + (L^* - L) < \epsilon.$$

Since the Darboux integral I and the Riemann sums $R = R(f, P, T)$ (with mesh P small) belong to $[L, U]$, an interval of length $< \epsilon$, we get $|R - I| < \epsilon$. Therefore f is Riemann integrable and its Riemann integral equals its Darboux integral I. $\square$

According to Theorem 20 and (2) we get

21 Riemann's Integrability Criterion

A bounded function is Riemann integrable if and only if
$$\forall \epsilon > 0 \; \exists P \; such \; that \; U(f, P) - L(f, P) < \epsilon.$$

Example Every continuous function $f : [a, b] \to \mathbb{R}$ is Riemann integrable. (See also Corollary 24 to the Riemann-Lebesgue Theorem, below.) Since $[a, b]$ is compact and f is continuous, f is uniformly continuous. See Theorem 42 in Chapter 2. Let $\epsilon > 0$ be given. Uniform continuity provides a $\delta > 0$ such that if $|t - s| < \delta$ then $|f(t) - f(s)| < \epsilon/(b-a)$. Choose any partition P with mesh $P < \delta$. On each partition interval $[x_{i-1}, x_i]$, we have $M_i - m_i \le \epsilon/(b-a)$. Thus

$$U - L = \sum_{i=1}^{n} (M_i - m_i)\Delta x_i \le \frac{\epsilon}{(b-a)} \sum \Delta x_i = \epsilon.$$

By Riemann's Integrability Criterion f is Riemann integrable.

Example The **characteristic function** (or **indicator function**) of a set $E \subset \mathbb{R}$, χ_E, takes value 1 at points of E and value 0 at points of E^c. See Figure 71. Some characteristic functions are Riemann integrable, while others aren't. Riemann's

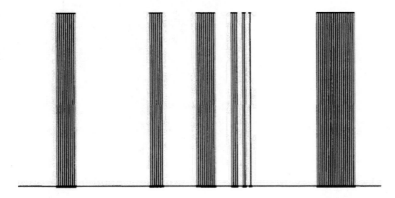

Figure 71 The region below the graph of a characteristic function

Integrability Criterion implies that the characteristic function of an interval (including the degenerate case that the interval is a point) is Riemann integrable. The integral of $\chi_{[a,b]}$ is $b - a$. A **step function** is a finite sum of constants times characteristic functions of intervals and is therefore Riemann integrable. A step function is a special type of **piecewise continuous** function, i.e., a function that is continuous except at finitely many points. See Figure 72. Bounded piecewise continuous functions are Riemann integrable. See Corollary 25 below.

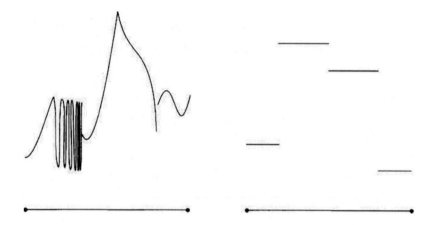

Figure 72 The graphs of a piecewise continuous function and a step function.

Example The characteristic function of $\mathbb{Q}$ is not integrable on $[a, b]$. It is defined as $\chi_{\mathbb{Q}}(x) = 1$ when $x \in \mathbb{Q}$ and $\chi_{\mathbb{Q}}(x) = 0$ when $x \notin \mathbb{Q}$. See Figure 73. Every lower sum

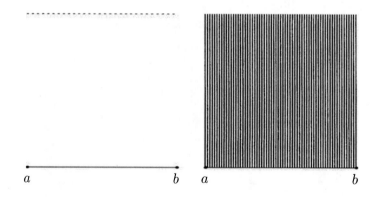

Figure 73 The graph of $\chi_{\mathbb{Q}}$ and the region below its graph

$L(\chi_{\mathbb{Q}}, P)$ is 0 and every upper sum is $b - a$. By Riemann's Integrability Criterion, $\chi_{\mathbb{Q}}$ is not integrable. Note that $\chi_{\mathbb{Q}}$ is discontinuous at every point, not merely at rational points.

The fact that $\chi_{\mathbb{Q}}$ fails to be Riemann integrable is actually a failing of Riemann integration theory, for the function $\chi_{\mathbb{Q}}$ is fairly tame. Its integral ought to exist and it ought to be 0, because the undergraph is just countably many line segments of height 1, and their area ought to be 0.

A handy consequence of Riemann's Integrability Criterion is the

22 Sandwich Principle $f : [a, b] \to \mathbb{R}$ *is Riemann integrable if, given* $\epsilon > 0$, *there are functions* $g, h \in \mathcal{R}$ *such that* $g \leq f \leq h$ *and* $\int_a^b h(x) - g(x)\, dx \leq \epsilon$.

Proof For any partition P it is clear that

$$L(g, P) \ \leq \ L(f, P) \ \leq \ U(f, P) \ \leq \ U(h, P).$$

Let $\epsilon > 0$ be given. Since g and h are Riemann integrable, there is a $\delta > 0$ such that if mesh $P < \delta$ then their Darboux sums differ from their integrals by $< \epsilon/3$, and $\int_a^b h(x) - g(x)\, dx \leq \epsilon/3$. Thus

$$\int_a^b g(x)\, dx - L(g, P) < \frac{\epsilon}{3} \quad \text{and} \quad U(h, P) - \int_a^b h(x)\, dx < \frac{\epsilon}{3},$$

from which it follows that

$$\int_a^b g(x)\, dx - \frac{\epsilon}{3} < L(g, P) \leq L(f, P) \leq U(f, P) \leq U(h, P) < \int_a^b h(x)\, dx + \frac{\epsilon}{3}.$$

Then $\int_a^b h(x)\, dx - \int_a^b g(x)\, dx = \int_a^b h(x) - g(x)\, dx \leq \epsilon/3$ gives $U(f, P) - L(f, P) < \epsilon$ and Riemann's Integrability Criterion implies that f is Riemann integrable. See Figure 74.

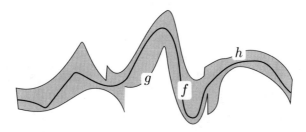

Figure 74 The graphs of g and h sandwich the graph of f.

Example Let $f : [0, 1] \to \mathbb{Q}$ be defined as $f(p/q) = 1/q$ when $p/q \in \mathbb{Q}$ is written in lowest terms, and $f(x) = 0$ when x is irrational. This is the **rational ruler function**.

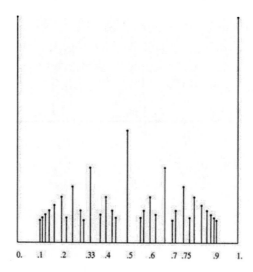

Figure 75 The graph of the rational ruler function and the region below its graph

Note that f is discontinuous at every $x \in \mathbb{Q}$ and is continuous at every $x \in \mathbb{Q}^c$. See Figure 75. It is Riemann integrable and its integral is zero. For, given $\epsilon > 0$, we can consider the degenerate step function

$$s(x) = \begin{cases} 1/q & \text{if } p/q \in \mathbb{Q} \cap [0,1] \text{ and } 1/q \geq \epsilon \\ 0 & \text{otherwise.} \end{cases}$$

Then f is sandwiched between the Riemann integrable functions $g = 0$ and

$$h(x) = \epsilon \chi_{[0,1]}(x) + s(x).$$

The integral of $h - g$ is ϵ, so the Sandwich Principle implies that $f \in \mathcal{R}$.

Example Zeno's staircase function $Z(x) = 1/2$ on the first half of $[0,1]$, $Z(x) = 3/4$ on the next quarter of $[0,1]$, and so on. See Figure 76. It is Riemann integrable and its integral is $2/3$. The function has infinitely many discontinuity points, one at each point $(2^k - 1)/2^k$. In fact, every monotone function is Riemann integrable.[†] See Corollary 26 below.

[†]To prove this directly is not hard. The key observation to make is that a monotone function is not much different from a continuous function. It has only jump discontinuities, and only countably many of them; given any $\epsilon > 0$, there are only finitely many at which the jump is $\geq \epsilon$. See Exercise 1.31.

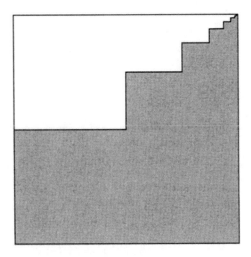

Figure 76 Zeno's staircase

These examples raise a natural question:

Exactly which functions are Riemann integrable?

To give an answer to the question, and for many other applications, the following concept is very useful. A set $Z \subset \mathbb{R}$ is a **zero set** if for each $\epsilon > 0$ there is a countable covering of Z by open intervals (a_i, b_i) such that

$$\sum_{i=1}^{\infty} b_i - a_i \leq \epsilon$$

The sum of the series is the **total length** of the covering. Think of zero sets as negligible. If a property holds for all points except those in a zero set then one says that the property holds **almost everywhere**, abbreviated "a.e."

23 Riemann-Lebesgue Theorem *A function $f : [a, b] \to \mathbb{R}$ is Riemann integrable if and only if it is bounded and its set of discontinuity points is a zero set.*

The set D of discontinuity points is exactly what its name implies,

$$D = \{x \in [a, b] : f \text{ is discontinuous at the point } x\}.$$

A function whose set of discontinuity points is a zero set is continuous almost everywhere. The Riemann-Lebesgue Theorem states that a function is Riemann integrable if and only if it is bounded and continuous almost everywhere.

Examples of zero sets are

(a) Every subset of a zero set.
(b) Every finite set.
(c) Every countable union of zero sets.
(d) Every countable set.
(e) The middle-thirds Cantor set.

(a) is clear. For if $Z_0 \subset Z$ where Z is a zero set, and if $\epsilon > 0$ is given, then there is some open covering of Z by intervals whose total length is $\leq \epsilon$; but the same collection of intervals covers Z_0, which shows that Z_0 is also a zero set.

(b) Let $Z = \{z_1, \ldots, z_n\}$ be a finite set and let $\epsilon > 0$ be given. The intervals $(z_i - \epsilon/2n, z_i + \epsilon/2n)$, for $i = 1, \ldots, n$, cover Z and have total length ϵ. Therefore Z is a zero set. In particular, the empty set and any single point are zero sets.

(c) This is a typical "$\epsilon/2^n$-argument." Let $Z_1, Z_2, \ldots$ be a sequence of zero sets and $Z = \bigcup Z_j$. We claim that Z is a zero set. Let $\epsilon > 0$ be given. The set Z_1 can be covered by countably many intervals (a_{i1}, b_{i1}) with total length $\sum(b_{i1} - a_{i1}) \leq \epsilon/2$. The set Z_2 can be covered by countably many intervals (a_{i2}, b_{i2}) with total length $\sum(b_{i2} - a_{i2}) \leq \epsilon/4$. In general, the set Z_j can be covered by countably many intervals (a_{ij}, b_{ij}) with total length

$$\sum_{i=1}^{\infty}(b_{ij} - a_{ij}) \leq \frac{\epsilon}{2^j}.$$

Since the countable union of countable sets is countable, the collection of all the intervals (a_{ij}, b_{ij}) is a countable covering of Z by open intervals, and the total length of all these intervals is

$$\sum_{j=1}^{\infty}\left(\sum_{i=1}^{\infty} b_{ij} - a_{ij}\right) \leq \sum_{j=1}^{\infty}\frac{\epsilon}{2^j} = \frac{\epsilon}{2} + \frac{\epsilon}{4} + \frac{\epsilon}{8} + \ldots = \epsilon.$$

Thus Z is a zero set and (c) is proved.

(d) This is implied by (b) and (c).

(e) Let $\epsilon > 0$ be given and choose $n \in \mathbb{N}$ such that $2^n/3^n < \epsilon$. The middle-thirds Cantor set C is contained inside 2^n closed intervals of length $1/3^n$, say $I_1, \ldots, I_{2^n}$. Enlarge each closed interval I_i to an open interval $(a_i, b_i) \supset I_i$ such that $b_i - a_i = \epsilon/2^n$. (Since $1/3^n < \epsilon/2^n$, and I_i has length $1/3^n$, this is possible.) The total length of these 2^n intervals (a_i, b_i) is ϵ. Thus C is a zero set.

It is nontrivial to prove that intervals are not zero sets. See Exercise 29.

In the proof of the Riemann-Lebesgue Theorem, it is useful to focus on the "size" of a discontinuity. A simple expression for this size is the **oscillation** of f at x,

$$\text{osc}_x(f) = \limsup_{t \to x} f(t) - \liminf_{t \to x} f(t).$$

Equivalently,

$$\text{osc}_x(f) = \lim_{r \to 0} \text{diam } f([x - r,\ x + r]).$$

(Of course, $r > 0$.) It is clear that f is continuous at x if and only if $\text{osc}_x(f) = 0$. It is also clear that if I is any interval containing x in its interior then

$$M_I - m_I \geq \text{osc}_x(f)$$

where M_I and m_I are the supremum and infimum of $f(t)$ as t varies in I. See Figure 77.

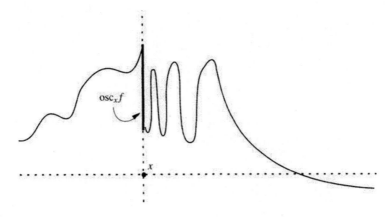

Figure 77 The oscillation of f at x

Proof of the Riemann-Lebesgue Theorem The set D of discontinuity points of $f : [a, b] \to [-M,\ M]$ naturally filters itself as the countable union

$$D = \bigcup_{k=1}^{\infty} D_k$$

where

$$D_k = \{x \in [a, b] : \text{osc}_x(f) \geq 1/k\}.$$

According to (a), (c) above, D is a zero set if and only if each D_k is a zero set.

Assume that f is Riemann integrable and let $\epsilon > 0$ and $k \in \mathbb{N}$ be given. By Theorem 20 there is a partition P such that

$$U - L = \sum (M_i - m_i)\Delta x_i < \epsilon/k.$$

We say that a P-interval I_i is "bad" if it contains a point of D_k in its interior. A bad interval has a fairly big f-variation, namely $M_i - m_i \geq 1/k$. Since $U - L = \sum(M_i - m_i)\Delta x_i < \epsilon/k$ is small, there cannot be too many bad intervals. (*This is the key insight in the estimates.*) More precisely,

$$\frac{\epsilon}{k} > U - L = \sum(M_i - m_i)\Delta x_i \geq \sum_{\text{bad}}(M_i - m_i)\Delta x_i \geq \frac{1}{k}\sum_{\text{bad}}\Delta x_i$$

implies (by canceling the factor $1/k$ from both sides of the inequality) the sum of the lengths of the bad intervals is $< \epsilon$. Thus, except for the finite set $D_k \cap P$, D_k is contained in finitely many open intervals whose total length is $< \epsilon$. Since finite sets are zero sets and ϵ is arbitrary, each D_k is a zero set. Therefore $D = \bigcup D_k$ is a zero set.

Conversely, assume that the discontinuity set D of $f : [a, b] \to [-M, M]$ is a zero set. Let $\epsilon > 0$ be given. By Riemann's Integrability Criterion, to prove that f is Riemann integrable it suffices to find P with $U(f, P) - L(f, P) < \epsilon$. Choose $k \in \mathbb{N}$ so that

$$\frac{1}{k} < \frac{\epsilon}{2(b-a)}.$$

Since D is a zero set, so is D_k and hence there is a countable covering $\mathcal{J}$ of D_k by open intervals $J_j = (a_j, b_j)$ with total length

$$\sum b_j - a_j \leq \frac{\epsilon}{4M}.$$

These J_j are "bad" intervals: The f-variation on each J_j is $\geq 1/k$. On the other hand, for each $x \in [a, b] \setminus D_k$ there is an open interval I_x containing x such that

$$\sup\{f(t) : t \in I_x\} - \inf\{f(t) : t \in I_x\} \quad < \quad 1/k.$$

These intervals I_x are a covering $\mathcal{I}$ of the good set $[a, b] \setminus D_k$. The union $\mathcal{V} = \mathcal{I} \cup \mathcal{J}$ is an open covering of $[a, b]$. Compactness of $[a, b]$ implies that $\mathcal{V}$ has a Lebesgue number $\lambda > 0$.

Let $P = \{x_0, \ldots, x_n\}$ be any partition of $[a, b]$ having mesh $P < \lambda$. We claim that $U(f, P) - L(f, P) < \epsilon$. Each P-interval I_i is contained wholly in some I_x or wholly in some J_j. (This is what Lebesgue numbers are good for.) Set

$$\mathbf{J} = \{i \in \{1, \ldots, n\} : I_i \text{ is contained in some bad interval } J_j\}.$$

See Figure 78. For some finite m, $J_1 \cup \cdots \cup J_m$ contains those P-intervals I_i with

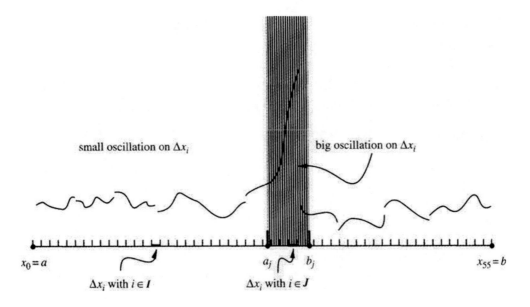

small oscillation on Δx_i big oscillation on Δx_i

$x_0 = a$ a_j b_j $x_{55} = b$

Δx_i with $i \in I$ Δx_i with $i \in J$

Figure 78 The P-intervals I_i with large oscillation have $i \in \mathbf{J}$ and are
potentially "bad."

$i \in \mathbf{J}$. Then

$$
\begin{aligned}
U - L &= \sum_{i=1}^{n}(M_i - m_i)\Delta x_i \\
&= \sum_{i \in \mathbf{J}}(M_i - m_i)\Delta x_i + \sum_{i \notin \mathbf{J}}(M_i - m_i)\Delta x_i \\
&\le \sum_{i \in \mathbf{J}} 2M\Delta x_i \quad + \sum_{i \notin \mathbf{J}} \Delta x_i / k \\
&\le 2M \sum_{j=1}^{m} b_j - a_j \quad + (b - a)/k \\
&< \frac{\epsilon}{2} + \frac{\epsilon}{2} = \epsilon.
\end{aligned}
$$

For the total length of the P-intervals I_i contained in the bad intervals $J_1, \ldots, J_m$ is no greater than $\sum b_j - a_j$. As remarked at the outset, Riemann's Integrability Criterion then implies that f is integrable. $\qquad \square$

The Riemann-Lebesgue Theorem has many consequences, ten of which we list as corollaries.

24 Corollary *Every continuous function is Riemann integrable, and so is every bounded piecewise continuous function.*

Proof The discontinuity set of a continuous function is empty, and is therefore a zero set. The discontinuity set of a piecewise continuous function is finite, and is therefore also a zero set. A continuous function defined on a compact interval $[a, b]$ is bounded. The piecewise continuous function was assumed to be bounded. By the Riemann-Lebesgue Theorem, both these functions are Riemann integrable.   □

25 Corollary *The characteristic function of $S \subset [a, b]$ is Riemann integrable if and only if the boundary of S is a zero set.*

Proof ∂S is the discontinuity set of χ_S. See also Exercise 5.44   □

26 Corollary *Every monotone function is Riemann integrable.*

Proof The set of discontinuities of a monotone function $f : [a, b] \to \mathbb{R}$ is countable and therefore is a zero set. (See Exercise 1.31.) Since f is monotone, its values lie in the interval between $f(a)$ and $f(b)$, so f is bounded. By the Riemann-Lebesgue Theorem, f is Riemann integrable.   □

27 Corollary *The product of Riemann integrable functions is Riemann integrable.*

Proof Let $f, g \in \mathcal{R}$ be given. They are bounded and their product is bounded. By the Riemann-Lebesgue Theorem their discontinuity sets, $D(f)$ and $D(g)$, are zero sets, and $D(f) \cup D(g)$ contains the discontinuity set of the product $f \cdot g$. Since the union of two zero sets is a zero set, the Riemann-Lebesgue Theorem implies that $f \cdot g$ is Riemann integrable.   □

28 Corollary *If $f : [a, b] \to [c, d]$ is Riemann integrable and $\phi : [c, d] \to \mathbb{R}$ is continuous, then the composite $\phi \circ f$ is Riemann integrable.*

Proof The discontinuity set of $\phi \circ f$ is contained in the discontinuity set of f, and therefore is a zero set. Since ϕ is continuous and $[c, d]$ is compact, $\phi \circ f$ is bounded. By the Riemann-Lebesgue Theorem, $\phi \circ f$ is Riemann integrable.   □

29 Corollary *If $f \in \mathcal{R}$ then $|f| \in \mathcal{R}$.*

Proof The function $\phi : y \mapsto |y|$ is continuous, so $x \mapsto |f(x)|$ is Riemann integrable according to Corollary 28.   □

30 Corollary *If $a < c < b$ and $f : [a, b] \to \mathbb{R}$ is Riemann integrable then its restrictions to $[a, c]$ and $[c, b]$ are Riemann integrable and*

$$\int_a^b f(x)\, dx = \int_a^c f(x)\, dx + \int_c^b f(x)\, dx.$$

Conversely, Riemann integrability on $[a, c]$ and $[c, b]$ implies Riemann integrability on $[a, b]$.

Proof See Figure 79. The union of the discontinuity sets for the restrictions of f to

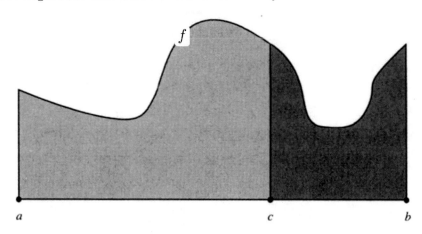

Figure 79 Additivity of the integral is equivalent to additivity of area.

the subintervals $[a, c]$, $[c, b]$ is the discontinuity set of f. The latter is a zero set if and only if the former two are, and so by the Riemann-Lebesgue Theorem, f is Riemann integrable if and only if its restrictions to $[a, c]$ and $[c, b]$ are.

Let $\chi_{[a,c]}$ and $\chi_{[c,b]}$ be the characteristic functions of $[a, c]$ and $[c, b]$. By Corollary 24 they are integrable, and by Corollary 27, so are the products $\chi_{[a,c]} \cdot f$ and $\chi_{[c,b]} \cdot f$. Since

$$f = \chi_{[a,c]} \cdot f + \chi_{(c,b]} \cdot f$$

the addition formula follows from linearity of the integral, Theorem 17. $\square$

31 Corollary *A Riemann integrable function $f : [a, b] \to [0, M]$ has integral zero if and only if $f(x) = 0$ almost everywhere.*

Proof Suppose $\int f = 0$ but $f(x_0) = c > 0$ at some continuity point x_0 of f. By continuity, $f(x) \geq c/2$ for all x in some small interval $I = [x_0 - \delta, x_0 + \delta]$. Then $f(x) \geq g(x) = (c/2)\chi_I(x)$ everywhere, so $\int f \geq \int g = c\delta > 0$, contradicting the assumption that f has integral zero. Thus, $f = 0$ at every continuity point. Riemann integrability implies that almost every x_0 is a continuity point, so $f = 0$ almost everywhere. See Figure 80.

Conversely, assume that $f(x) = 0$ almost everywhere. Then every partition interval $I_i = [x_{i-1}, x_i]$ of every partition of $[a, b]$ contains points at which $f = 0$, so $m_i = 0$

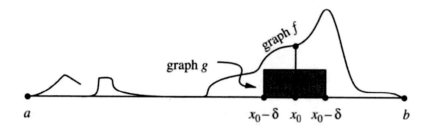

Figure 80 The shaded rectangle prevents the integral of f being zero.

where $m_i = \inf\{f(t) : t \in I_i\}$. Thus, $L(f, P) \equiv 0$. Riemann integrability implies that $L(f, P) \to \int f$ as $\|P\| \to 0$, so $\int f = 0$. □

Corollary 28 and Exercises 33, 35, 47, 49 deal with the way that Riemann integrability behaves under composition. If $f \in \mathcal{R}$ and ϕ is continuous then $\phi \circ f \in \mathcal{R}$, although the composition in the other order, $f \circ \phi$, may fail to be integrable. Continuity is too weak a hypothesis for such a "change of variable." See Exercise 35. In particular, the composite of Riemann integrable functions may fail to be Riemann integrable. See Exercise 33. However, we do have the following result when the change of variables is sufficiently nice.

32 Corollary *If f is Riemann integrable and ψ is a homeomorphism whose inverse satisfies a Lipschitz condition then $f \circ \psi$ is Riemann integrable.*

Proof More precisely, we assume that $f : [a, b] \to \mathbb{R}$ is Riemann integrable, ψ bijects $[c, d]$ onto $[a, b]$, $\psi(c) = a$, $\psi(d) = b$, and for some constant K and all $s, t \in [a, b]$ we have

$$|\psi^{-1}(s) - \psi^{-1}(t)| \leq K|s - t|.$$

We then assert that $f \circ \psi$ is a Riemann integrable function $[c, d] \to \mathbb{R}$. Clearly $f \circ \psi$ is bounded.

Let D be the set of discontinuity points of f. Then $D' = \psi^{-1}(D)$ is the set of discontinuity points of $f \circ \psi$. Let $\epsilon > 0$ be given. There is an open covering of D by intervals (a_i, b_i) whose total length is $\leq \epsilon/K$. The homeomorphic intervals $(a_i', b_i') = \psi^{-1}(a_i, b_i)$ cover D' and have total length

$$\sum b_i' - a_i' \leq \sum K(b_i - a_i) \leq \epsilon.$$

Therefore D' is a zero set and it follows from the Riemann-Lebesgue Theorem that $f \circ \psi$ is integrable. □

33 Corollary *If $f \in \mathcal{R}$ and $\psi : [c,d] \to [a,b]$ is a C^1 diffeomorphism then $f \circ \psi$ is Riemann integrable.*

Proof The hypothesis that ψ is a C^1 diffeomorphism means that it is a continuously differentiable homeomorphism whose inverse is also continuously differentiable. By the Mean Value Theorem, for all $s, t \in [a,b]$ we have

$$\left| \psi^{-1}(s) - \psi^{-1}(t) \right| \leq K \left| s - t \right|$$

where $K = \max\limits_{x \in [a,b]} \left| (\psi^{-1})'(x) \right|$. By Corollary 32, $f \circ \psi$ is Riemann integrable.   □

Versions of the preceding theorem and corollary remain true without the hypotheses that ψ bijects. The proofs are harder because ψ can fold infinitely often. See Exercises 42 and 44.

In calculus you learn that the derivative of the integral is the integrand. This we now prove.

34 Fundamental Theorem of Calculus *If $f : [a,b] \to \mathbb{R}$ is Riemann integrable then its indefinite integral*

$$F(x) = \int_a^x f(t) \, dt$$

is a continuous function of x. The derivative of $F(x)$ exists and equals $f(x)$ at every point x at which f is continuous.

Proof #1 Obvious from Figure 81.   □

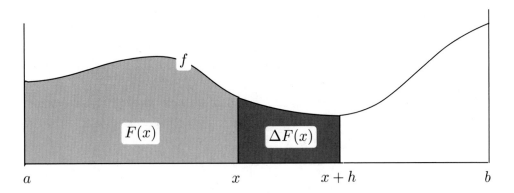

Figure 81 Why *does* this picture give a proof of the Fundamental Theorem of Calculus?

Proof #2 Since f is Riemann integrable, it is bounded; say $|f(x)| \leq M$ for all x. By Corollary 30 we have

$$|F(y) - F(x)| = \left| \int_x^y f(t)\,dt \right| \leq M|y - x|.$$

Therefore F is continuous. Given $\epsilon > 0$, choose $\delta < \epsilon/M$, and observe that $|y - x| < \delta$ implies that $|F(y) - F(x)| < M\delta < \epsilon$. In exactly the same way, if f is continuous at x then

$$\frac{F(x + h) - F(x)}{h} = \frac{1}{h} \int_x^{x+h} f(t)\,dt \;\; \rightarrow \;\; f(x)$$

as $h \to 0$. For if

$$m(x, h) = \inf\{f(s) : |s - x| \leq |h|\} \qquad M(x, h) = \sup\{f(s) : |s - x| \leq |h|\}$$

then

$$
\begin{aligned}
m(x, h) &= \frac{1}{h} \int_x^{x+h} m(x, h)\,dt \;\; \leq \;\; \frac{1}{h} \int_x^{x+h} f(t)\,dt \\
&\leq \;\; \frac{1}{h} \int_x^{x+h} M(x, h)\,dt \;\; = \;\; M(x, h).
\end{aligned}
$$

When f is continuous at x, $m(x, h)$ and $M(x, h)$ converge to $f(x)$ as $h \to 0$, and so must the integral sandwiched between them,

$$\frac{1}{h} \int_x^{x+h} f(t)\,dt \to f(x).$$

(If $h < 0$ then $\frac{1}{h} \int_x^{x+h} f(t)\,dt$ is interpreted as $-\frac{1}{h} \int_{x+h}^x f(t)\,dt$.) $\qquad\square$

35 Corollary *The derivative of an indefinite Riemann integral exists almost everywhere and equals the integrand almost everywhere.*

Proof Assume that $f : [a, b] \to \mathbb{R}$ is Riemann integrable and $F(x)$ is its indefinite integral. By the Riemann-Lebesgue Theorem, f is continuous almost everywhere, and by the Fundamental Theorem of Calculus, $F'(x)$ exists and equals $f(x)$ wherever f is continuous. $\qquad\square$

A second version of the Fundamental Theorem of Calculus concerns antiderivatives. If one function is the derivative of another, the second function is an **antiderivative** of the first.

Note When G is an antiderivative of $g : [a, b] \to \mathbb{R}$, we have

$$G'(x) = g(x)$$

for *every* $x \in [a, b]$, not merely for almost every $x \in [a, b]$.

36 Corollary *Every continuous function has an antiderivative.*

Proof Assume that $f : [a, b] \to \mathbb{R}$ is continuous. By the Fundamental Theorem of Calculus, the indefinite integral $F(x)$ has a derivative everywhere, and $F'(x) = f(x)$ everywhere. $\square$

Some discontinuous functions have an antiderivative and others don't. Surprisingly, the wildly oscillating function

$$f(x) = \begin{cases} 0 & \text{if } x \leq 0 \\ \sin \dfrac{\pi}{x} & \text{if } x > 0 \end{cases}$$

has an antiderivative, but the jump function

$$g(x) = \begin{cases} 0 & \text{if } x \leq 0 \\ 1 & \text{if } x > 0 \end{cases}$$

does not. See Exercise 40.

37 Antiderivative Theorem *An antiderivative of a Riemann integrable function, if it exists, differs from the indefinite integral by a constant.*

Proof We assume that $f : [a, b] \to \mathbb{R}$ is Riemann integrable, that G is an antiderivative of f, and we assert that for all $x \in [a, b]$ we have

$$G(x) = \int_a^x f(t) \, dt + C,$$

where C is a constant. (In fact, $C = G(a)$.) Partition $[a, x]$ as

$$a = x_0 < x_1 < \ldots < x_n = x,$$

and choose $t_k \in [x_{k-1}, x_k]$ such that

$$G(x_k) - G(x_{k-1}) = G'(t_k)\Delta x_k.$$

Such a t_k exists by the Mean Value Theorem applied to the differentiable function G. Telescoping gives

$$G(x) - G(a) \;=\; \sum_{k=1}^{n} G(x_k) - G(x_{k-1}) \;=\; \sum_{k=1}^{n} f(t_k)\Delta x_k,$$

which is a Riemann sum for f on the interval $[a, x]$. Since f is Riemann integrable, the Riemann sum converges to $F(x)$ as the mesh of the partition tends to zero. This gives $G(x) - G(a) = F(x)$ as claimed. $\qquad\square$

38 Corollary *Standard integral formulas, such as*

$$\int_a^b x^2 \, dx = \frac{b^3 - a^3}{3},$$

are valid.

Proof Every integral formula is actually a derivative formula, and the Antiderivative Theorem converts derivative formulas to integral formulas. $\qquad\square$

Example The **logarithm function** is defined as the integral,

$$\log x = \int_1^x \frac{1}{t} \, dt.$$

Since the integrand $1/t$ is well defined and continuous when $t > 0$, $\log x$ is well defined and differentiable for $x > 0$. Its derivative is $1/x$. By the way, as is standard in post-calculus vocabulary, $\log x$ refers to the natural logarithm, not to the base-10 logarithm. See also Exercise 16.

    An antiderivative of f has $G'(x) = f(x)$ everywhere, and differs from the indefinite integral $F(x)$ by a constant. But what if we assume instead that $H'(x) = f(x)$ *almost* everywhere? Should this not also imply $H(x)$ differs from $F(x)$ by a constant? Surprisingly, the answer is "no."

37 Theorem *There exists a continuous function $H : [0, 1] \to \mathbb{R}$ whose derivative exists and equals zero almost everywhere, but which is not constant.*

Proof The counterexample is the **Devil's staircase function**, also called the **Cantor function**. Its graph is shown in Figure 82 and it is defined as follows.

    Each $x \in [0, 1]$ has a base-3 expansion $(.\omega_1\omega_2\omega_3\ldots)_3$ where

$$x = \sum_{i=1}^{\infty} \frac{\omega_i}{3^i}.$$

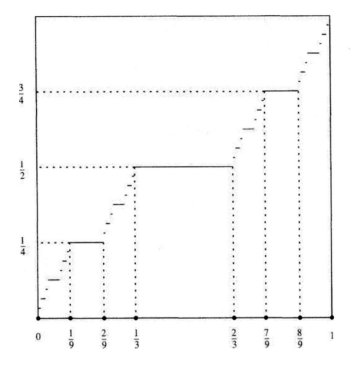

Figure 82 The Devil's staircase

Each ω_i is 0, 1, or 2. If $x \in C$, the standard Cantor set constructed in Chapter 2, then x has a unique expansion in which each ω_i equals 0 or 2. The function H sends $x \in C$ to

$$H(x) \; = \; \sum_{i=1}^{\infty} \frac{\omega_i/2}{2^i}.$$

H has equal values at the endpoints of the discarded gap intervals and so we extend H to them by letting it be constant on each. This accounts for the steps in its graph.

There are two things to check – the definition of H makes sense and H has the properties asserted. Continuity of the map $H : C \to [0,1]$ is simple. As we showed in Chapter 2, C is the nested intersection $\bigcap C^n$ where C^n is the disjoint union of 2^n intervals of length $1/3^n$, the endpoints of which are fractions with denominator 3^n. Between the intervals C_α in C^n there are open discarded intervals of length $\geq 1/3^n$. Let $\epsilon > 0$ be given, choose n with $1/2^n < \epsilon$, and take $\delta = 1/3^n$. If $x, x' \in C$ have $|x - x'| < \delta = 1/3^n$ then they lie in a common interval C_α in C^n. For the distance between different intervals C_α, C_β in C^n is at least $1/3^n$. Therefore the base-3 expansion of x and x' agree for the first n terms, which implies $|H(x) - H(x')| \leq \sum_{j=n+1}^{\infty} 1/2^j < \epsilon$ and gives continuity on C.

At stage n in the Cantor set construction we discard the open middle third of an interval $C_\alpha = [\ell_\alpha, \ell_\alpha + 1/3^n]$, where the left endpoint is

$$\ell_\alpha = \sum_{i=1}^{n} \frac{\alpha_i}{3^i} = (.\alpha_1\,\alpha_2\,\ldots\,\alpha_n)_3.$$

and each α_i is 0 or 2. Thus the discarded interval is

$$(\ell_\alpha + 1/3^{n+1},\ \ell_\alpha + 2/3^{n+1}) = ((.\alpha\,1)_3,\ (.\alpha\,2)_3) = ((.\alpha\,0\,\overline{2})_3,\ (.\alpha\,2)_3)$$

since $1/3^{n+1} = \sum_{j=n+2}^{\infty} 2/3^j$. This expresses both endpoints base-3 using only the numerals 0 and 2. Evaluating H on them gives equal value:

$$H(\ell_\alpha + 1/3^{n+1}) = H((.\alpha\,0\,\overline{2})_3) = \sum_{i=1}^{n} \frac{\alpha_i/2}{2^i} + \frac{0}{2^{n+1}} + \sum_{j=n+2}^{\infty} \frac{1}{2^j}$$

$$H(\ell_\alpha + 2/3^{n+1}) = H((.\alpha\,2)_3) = \sum_{i=1}^{n} \frac{\alpha_i/2}{2^i} + \frac{1}{2^{n+1}}.$$

This verifies that the definition of H being constant on the discarded intervals makes sense and completes the proof that H is continuous on $[0, 1]$.

It is clear that $H(0) = 0$ and

$$H(1) = H((.\overline{2})_3) = \sum_{i=1}^{\infty} \frac{2/2}{2^i} = 1.$$

Thus H is surjective. If $x, x' \in C$ and $x < x'$ then it is also clear that $H(x) \leq H(x')$, which implies that H is nondecreasing on $[0, 1]$. Since H is constant on the complement of the Cantor set, its derivative exists and is zero almost everywhere. $\square$

A yet more pathological example is a *strictly* monotone, continuous function J whose derivative is almost everywhere zero. Its graph is a sort of **Devil's ski slope**, almost everywhere level but also everywhere downhill. To construct J, start with H and extend it to a function $\widehat{H} : \mathbb{R} \to \mathbb{R}$ by setting $\widehat{H}(x + n) = H(x) + n$ for all $n \in \mathbb{Z}$ and all $x \in [0, 1]$. Then set

$$J(x) = \sum_{k=0}^{\infty} \frac{\widehat{H}(3^k x)}{4^k}.$$

The values of $\widehat{H}(3^k x)$ for $x \in [0, 1]$ are $\leq 3^k$, which is much smaller than the denominator 4^k. Thus the series converges and $J(x)$ is well defined. According to the Weierstrass M-test, proved in the next chapter, J is continuous. Since $\widehat{H}(3^k x)$

strictly increases for any pair of points at distance $> 1/3^k$ apart, and this fact is preserved when we take sums, J strictly increases. The proof that $J'(x) = 0$ almost everywhere requires deeper theory. See Exercise 48 on page 456.

Next, we justify two common methods of integration.

38 Integration by Substitution *If $f \in \mathcal{R}$ and $g : [c, d] \to [a, b]$ is a continuously differentiable bijection with $g' > 0$ (g is a C^1 diffeomorphism) then*

$$\int_a^b f(y)\,dy \;=\; \int_c^d f(g(x))g'(x)\,dx.$$

Proof The first integral exists by assumption. By Corollary 33 the composite $f \circ g$ is Riemann integrable. Since g' is continuous, the second integral exists by Corollary 27. To show that the two integrals are equal we resort again to Riemann sums. Let P partition the interval $[c, d]$ as

$$c = x_0 < x_1 < \cdots < x_n = d$$

and choose $t_k \in [x_{k-1}, x_k]$ such that

$$g(x_k) - g(x_{k-1}) \;=\; g'(t_k)\Delta x_k.$$

The Mean Value Theorem ensures that such a t_k exists. Since g is a diffeomorphism we have a partition Q of the interval $[a, b]$

$$a = y_0 < y_1 < \ldots < y_n = b$$

where $y_k = g(x_k)$, and mesh $P \to 0$ implies that mesh $Q \to 0$. Set $s_k = g(t_k)$. This gives two equal Riemann sums

$$\sum_{k=1}^n f(s_k)\Delta y_k \;=\; \sum_{k=1}^n f(g(t_k))g'(t_k)\Delta x_k$$

which converge to the integrals $\int_a^b f(y)\,dy$ and $\int_c^d f(g(t))g'(t)\,dt$ as mesh $P \to 0$. Since the limits of equals are equal, the integrals are equal. $\qquad\square$

Actually, it is sufficient to assume that $g' \in \mathcal{R}$.

39 Integration by Parts *If $f, g : [a, b] \to \mathbb{R}$ are differentiable and $f', g' \in \mathcal{R}$ then*

$$\int_a^b f(x)g'(x)\,dx \;=\; f(b)g(b) - f(a)g(a) - \int_a^b f'(x)g(x)\,dx.$$

Proof Differentiability implies continuity implies integrability, so $f, g \in \mathcal{R}$. Since the product of Riemann integrable functions is Riemann integrable, $f'g$ and fg' are Riemann integrable. By the Leibniz Rule, $(fg)'(x) = f(x)g'(x) + f'(x)g(x)$ everywhere. That is, fg is an antiderivative of $f'g + fg'$. The Antiderivative Theorem states that fg differs from the indefinite integral of $f'g + fg'$ by a constant. That is, for all $t \in [a, b]$ we have

$$
f(t)g(t) - f(a)g(a) = \int_a^t f'(x)g(x) + f(x)g'(x)\,dx
$$
$$
= \int_a^t f'(x)g(x)\,dx + \int_a^t f(x)g'(x)\,dx.
$$

Setting $t = b$ gives the result. □

Improper Integrals

Assume that $f : [a, b) \to \mathbb{R}$ is Riemann integrable when restricted to any closed subinterval $[a, c] \subset [a, b)$. You may imagine that $f(x)$ has some unpleasant behavior as $x \to b$, such as $\limsup_{x \to b} |f(x)| = \infty$ and/or $b = \infty$. See Figure 83.

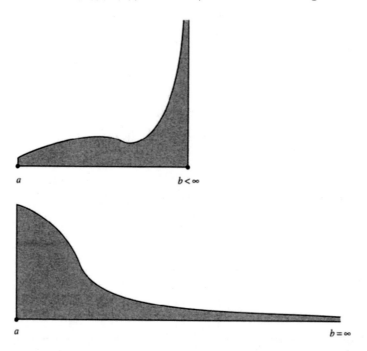

Figure 83 The improper integral converges if and only if the total undergraph area is finite.

If the limit of $\int_a^c f(x)\,dx$ exists (and is a real number) as $c \to b$ then it is natural to define it as the **improper Riemann integral**

$$\int_a^b f(x)\,dx \;=\; \lim_{c \to b} \int_a^c f(x)\,dx.$$

In order that the two-sided improper integral exists for a function $f : (a,b) \to \mathbb{R}$ it is natural to fix some point $m \in (a,b)$ and require that both improper integrals $\int_a^m f(x)\,dx$ and $\int_m^b f(x)\,dx$ exist. Their sum is the improper integral $\int_a^b f(x)\,dx$. With some ingenuity you can devise a function $f : \mathbb{R} \to \mathbb{R}$ whose improper integral $\int_{-\infty}^\infty f(x)\,dx$ exists despite the fact that f is unbounded at both $\pm\infty$. See Exercise 55.

3 Series

A series is a formal sum $\sum a_k$ where the terms a_k are real numbers. The nth **partial sum** of the series is

$$A_n = a_0 + a_1 + a_2 + \cdots + a_n.$$

The series **converges** to a real number A if $A_n \to A$ as $n \to \infty$, and we write

$$A \;=\; \sum_{k=0}^{\infty} a_k.$$

A series that does not converge **diverges**. The basic question to ask about a series is: Does it converge or diverge?

For example, if λ is a constant and $|\lambda| < 1$ then the **geometric series**

$$\sum_{k=0}^{\infty} \lambda^k \;=\; 1 + \lambda + \cdots + \lambda^n + \ldots$$

converges to $1/(1-\lambda)$. For its partial sums are

$$\Lambda_n \;=\; 1 + \lambda + \lambda^2 + \cdots + \lambda^n \;=\; \frac{1 - \lambda^{n+1}}{1 - \lambda}$$

and $\lambda^{n+1} \to 0$ as $n \to \infty$. On the other hand, if $|\lambda| \geq 1$ then the series $\sum \lambda^k$ diverges.

Let $\sum a_n$ be a series. The Cauchy Convergence Criterion from Chapter 1 applied to its sequence of partial sums yields the **CCC for series**

$$\sum a_k \quad \text{converges if and only if}$$

$$\forall \epsilon > 0 \; \exists N \text{ such that } m, n \geq N \;\Rightarrow\; \left| \sum_{k=m}^{n} a_k \right| < \epsilon.$$

One immediate consequence of the CCC is that no finite number of terms affects convergence of a series. Rather, it is the **tail** of the series, the terms a_k with k large, that determines convergence or divergence. Likewise, whether the series leads off with a term of index $k = 0$ or $k = 1$, etc. is irrelevant.

A second consequence of the CCC is that if a_k does not converge to zero as $k \to \infty$ then $\sum a_k$ does not converge. For Cauchyness of the partial sum sequence (A_n) implies that $a_n = A_n - A_{n-1}$ becomes small when $n \to \infty$. If $|\lambda| \geq 1$ then the geometric series $\sum \lambda^k$ diverges since its terms do not converge to zero. The **harmonic series**

$$\sum_{k=1}^{\infty} \frac{1}{k} = 1 + \frac{1}{2} + \frac{1}{3} + \cdots$$

gives an example that a series can diverge even though its terms do tend to zero. See below.

Series theory has a large number of convergence tests. All boil down to the following result.

40 Comparison Test *If a series $\sum b_k$ **dominates** a series $\sum a_k$ in the sense that for all sufficiently large k we have $|a_k| \leq b_k$ then convergence of $\sum b_k$ implies convergence of $\sum a_k$.*

Proof Given $\epsilon > 0$, convergence of $\sum b_k$ implies there is a large N such that for all $m, n \geq N$ we have $\sum_{k=m}^{n} b_k < \epsilon$. Thus

$$\left| \sum_{k=m}^{n} a_k \right| \leq \sum_{k=m}^{n} |a_k| \leq \sum_{k=m}^{n} b_k < \epsilon$$

and convergence of $\sum a_k$ follows from the CCC.   □

Example The series $\sum \sin(k)/2^k$ converges since it is dominated by the geometric series $\sum 1/2^k$.

A series $\sum a_k$ converges **absolutely** if $\sum |a_k|$ converges. The comparison test shows that absolute convergence implies convergence. A series that converges but not absolutely converges **conditionally**. That is, $\sum a_k$ converges and $\sum |a_k|$ diverges. See below.

Series and integrals are both infinite sums. You can imagine a series as an improper integral in which the integration variable is an integer,

$$\sum_{k=0}^{\infty} a_k = \int_{\mathbb{N}} a_k \, dk.$$

More precisely, given a series $\sum a_k$, define $f : [0, \infty) \to \mathbb{R}$ by setting

$$f(x) = a_k \quad \text{if} \quad k - 1 < x \le k.$$

See Figure 84.

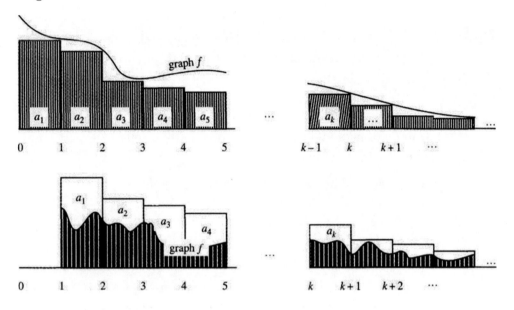

Figure 84 The pictorial proof of the integral test

Then

$$\sum_{k=0}^{\infty} a_k = \int_0^{\infty} f(x)\, dx.$$

The series converges if and only if the improper integral does. The natural interpretation of this picture is the

41 Integral Test *Suppose that $\int_0^{\infty} f(x)\, dx$ is a given improper integral and $\sum a_k$ is a given series.*

(a) If $|a_k| \le f(x)$ for all sufficiently large k and all $x \in (k-1, k]$ then convergence of the improper integral implies convergence of the series.

(b) If $|f(x)| \le a_k$ for all sufficiently large k and all $x \in [k, k+1)$ then divergence of the improper integral implies divergence of the series .

Proof (a) For some large N_0 and all $N \ge N_0$ we have

$$\sum_{k=N_0+1}^{N} |a_k| \quad \le \quad \int_{N_0}^{N} f(x)\, dx \quad \le \quad \int_0^{\infty} f(x)\, dx,$$

which is a finite real number. An increasing, bounded sequence converges to a limit, so the tail of the series $\sum |a_k|$ converges and the whole series $\sum |a_k|$ converges. Absolute convergence implies convergence.

The proof of (b) is left as Exercise 58.   □

Example The **p-series** $\sum 1/k^p$ converges when $p > 1$ and diverges when $p \leq 1$.

<u>Case 1.</u> $p > 1$. By the Fundamental Theorem of Calculus and differentiation,

$$\int_1^b \frac{1}{x^p}\, dx = \frac{b^{1-p} - 1}{1 - p} \to \frac{1}{p - 1}$$

as $b \to \infty$. The improper integral converges and dominates the p-series, which implies convergence of the series by the integral test.

<u>Case 2.</u> $p \leq 1$. The p-series dominates the improper integral

$$\int_1^b \frac{1}{x^p}\, dx = \begin{cases} \log b & \text{if} \quad p = 1 \\ \dfrac{b^{1-p} - 1}{1 - p} & \text{if} \quad p < 1. \end{cases}$$

As $b \to \infty$, these quantities blow up, and the integral test implies divergence of the series. When $p = 1$ we have the harmonic series, which we have just shown to diverge.

The **exponential growth rate** of the series $\sum a_k$ is

$$\alpha = \limsup_{k \to \infty} \sqrt[k]{|a_k|}.$$

Example $\sum \alpha^k$ has exponential growth rate α.

42 Root Test *Let α be the exponential growth rate of a series $\sum a_k$. If $\alpha < 1$ then the series converges, if $\alpha > 1$ then the series diverges, and if $\alpha = 1$ then the root test is inconclusive.*

Proof If $\alpha < 1$ then we fix a constant β with

$$\alpha < \beta < 1.$$

Then for all large k we have $|a_k|^{1/k} \leq \beta$; i.e., $|a_k| \leq \beta^k$, which gives convergence of $\sum a_k$ by comparison to the geometric series $\sum \beta^k$.

If $\alpha > 1$, choose β with $1 < \beta < \alpha$. Then $|a_k| \geq \beta^k$ for infinitely many k. Since the terms a_k do not converge to 0, the series diverges.

To show that the root test is inconclusive when $\alpha = 1$, we must find two series, one convergent and the other divergent, both having exponential growth rate $\alpha = 1$. The examples are p-series. We have

$$\log\left(\frac{1}{k^p}\right)^{1/k} = \frac{-p\log(k)}{k} \sim \frac{-p\log(x)}{x} \sim \frac{-p/x}{1} \sim 0$$

by L'Hôpital's Rule as $k = x \to \infty$. Therefore $\alpha = \lim_{k\to\infty}(1/k^p)^{1/k} = 1$ for all p-series. Since the square series $\sum 1/k^2$ converges and the harmonic series $\sum 1/k$ diverges the root test is inconclusive when $\alpha = 1$. $\qquad\square$

43 Ratio Test *Let the ratio between successive terms of the series $\sum a_k$ be $r_k = |a_{k+1}/a_k|$, and set*

$$\limsup_{k\to\infty} r_k = \rho \qquad \liminf_{k\to\infty} r_k = \lambda.$$

If $\rho < 1$ then the series converges, if $\lambda > 1$ then the series diverges, and otherwise the ratio test is inconclusive.

Proof If $\rho < 1$, choose β with $\rho < \beta < 1$. For all $k \geq$ some K, $|a_{k+1}/a_k| < \beta$; i.e.,

$$|a_k| \leq \beta^{k-K}|a_K| = C\beta^k$$

where $C = \beta^{-K}|a_K|$ is a constant. Convergence of $\sum a_k$ follows from comparison with the geometric series $\sum C\beta^k$. If $\lambda > 1$, choose β with $1 < \beta < \lambda$. Then $|a_k| \geq \beta^k/C$ for all large k, and $\sum a_k$ diverges because its terms do not converge to 0. Again the p-series all have ratio limit $\rho = \lambda = 1$ and demonstrate the inconclusiveness of the ratio test when $\rho = 1$ or $\lambda = 1$. $\qquad\square$

Although it is usually easier to apply the ratio test than the root test, the latter has a strictly wider scope. See Exercises 61 and 65.

Conditional Convergence

If (a_k) is a decreasing sequence in $\mathbb{R}$ that converges to 0 then its **alternating series**

$$\sum(-1)^{k+1}a_k = a_1 - a_2 + a_3 - \ldots$$

converges. For

$$A_{2n} = (a_1 - a_2) + (a_3 - a_4) + \ldots (a_{2n-1} - a_{2n})$$

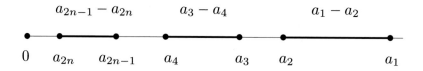

Figure 85 The pictorial proof of alternating convergence

and $a_{k-1} - a_k$ is the length of the the interval $I_k = (a_k, a_{k-1})$. The intervals I_k are disjoint, so the sum of their lengths is at most the length of $(0, a_1)$, namely a_1. See Figure 85.

The sequence (A_{2n}) is increasing and bounded, so $\lim_{n\to\infty} A_{2n}$ exists. The partial sum A_{2n+1} differs from A_{2n} by a_{2n+1}, a quantity that converges to 0 as $n \to \infty$, so

$$\lim_{n\to\infty} A_{2n} = \lim_{n\to\infty} A_{2n+1}$$

and the alternating series converges.

When $a_k = 1/k$ we have the **alternating harmonic series**,

$$\sum_{k=1}^{\infty} \frac{(-1)^{k+1}}{k} = 1 - \frac{1}{2} + \frac{1}{3} - \frac{1}{4} + \cdots$$

which we have just shown is convergent.

Series of Functions

A series of functions is of the form

$$\sum_{k=0}^{\infty} f_k(x),$$

where each term $f_k : (a, b) \to \mathbb{R}$ is a function. For example, in a power series

$$\sum c_k x^k$$

the functions are monomials $c_k x^k$. (The coefficients c_k are constants and x is a real variable.) If you think of $\lambda = x$ as a variable then the geometric series is a power series whose coefficients are 1, namely $\sum x^k$. Another example of a series of functions is a **Fourier series**

$$\sum a_k \sin(kx) + b_k \cos(kx).$$

44 Radius of Convergence Theorem *If $\sum c_k x^k$ is a power series then there is a unique R with $0 \leq R \leq \infty$, its **radius of convergence**, such that the series converges whenever $|x| < R$, and diverges whenever $|x| > R$. Moreover R is given by the formula*

$$R = \frac{1}{\limsup_{k \to \infty} \sqrt[k]{|c_k|}}.$$

Proof Apply the root test to the series $\sum c_k x^k$. Then

$$\limsup_{k \to \infty} \sqrt[k]{|c_k x^k|} = |x| \limsup_{k \to \infty} \sqrt[k]{|c_k|} = \frac{|x|}{R}.$$

If $|x| < R$ the root test gives convergence. If $|x| > R$ it gives divergence.   □

There are power series with any given radius of convergence, $0 \leq R \leq \infty$. The series $\sum k^k x^k$ has $R = 0$. The series $\sum x^k / \sigma^k$ has $R = \sigma$ for $0 < \sigma < \infty$. The series $\sum x^k / k!$ has $R = \infty$. Eventually, we show that a function defined by a power series is analytic: It has all derivatives at all points and it can be expanded as a Taylor series at each point inside its radius of convergence, not merely at $x = 0$. See Section 6 in Chapter 4.

Exercises

1. Assume that $f : \mathbb{R} \to \mathbb{R}$ satisfies $|f(t) - f(x)| \leq |t - x|^2$ for all t, x. Prove that f is constant.

2. A function $f : (a, b) \to \mathbb{R}$ satisfies a **Hölder condition** of order α if $\alpha > 0$, and for some constant H and all $u, x \in (a, b)$ we have

$$|f(u) - f(x)| \leq H|u - x|^\alpha$$

   The function is said to be α-Hölder, with α-Hölder constant H. (The terms "Lipschitz function of order α" and "α-Lipschitz function" are sometimes used with the same meaning.)

   (a) Prove that an α-Hölder function defined on (a, b) is uniformly continuous and infer that it extends uniquely to a continuous function defined on $[a, b]$. Is the extended function α-Hölder?

   (b) What does α-Hölder continuity mean when $\alpha = 1$?

   (c) Prove that α-Hölder continuity when $\alpha > 1$ implies that f is constant.

3. Assume that $f : (a, b) \to \mathbb{R}$ is differentiable.

   (a) If $f'(x) > 0$ for all x, prove that f is strictly monotone increasing.

   (b) If $f'(x) \geq 0$ for all x, what can you prove?

4. Prove that $\sqrt{n + 1} - \sqrt{n} \to 0$ as $n \to \infty$.

5. Assume that $f : \mathbb{R} \to \mathbb{R}$ is continuous, and for all $x \neq 0$, $f'(x)$ exists. If $\lim_{x \to 0} f'(x) = L$ exists, does it follow that $f'(0)$ exists? Prove or disprove.

6. In L'Hôpital's Rule, replace the interval (a, b) with the half-line (a, ∞) and interpret "x tends to b" as "$x \to \infty$." Show that if f/g tends to $0/0$ and f'/g' tends to L then f/g tends to L also. Prove that this continues to hold when $L = \infty$ in the sense that if $f'/g' \to \infty$ then $f/g \to \infty$.

7. In L'Hôpital's Rule, replace the assumption that f/g tends to $0/0$ with the assumption that it tends to ∞/∞. If f'/g' tends to L, prove that f/g tends to L also. [Hint: Think of a rear guard instead of an advance guard.] [Query: Is there a way to deduce the ∞/∞ case from the $0/0$ case? Naïvely taking reciprocals does not work.]

8. (a) Draw the graph of a continuous function defined on $[0, 1]$ that is differentiable on the interval $(0, 1)$ but not at the endpoints.

   (b) Can you find a formula for such a function?

   (c) Does the Mean Value Theorem apply to such a function?

9. Assume that $f : \mathbb{R} \to \mathbb{R}$ is differentiable.

   (a) If there is an $L < 1$ such that for each $x \in \mathbb{R}$ we have $f'(x) < L$, prove that there exists a unique point x such that $f(x) = x$. [x is a fixed point for f.]

   (b) Show by example that (a) fails if $L = 1$.

10. Concoct a function $f : \mathbb{R} \to \mathbb{R}$ with a discontinuity of the second kind at $x = 0$ such that f does not have the intermediate value property there. Infer that it is incorrect to assert that functions without jumps are Darboux continuous.

*11. Let $f : (a, b) \to \mathbb{R}$ be given.

(a) If $f''(x)$ exists, prove that

$$\lim_{h \to 0} \frac{f(x - h) - 2f(x) + f(x + h)}{h^2} = f''(x).$$

(b) Find an example that this limit can exist even when $f''(x)$ fails to exist.

*12. Assume that $f : (-1, 1) \to \mathbb{R}$ and $f'(0)$ exists. If $\alpha_n, \beta_n \to 0$ as $n \to \infty$, define the difference quotient

$$D_n = \frac{f(\beta_n) - f(\alpha_n)}{\beta_n - \alpha_n}.$$

(a) Prove that $\lim_{n \to \infty} D_n = f'(0)$ under each of the following conditions.

(i) $\alpha_n < 0 < \beta_n$.

(ii) $0 < \alpha_n < \beta_n$ and $\dfrac{\beta_n}{\beta_n - \alpha_n} \leq M$.

(iii) $f'(x)$ exists and is continuous for all $x \in (-1, 1)$.

(b) Set $f(x) = x^2 \sin(1/x)$ for $x \neq 0$ and $f(0) = 0$. Observe that f is differentiable everywhere in $(-1, 1)$ and $f'(0) = 0$. Find α_n, β_n that tend to 0 in such a way that D_n converges to a limit unequal to $f'(0)$.

13. Assume that f and g are r^{th} order differentiable functions $(a, b) \to \mathbb{R}$, $r \geq 1$. Prove the **Higher-Order Leibniz Product Rule** for the function $f \cdot g$,

$$(f \cdot g)^{(r)}(x) = \sum_{k=0}^{r} \binom{r}{k} f^{(k)}(x) \cdot g^{(r-k)}(x).$$

where $\binom{r}{k} = r!/(k!(r - k)!)$ is the binomial coefficient, r choose k. [Hint: Induction.]

14. For each $r \geq 1$, find a function that is C^r but not C^{r+1}.

15. Define $f(x) = x^2$ if $x < 0$ and $f(x) = x + x^2$ if $x \geq 0$. Differentiation gives $f''(x) \equiv 2$. This is bogus. Why?

16. $\log x$ is defined to be $\int_1^x 1/t \, dt$ for $x > 0$. Using only the mathematics explained in this chapter,

(a) Prove that log is a smooth function.

(b) Prove that $\log(xy) = \log x + \log y$ for all $x, y > 0$. [Hint: Fix y and define $f(x) = \log(xy) - \log x - \log y$. Show that $f(x) \equiv 0$.]

(c) Prove that log is strictly monotone increasing and its range is all of $\mathbb{R}$.

17. Define $e : \mathbb{R} \to \mathbb{R}$ by

$$e(x) = \begin{cases} e^{-1/x} & \text{if } x > 0 \\ 0 & \text{if } x \leq 0 \end{cases}$$

(a) Prove that e is smooth; that is, e has derivatives of all orders at all points x. [Hint: L'Hôpital and induction. Feel free to use the standard differentiation formulas about e^x from calculus.]

(b) Is e analytic?

(c) Show that the **bump function**

$$\beta(x) = e^2 e(1 - x) \cdot e(x + 1)$$

is smooth, identically zero outside the interval $(-1, 1)$, positive inside the interval $(-1, 1)$, and takes value 1 at $x = 0$.[†] (e^2 is the square of the base of the natural logarithms, while $e(x)$ is the function just defined. Apologies to the abused notation.)

(d) For $|x| < 1$ show that

$$\beta(x) = e^{2x^2/(x^2-1)}.$$

Bump functions have wide use in smooth function theory and differential topology. The graph of β looks like a bump. See Figure 86.

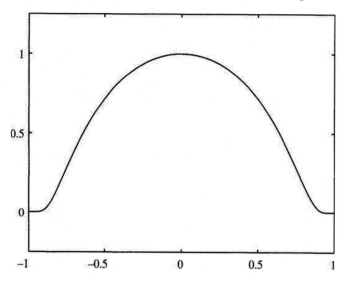

Figure 86 The graph of the bump function β

**18. Let L be any closed set in $\mathbb{R}$. Prove that there is a smooth function $f : \mathbb{R} \to [0, 1]$ such that $f(x) = 0$ if and only if $x \in L$. To put it another way, every closed set in $\mathbb{R}$ is the zero locus of some smooth function. [Hint: Use Exercise 17(c).]

[†]The **support** of a function is the closure of the set of points at which the function is nonzero. The support of β is $[-1, 1]$.

19. Recall that the oscillation of an arbitrary function $f : [a, b] \to \mathbb{R}$ at x is

$$\operatorname{osc}_x f = \limsup_{t \to x} f(t) - \liminf_{t \to x} f(t)$$

In the proof of the Riemann-Lebesgue Theorem D_k refers to the set of points with oscillation $\geq 1/k$.

(a) Prove that D_k is closed.

(b) Infer that the discontinuity set of f is a countable union of closed sets. (This is called an $\boldsymbol{F_\sigma}$-**set**.)

(c) Infer from (b) that the set of continuity points is a countable intersection of open sets. (This is called a $\boldsymbol{G_\delta}$-set).)

*20. Baire's Theorem (page 256) asserts that if a complete metric space is the countable union of closed subsets then at least one of them has nonempty interior. Use Baire's Theorem to show that the set of irrational numbers is not the countable union of closed subsets of $\mathbb{R}$.

21. Use Exercises 19 and 20 to show there is no function $f : \mathbb{R} \to \mathbb{R}$ which is discontinuous at every irrational number and continuous at every rational number.

**22. Show that there exists a subset S of the middle-thirds Cantor set which is never the discontinuity set of a function $f : \mathbb{R} \to \mathbb{R}$. Infer that some zero sets are never discontinuity sets of Riemann integrable functions. [Hint: How many subsets of C are there? How many can be countable unions of closed sets?]

23. Suppose that $f_n : [a, b] \to \mathbb{R}$ is a sequence of continuous functions that converges pointwise to a limit function $f : [a, b] \to \mathbb{R}$. Such an f is said to be of **Baire class 1. (Pointwise convergence is discussed in the next chapter. It means what it says: For each x, $f_n(x)$ converges to $f(x)$ as $n \to \infty$. Continuous functions are considered to be of Baire class 0, and in general a Baire class r function is the pointwise limit of a sequence of Baire class $r - 1$ functions. Strictly speaking, it should not be of Baire class $r - 1$ itself, but for simplicity I include continuous functions among Baire class 1 functions. It is an interesting fact that for every r there are Baire class r functions not of Baire class $r - 1$. You might consult *A Primer of Real Functions* by Ralph Boas.) Prove that the set D_k of discontinuity points with oscillation $\geq 1/k$ is nowhere dense, as follows. To arrive at a contradiction, suppose that D_k is dense in some interval $(\alpha, \beta) \subset [a, b]$. By Exercise 19, D_k is closed, so it contains (α, β). Cover $\mathbb{R}$ by countably many intervals (a_ℓ, b_ℓ) of length $< 1/k$ and set

$$H_\ell = f^{\mathrm{pre}}(a_\ell, b_\ell).$$

(a) Why does $\bigcup_\ell H_\ell = [a, b]$?

(b) Show that no H_ℓ contains a subinterval of (α, β).

(c) Why are

$$F_{\ell m n} = \{x \in [a,b] : a_\ell + \frac{1}{m} \leq f_n(x) \leq b_\ell - \frac{1}{m}\}$$
$$E_{\ell m N} = \bigcap_{n \geq N} F_{\ell m n}$$

closed?

(d) Show that

$$H_\ell = \bigcup_{m, N \in \mathbb{N}} E_{\ell m N}.$$

(e) Use (a) and Baire's Theorem (page 243) to deduce that some $E_{\ell m N}$ contains a subinterval of (α, β).

(f) Why does (e) contradict (b) and complete the proof that D_k is nowhere dense?

24. Combine Exercises 19, 23, and Baire's Theorem to show that a Baire class 1 function has a dense set of continuity points.

25. Suppose that $g : [a,b] \to \mathbb{R}$ is differentiable.

(a) Prove that g' is of Baire class 1. [Hint: Extend g to a differentiable function defined on a larger interval and consider

$$f_n(x) = \frac{g(x + 1/n) - g(x)}{1/n}$$

for $x \in [a,b]$. Is $f_n(x)$ continuous? Does $f_n(x)$ converge pointwise to $g'(x)$ as $n \to \infty$?]

(b) Infer from Exercise 24 that *a derivative cannot be everywhere discontinuous*. It must be continuous on a dense subset of its domain of definition.

26. Redefine Riemann and Darboux integrability using dyadic partitions.

(a) Prove that the integrals are unaffected.

(b) Infer that Riemann's integrability criterion can be restated in terms of dyadic partitions.

(c) Repeat the analysis using only partitions of $[a,b]$ into subintervals of length $(b-a)/n$.

27. In many calculus books, the definition of the integral is given as

$$\int_a^b f(x)\, dx = \lim_{n \to \infty} \sum_{k=1}^n f(x_k^*) \frac{b-a}{n}$$

where x_k^* is the midpoint of the kth interval of $[a,b]$ having length $(b-a)/n$, namely

$$[a + (k-1)(b-a)/n, \ a + k(b-a)/n].$$

See Stewart's *Calculus with Early Transcendentals*, for example.

(a) If f is continuous, show that the calculus-style limit exists and equals the Riemann integral of f. [Hint: This is a one-liner.]

(b) Show by example that the calculus-style limit can exist for functions which are not Riemann integrable.

(c) Infer that the calculus-style definition of the integral is inadequate for real analysis.

28. Suppose that $Z \subset \mathbb{R}$. Prove that the following are equivalent.

(i) Z is a zero set.

(ii) For each $\epsilon > 0$ there is a countable covering of Z by closed intervals $[a_i, b_i]$ with total length $\sum b_i - a_i < \epsilon$.

(iii) For each $\epsilon > 0$ there is a countable covering of Z by sets S_i with total diameter $\sum \operatorname{diam} S_i < \epsilon$.

*29. Prove that the interval $[a, b]$ is not a zero set.

(a) Explain why the following observation is not a solution to the problem: "Every open interval that contains $[a, b]$ has length $> b - a$."

(b) Instead, suppose there is a "bad" covering of $[a, b]$ by open intervals $\{I_i\}$ whose total length is $< b - a$, and justify the following steps.

(i) It is enough to deal with finite bad coverings.

(ii) Let $\mathcal{B} = \{I_1, \ldots, I_n\}$ be a bad covering such that n is minimal among all bad coverings.

(iii) Show that no bad covering has $n = 1$ so we have $n \geq 2$.

(iv) Show that it is no loss of generality to assume $a \in I_1$ and $I_1 \cap I_2 \neq \emptyset$.

(v) Show that $I = I_1 \cup I_2$ is an open interval and $|I| < |I_1| + |I_2|$.

(vi) Show that $\mathcal{B}' = \{I, I_3, \ldots, I_n\}$ is a bad covering of $[a, b]$ with fewer intervals, a contradiction to minimality of n.

30. The standard **middle-quarters Cantor set** Q is formed by removing the middle quarter from $[0, 1]$, then removing the middle quarter from each of the remaining two intervals, then removing the middle quarter from each of the remaining four intervals, and so on.

(a) Prove that Q is a zero set.

(b) Formulate the natural definition of the middle β-Cantor set.

(c) Is it also a zero set? Prove or disprove.

*31. Define a Cantor set by removing from $[0, 1]$ the middle interval of length $1/4$. From the remaining two intervals F^1 remove the middle intervals of length $1/16$. From the remaining four intervals F^2 remove the middle intervals of length $1/64$, and so on. At the nth step in the construction F^n consists of 2^n subintervals of F^{n-1}.

(a) Prove that $F = \bigcap F^n$ is a Cantor set but not a zero set. It is referred to as a **fat Cantor set**.

(b) Infer that being a zero set is not a topological property: If two sets are homeomorphic and one is a zero set then the other need not be a zero set. [Hint: To get a sense of this fat Cantor set, calculate the total length of the intervals which comprise its complement. See Figure 52 and Exercise 35.]

32. Consider the characteristic function of the dyadic rational numbers, $f(x) = 1$ if $x = k/2^n$ for some $k \in \mathbb{Z}$ and $n \in \mathbb{N}$, and $f(x) = 0$ otherwise.

    (a) What is its set of discontinuities?

    (b) At which points is its oscillation $\geq \epsilon$?

    (c) Is it integrable? Explain, both by the Riemann-Lebesgue Theorem and directly from the definition.

    (d) Consider the **dyadic ruler function** $g(x) = 1/2^n$ if $x = k/2^n$ and $g(x) = 0$ otherwise. Graph it and answer the questions posed in (a), (b), (c).

33. (a) Prove that the characteristic function f of the middle-thirds Cantor set C is Riemann integrable but the characteristic function g of the fat Cantor set F (Exercise 31) is not.

    (b) Why is there a homeomorphism $h : [0, 1] \to [0, 1]$ sending C onto F?

    (c) Infer that the composite of Riemann integrable functions need not be Riemann integrable. How is this example related to Corollaries 28 and 32 of the Riemann-Lebesgue Theorem? See also Exercise 35.

*34. Assume that $\psi : [a, b] \to \mathbb{R}$ is continuously differentiable. A **critical point** of ψ is an x such that $\psi'(x) = 0$. A **critical value** is a number y such that for at least one critical point x we have $y = \psi(x)$.

    (a) Prove that the set of critical values is a zero set. (This is the **Morse-Sard Theorem** in dimension one.)

    (b) Generalize this to continuously differentiable functions $\mathbb{R} \to \mathbb{R}$.

*35. Let $F \subset [0, 1]$ be the fat Cantor set from Exercise 31, and define

$$\psi(x) = \int_0^x \text{dist}(t, F) \, dt$$

where $\text{dist}(t, F)$ refers to the minimum distance from t to F.

    (a) Why is ψ a continuously differentiable homeomorphism from $[0, 1]$ onto $[0, L]$ where $L = \psi(1)$?

    (b) What is the set of critical points of ψ? (See Exercise 34.)

    (c) Why is $\psi(F)$ a Cantor set of zero measure?

    (d) Let f be the characteristic function of $\psi(F)$. Why is f Riemann integrable but $f \circ \psi$ not?

    (e) What is the relation of (d) to Exercise 33?

    See also Exercise 6.77.

36. Generalizing Exercise 1.31, we say that $f : (a, b) \to \mathbb{R}$ has a **jump discontinuity** (or a discontinuity of the **first kind**) at $c \in (a, b)$ if

$$f(c^-) = \lim_{x \to c^-} f(x) \quad \text{and} \quad f(c^+) = \lim_{x \to c^+} f(x)$$

exist, but are either unequal or are unequal to $f(c)$. (The three quantities exist and are equal if and only if f is continuous at c.) An **oscillating discontinuity** (or a discontinuity of the **second kind** is any nonjump discontinuity.

(a) Show that $f : \mathbb{R} \to \mathbb{R}$ has at most countably many jump discontinuities.

(b) What about the function

$$f(x) = \begin{cases} \sin \dfrac{1}{x} & \text{if } x > 0 \\ 0 & \text{if } x \le 0? \end{cases}$$

(c) What about the characteristic function of the rationals?

37. Suppose that $f : \mathbb{R} \to [-M, M]$ has no jump discontinuities. Does f have the intermediate value property? (Proof or counterexample.)

**38. Recall that $\mathcal{P}(S) = 2^S$ is the power set of S, the collection of all subsets of S, and $\mathcal{R}$ is the set of Riemann integrable functions $f : [a, b] \to \mathbb{R}$.

(a) Prove that the cardinality of $\mathcal{R}$ is the same as the cardinality of $\mathcal{P}(\mathbb{R})$, which is greater than the cardinality of $\mathbb{R}$.

(b) Call two functions in $\mathcal{R}$ **integrally equivalent** if they differ only on a zero set. Prove that the collection of integral equivalence classes of $\mathcal{R}$ has the same cardinality as $\mathbb{R}$, namely $2^{\mathbb{N}}$.

(c) Is it better to count Riemann integrable functions or integral equivalence classes of Riemann integrable functions?

(d) Show that $f, g \in \mathcal{R}$ are integrally equivalent if and only if the integral of $|f - g|$ is zero.

39. Consider the characteristic functions $f(x)$ and $g(x)$ of the intervals $[1, 4]$ and $[2, 5]$. The derivatives f' and g' exist almost everywhere. The integration-by-parts formula says that

$$\int_0^3 f(x)g'(x)\, dx = f(3)g(3) - f(0)g(0) - \int_0^3 f'(x)g(x)\, dx.$$

But both integrals are zero, while $f(3)g(3) - f(0)g(0) = 1$. Where is the error?

40. Set

$$f(x) = \begin{cases} 0 & \text{if } x \le 0 \\ \sin \dfrac{\pi}{x} & \text{if } x > 0 \end{cases} \quad \text{and} \quad g(x) = \begin{cases} 0 & \text{if } x \le 0 \\ 1 & \text{if } x > 0. \end{cases}$$

Prove that f has an antiderivative but g does not.

41. Show that any two antiderivatives of a function differ by a constant. [Hint: This is a one-liner.]

42. Suppose that $\psi : [c, d] \to [a, b]$ is continuous and for every zero set $Z \subset [a, b]$, $\psi^{\mathrm{pre}}(Z)$ is a zero set in $[c, d]$.

(a) If f is Riemann integrable, prove that $f \circ \psi$ is Riemann integrable.

(b) Derive Corollary 32 from (a).

43. Let $\psi(x) = x \sin 1/x$ for $0 < x \le 1$ and $\psi(0) = 0$.
    (a) If $f : [-1, 1] \to \mathbb{R}$ is Riemann integrable, prove that $f \circ \psi$ is Riemann integrable.
    (b) What happens for $\psi(x) = \sqrt{x} \sin 1/x$?
*44. Assume that $\psi : [c, d] \to [a, b]$ is continuously differentiable.
    (a) If the critical points of ψ form a zero set in $[c, d]$ and f is Riemann integrable on $[a, b]$ prove that $f \circ \psi$ is Riemann integrable on $[c, d]$.
    (b) Conversely, prove that if $f \circ \psi$ is Riemann integrable for each Riemann integrable f on $[a, b]$, then the critical points of ψ form a zero set. [Hint: Think in terms of Exercise 34.]
    (c) Prove (a) and (b) under the weaker assumption that ψ is continuously differentiable except at finitely many points of $[c, d]$.
    (d) Derive part (a) of Exercise 35 from (c).
    (e) Weaken the assumption further to ψ being continuously differentiable on an open subset of $[c, d]$ whose complement is a zero set.

    Remark The following assertion, to be proved in Chapter 6, is related to the preceding exercises. If $f : [a, b] \to \mathbb{R}$ satisfies a Lipschitz condition or is monotone then the set of points at which $f'(x)$ fails to exist is a zero set. Thus: "A Lipschitz function is differentiable almost everywhere," which is **Rademacher's Theorem** in dimension 1, and a "monotone function is almost everywhere differentiable," which is the last theorem in Lebesgue's book, *Leçons sur l'intégration.* See Theorem 6.57 and Corollary 6.59.

45. (a) Define the oscillation for a function from one metric space to another, $f : M \to N$.
    (b) Is it true that f is continuous at a point if and only if its oscillation is zero there? Prove or disprove.
    (c) Is the set of points at which the oscillation of f is $\ge 1/k$ closed in M? Prove or disprove.
46. (a) Prove that the integral of the Zeno's staircase function described on page 174 is $2/3$.
    (b) What about the Devil's staircase?
47. In the proof of Corollary 28 of the Riemann-Lebesgue Theorem, it is asserted that when ϕ is continuous the discontinuity set of $\phi \circ f$ is contained in the discontinuity set of f.
    (a) Prove this.
    (b) Give an example where the inclusion is not an equality.
    (c) Find a sufficient condition on ϕ so that $\phi \circ f$ and f have equal discontinuity sets for all $f \in \mathcal{R}$
    (d) Is your condition necessary too?

48. Assume that $f \in \mathcal{R}$ and for some $m > 0$ we have $|f(x)| \geq m$ for all $x \in [a, b]$. Prove that the reciprocal of f, $1/f(x)$, also belongs to $\mathcal{R}$. If $f \in \mathcal{R}$, $|f(x)| > 0$, but no $m > 0$ is an underbound for $|f|$, prove that the reciprocal of f is not Riemann integrable.

49. Corollary 28 to the Riemann-Lebesgue Theorem asserts that if $f \in \mathcal{R}$ and ϕ is continuous, then $\phi \circ f \in \mathcal{R}$. Show that piecewise continuity cannot replace continuity. [Hint: Take f to be a ruler function and ϕ to be a characteristic function.]

**50. Assume that $f : [a, b] \to [c, d]$ is a Riemann integrable bijection. Is the inverse bijection also Riemann integrable? Prove or disprove.

51. If f, g are Riemann integrable on $[a, b]$, and $f(x) < g(x)$ for all $x \in [a, b]$, prove that $\int_a^b f(x)\, dx < \int_a^b g(x)\, dx$. (Note the *strict* inequality.)

52. Let $f : [a, b] \to \mathbb{R}$ be given. Prove or give counterexamples to the following assertions.
    (a) $f \in \mathcal{R} \Rightarrow |f| \in \mathcal{R}$.
    (b) $|f| \in \mathcal{R} \Rightarrow f \in \mathcal{R}$.
    (c) $f \in \mathcal{R}$ and $|f(x)| \geq c > 0$ for all $x \Rightarrow 1/f \in \mathcal{R}$.
    (d) $f \in \mathcal{R} \Rightarrow f^2 \in \mathcal{R}$.
    (e) $f^2 \in \mathcal{R} \Rightarrow f \in \mathcal{R}$.
    (f) $f^3 \in \mathcal{R} \Rightarrow f \in \mathcal{R}$.
    (g) $f^2 \in \mathcal{R}$ and $f(x) \geq 0$ for all $x \Rightarrow f \in \mathcal{R}$.
       [Here f^2 and f^3 refer to the functions $f(x) \cdot f(x)$ and $f(x) \cdot f(x) \cdot f(x)$, not the iterates.]

53. Given $f, g \in \mathcal{R}$, prove that $\max(f, g)$ and $\min(f, g)$ are Riemann integrable, where $\max(f, g)(x) = \max(f(x), g(x))$ and $\min(f, g)(x) = \min(f(x), g(x))$.

54. Assume that $f, g : [0, 1] \to \mathbb{R}$ are Riemann integrable and $f(x) = g(x)$ except on the middle-thirds Cantor set C.
    (a) Prove that f and g have the same integral.
    (b) Is the same true if $f(x) = g(x)$ except for $x \in \mathbb{Q}$?
    (c) How is this related to the fact that the characteristic function of $\mathbb{Q}$ is not Riemann integrable?

55. Invent a continuous function $f : \mathbb{R} \to \mathbb{R}$ whose improper integral is zero, but which is unbounded as $x \to -\infty$ and $x \to \infty$. [Hint: f is far from monotone.]

56. Assume that $f : \mathbb{R} \to \mathbb{R}$ and that the restriction of f to each closed interval is Riemann integrable.
    (a) Formulate the concepts of conditional and absolute convergence of the improper Riemann integral of f.
    (b) Find an example that distinguishes them.

57. Construct a function $f : [-1, 1] \to \mathbb{R}$ such that

$$\lim_{r \to 0} \left(\int_{-1}^{-r} f(x)\, dx + \int_{r}^{1} f(x)\, dx \right)$$

exists (and is a finite real number) but the improper integral $\int_{-1}^{1} f(x)\, dx$ does not exist. Do the same for a function $g : \mathbb{R} \to \mathbb{R}$ such that

$$\lim_{R \to \infty} \int_{-R}^{R} f(x)\, dx$$

exists but the improper integral $\int_{-\infty}^{\infty} g(x)\, dx$ fails to exist. [Hint: The functions are not symmetric across 0.]

58. Let $f : [0, \infty) \to [0, \infty)$ and $\sum a_k$ be given. Assume that for all sufficiently large k and all $x \in [k, k+1)$ we have $f(x) \leq a_k$. Prove that divergence of the improper integral $\int_0^\infty f(x)\, dx$ implies divergence of $\sum a_k$.

59. Prove that if $a_n \geq 0$ and $\sum a_n$ converges then $\sum (\sqrt{a_n})/n$ converges.

60. (a) If $\sum a_n$ converges and (b_n) is monotonic and bounded, prove that $\sum a_n b_n$ converges.

   (b) If the monotonicity condition is dropped, or replaced by the assumption that $\lim_{n \to \infty} b_n = 0$, find a counterexample to convergence of $\sum a_n b_n$.

61. Find an example of a series of positive terms that converges despite the fact that $\limsup_{n \to \infty} |a_{n+1}/a_n| = \infty$. Infer that ρ cannot replace λ in the divergence half of the ratio test.

62. Prove that if the terms of a sequence decrease monotonically, $a_1 \geq a_2 \geq \cdots$, and converge to 0 then the series $\sum a_k$ converges if and only if the associated dyadic series

$$a_1 + 2a_2 + 4a_4 + 8a_8 + \cdots = \sum 2^k a_{2^k}$$

converges. (I call this the **block test** because it groups the terms of the series in blocks of length 2^{k-1}.)

63. Prove that $\sum 1/k(\log(k))^p$ converges when $p > 1$ and diverges when $p \leq 1$. Here $k = 2, 3, \ldots$. [Hint: Integral test or block test.]

64. Concoct a series $\sum a_k$ such that $(-1)^k a_k > 0$, $a_k \to 0$, but the series diverges.

65. Compare the root and ratio tests.

   (a) Show that if a series has exponential growth rate ρ then it has ratio $\limsup \rho$. Infer that the ratio test is subordinate to the root test.

   (b) Concoct a series such that the root test is conclusive but the ratio test is not. Infer that the root test has strictly wider scope than the ratio test.

66. Show that there is no simple comparison test for conditionally convergent series:

   (a) Find two series $\sum a_k$ and $\sum b_k$ such that $\sum b_k$ converges conditionally, $a_k/b_k \to 1$ as $k \to \infty$, and $\sum a_k$ diverges.

(b) Why is this impossible if the series $\sum b_k$ is absolutely convergent?

67. An **infinite product** is an expression $\prod c_k$ where $c_k > 0$. The nth **partial product** is $C_n = c_1 \cdots c_n$. If C_n converges to a limit $C \neq 0$ then the product converges to C. Write $c_k = 1 + a_k$. If each $a_k \geq 0$ or each $a_k \leq 0$ prove that $\sum a_k$ converges if and only if $\prod c_k$ converges. [Hint: Take logarithms.]

68. Show that conditional convergence of the series $\sum a_k$ and the product $\prod(1+a_k)$ can be unrelated to each other:

   (a) Set $a_k = (-1)^k/\sqrt{k}$. The series $\sum a_k$ converges but the corresponding product $\prod(1 + a_k)$ diverges. [Hint: Group the terms in the product two at a time.]

   (b) When k is odd, set $b_k = 1/\sqrt{k}$ and $b_{k+1} = 1/k - 1/\sqrt{k}$. The series $\sum b_k$ diverges while the corresponding product $\prod_{k \geq 2}(1 + b_k)$ converges.

69. Consider a series $\sum a_k$ and **rearrange** its terms by some bijection $\beta : \mathbb{N} \to \mathbb{N}$, forming a new series $\sum a_{\beta(k)}$. The rearranged series converges if and only if the partial sums $a_{\beta(1)} + \ldots + a_{\beta(n)}$ converge to a limit as $n \to \infty$.

   (a) Prove that every rearrangement of a convergent series of nonnegative terms converges – and converges to the same sum as the original series.

   (b) Do the same for absolutely convergent series.

*70. Let $\sum a_k$ be given.

   (a) If $\sum a_k$ converges conditionally, prove that rearrangement totally alters its convergence in the sense that some rearrangements $\sum b_k$ of $\sum a_k$ diverge to $+\infty$, others diverge to $-\infty$, and others converge to any given real number.

   (b) Infer that a series is absolutely convergent if and only if every rearrangement converges. (The fact that rearrangement radically alters conditional convergence shows that although finite addition is commutative, infinite addition (i.e., summing a series) is not.)

**71. Suppose that $\sum a_k$ converges conditionally. If $\sum b_k$ is a rearrangement of $\sum a_k$, let Y be the set of subsequential limits of (B_n) where B_n is the nth partial sum of $\sum b_k$. That is, $y \in Y$ if and only if some $B_{n_\ell} \to y$ as $\ell \to \infty$.

   (a) Prove that Y is closed and connected.

   (b) If Y is compact and nonempty, prove that $\sum b_k$ converges to Y in the sense that $d_H(Y_n, Y) \to 0$ as $n \to \infty$, where d_H is the Hausdorff metric on the space of compact subsets of $\mathbb{R}$ and Y_n is the closure of $\{B_m : m \geq n\}$. See Exercise 2.147.

   (c) Prove that each closed and connected subset of $\mathbb{R}$ is the set of subsequential limits of some rearrangement of $\sum a_k$.

   The article, "The Remarkable Theorem of Lévy and Steinitz" by Peter Rosenthal in the *American Math Monthly* of April 1987 deals with some of these issues, including the higher dimensional situation.

72. Absolutely convergent series can be multiplied in a natural way, the result being their **Cauchy product,

$$\left(\sum_{i=0}^{\infty} a_i\right)\left(\sum_{j=0}^{\infty} b_j\right) = \sum_{k=0}^{\infty} c_k$$

where $c_k = a_0 b_k + a_1 b_{k-1} + \cdots + a_k b_0$.

(a) Prove that $\sum c_k$ converges absolutely.

(b) Formulate some algebraic laws for such products (commutativity, distributivity, and so on). Prove two of them.
[Hint for (a): Write the products $a_i b_j$ in an $\infty \times \infty$ matrix array M, and let A_n, B_n, C_n be the nth partial sums of $\sum a_i$, $\sum b_j$, $\sum c_k$. You are asked to prove that $(\lim A_n)(\lim B_n) = \lim C_n$. The product of the limits is the limit of the products. The product $A_n B_n$ is the sum of all the $a_i b_j$ in the $n \times n$ corner submatrix of M and c_n is the sum of its antidiagonal. Now estimate $A_n B_n - C_n$. Alternately, assume that $a_n, b_n \geq 0$ and draw a rectangle R with edges A, B. Observe that R is the union of rectangles R_{ij} with edges a_i, b_j.]

**73. With reference to Exercise 72,

(a) Reduce the hypothesis that both series $\sum a_i$ and $\sum b_j$ are absolutely convergent to merely one being absolutely convergent and the other convergent. (Exercises 72 and 73(a) are known as **Mertens' Theorem**.)

(b) Find an example to show that the Cauchy product of two conditionally convergent series may diverge.

74. The **Riemann ζ-function is defined to be $\zeta(s) = \sum_{n=1}^{\infty} n^{-s}$ where $s > 1$. It is the sum of the p-series when $p = s$. Establish **Euler's product formula**,

$$\zeta(s) = \prod_{k=1}^{\infty} \frac{1}{1 - p_k^{-s}}$$

where p_k is the kth prime number. Thus, $p_1 = 2$, $p_2 = 3$, and so on. Prove that the infinite product converges. [Hint: Each factor in the infinite product is the sum of a geometric series $1 + p_k^{-s} + (p_k^{-s})^2 + \ldots$. Replace each factor by its geometric series and write out the nth partial product. Apply Mertens' Theorem, collect terms, and recall that every integer has a unique prime factorization.]

4

Function Spaces

1 Uniform Convergence and $C^0[a, b]$

Points converge to a limit if they get physically closer and closer to it. What about a sequence of functions? When do functions converge to a limit function? What should it mean that they get closer and closer to a limit function? The simplest idea is that a sequence of functions f_n converges to a limit function f if for each x, the values $f_n(x)$ converge to $f(x)$ as $n \to \infty$. This is called **pointwise convergence**: A sequence of functions $f_n : [a, b] \to \mathbb{R}$ **converges pointwise** to a limit function $f : [a, b] \to \mathbb{R}$ if for each $x \in [a, b]$ we have

$$\lim_{n \to \infty} f_n(x) = f(x).$$

The function f is the **pointwise limit** of the sequence (f_n) and we write

$$f_n \to f \quad \text{or} \quad \lim_{n \to \infty} f_n = f.$$

Note that the limit refers to $n \to \infty$, not to $x \to \infty$. The same definition applies to functions from one metric space to another.

The requirement of uniform convergence is stronger. The sequence of functions $f_n : [a, b] \to \mathbb{R}$ **converges uniformly** to the limit function $f : [a, b] \to \mathbb{R}$ if for each $\epsilon > 0$ there is an N such that for all $n \geq N$ and all $x \in [a, b]$,

(1) $$|f_n(x) - f(x)| < \epsilon.$$

© Springer International Publishing Switzerland 2015
C.C. Pugh, *Real Mathematical Analysis*, Undergraduate Texts
in Mathematics, DOI 10.1007/978-3-319-17771-7_4

The function f is the **uniform limit** of the sequence (f_n) and we write

$$f_n \rightrightarrows f \quad \text{or} \quad \underset{n \to \infty}{\text{unif}\lim} f_n = f.$$

Your intuition about uniform convergence is crucial. Draw a tube V of vertical radius ϵ around the graph of f. For n large, the graph of f_n must lie wholly in V. See Figure 87. Absorb this picture!

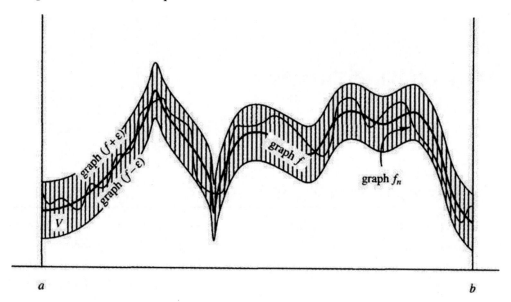

Figure 87 The graph of f_n is contained in the ϵ-tube around the graph of
f.

It is clear that uniform convergence implies pointwise convergence. The difference between the two definitions is apparent in the following standard example.

Example Define $f_n : (0, 1) \to \mathbb{R}$ by $f_n(x) = x^n$. For each $x \in (0, 1)$ it is clear that $f_n(x) \to 0$. The functions converge pointwise to the zero function as $n \to \infty$. They do not converge uniformly. For if $\epsilon = 1/10$ then the point $x_n = \sqrt[n]{1/2}$ is sent by f_n to $1/2$ and thus not all points x satisfy (1) when n is large. The graph of f_n fails to lie in the ϵ-tube V. See Figure 88.

The lesson to draw is that pointwise convergence of a sequence of functions is frequently too weak a concept. Gravitating toward uniform convergence we ask the natural question:

Which properties of functions are
preserved under uniform convergence?

The answers are found in Theorem 1, Exercise 4, Theorem 6, and Theorem 9. Uniform limits preserve continuity, uniform continuity, integrability, and – with an additional hypothesis – differentiability.

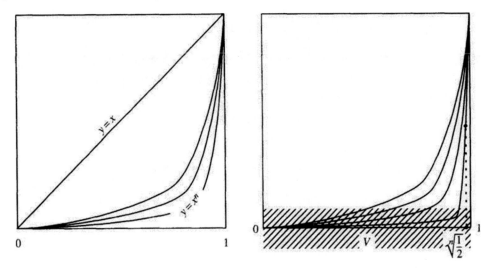

Figure 88 Non-uniform, pointwise convergence

1 Theorem *If $f_n \rightrightarrows f$ and each f_n is continuous at x_0 then f is continuous at x_0.*

In other words,

The uniform limit of continuous functions is continuous.

Proof For simplicity, assume that the functions have domain $[a, b]$ and target $\mathbb{R}$. (See also Section 8 and Exercise 2.) Let $\epsilon > 0$ and $x_0 \in [a, b]$ be given. There is an N such that for all $n \geq N$ and all $x \in [a, b]$ we have

$$|f_n(x) - f(x)| < \frac{\epsilon}{3}.$$

The function f_N is continuous at x_0 and so there is a $\delta > 0$ such that $|x - x_0| < \delta$ implies

$$|f_N(x) - f_N(x_0)| < \frac{\epsilon}{3}.$$

Thus, if $|x - x_0| < \delta$ then

$$
\begin{aligned}
|f(x) - f(x_0)| &\leq |f(x) - f_N(x)| + |f_N(x) - f_N(x_0)| + |f_N(x_0) - f(x_0)| \\
&\leq \frac{\epsilon}{3} + \frac{\epsilon}{3} + \frac{\epsilon}{3} = \epsilon,
\end{aligned}
$$

which completes the proof that f is continuous at x_0. $\qquad\qquad\square$

Without uniform convergence the theorem fails. For example, we can define $f_n : [0, 1] \to \mathbb{R}$ as before, $f_n(x) = x^n$. Then $f_n(x)$ converges pointwise to the function

$$f(x) = \begin{cases} 0 & \text{if } 0 \le x < 1 \\ 1 & \text{if } x = 1. \end{cases}$$

The function f is not continuous and the convergence is not uniform. What about the converse? If the limit and the functions are continuous, does pointwise convergence imply uniform convergence? The answer is "no," as is shown by x^n on $(0, 1)$. But what if the functions have a compact domain of definition, $[a, b]$? The answer is still "no."

Example John Kelley refers to this as the **growing steeple**,

$$f_n(x) = \begin{cases} n^2 x & \text{if } 0 \le x \le \dfrac{1}{n} \\[2mm] 2n - n^2 x & \text{if } \dfrac{1}{n} \le x \le \dfrac{2}{n} \\[2mm] 0 & \text{if } \dfrac{2}{n} \le x \le 1. \end{cases}$$

See Figure 89.

Then $\lim\limits_{n\to\infty} f_n(x) = 0$ for each x, and f_n converges pointwise to the function $f = 0$. Even if the functions have compact domain of definition, and are uniformly bounded and uniformly continuous, pointwise convergence does not imply uniform convergence. For an example, just multiply the growing steeple functions by $1/n$.

The *natural* way to view uniform convergence is in a function space. Let $C_b = C_b([a, b], \mathbb{R})$ denote the set of all bounded functions $[a, b] \to \mathbb{R}$. The elements of C_b are functions f, g, etc. Each is bounded. Define the **sup norm** on C_b as

$$\|f\| \;=\; \sup\{|f(x)| : x \in [a, b]\}.$$

The sup norm satisfies the norm axioms discussed in Chapter 1, page 28.

$$\|f\| \ge 0 \text{ and } \|f\| = 0 \text{ if and only if } f = 0$$
$$\|cf\| = |c|\|f\|$$
$$\|f + g\| \le \|f\| + \|g\|.$$

As we observed in Chapter 2, any norm defines a metric. In the case at hand,

$$d(f, g) \;=\; \sup\{|f(x) - g(x)| : x \in [a, b]\}$$

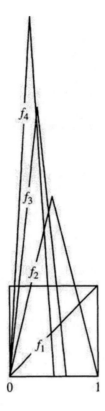

Figure 89 The sequence of functions converges pointwise to the zero function, but not uniformly.

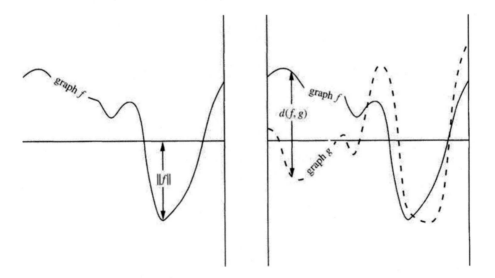

Figure 90 The sup norm of f and the sup distance between f and g

is the corresponding metric on C_b. See Figure 90. To distinguish the norm $\|f\| = \sup|f(x)|$ from other norms on C_b we sometimes write $\|f\|_{\sup}$ for the sup norm.

The thing to remember is that C_b is a metric space whose *elements* are functions. Ponder this.

2 Theorem *Convergence with respect to the sup metric d is equivalent to uniform convergence.*

Proof If $d(f_n, f) \to 0$ then $\sup\{|f_nx - fx| : x \in [a,b]\} \to 0$, so $f_n \rightrightarrows f$, and conversely.   □

3 Theorem *C_b is a complete metric space.*

Proof Let (f_n) be a Cauchy sequence in C_b. For each individual $x_0 \in [a,b]$ the values $f_n(x_0)$ form a Cauchy sequence in $\mathbb{R}$ since

$$|f_n(x_0) - f_m(x_0)| \leq \sup\{|f_n(x) - f_m(x)| : x \in [a,b]\} = d(f_n, f_m).$$

Thus, for each $x \in [a,b]$,

$$\lim_{n \to \infty} f_n(x)$$

exists. Define this limit to be $f(x)$. It is clear that f_n converges pointwise to f. In fact, the convergence is uniform. For let $\epsilon > 0$ be given. Since (f_n) is a Cauchy sequence with respect to d, there exists N such that $m, n \geq N$ imply

$$d(f_n, f_m) < \frac{\epsilon}{2}.$$

Also, since f_n converges pointwise to f, for each $x \in [a,b]$ there exists an $m = m(x) \geq N$ such that

$$|f_m(x) - f(x)| < \frac{\epsilon}{2}.$$

If $n \geq N$ and $x \in [a,b]$ then

$$
\begin{aligned}
|f_n(x) - f(x)| &\leq |f_n(x) - f_{m(x)}(x)| + |f_{m(x)}(x) - f(x)| \\
&< \frac{\epsilon}{2} + \frac{\epsilon}{2} = \epsilon.
\end{aligned}
$$

Hence $f_n \rightrightarrows f$. The function f is bounded. For f_N is bounded and for all x we have $|f_N(x) - f(x)| < \epsilon$. Thus $f \in C_b$. By Theorem 2, uniform convergence implies d-convergence, $d(f_n, f) \to 0$, and the Cauchy sequence (f_n) converges to a limit in the metric space C_b.   □

The preceding proof is subtle. The uniform inequality $d(f_n, f) < \epsilon$ is derived by nonuniform means – for each x we make a separate estimate using an $m(x)$ depending nonuniformly on x. It is a case of the ends justifying the means.

Let $C^0 = C^0([a, b], \mathbb{R})$ denote the set of continuous functions $[a, b] \to \mathbb{R}$. Each $f \in C^0$ belongs to C_b since a continuous function defined on a compact domain is bounded. That is, $C^0 \subset C_b$.

4 Corollary C^0 *is a closed subset of* C_b. *It is a complete metric space.*

Proof Theorem 1 implies that a limit in C_b of a sequence of functions in C^0 lies in C^0. That is, C^0 is closed in C_b. A closed subset of a complete space is complete. $\square$

Just as it is reasonable to discuss the convergence of a sequence of functions we can also discuss the convergence of a series of functions $\sum f_k$. Merely consider the nth partial sum

$$F_n(x) = \sum_{k=0}^{n} f_k(x).$$

It is a function. If the sequence of functions (F_n) converges to a limit function F then the series converges, and we write

$$F(x) = \sum_{k=0}^{\infty} f_k(x).$$

If the sequence of partial sums converges uniformly then we say the series **converges uniformly**. If the series of absolute values $\sum |f_k(x)|$ converges then the series $\sum f_k$ converges **absolutely**.

5 Weierstrass M-test *If* $\sum M_k$ *is a convergent series of constants and if* $f_k \in C_b$ *satisfies* $\|f_k\| \leq M_k$ *for all* k *then* $\sum f_k$ *converges uniformly and absolutely.*

Proof If $n > m$ then the partial sums of the series of absolute values telescope as

$$d(F_n, F_m) \leq d(F_n, F_{n-1}) + \cdots + d(F_{m+1}, F_m)$$
$$= \sum_{k=m+1}^{n} \|f_k\| \leq \sum_{k=m+1}^{n} M_k.$$

Since $\sum M_k$ converges, the last sum is $< \epsilon$ when m, n are large. Thus (F_n) is Cauchy in C_b, and by Theorem 3 it converges uniformly. $\square$

Next we ask how integrals and derivatives behave with respect to uniform convergence. Integrals behave better than derivatives.

6 Theorem *The uniform limit of Riemann integrable functions is Riemann integrable, and the limit of the integrals is the integral of the limit,*

$$\lim_{n \to \infty} \int_a^b f_n(x)\, dx \;=\; \int_a^b \operatorname{unif} \lim_{n \to \infty} f_n(x)\, dx.$$

In other words, $\mathcal{R}$, the set of Riemann integrable functions defined on $[a, b]$, is a closed subset of C_b and the integral functional $f \mapsto \int_a^b f(x)\, dx$ is a continuous map from $\mathcal{R}$ to $\mathbb{R}$. This extends the regularity hierarchy to

$$C_b \supset \mathcal{R} \supset C^0 \supset C^1 \supset \cdots \supset C^\infty \supset C^\omega.$$

Theorem 6 gives the simplest condition under which the operations of taking limits and integrals commute.

Proof Let $f_n \in \mathcal{R}$ be given and assume that $f_n \rightrightarrows f$ as $n \to \infty$. By the Riemann-Lebesgue Theorem, f_n is bounded and there is a zero set Z_n such that f_n is continuous at each $x \in [a, b] \setminus Z_n$. Theorem 1 implies that f is continuous at each $x \in [a, b] \setminus \bigcup Z_n$, while Theorem 3 implies that f is bounded. Since $\bigcup Z_n$ is a zero set, the Riemann-Lebesgue Theorem implies that $f \in \mathcal{R}$. Finally

$$\left| \int_a^b f(x)\, dx - \int_a^b f_n(x)\, dx \right| = \left| \int_a^b f(x) - f_n(x)\, dx \right|$$

$$\leq \int_a^b |f(x) - f_n(x)|\, dx \;\leq\; d(f, f_n)(b - a) \to 0$$

as $n \to \infty$. Hence the integral of the limit is the limit of the integrals. $\qquad\square$

7 Corollary *If $f_n \in \mathcal{R}$ and $f_n \rightrightarrows f$ then the indefinite integrals converge uniformly,*

$$\int_a^x f_n(t)\, dt \;\rightrightarrows\; \int_a^x f(t)\, dt.$$

Proof As above,

$$\left| \int_a^x f(t)dt - \int_a^x f_n(t)dt \right| \;\leq\; d(f_n, f)(x - a) \;\leq\; d(f_n, f)(b - a) \to 0$$

when $n \to \infty$. $\qquad\square$

8 Term by Term Integration Theorem *A uniformly convergent series of integrable functions $\sum f_k$ can be integrated term-by-term in the sense that*

$$\int_a^b \sum_{k=0}^\infty f_k(x)\,dx \;=\; \sum_{k=0}^\infty \int_a^b f_k(x)\,dx.$$

Proof The sequence of partial sums F_n converges uniformly to $\sum f_k$. Each F_n belongs to $\mathcal{R}$ since it is the finite sum of members of $\mathcal{R}$. According to Theorem 6,

$$\sum_{k=0}^n \int_a^b f_k(x)\,dx \;=\; \int_a^b F_n(x)\,dx \to \int_a^b \sum_{k=0}^\infty f_k(x)\,dx.$$

This shows that the series $\sum \int_a^b f_k(x)\,dx$ converges to $\int_a^b \sum f_k(x)\,dx$. $\qquad\square$

9 Theorem *The uniform limit of a sequence of differentiable functions is differentiable provided that the sequence of derivatives also converges uniformly.*

Proof We suppose that $f_n : [a, b] \to \mathbb{R}$ is differentiable for each n and that $f_n \rightrightarrows f$ as $n \to \infty$. Also we assume that $f_n' \rightrightarrows g$ for some function g. Then we show that f is differentiable and in fact $f' = g$.

We first prove the theorem with a major loss of generality – we assume that each f_n' is continuous. Then $f_n', g \in \mathcal{R}$ and we can apply the Fundamental Theorem of Calculus and Corollary 7 to write

$$f_n(x) \;=\; f_n(a) + \int_a^x f_n'(t)\,dt \quad \rightrightarrows \quad f(a) + \int_a^x g(t)\,dt.$$

Since $f_n \rightrightarrows f$ we see that $f(x) = f(a) + \int_a^x g(t)\,dt$ and, again by the Fundamental Theorem of Calculus, $f' = g$.

In the general case the proof is harder. Fix some $x \in [a, b]$ and define

$$\phi_n(t) \;=\; \begin{cases} \dfrac{f_n(t) - f_n(x)}{t - x} & \text{if } t \neq x \\ f_n'(x) & \text{if } t = x \end{cases}$$

$$\phi(t) \;=\; \begin{cases} \dfrac{f(t) - f(x)}{t - x} & \text{if } t \neq x \\ g(x) & \text{if } t = x. \end{cases}$$

Each function ϕ_n is continuous since $\phi_n(t)$ converges to $f_n'(x)$ as $t \to x$. Also it is clear that ϕ_n converges pointwise to ϕ as $n \to \infty$. We claim the convergence is uniform. For any m, n the Mean Value Theorem applied to the function $f_m - f_n$ gives

$$\phi_m(t) - \phi_n(t) = \frac{(f_m(t) - f_n(t)) - (f_m(x) - f_n(x))}{t - x} = f_m'(\theta) - f_n'(\theta)$$

for some θ between t and x. Since $f_n' \rightrightarrows g$ the difference $f_m' - f_n'$ tends uniformly to 0 as $m, n \to \infty$. Thus (ϕ_n) is Cauchy in C^0. Since C^0 is complete, ϕ_n converges uniformly to a limit function ψ, and ψ is continuous. As already remarked, the pointwise limit of ϕ_n is ϕ, and so $\psi = \phi$. Continuity of $\psi = \phi$ implies that $g(x) = f'(x)$. □

10 Theorem *A uniformly convergent series of differentiable functions can be differentiated term-by-term, provided that the derivative series converges uniformly,*

$$\left(\sum_{k=0}^{\infty} f_k(x) \right)' = \sum_{k=0}^{\infty} f_k'(x).$$

Proof Apply Theorem 9 to the sequence of partial sums. □

Note that Theorem 9 fails if we forget to assume the derivatives converge. For example, consider the sequence of functions $f_n : [-1, 1] \to \mathbb{R}$ defined by

$$f_n(x) = \sqrt{x^2 + \frac{1}{n}}.$$

See Figure 91. The functions converge uniformly to $f(x) = |x|$, a nondifferentiable function. The derivatives converge pointwise but not uniformly. Worse examples are easy to imagine. In fact, a sequence of everywhere differentiable functions can converge uniformly to a nowhere differentiable function. See Sections 4 and 7. It is one of the miracles of the complex numbers that a uniform limit of complex differentiable functions is complex differentiable, and automatically the sequence of derivatives converges uniformly to a limit. Real and complex analysis diverge radically on this point.

2 Power Series

As another application of the Weierstrass M-test we say a little more about the power series $\sum c_k x^k$. A power series is a special type of series of functions, the functions

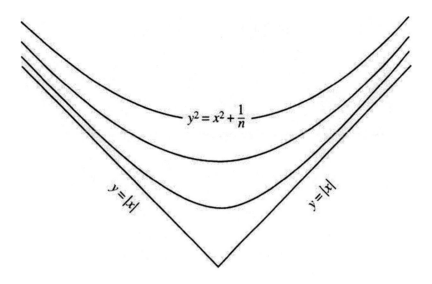

Figure 91 The uniform limit of differentiable functions need not be differentiable.

being constant multiples of powers of x. As explained in Section 3 of Chapter 3, its radius of convergence is

$$R = \frac{1}{\limsup\limits_{k \to \infty} \sqrt[k]{|c_k|}}.$$

Its interval of convergence is $(-R, R)$. If $x \in (-R, R)$, the series converges and defines a function $f(x) = \sum c_k x^k$, while if $x \notin [-R, R]$ the series diverges. More is true on compact subintervals of $(-R, R)$.

11 Theorem *If $r < R$ then the power series converges uniformly and absolutely on the interval $[-r, r]$.*

Proof Choose β with $r < \beta < R$. For all large k, $\sqrt[k]{|c_k|} < 1/\beta$ since $\beta < R$. Thus, if $|x| \le r$ then

$$|c_k x^k| \le \left(\frac{r}{\beta}\right)^k.$$

These are terms in a convergent geometric series and according to the M-test $\sum c_k x^k$ converges uniformly when $x \in [-r, r]$. $\qquad\square$

12 Theorem *A power series can be integrated and differentiated term-by-term on its interval of convergence.*

For $f(x) = \sum c_k x^k$ and $|x| < R$ this means

$$\int_0^x f(t)\, dt = \sum_{k=0}^{\infty} \frac{c_k}{k+1} x^{k+1} \quad \text{and} \quad f'(x) = \sum_{k=1}^{\infty} k c_k x^{k-1}.$$

Proof The radius of convergence of the integral series is determined by the exponential growth rate of its coefficients,

$$\limsup_{k \to \infty} \sqrt[k]{\left| \frac{c_{k-1}}{k} \right|} = \limsup_{k \to \infty} \left(|c_{k-1}|^{1/(k-1)} \right)^{(k-1)/k} \left(\frac{1}{k} \right)^{1/k}.$$

Since $(k-1)/k \to 1$ and $k^{-1/k} \to 1$ as $k \to \infty$, we see that the integral series has the same radius of convergence R as the original series. According to Theorem 8, term-by-term integration is valid when the series converges uniformly, and by Theorem 11, the integral series does converge uniformly on every closed interval $[-r, r]$ contained in $(-R, R)$.

A similar calculation for the derivative series shows that its radius of convergence too is R. Term-by-term differentiation is valid provided the series and the derivative series converge uniformly. Since the radius of convergence of the derivative series is R, the derivative series does converge uniformly on every $[-r, r] \subset (-R, R)$.   □

13 Theorem *Analytic functions are smooth, i.e., $C^\omega \subset C^\infty$.*

Proof An analytic function f is defined by a convergent power series. According to Theorem 12, the derivative of f is given by a convergent power series with the same radius of convergence, so repeated differentiation is valid, and we see that f is indeed smooth.   □

The general smooth function is not analytic, as is shown by the example

$$e(x) = \begin{cases} e^{-1/x} & \text{if } x > 0 \\ 0 & \text{if } x \le 0 \end{cases}$$

on page 149. Near $x = 0$, $e(x)$ cannot be expressed as a convergent power series.

Power series provide a clean and unambiguous way to define functions, especially trigonometric functions. The usual definitions of sine, cosine, etc. involve angles and circular arc length, and these concepts seem less fundamental than the functions being defined. To avoid circular reasoning, as it were, we declare that by definition

$$\exp x = \sum_{k=0}^{\infty} \frac{x^k}{k!} \quad \sin x = \sum_{k=0}^{\infty} \frac{(-1)^k x^{2k+1}}{(2k+1)!} \quad \cos x = \sum_{k=0}^{\infty} \frac{(-1)^k x^{2k}}{(2k)!}.$$

We then must prove that these functions have the properties we know and love from calculus. All three series are easily seen to have radius of convergence $R = \infty$. Theorem 12 justifies term-by-term differentiation, yielding the usual formulas,

$$\exp'(x) = \exp x \quad \sin'(x) = \cos x \quad \cos'(x) = -\sin x.$$

The logarithm has already been defined as the indefinite integral $\int_1^x 1/t \, dt$. We claim that if $|x| < 1$ then $\log(1 + x)$ is given as the power series

$$\log(1 + x) = \sum_{k=1}^{\infty} \frac{(-1)^{k+1}}{k} x^k.$$

To check this, we merely note that its derivative is the sum of a geometric series,

$$(\log(1 + x))' = \frac{1}{x + 1} = \frac{1}{1 - (-x)} = \sum_{k=0}^{\infty} (-x)^k = \sum_{k=0}^{\infty} (-1)^k x^k.$$

The last is a power series with radius of convergence 1. Since term by term integration of a power series inside its radius of convergence is legal, we integrate both sides of the equation and get the series expression for $\log(1 + x)$ as claimed.

The functions e^x and $1/(1+x^2)$ both have perfectly smooth graphs, but the power series for e^x has radius of convergence ∞ while that of $1/(1 + x^2)$ is 1. Why is this? What goes "wrong" at radius 1? The function $1/(1+x^2)$ doesn't blow up or have bad behavior at $x = \pm 1$ like $\log(1 + x)$ does. It's because of $\mathbb{C}$. The denominator $1 + x^2$ equals 0 when $x = \pm\sqrt{-1}$. The bad behavior in $\mathbb{C}$ wipes out the good behavior in $\mathbb{R}$.

3 Compactness and Equicontinuity in C^0

The Heine-Borel theorem states that a closed and bounded set in $\mathbb{R}^m$ is compact. On the other hand, closed and bounded sets in C^0 are rarely compact. Consider, for example, the closed unit ball

$$\mathcal{B} = \{f \in C^0([0, 1], \mathbb{R}) : \|f\| \leq 1\}.$$

To see that $\mathcal{B}$ is not compact we look again at the sequence $f_n(x) = x^n$. It lies in $\mathcal{B}$. Does it have a subsequence that converges (with respect to the metric d of C^0) to a limit in C^0? No. For if f_{n_k} converges to f in C^0 then $f(x) = \lim_{k \to \infty} f_{n_k}(x)$. Thus $f(x) = 0$ if $x < 1$ and $f(1) = 1$, but this function f does not belong to C^0. The cause of the problem is the fact that C^0 is infinite-dimensional. In fact it can be shown that if V is a vector space with a norm then its closed unit ball is compact if and only if the space is finite-dimensional. The proof is not especially hard.

Nevertheless, we want to have theorems that guarantee certain closed and bounded subsets of C^0 are compact. For we want to extract a convergent subsequence of functions from a given sequence of functions. The simple condition that lets us go ahead is equicontinuity. A sequence of functions (f_n) in C^0 is **equicontinuous** if

$$\forall \epsilon > 0 \ \exists \delta > 0 \text{ such that}$$
$$|s - t| < \delta \text{ and } n \in \mathbb{N} \quad \Rightarrow \quad |f_n(s) - f_n(t)| < \epsilon.$$

The functions f_n are *equally continuous*. The δ depends on ϵ but it does not depend on n. Roughly speaking, the graphs of all the f_n are similar. For total clarity, the concept might better be labeled uniform equicontinuity, in contrast to **pointwise equicontinuity**, which requires

$$\forall \epsilon > 0 \ \text{ and } \forall x \in [a, b] \ \exists \delta > 0 \text{ such that}$$
$$|x - t| < \delta \text{ and } n \in \mathbb{N} \quad \Rightarrow \quad |f_n(x) - f_n(t)| < \epsilon.$$

The definitions work equally well for sets of functions, not only sequences of functions. The set $\mathcal{E} \subset C^0$ is equicontinuous if

$$\forall \epsilon > 0 \ \exists \delta > 0 \text{ such that}$$
$$|s - t| < \delta \text{ and } f \in \mathcal{E} \quad \Rightarrow \quad |f(s) - f(t)| < \epsilon.$$

The crucial point is that δ does not depend on the particular $f \in \mathcal{E}$. It is valid for all $f \in \mathcal{E}$ simultaneously. To picture equicontinuity of a family $\mathcal{E}$, imagine the graphs. Their shapes are uniformly controlled. Note that any finite number of continuous functions $[a, b] \to \mathbb{R}$ forms an equicontinuous family so Figures 92 and 93 are only suggestive.

The basic theorem about equicontinuity is the

14 Arzelà-Ascoli Theorem *Every bounded equicontinuous sequence of functions in $C^0([a, b], \mathbb{R})$ has a uniformly convergent subsequence.*

Think of this as a compactness result. If (f_n) is the sequence of equicontinuous functions, the theorem amounts to asserting that the closure of the set $\{f_n : n \in \mathbb{N}\}$ is compact. Any compact metric space serves just as well as $[a, b]$, and the target space $\mathbb{R}$ can also be more general. See Section 8.

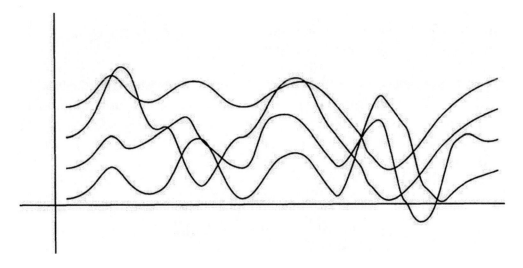

Figure 92 Equicontinuity

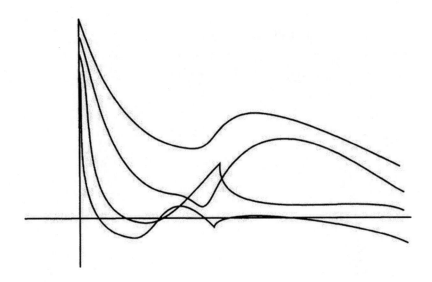

Figure 93 Nonequicontinuity

15 Lemma *If (f_k) is a subsequence of (g_n) then for each k we have $f_k = g_r$ for some $r \geq k$.*

Proof By definition of what a subsequence is, $f_k = g_{n_k}$ for some n_k such that $1 \leq n_1 < n_2 < \cdots < n_k$. Hence $r = n_k \geq k$. $\hfill\square$

Proof of the Arzelà-Ascoli Theorem $[a, b]$ has a countable dense subset $D = \{d_1, d_2, \ldots\}$. For instance we could take $D = \mathbb{Q} \cap [a, b]$. Boundedness of (f_n) means that for some constant M, all $x \in [a, b]$, and all $n \in \mathbb{N}$ we have $|f_n(x)| \leq M$. Thus $(f_n(d_1))$ is a bounded sequence of real numbers. Bolzano-Weierstrass implies that some subsequence of it converges to a limit in $\mathbb{R}$, say

$$f_{1,k}(d_1) \to y_1 \text{ as } k \to \infty.$$

The subsequence $(f_{1,k})$ evaluated at the point d_2 is also a bounded sequence in $\mathbb{R}$, and there exists a sub-subsequence $(f_{2,k})$ such that $f_{2,k}(d_2)$ converges to a limit in $\mathbb{R}$, say $f_{2,k}(d_2) \to y_2$ as $k \to \infty$. The sub-subsequence evaluated at d_1 still converges to y_1. Continuing in this way gives a nested family of subsequences $(f_{m,k})$ such that

$$(f_{m,k}) \text{ is a subsequence of } (f_{m-1,k})$$
$$j \leq m \quad \Rightarrow \quad f_{m,k}(d_j) \to y_j \text{ as } k \to \infty.$$

Now consider the diagonal subsequence $(g_m) = (f_{m,m})$. We claim that it converges uniformly to a limit, which will complete the proof. First we show it converges pointwise on D. Fix any $j \in \mathbb{N}$ and look at $m \gg j$. Lemma 15 implies that $f_{m,m} = f_{m-1,r_1}$ for some $r_1 \geq m$. Applying the lemma again, we see that $f_{m-1,r_1} = f_{m-2,r_2}$ for some $r_2 \geq r_1 \geq m$. Repetition gives

$$f_{m,m} = f_{m-1,r_1} = f_{m-2,r_2} = \cdots = f_{j,r}$$

for some $r = r_{m-j} \geq \cdots \geq r_2 \geq r_1 \geq m$. Since $r \geq m$ this gives

$$g_m(d_j) = f_{m,m}(d_j) = f_{j,r}(d_j) \to y_j$$

as $m \to \infty$.

We claim that $g_m(x)$ converges also at the other points $x \in [a, b]$ and that the convergence is uniform. It suffices to show that (g_m) is a Cauchy sequence in C^0.

Let $\epsilon > 0$ be given. Equicontinuity gives a $\delta > 0$ such that for all $s, t \in [a, b]$ we have

$$|s - t| < \delta \quad \Rightarrow \quad |g_m(s) - g_m(t)| < \frac{\epsilon}{3}.$$

Choose J large enough that every $x \in [a, b]$ lies in the δ-neighborhood of some d_j with $j \leq J$. Since D is dense and $[a, b]$ is compact, this is possible. See Exercise 19. Since $\{d_1, \ldots, d_J\}$ is a finite set and $g_m(d_j)$ converges for each d_j, there is an N such that for all $\ell, m \geq N$ and all $j \leq J$,

$$|g_m(d_j) - g_\ell(d_j)| < \frac{\epsilon}{3}.$$

If $\ell, m \geq N$ and $x \in [a, b]$, choose d_j with $|d_j - x| < \delta$ and $j \leq J$. Then

$$
\begin{aligned}
|g_m(x) - g_\ell(x)| &\leq |g_m(x) - g_m(d_j)| + |g_m(d_j) - g_\ell(d_j)| + |g_\ell(d_j) - g_\ell(x)| \\
&\leq \frac{\epsilon}{3} + \frac{\epsilon}{3} + \frac{\epsilon}{3} = \epsilon.
\end{aligned}
$$

Hence (g_m) is Cauchy in C^0, it converges in C^0, and the proof is complete. $\qquad \square$

Part of the preceding development can be isolated as the

16 Arzelà-Ascoli Propagation Theorem *Pointwise convergence of an equicontinuous sequence of functions on a dense subset of the domain **propagates** to uniform convergence on the whole domain.*

Proof This is the $\epsilon/3$ part of the proof. $\qquad \square$

The example cited over and over again in the equicontinuity world is the following.

17 Corollary *Assume that $f_n : [a, b] \to \mathbb{R}$ is a sequence of differentiable functions whose derivatives are uniformly bounded. If for one point x_0, the sequence $(f_n(x_0))$ is bounded as $n \to \infty$ then the sequence (f_n) has a subsequence that converges uniformly on the whole interval $[a, b]$.*

Proof Let M be a bound for the derivatives $|f_n'(x)|$, valid for all $n \in \mathbb{N}$ and all $x \in [a, b]$. Equicontinuity of (f_n) follows from the Mean Value Theorem:

$$|s - t| < \delta \quad \Rightarrow \quad |f_n(s) - f_n(t)| = |f_n'(\theta)| \, |s - t| \leq M\delta$$

for some θ between s and t. Thus, given $\epsilon > 0$, the choice $\delta = \epsilon/(M + 1)$ shows that (f_n) is equicontinuous.

Let C be a bound for $|f_n(x_0)|$, valid for all $n \in \mathbb{N}$. Then

$$
\begin{aligned}
|f_n(x)| &\leq |f_n(x) - f_n(x_0)| + |f_n(x_0)| \leq M|x - x_0| + C \\
&\leq M|b - a| + C
\end{aligned}
$$

shows that the sequence (f_n) is bounded in C^0. The Arzelà-Ascoli theorem then supplies the uniformly convergent subsequence. $\qquad \square$

Two other consequences of the same type are fundamental theorems in the fields of ordinary differential equations and complex variables.

(a) A sequence of solutions to a continuous ordinary differential equation in $\mathbb{R}^m$ has a subsequence that converges to a limit, and that limit is also a solution of the ODE.

(b) A sequence of complex analytic functions that converges pointwise, converges uniformly (on compact subsets of the domain of definition) and the limit is complex analytic.

Finally, we give a topological interpretation of the Arzelà-Ascoli theorem.

18 Heine-Borel Theorem in a Function Space *A subset $\mathcal{E} \subset C^0$ is compact if and only if it is closed, bounded, and equicontinuous.*

Proof Assume that $\mathcal{E}$ is compact. By Theorem 2.65, it is closed and totally bounded. This means that given $\epsilon > 0$ there is a finite covering of $\mathcal{E}$ by neighborhoods in C^0 having radius $\epsilon/3$, say $\mathcal{N}_{\epsilon/3}(f_k)$, with $k = 1, \ldots, n$. Each f_k is uniformly continuous so there is a $\delta > 0$ such that

$$|s - t| < \delta \quad \Rightarrow \quad |f_k(s) - f_k(t)| < \frac{\epsilon}{3}.$$

If $f \in \mathcal{E}$ then for some k we have $f \in \mathcal{N}_{\epsilon/3}(f_k)$, and $|s - t| < \delta$ implies

$$\begin{aligned} |f(s) - f(t)| &\leq |f(s) - f_k(s)| + |f_k(s) - f_k(t)| + |f_k(t) - f(t)| \\ &< \frac{\epsilon}{3} + \frac{\epsilon}{3} + \frac{\epsilon}{3} = \epsilon \end{aligned}$$

Thus $\mathcal{E}$ is equicontinuous.

Conversely, assume that $\mathcal{E}$ is closed, bounded, and equicontinuous. If (f_n) is a sequence in $\mathcal{E}$ then by the Arzelà-Ascoli theorem, some subsequence (f_{n_k}) converges uniformly to a limit. The limit lies in $\mathcal{E}$ since $\mathcal{E}$ is closed. Thus $\mathcal{E}$ is compact. $\square$

4 Uniform Approximation in C^0

Given a continuous but nondifferentiable function f, we often want to make it smoother by a small perturbation. We want to approximate f in C^0 by a smooth function g. The ultimately smooth function is a polynomial, and the first thing we prove is a polynomial approximation result.

19 Weierstrass Approximation Theorem *The set of polynomials is dense in $C^0([a, b], \mathbb{R})$.*

Density means that for each $f \in C^0$ and each $\epsilon > 0$ there is a polynomial function $p(x)$ such that for all $x \in [a, b]$,

$$|f(x) - p(x)| < \epsilon.$$

There are several proofs of this theorem, and although they appear quite different from each other, they share a common thread: The approximating function is built from f by sampling the values of f and recombining them in some clever way. It is no loss of generality to assume that the interval $[a, b]$ is $[0, 1]$. We do so.

Proof #1 For each $n \in \mathbb{N}$, consider the sum

$$p_n(x) = \sum_{k=0}^{n} \binom{n}{k} c_k x^k (1 - x)^{n-k},$$

where $c_k = f(k/n)$ and $\binom{n}{k}$ is the binomial coefficient $n!/k!(n-k)!$. Clearly p_n is a polynomial. It is called a **Bernstein polynomial**. We claim that the nth Bernstein polynomial converges uniformly to f as $n \to \infty$. The proof relies on two formulas about how the functions

$$r_k(x) = \binom{n}{k} x^k (1 - x)^{n-k}$$

whose graphs are shown in Figure 94 behave. They are

(2)
$$\sum_{k=0}^{n} r_k(x) = 1$$

(3)
$$\sum_{k=0}^{n} (k - nx)^2 r_k(x) = nx(1 - x).$$

In terms of the functions r_k we write

$$p_n(x) = \sum_{k=0}^{n} c_k r_k(x) \qquad f(x) = \sum_{k=0}^{n} f(x) r_k(x).$$

Then we divide the sum $p_n - f = \sum (c_k - f) r_k$ into the terms where k/n is near x, and other terms where k/n is far from x. More precisely, given $\epsilon > 0$ we use uniform continuity of f on $[0, 1]$ to find $\delta > 0$ such that $|t - s| < \delta$ implies $|f(t) - f(s)| < \epsilon/2$. Then we set

$$K_1 = \{k \in \{0, \ldots, n\} : \left|\frac{k}{n} - x\right| < \delta\} \text{ and } K_2 = \{0, \ldots, n\} \setminus K_1.$$

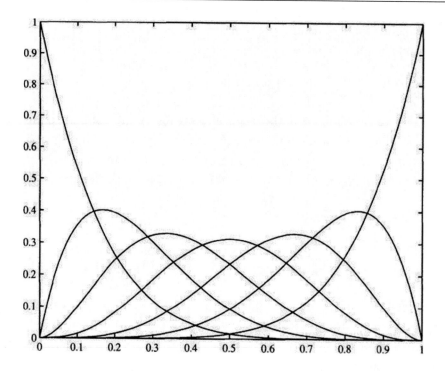

Figure 94 The seven basic Bernstein polynomials of degree 6,
$\binom{6}{k} x^k (1-x)^{6-k}, k = 0, \ldots, 6$

This gives

$$
\begin{aligned}
|p_n(x) - f(x)| &\leq \sum_{k=0}^{n} |c_k - f(x)| r_k(x) \\
&= \sum_{k \in K_1} |c_k - f(x)| r_k(x) + \sum_{k \in K_2} |c_k - f(x)| r_k(x).
\end{aligned}
$$

The factors $|c_k - f(x)|$ in the first sum are less than $\epsilon/2$ since $c_k = f(k/n)$ and k/n differs from x by less than δ. Since the sum of all the terms r_k is 1 and the terms are nonnegative, the first sum is less than $\epsilon/2$. To estimate the second sum, use (3) to write

$$
\begin{aligned}
nx(1-x) &= \sum_{k=0}^{n} (k - nx)^2 r_k(x) \geq \sum_{k \in K_2} (k - nx)^2 r_k(x) \\
&\geq \sum_{k \in K_2} (n\delta)^2 r_k(x),
\end{aligned}
$$

since $k \in K_2$ implies that $(k - nx)^2 \geq (n\delta)^2$. This implies that

$$\sum_{k \in K_2} r_k(x) \leq \frac{nx(1-x)}{(n\delta)^2} \leq \frac{1}{4n\delta^2}$$

since $\max x(1-x) = 1/4$ as x varies in $[0,1]$. The factors $|c_k - f(x)|$ in the second sum are at most $2M$ where $M = \|f\|$. Thus the second sum is

$$\sum_{k \in K_2} |c_k - f(x)| r_k(x) \leq \frac{M}{2n\delta^2} \leq \frac{\epsilon}{2}$$

when n is large, completing the proof that $|p_n(x) - f(x)| < \epsilon$ when n is large.

It remains to check the identities (2) and (3). The binomial coefficients satisfy

$$(4) \qquad (x+y)^n = \sum_{k=0}^{n} \binom{n}{k} x^k y^{n-k},$$

which becomes (2) if we set $y = 1 - x$. On the other hand, if we fix y and differentiate (4) with respect to x once, and then again, we get

$$(5) \qquad n(x+y)^{n-1} = \sum_{k=0}^{n} \binom{n}{k} k x^{k-1} y^{n-k},$$

$$(6) \qquad n(n-1)(x+y)^{n-2} = \sum_{k=0}^{n} \binom{n}{k} k(k-1) x^{k-2} y^{n-k}.$$

Note that the bottom term in (5) and the bottom two terms in (6) are 0. Multiplying (5) by x and (6) by x^2 and then setting $y = 1 - x$ in both equations gives

$$(7) \qquad nx = \sum_{k=0}^{n} \binom{n}{k} k x^k (1-x)^{n-k} = \sum_{k=0}^{n} k r_k(x),$$

$$(8) \qquad n(n-1)x^2 = \sum_{k=0}^{n} \binom{n}{k} k(k-1) x^k (1-x)^{n-k} = \sum_{k=0}^{n} k(k-1) r_k(x).$$

The last sum is $\sum k^2 r_k(x) - \sum k r_k(x)$. Hence (7) and (8) become

$$(9) \qquad \sum_{k=0}^{n} k^2 r_k(x) = n(n-1)x^2 + \sum_{k=0}^{n} k r_k(x) = n(n-1)x^2 + nx.$$

Using (2), (7), and (9) we get

$$\sum_{k=0}^{n}(k - nx)^2 r_k(x)$$

$$= \sum_{k=0}^{n} k^2 r_k(x) - 2nx \sum_{k=0}^{n} kr_k(x) + (nx)^2 \sum_{k=0}^{n} r_k(x)$$

$$= n(n-1)x^2 + nx - 2(nx)^2 + (nx)^2$$

$$= -nx^2 + nx = nx(1 - x),$$

as claimed in (3). $\square$

Proof #2 Let $f \in C^0([0,1], \mathbb{R})$ be given and let $g(x) = f(x) - (mx + b)$ where

$$m = \frac{f(1) - f(0)}{1} \quad \text{and} \quad b = f(0).$$

Then $g \in C^0$ and $g(0) = 0 = g(1)$. If we can approximate g arbitrarily well by polynomials, then the same is true of f since $mx + b$ is a polynomial. In other words it is no loss of generality to assume that $f(0) = f(1) = 0$ in the first place. We do so. Also, we extend f to all of $\mathbb{R}$ by defining $f(x) = 0$ for all $x \in \mathbb{R} \setminus [0,1]$. Then we consider a function

$$\beta_n(t) = b_n(1 - t^2)^n \quad -1 \le t \le 1,$$

where the constant b_n is chosen so that $\int_{-1}^{1} \beta_n(t)\,dt = 1$. As shown in Figure 95, β_n is a kind of polynomial bump function. For $0 \le x \le 1$, set

$$P_n(x) = \int_{-1}^{1} f(x + t)\beta_n(t)\,dt.$$

This is a weighted average of the values of f using the weight function β_n. We claim that P_n is a polynomial and $P_n(x) \rightrightarrows f(x)$ as $n \to \infty$.

To check that P_n is a polynomial we use a change of variables, $u = x + t$. Then, for $0 \le u \le 1$ we have

$$P_n(x) = \int_{x-1}^{x+1} f(u)\beta_n(u - x)\,du = \int_{0}^{1} f(u)\beta_n(u - x)\,du$$

since $f = 0$ outside $[0,1]$. The function $\beta_n(u-x) = b_n(1-(u-x)^2)^n$ is a polynomial in x whose coefficients are polynomials in u. The powers of x pull out past the integral and we are left with these powers of x multiplied by numbers, namely, the integrals of the polynomials in u times $f(u)$. In other words, by merely inspecting the last formula, it becomes clear that $P_n(x)$ is a polynomial in x.

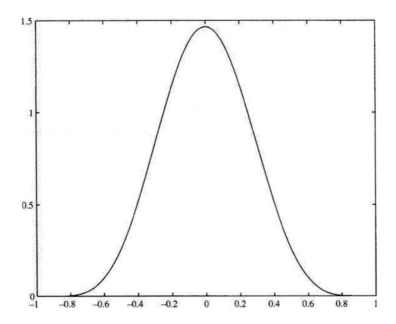

Figure 95 The graph of the function $\beta_6(t) = 1.467(1 - t^2)^6$

To check that $P_n \rightrightarrows f$ as $n \to \infty$, we need to estimate $\beta_n(t)$. We claim that if $\delta > 0$ then

(10) $$\beta_n(t) \rightrightarrows 0 \text{ as } n \to \infty \text{ and } \delta \leq |t| \leq 1.$$

This is "clear" from Figure 95. Proceeding more rigorously and using the definition of β_n as $\beta_n(t) = b_n(1 - t^2)^n$, we have

$$1 = \int_{-1}^{1} \beta_n(t) \, dt \geq \int_{-1/\sqrt{n}}^{1/\sqrt{n}} b_n(1 - t^2)^n \, dt \geq b_n \frac{2}{\sqrt{n}} (1 - \frac{1}{n})^n.$$

Since $1/e = \lim_{n \to \infty} (1 - 1/n)^n$, we see that for some constant c and all n,

$$b_n \leq c\sqrt{n}.$$

See also Exercise 31. Hence, if $\delta \leq |t| \leq 1$ then

$$\beta_n(t) = b_n(1 - t^2)^n \leq c\sqrt{n}(1 - \delta^2)^n \to 0 \text{ as } n \to \infty,$$

due to the fact that $\sqrt{n}$ tends to ∞ more slowly than $(1 - \delta^2)^{-n}$ does as $n \to \infty$. This proves (10).

From (10) we deduce that $P_n \rightrightarrows f$ as follows. Let $\epsilon > 0$ be given. Uniform continuity of f gives $\delta > 0$ such that $|t| < \delta$ implies $|f(x+t) - f(x)| < \epsilon/2$. Since β_n has integral 1 on $[-1, 1]$ we have

$$
\begin{aligned}
|P_n(x) - f(x)| &= \left| \int_{-1}^{1} (f(x+t) - f(x))\beta_n(t)\, dt \right| \\
&\leq \int_{-1}^{1} |f(x+t) - f(x)|\beta_n(t)\, dt \\
&= \int_{|t|<\delta} |f(x+t) - f(x)|\beta_n(t)\, dt + \int_{|t|\geq\delta} |f(x+t) - f(x)|\beta_n(t)\, dt.
\end{aligned}
$$

The first integral is less than $\epsilon/2$, while the second is at most $2M \int_{|t|\geq\delta} \beta_n(t)\, dt$. By (10), the second integral is less than $\epsilon/2$ when n is large. Thus $P_n \rightrightarrows f$ as claimed. $\square$

Next we see how to extend this result to functions defined on a compact metric space M instead of merely on an interval. A subset $\mathcal{A}$ of $C^0 M = C^0(M, \mathbb{R})$ is a **function algebra** if it is closed under addition, scalar multiplication, and function multiplication. That is, if $f, g \in \mathcal{A}$ and c is a constant then $f + g, cf$, and $f \cdot g$ belong to $\mathcal{A}$. For example, the set of polynomials is a function algebra. The function algebra **vanishes at a point** p if $f(p) = 0$ for all $f \in \mathcal{A}$. For example, the function algebra of all polynomials with zero constant term vanishes at $x = 0$. The function algebra **separates points** if for each pair of distinct points $p_1, p_2 \in M$ there is a function $f \in \mathcal{A}$ such that $f(p_1) \neq f(p_2)$. For example, the function algebra of all trigonometric polynomials separates points of $[0, 2\pi)$ and vanishes nowhere.

20 Stone-Weierstrass Theorem *If M is a compact metric space and $\mathcal{A}$ is a function algebra in $C^0 M$ that vanishes nowhere and separates points then $\mathcal{A}$ is dense in $C^0 M$.*

Although the Weierstrass Approximation Theorem is a special case of the Stone-Weierstrass Theorem, the proof of the latter does not stand on its own; it depends crucially on the former. We also need two lemmas.

21 Lemma *If $\mathcal{A}$ vanishes nowhere and separates points then there exists $f \in \mathcal{A}$ with specified values at any pair of distinct points.*

Proof Given distinct points $p_1, p_2 \in M$, and given constants c_1, c_2, we seek a function $f \in \mathcal{A}$ such that $f(p_1) = c_1$ and $f(p_2) = c_2$.

Since $\mathcal{A}$ vanishes nowhere there exist $g_1, g_2 \in \mathcal{A}$ such that $g_1(p_1) \neq 0 \neq g_2(p_2)$. Then $g = g_1^2 + g_2^2$ belongs to $\mathcal{A}$ and vanishes at neither p_1 nor p_2. Since $\mathcal{A}$ separates points there exists $h \in \mathcal{A}$ with different values at p_1 and p_2. Consider the matrix

$$H = \begin{bmatrix} a & ab \\ c & cd \end{bmatrix} = \begin{bmatrix} g(p_1) & g(p_1)h(p_1) \\ g(p_2) & g(p_2)h(p_2) \end{bmatrix}.$$

By construction $a, c \neq 0$ and $b \neq d$. Hence $\det H = acd - abc = ac(d - b) \neq 0$, H has rank 2, and the linear equations

$$\begin{aligned} a\xi + ab\eta &= c_1 \\ c\xi + cd\eta &= c_2 \end{aligned}$$

have a solution (ξ, η). Then $f = \xi g + \eta g h$ belongs to $\mathcal{A}$ and $f(p_1) = c_1, f(p_2) = c_2$. $\square$

22 Lemma *The closure of a function algebra in $C^0 M$ is a function algebra.*

Proof Clear enough. $\qquad\qquad\qquad\qquad\qquad\qquad\qquad\qquad\qquad\qquad\qquad\qquad\qquad$ $\square$

Proof of the Stone-Weierstrass Theorem Let $\mathcal{A}$ be a function algebra in $C^0 M$ that vanishes nowhere and separates points. We must show that $\mathcal{A}$ is dense in $C^0 M$. Given $F \in C^0 M$ and $\epsilon > 0$, we must find $G \in \mathcal{A}$ such that for all $x \in M$ we have

(11) $$F(x) - \epsilon \; < \; G(x) \; < \; F(x) + \epsilon.$$

First we observe that

(12) $$f \in \overline{\mathcal{A}} \quad \Rightarrow \quad |f| \in \overline{\mathcal{A}}$$

where $\overline{\mathcal{A}}$ denotes the closure of $\mathcal{A}$ in $C^0 M$. Let $\epsilon > 0$ be given. According to the Weierstrass Approximation Theorem, there exists a polynomial $p(y)$ such that

(13) $$\sup\{|p(y) - |y|| : |y| \le \|f\|\} < \frac{\epsilon}{2}$$

After all, $|y|$ is a continuous function defined on the interval $[-\|f\|, \|f\|]$. The constant term of $p(y)$ is at most $\epsilon/2$ since $|p(0) - |0|| < \epsilon/2$. Let $q(y) = p(y) - p(0)$. Then $q(y)$ is a polynomial with zero constant term and (13) becomes

(14) $$|q(y) - |y|| < \epsilon$$

for all $y \in [-\|f\|, \|f\|]$. Write $q(y) = a_1 y + a_2 y^2 + \cdots + a_n y^n$ and

$$g = a_1 f + a_2 f^2 + \cdots + a_n f^n.$$

(Here, f^n denotes $f \cdot f \cdots f$.) Lemma 22 states that $\overline{\mathcal{A}}$ is an algebra, so $g \in \overline{\mathcal{A}}$.[†]
Besides, if $x \in M$ and $y = f(x)$ then

$$|g(x) - |f(x)|| = |q(y) - |y|| < \epsilon.$$

Hence $|f| \in \overline{\overline{\mathcal{A}}} = \overline{\mathcal{A}}$ as claimed in (12).

Next we observe that if f, g belong to $\overline{\mathcal{A}}$, then $\max(f, g)$ and $\min(f, g)$ also belong
to $\overline{\mathcal{A}}$. For

$$\begin{aligned}
\max(f, g) &= \frac{f + g}{2} + \frac{|f - g|}{2} \\
\min(f, g) &= \frac{f + g}{2} - \frac{|f - g|}{2}.
\end{aligned}$$

Repetition shows that the maximum and minimum of any finite number of functions
in $\overline{\mathcal{A}}$ also belongs to $\overline{\mathcal{A}}$.

Now we return to (11). Let $F \in C^0 M$ and $\epsilon > 0$ be given. We are trying to find
$G \in \overline{\mathcal{A}}$ whose graph lies in the ϵ-tube around the graph of F. Fix any distinct points
$p, q \in M$. According to Lemma 21, we can find a function in $\mathcal{A}$ with specified values
at p, q, so there exists $H_{pq} \in \mathcal{A}$ that satisfies

$$H_{pq}(p) = F(p) \quad \text{and} \quad H_{pq}(q) = F(q).$$

Fix p and let q vary. Each $q \in M$ has a neighborhood U_q such that

(15) $x \in U_q \quad \Rightarrow \quad F(x) - \epsilon < H_{pq}(x).$

For $H_{pq}(x) - F(x) + \epsilon$ is a continuous function of x which is positive at $x = q$. See
Figure 96.

Compactness of M implies that finitely many of these neighborhoods U_q cover
M, say $U_{q_1}, \ldots, U_{q_n}$. Define

$$G_p(x) = \max(H_{pq_1}(x), \ldots, H_{pq_n}(x)).$$

Then $G_p \in \overline{\mathcal{A}}$ and, as shown in Figure 97, for all $x \in M$ we have

(16) $G_p(p) = F(p) \quad \text{and} \quad F(x) - \epsilon < G_p(x).$

Continuity implies that each p has a neighborhood V_p such that

(17) $x \in V_p \quad \Rightarrow \quad G_p(x) < F(x) + \epsilon.$

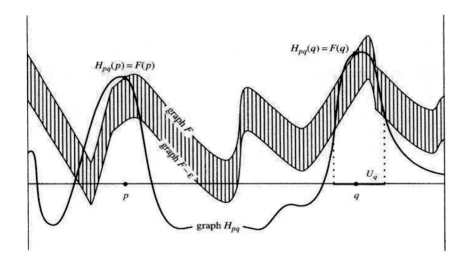

Figure 96 For all x in a neighborhood of q we have $H_{pq}(x) > F(x) - \epsilon$.

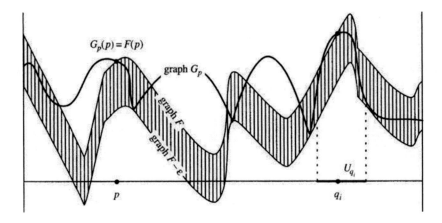

Figure 97 G_p is the maximum of $H_{pq_i}, i = 1, \ldots, n$.

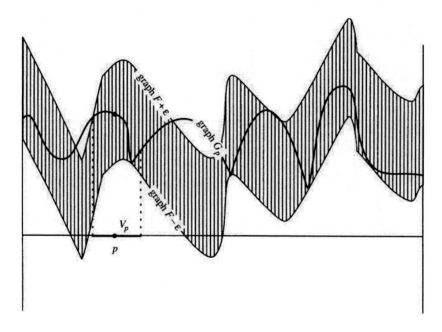

Figure 98 $G_p(p) = F(p)$ and $G_p > F - \epsilon$ everywhere.

See Figure 98.

By compactness, finitely many of these neighborhoods cover M, say $V_{p_1}, \ldots, V_{p_m}$. Set

$$G(x) = \min(G_{p_1}(x), \ldots, G_{p_m}(x)).$$

We know that $G \in \overline{\mathcal{A}}$ and (16), (17) imply (11). See Figure 99. $\quad\square$

23 Corollary *Any 2π-periodic continuous function of $x \in \mathbb{R}$ can be uniformly approximated by a **trigonometric polynomial***

$$T(x) \; = \; a_0 + \sum_{k=1}^{n} a_k \cos kx + \sum_{k=1}^{n} b_k \sin kx.$$

Proof Think of $[0, 2\pi)$ parameterizing the circle S^1 by $x \mapsto (\cos x, \sin x)$. The circle is compact, and 2π-periodic continuous functions on $\mathbb{R}$ become continuous functions on S^1. The trigonometric polynomials on S^1 form an algebra $\mathcal{T} \subset C^0 S^1$ that vanishes nowhere and separates points. The Stone-Weierstrass Theorem implies that $\mathcal{T}$ is dense in $C^0 S^1$. $\quad\square$

[†]Since a function algebra need not contain constant functions, it was important that q has no constant term. One should not expect that $g = a_0 + a_1 f + \cdots + a_n f^n$ belongs to $\overline{\mathcal{A}}$.

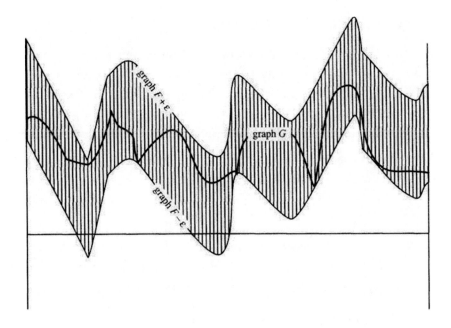

Figure 99 The graph of G lies in the ϵ-tube around the graph of F.

Here is a typical application of the Stone-Weierstrass Theorem: Consider a continuous vector field $F : \Delta \to \mathbb{R}^2$ where Δ is the closed unit disc in the plane, and suppose that we want to approximate F by a vector field that vanishes (equals zero) at most finitely often. A simple way to do so is to approximate F by a polynomial vector field G. Real polynomials in two variables are finite sums

$$P(x, y) = \sum_{i,j=0}^{n} c_{ij} x^i y^j$$

where the c_{ij} are constants. They form a function algebra $\mathcal{A}$ in $C^0(\Delta, \mathbb{R})$ that separates points and vanishes nowhere. By the Stone-Weierstrass Theorem, $\mathcal{A}$ is dense in C^0, so we can approximate the components of $F = (F_1, F_2)$ by polynomials

$$F_1 \overset{\cdot}{=} P \qquad F_2 \overset{\cdot}{=} Q.$$

(The symbol $\overset{\cdot}{=}$ indicates "almost equal.") The vector field (P, Q) then approximates F. Changing the coefficients of P by a small amount ensures that P and Q have no common polynomial factor and F vanishes at most finitely often.

5 Contractions and ODEs

Fixed-point theorems are of great use in the applications of analysis, including the basic theory of vector calculus such as the general implicit function theorem. If $f : M \to M$ and for some $p \in M$ we have $f(p) = p$ then p is a **fixed-point** of f. When must f have a fixed-point? This question has many answers, and the two most famous are given in the next two theorems.

Let M be a metric space. A **contraction** of M is a mapping $f : M \to M$ such that for some constant $k < 1$ and all $x, y \in M$ we have

$$d(fx, fy) \leq kd(x, y).$$

24 Banach Contraction Principle *Suppose that $f : M \to M$ is a contraction and the metric space M is complete. Then f has a unique fixed-point p and for any $x \in M$, the iterate[†] $f^n(x) = f \circ f \circ \cdots \circ f(x)$ converges to p as $n \to \infty$.*

Brouwer Fixed-Point Theorem *Suppose that $f : B^m \to B^m$ is continuous where B^m is the closed unit ball in $\mathbb{R}^m$. Then f has a fixed-point $p \in B^m$.*

The proof of the first result is fairly easy, the second not. See Figure 100 to picture a contraction and Section 10 of Chapter 5 for a proof of the Brouwer theorem.

Proof #1 of the Banach Contraction Principle Beautiful, simple, and dynamical! See Figure 100. Choose any $x_0 \in M$ and define $x_n = f^n(x_0)$. We claim that for all $n \in \mathbb{N}$ we have

$$(18) \qquad\qquad d(x_n, x_{n+1}) \leq k^n d(x_0, x_1).$$

This is easy:

$$\begin{aligned} d(x_n, x_{n+1}) &= d(f(x_{n-1}), f(x_n)) \leq kd(x_{n-1}, x_n) \leq k^2 d(x_{n-2}, x_{n-1}) \\ &\leq \ \ldots \ \leq k^n d(x_0, x_1). \end{aligned}$$

From this and a geometric series type of estimate, it follows that the sequence (x_n) is Cauchy. For let $\epsilon > 0$ be given. Choose N large enough that

$$(19) \qquad\qquad \frac{k^N}{1-k} d(x_0, x_1) < \epsilon.$$

[†]Note the abuse of notation. In the proof of the Stone-Weierstrass Theorem, $f^n(x)$ denotes the nth power of the real number $f(x)$, while here f^n denotes the composition of f with itself n times. Deal with it!

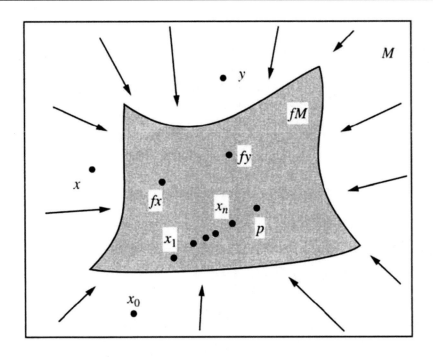

Figure 100 f contracts M toward the fixed-point p.

Note that (19) needs the hypothesis $k < 1$. If $N \leq m \leq n$ then (18) gives

$$
\begin{aligned}
d(x_m, x_n) &\leq d(x_m, x_{m+1}) + d(x_{m+1}, x_{m+2}) + \ldots + d(x_{n-1}, x_n) \\
&\leq k^m d(x_0, x_1) + k^{m+1} d(x_0, x_1) + \ldots + k^{n-1} d(x_0, x_1) \\
&= k^m (1 + k + \ldots + k^{n-m-1}) d(x_0, x_1) \\
&\leq k^N \sum_{\ell=0}^{\infty} k^\ell d(x_0, x_1) = \frac{k^N}{1-k} d(x_0, x_1) < \epsilon.
\end{aligned}
$$

Thus (x_n) is Cauchy. Since M is complete, x_n converges to some $p \in M$ as $n \to \infty$. Let $\epsilon > 0$ be given. For large n, the points x_n and x_{n+1} lie in the ϵ-neighborhood of p. Since $f(x_n) = x_{n+1}$, the map f moves x_n a distance $< 2\epsilon$, and since ϵ is arbitrarily small, continuity of f implies f does not move p at all. It is a fixed-point of f. Uniqueness of the fixed-point is immediate. After all, how can two points simultaneously stay fixed and move closer together? $\square$

Proof #2 of the Banach Contraction Principle Choose any point $x_0 \in M$ and choose r_0 so large that $f(\overline{M_{r_0}(x_0)}) \subset M_{r_0}(x_0)$. Let $B_0 = \overline{M_{r_0}(x_0)}$ and $B_n = \overline{f^n(B_{n-1})}$. The diameter of B_n is at most $k^n \operatorname{diam}(B_0)$, and this tends to 0 as $n \to \infty$.

The sets B_n nest downward as $n \to \infty$ and f sends B_n inside B_{n+1}. Since M is complete, this implies that $\bigcap B_n$ is a single point, say p, and $f(p) = p$.   □

Proof of Brouwer's Theorem in Dimension One The closed unit 1-ball is the interval $[-1, 1]$ in $\mathbb{R}$. If $f : [-1, 1] \to [-1, 1]$ is continuous then so is $g(x) = x - f(x)$. At the endpoints ± 1, we have $g(-1) \leq 0 \leq g(1)$. By the Intermediate Value Theorem, there is a point $p \in [-1, 1]$ such that $g(p) = 0$. That is, $f(p) = p$.   □

The proof in higher dimensions is harder. One proof is a consequence of the general Stokes' Theorem, and is given in Chapter 5. Another depends on algebraic topology, a third on differential topology.

Ordinary Differential Equations

The qualitative theory of ordinary differential equations (ODEs) begins with the basic existence/uniqueness theorem, Picard's Theorem. Throughout, U is an open subset of m-dimensional Euclidean space $\mathbb{R}^m$.

A **vector ODE** on U is given as m simultaneous scalar equations

$$x_1' = f_1(x_1, x_2, \ldots, x_m)$$
$$x_2' = f_2(x_1, x_2, \ldots, x_m)$$
$$\ldots$$
$$x_m' = f_m(x_1, x_2, \ldots, x_m)$$

where each f_i is a function from U to $\mathbb{R}$. One seeks m real-valued functions $x_1(t), \ldots, x_m(t)$ such that

$$\frac{dx_1(t)}{dt} = f_1(x_1(t), x_2(t), \ldots, x_m(t))$$
$$\frac{dx_2(t)}{dt} = f_2(x_1(t), x_2(t), \ldots, x_m(t))$$
$$\ldots$$
$$\frac{dx_m(t)}{dt} = f_m(x_1(t), x_2(t), \ldots, x_m(t))$$

hold identically and simultaneously. The functions $x_1(t), \ldots, x_m(t)$ are said to **solve** the ODE with **initial condition**

$$(x_1(0), x_2(0), \ldots, x_m(0)).$$

The ODE can be expressed geometrically as follows. The m real-valued functions f_i can be combined into a vector function $F(x) = (f_1(x), \ldots, f_m(x))$ where $x = (x_1, \ldots, x_m)$. Thus F is a **vector field** on U, and we seek a **trajectory** of F, that is, a curve $\gamma : (a, b) \to U$ such that $a < 0 < b$ and for all $t \in (a, b)$ we have

$$(20) \qquad\qquad \gamma'(t) = F(\gamma(t)) \quad \text{and} \quad \gamma(0) = p.$$

The components of γ are the functions $x_i(t)$ that solve the ODE and p is their initial condition. I contend that this geometric view of an ODE as a vector field is the best way to get intuition about it. See Figure 101.

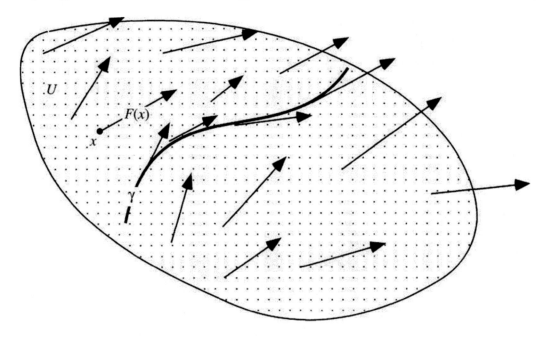

Figure 101 γ is always tangent to the vector field F.

We think of the vector field F defining at each $x \in U$ a vector $F(x)$ whose foot lies at x and to which γ must be tangent. The vector $\gamma'(t)$ is $(\gamma_1'(t), \ldots, \gamma_m'(t))$ where $\gamma_1, \ldots, \gamma_m$ are the components of γ. The trajectory $\gamma(t)$ describes how a particle travels with prescribed velocity F. At each time t, $\gamma(t)$ is the position of the particle; its velocity there is exactly the vector F at that point. Intuitively, trajectories should exist because particles do move.

The contraction principle gives a way to find trajectories of vector fields, or what is the same thing, to solve ODEs. We will assume that F satisfies a **Lipschitz**

condition – there is a constant L such that for all points $x, y \in U$ we have

$$|F(x) - F(y)| \leq L|x - y|.$$

Here, $|\ |$ refers to the Euclidean length of a vector. F, x, y are all vectors in $\mathbb{R}^m$. It follows that F is continuous. The Lipschitz condition is stronger than continuity, but still fairly mild. Any differentiable vector field with a bounded derivative is Lipschitz.

25 Picard's Theorem *Given $p \in U$ there exists an F-trajectory $\gamma(t)$ in U through p. This means that $\gamma : (a, b) \to U$ solves (20). Locally, γ is unique.*

To prove Picard's Theorem it is convenient to reexpress (20) as an integral equation; to do this we make a brief digression about vector-valued integrals. Let's recall four key facts about integrals of real-valued functions of a real variable, $y = f(x)$, $a \leq x \leq b$.

(a) $\int_a^b f(x)\, dx$ is approximated by Riemann sums $R = \sum f(t_k) \Delta x_k$.
(b) Continuous functions are integrable.
(c) If $f'(x)$ exists and is continuous then $\int_a^b f'(x)\, dx = f(b) - f(a)$.
(d) $\left| \int_a^b f(x)\, dx \right| \leq M(b - a)$ where $M = \sup |f(x)|$.

The Riemann sum R in (a) has $a = x_0 \leq \cdots \leq x_{k-1} \leq t_k \leq x_k \leq \cdots \leq x_n = b$ and all the $\Delta x_k = x_k - x_{k-1}$ are small.

Given a continuous vector-valued function of a real variable

$$f(x) = (f_1(x), \ldots, f_m(x)),$$

$a \leq x \leq b$, we define its integral componentwise as the vector of integrals

$$\int_a^b f(x)\, dx = \left(\int_a^b f_1(x)\, dx, \quad \ldots \quad, \int_a^b f_m(x)\, dx \right).$$

Corresponding to (a) - (d) are the following:

(a′) $\int_a^b f(x)\, dx$ is approximated by $R = (R_1, \ldots, R_m)$, with R_j a Riemann sum for f_j.
(b′) Continuous vector-valued functions are integrable.
(c′) If $f'(x)$ exists and is continuous, then $\int_a^b f'(x)\, dx = f(b) - f(a)$.
(d′) $\left| \int_a^b f(x)\, dx \right| \leq M(b - a)$ where $M = \sup |f(x)|$.

$(a'), (b')$, and (c') are clear enough. To check (d') we write

$$R = \sum R_j e_j = \sum_j \sum_k f_j(t_k) \Delta x_k e_j$$
$$= \sum_k \sum_j f_j(t_k) e_j \Delta x_k = \sum_k f(t_k) \Delta x_k$$

where $e_1, \ldots, e_m$ is the standard vector basis for $\mathbb{R}^m$. Thus,

$$|R| \leq \sum_k |f(t_k)| \Delta x_k \leq \sum_k M \Delta x_k = M(b-a).$$

By (a'), R approximates the integral, which implies (d'). (Note that a weaker inequality with M replaced by $\sqrt{m} M$ follows immediately from (d). This weaker inequality would suffice for most of what we do but it is inelegant.)

Now consider the following integral version of (20),

$$(21) \qquad\qquad \gamma(t) = p + \int_0^t F(\gamma(s)) \, ds.$$

A solution of (21) is by definition any continuous curve $\gamma : (a, b) \to U$ for which (21) holds identically in $t \in (a, b)$. By (b') any solution of (21) is automatically differentiable and its derivative is $F(\gamma(t))$. That is, every solution of (21) solves (20). The converse is also clear, so solving (20) is equivalent to solving (21) for a continuous function $\gamma(t)$.

Proof of Picard's Theorem Since F is continuous, there exist a compact neighborhood $N = \overline{N}_r(p) \subset U$ and a constant M such that $|F(x)| \leq M$ for all $x \in N$. Choose $\tau > 0$ such that

$$(22) \qquad\qquad \tau M \leq r \quad \text{and} \quad \tau L < 1.$$

Consider the set $\mathcal{C}$ of all continuous functions $\gamma, \sigma : [-\tau, \tau] \to N$. With respect to the metric

$$d(\gamma, \sigma) = \sup\{|\gamma(t) - \sigma(t)| : t \in [-\tau, \tau]\}$$

the set $\mathcal{C}$ is a complete metric space. Given $\gamma \in \mathcal{C}$, define a new curve $\Phi(\gamma)$ as

$$\Phi(\gamma)(t) = p + \int_0^t F(\gamma(s)) \, ds.$$

Solving (21) is the same as finding γ such that $\Phi(\gamma) = \gamma$. That is, we seek a fixed point of Φ.

We just need to show that Φ is a contraction of $\mathcal{C}$. Does Φ send $\mathcal{C}$ into itself? Given $\gamma \in \mathcal{C}$ we see that $\Phi(\gamma)(t)$ is a continuous (in fact differentiable) vector-valued function of t and that by (22),

$$|\Phi(\gamma)(t) - p| = \left| \int_0^t F(\gamma(s))\, ds \right| \leq \tau M \leq r.$$

Therefore, Φ does send $\mathcal{C}$ into itself. Φ contracts $\mathcal{C}$ because

$$
\begin{aligned}
d(\Phi(\gamma), \Phi(\sigma)) &= \sup_t \left| \int_0^t F(\gamma(s)) - F(\sigma(s))\, ds \right| \\
&\leq \tau \sup_s |F(\gamma(s)) - F(\sigma(s))| \\
&\leq \tau \sup_s L|\gamma(s) - \sigma(s)| \leq \tau L d(\gamma, \sigma)
\end{aligned}
$$

and $\tau L < 1$ by (22). Therefore Φ has a fixed-point γ, and $\Phi(\gamma) = \gamma$ implies that $\gamma(t)$ solves (21), which implies that γ is differentiable and solves (20).

Any other solution $\sigma(t)$ of (20) defined on the interval $[-\tau, \tau]$ also solves (21) and is a fixed-point of Φ, $\Phi(\sigma) = \sigma$. Since a contraction mapping has a unique fixed-point, $\gamma = \sigma$, which is what local uniqueness means. $\qquad \square$

The F-trajectories define a **flow** in the following way: To avoid the possibility that trajectories cross the boundary of U (they "escape from U") or become unbounded in finite time (they "escape to infinity") we assume that U is all of $\mathbb{R}^m$. Then trajectories can be defined for all time $t \in \mathbb{R}$. Let $\gamma(t, p)$ denote the trajectory through p. Imagine all points $p \in \mathbb{R}^m$ moving *in unison* along their trajectories as t increases. They are leaves on a river, motes in a breeze. The point $p_1 = \gamma(t_1, p)$ at which p arrives after time t_1 moves according to $\gamma(t, p_1)$. Before p arrives at p_1, however, p_1 has already gone elsewhere. This is expressed by the flow equation

$$\gamma(t, p_1) = \gamma(t + t_1, p).$$

See Figure 102.

The flow equation is true because as functions of t both sides of the equation are F-trajectories through p_1, and the F-trajectory through a point is locally unique. It is revealing to rewrite the flow equation with different notation. Setting $\varphi_t(p) = \gamma(t, p)$ gives

$$\varphi_{t+s}(p) = \varphi_t(\varphi_s(p)) \text{ for all } t, s \in \mathbb{R}.$$

φ_t is called the **t-advance map**. It specifies where each point moves after time t. See Figure 103. The flow equation states that $t \mapsto \varphi_t$ is a group homomorphism from

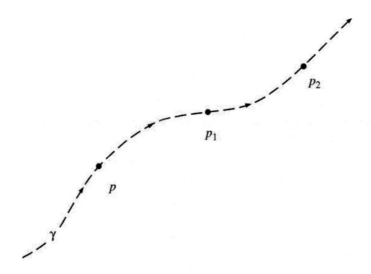

Figure 102 The time needed to flow from from p to p_2 is the sum of the times needed to flow from p to p_1 and from p_1 to p_2.

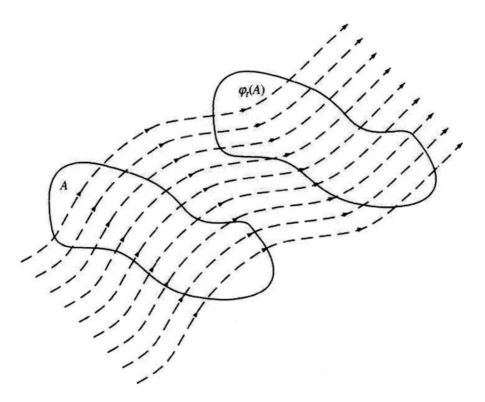

Figure 103 The t-advance map shows how a set A flows to a set $\varphi_t(A)$.

$\mathbb{R}$ into the group of motions of $\mathbb{R}^m$. In fact each φ_t is a homeomorphism of $\mathbb{R}^m$ onto itself and its inverse is φ_{-t}. For $\varphi_{-t} \circ \varphi_t = \varphi_0$ and φ_0 is the time-zero map where nothing moves at all, $\varphi_0 = $ identity map.

6* Analytic Functions

Recall from Chapter 3 that a function $f : (a, b) \to \mathbb{R}$ is **analytic** if it can be expressed locally as a power series. For each $x \in (a, b)$ there exists a convergent power series $\sum c_k h^k$ such that for all $x + h$ near x we have

$$f(x + h) = \sum_{k=0}^{\infty} c_k h^k.$$

As we have shown previously, every analytic function is smooth but not every smooth function is analytic. In this section we give a necessary and sufficient condition that a smooth function be analytic. It involves the speed with which the rth derivative grows as $r \to \infty$.

Let $f : (a, b) \to \mathbb{R}$ be smooth. The **Taylor series** for f at $x \in (a, b)$ is

$$\sum_{k=0}^{\infty} \frac{f^{(k)}(x)}{k!} h^k.$$

Let $I = [x - \sigma, \ x + \sigma]$ be a subinterval of (a, b), $\sigma > 0$, and denote by M_r the maximum of $|f^{(r)}(t)|$ for $t \in I$. The **derivative growth rate** of f on I is

$$\alpha = \limsup_{r \to \infty} \sqrt[r]{\frac{M_r}{r!}}.$$

Clearly, $\sqrt[r]{|f^{(r)}(x)|/r!} \leq \sqrt[r]{M_r/r!}$, so the radius of convergence

$$R = \frac{1}{\displaystyle\limsup_{r \to \infty} \sqrt[r]{\frac{|f^{(r)}(x)|}{r!}}}$$

of the Taylor series at x satisfies

$$\frac{1}{\alpha} \leq R.$$

In particular, if α is finite the radius of convergence of the Taylor series is positive.

26 Theorem *If $\alpha\sigma < 1$ then the Taylor series converges uniformly to f on the interval I.*

Proof Choose $\delta > 0$ such that $(\alpha + \delta)\sigma < 1$. The Taylor remainder formula from Chapter 3, applied to the $(r-1)^{\text{st}}$-order remainder, gives

$$f(x+h) - \sum_{k=0}^{r-1} \frac{f^{(k)}(x)}{k!} h^k \;=\; \frac{f^{(r)}(\theta)}{r!} h^r$$

for some θ between x and $x + h$. Thus, for r large we have

$$\left| f(x+h) - \sum_{k=0}^{r-1} \frac{f^{(k)}(x)}{k!} h^k \right| \;\leq\; \frac{M_r}{r!}\sigma^r \;=\; \left(\left(\frac{M_r}{r!} \right)^{1/r} \sigma \right)^r \;\leq\; ((\alpha+\delta)\sigma)^r.$$

Since $(\alpha + \delta)\sigma < 1$, the Taylor series converges uniformly to $f(x+h)$ on I. $\square$

27 Theorem *If f is expressed as a convergent power series $f(x+h) = \sum c_k h^k$ with radius of convergence $R > \sigma$ then f has bounded derivative growth rate on I.*

The proof of Theorem 27 uses two estimates about the growth rate of factorials. If you know Stirling's formula they are easy, but we prove them directly.

$$(23) \qquad\qquad\qquad\qquad \lim_{r\to\infty} \sqrt[r]{\frac{r^r}{r!}} = e$$

$$(24) \qquad\qquad 0 < \lambda < 1 \quad \Rightarrow \quad \limsup_{r\to\infty} \sqrt[r]{\sum_{k=r}^{\infty} \binom{k}{r} \lambda^k} < \infty.$$

Taking logarithms, applying the integral test, and ignoring terms that tend to zero as $r \to \infty$ gives

$$
\begin{aligned}
\frac{1}{r}(\log r^r - \log r!) &= \log r - \frac{1}{r}(\log r + \log(r-1) + \cdots + \log 1) \\
\sim \log r - \frac{1}{r}\int_1^r \log x \, dx &= \log r - \frac{1}{r}(x\log x - x)\Big|_1^r \\
&= 1 - \frac{1}{r},
\end{aligned}
$$

which tends to 1 as $r \to \infty$. This proves (23).

To prove (24) we write $\lambda = e^{-\mu}$ for $\mu > 0$, and reason similarly:

$$
\begin{aligned}
\sum_{k=r}^{\infty} \binom{k}{r} \lambda^k &= \sum_{k=r}^{\infty} \frac{k(k-1)(k-2)\ldots(k-r+1)}{r!} e^{-k\mu} \\
&\leq \frac{1}{r!} \sum_{k=r}^{\infty} k^r e^{-k\mu} \quad \sim \quad \frac{1}{r!} \int_r^{\infty} x^r e^{-\mu x}\, dx \\
&= \frac{-1}{r!} e^{-\mu x} \left(\frac{x^r}{\mu} + \frac{r x^{r-1}}{\mu^2} + \frac{r(r-1)x^{r-2}}{\mu^3} + \cdots + \frac{r!}{\mu^{r+1}} \right) \Bigg|_r^{\infty} \\
&\leq \frac{1}{r!} e^{-\mu r} (r+1) r^r \left(\frac{1}{\min(1,\mu)} \right)^{r+1} .
\end{aligned}
$$

According to (23) the rth root of this quantity tends to $e^{1-\mu}/\min(1,\mu)$ as $r \to \infty$, completing the proof of (24).

Proof of Theorem 27 By assumption the power series $\sum c_k h^k$ has radius of convergence R and $\sigma < R$. Since $1/R$ is the lim sup of $\sqrt[k]{|c_k|}$ as $k \to \infty$, there is a number $\lambda < 1$ such that for all large k we have $|c_k \sigma^k| \leq \lambda^k$. Differentiating the series term by term with $|h| \leq \sigma$ gives

$$
\begin{aligned}
|f^{(r)}(x+h)| &\leq \sum_{k=r}^{\infty} k(k-1)(k-2)\ldots(k-r+1)|c_k h^{k-r}| \\
&\leq \frac{r!}{\sigma^r} \sum_{k=r}^{\infty} \binom{k}{r} |c_k \sigma^k| \leq \frac{r!}{\sigma^r} \sum_{k=r}^{\infty} \binom{k}{r} \lambda^k
\end{aligned}
$$

for r large. Thus,

$$
M_r = \sup_{|h| \leq \sigma} |f^{(r)}(x+h)| \leq \frac{r!}{\sigma^r} \sum_{k=r}^{\infty} \binom{k}{r} \lambda^k .
$$

According to (24),

$$
\alpha = \limsup_{r \to \infty} \sqrt[r]{\frac{M_r}{r!}} \leq \frac{1}{\sigma} \limsup_{r \to \infty} \sqrt[r]{\sum_{k=r}^{\infty} \binom{k}{r} \lambda^k} < \infty,
$$

and f has bounded derivative growth rate on I. $\qquad \square$

From Theorems 26 and 27 we deduce the main result of this section.

28 Analyticity Theorem *A smooth function is analytic if and only if it has locally bounded derivative growth rate.*

Proof Assume that $f : (a, b) \to \mathbb{R}$ is smooth and has locally bounded derivative growth rate. Then $x \in (a, b)$ has a neighborhood N on which the derivative growth rate α is finite. Choose $\sigma > 0$ such that $I = [x - \sigma, x + \sigma] \subset N$ and $\alpha\sigma < 1$. We infer from Theorem 26 that the Taylor series for f at x converges uniformly to f on I. Hence f is analytic.

Conversely, assume that f is analytic and let $x \in (a, b)$ be given. There is a power series $\sum c_k h^k$ that converges to $f(x+h)$ for all h in some interval $(-R, R)$ with $R > 0$. Choose σ with $0 < \sigma < R$. We infer from Theorem 27 that f has bounded derivative growth rate on I. $\qquad\square$

29 Corollary *A smooth function is analytic if its derivatives are uniformly bounded.*

An example of such a function is $f(x) = \sin x$.

Proof If $|f^{(r)}(\theta)| \leq M$ for all r and θ then the derivative growth rate of f is bounded. In fact, $\alpha = 0$ and $R = \infty$. $\qquad\square$

30 Taylor's Theorem *If $f(x) = \sum c_k x^k$ and the power series has radius of convergence R then f is analytic on $(-R, R)$.*

Proof The function f is smooth, and by Theorem 27 it has bounded derivative growth rate on each compact interval $I \subset (-R, R)$. Hence it is analytic. $\qquad\square$

Taylor's Theorem states that not only can f be expanded as a convergent power series at $x = 0$, but also at any other point $x_0 \in (-R, R)$. Other proofs of Taylor's theorem rely more heavily on series manipulations and Mertens' theorem (Exercise 73 in Chapter 3).

The concept of analyticity extends immediately to complex functions. A function $f : D \to \mathbb{C}$ is **complex analytic** if D is an open subset of $\mathbb{C}$ and for each $z \in D$ there is a power series

$$\sum c_k \zeta^k$$

such that for all $z + \zeta$ near z,

$$f(z + \zeta) = \sum_{k=0}^{\infty} c_k \zeta^k.$$

The coefficients c_k are complex and so is the variable ζ. Convergence occurs on a disc of radius R. This lets us define e^z, $\log z$, $\sin z$, $\cos z$ for the complex number z

by setting

$$e^z = \sum_{k=0}^{\infty} \frac{z^k}{k!} \qquad \log(1+z) = \sum_{k=1}^{\infty} \frac{(-1)^{k+1} z^k}{k} \text{ when } |z| < 1$$

$$\sin z = \sum_{k=0}^{\infty} \frac{(-1)^k z^{2k+1}}{(2k+1)!} \qquad \cos z = \sum_{k=0}^{\infty} \frac{(-1)^k z^{2k}}{(2k)!}.$$

It is enlightening and reassuring to derive formulas such as

$$e^{i\theta} = \cos\theta + i\sin\theta$$

directly from these definitions. (Just plug in $z = i\theta$ and use the equations $i^2 = -1, i^3 = -i, i^4 = 1$, etc.) A key formula to check is $e^{z+w} = e^z e^w$. One proof involves a manipulation of product series; a second merely uses analyticity. Another formula is $\log(e^z) = z$.

There are many natural results about real analytic functions that can be proved by direct power series means; e.g., the sum, product, reciprocal, composite, and inverse function of analytic functions are analytic. Direct proofs, like those for the Analyticity Theorem above, involve major series manipulations. The use of complex variables leads to greatly simplified proofs of these real variable theorems, thanks to the following fact.

Real analyticity propagates to complex analyticity and

complex analyticity is equivalent to complex differentiability.[†]

For it is relatively easy to check that the composition, etc., of complex differentiable functions is complex differentiable.

The analyticity concept extends even beyond $\mathbb{C}$. You may already have seen such an extension when you studied the vector linear ODE

$$x' = Ax$$

in calculus. A is a given $m \times m$ matrix and the unknown solution $x = x(t)$ is a vector function of t, on which an initial condition $x(0) = x_0$ is usually imposed. A

[†]A function $f : D \to \mathbb{C}$ is **complex differentiable** or **holomorphic** if D is an open subset of $\mathbb{C}$ and for each $z \in D$, the limit of

$$\frac{\Delta f}{\Delta z} = \frac{f(z + \Delta z) - f(z)}{\Delta z}$$

exists as $\Delta z \to 0$ in $\mathbb{C}$. The limit, if it exists, is a complex number.

vector ODE is equivalent to m coupled, scalar, linear ODEs. The solution $x(t)$ can be expressed as

$$x(t) = e^{tA} x_0$$

where

$$e^{tA} = \lim_{n\to\infty} \left(I + tA + \frac{1}{2!}(tA)^2 + \cdots + \frac{1}{n!}(tA)^n \right) = \sum_{k=0}^{\infty} \frac{t^k}{k!} A^k.$$

I is the $m \times m$ identity matrix. View this series as a power series with k^{th} coefficient $t^k/k!$ and variable A. (A is a matrix variable!) The limit exists in the space of all $m \times m$ matrices, and its product with the constant vector x_0 does indeed give a vector function of t that solves the original linear ODE.

The previous series defines the exponential of a matrix as $e^A = \sum A^k/k!$. You might ask yourself – is there such a thing as the logarithm of a matrix? A function that assigns to a matrix its matrix logarithm? A power series that expresses the matrix logarithm? What about other analytic functions? Is there such a thing as the sine of a matrix? What about inverting a matrix? Is there a power series that expresses matrix inversion? Are formulas such as $\log A^2 = 2 \log A$ true? These questions are explored in nonlinear functional analysis.

A terminological point on which to insist is that the word "analytic" be defined as "locally power series expressible." In the complex case, some mathematicians define complex analyticity as complex differentiability, and although complex differentiability turns out to be equivalent to local expressibility as a complex power series, this is a very special feature of $\mathbb{C}$. In fact it is responsible for every distinction between real and complex analysis. For cross-theory consistency, then, one should use the word "analytic" to mean local power series expressible, and use "differentiable" to mean differentiable. Why confound the two ideas?

7* Nowhere Differentiable Continuous Functions

Although many continuous functions, such as $|x|$, $\sqrt[3]{x}$, and $x \sin(1/x)$ fail to be differentiable at a few points, it is quite surprising that there can exist a function which is everywhere continuous but nowhere differentiable.

31 Theorem *There exists a continuous function $f : \mathbb{R} \to \mathbb{R}$ that has a derivative at no point whatsoever.*

Proof The construction is due to Weierstrass. The letters k, m, n denote integers. Start with a **sawtooth function** $\sigma_0 : \mathbb{R} \to \mathbb{R}$ defined as

$$\sigma_0(x) = \begin{cases} x - 2n & \text{if } 2n \le x \le 2n+1 \\ 2n + 2 - x & \text{if } 2n+1 \le x \le 2n+2. \end{cases}$$

σ_0 is periodic with period 2; if $t = x + 2m$ then $\sigma_0(t) = \sigma_0(x)$. The compressed sawtooth function

$$\sigma_k(x) = \left(\frac{3}{4}\right)^k \sigma_0(4^k x)$$

has period $\pi_k = 2/4^k$. If $t = x + m\pi_k$ then $\sigma_k(t) = \sigma_k(x)$. See Figure 104.

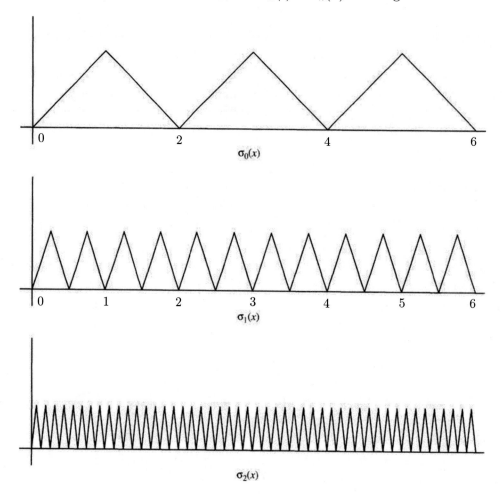

Figure 104 The graphs of the sawtooth function and two compressed sawtooth functions

According to the M-test, the series $\sum \sigma_k(x)$ converges uniformly to a limit f, and

$$f(x) = \sum_{k=0}^{\infty} \sigma_k(x)$$

is continuous. We claim that f is nowhere differentiable. Fix an arbitrary point x, and set $\delta_n = 1/2 \cdot 4^n$. We will show that

$$\frac{\Delta f}{\Delta x} = \frac{f(x \pm \delta_n) - f(x)}{\delta_n}$$

does not converge to a limit as $\delta_n \to 0$, and thus that $f'(x)$ does not exist. The quotient is

$$\frac{\Delta f}{\Delta x} = \sum_{k=0}^{\infty} \frac{\sigma_k(x \pm \delta_n) - \sigma_k(x)}{\delta_n}.$$

There are three types of terms in the series, $k > n$, $k = n$, and $k < n$. If $k > n$ then $\sigma_k(x \pm \delta_n) - \sigma_k(x) = 0$. For δ_n is an integer multiple of the period of σ_k,

$$\delta_n = \frac{1}{2 \cdot 4^n} = 4^{k-(n+1)} \cdot \frac{2}{4^k} = 4^{k-(n+1)} \cdot \pi_k.$$

Thus the infinite series expression for $\Delta f/\Delta x$ reduces to a sum of $n+1$ terms

$$\frac{\Delta f}{\Delta x} = \frac{\sigma_n(x \pm \delta_n) - \sigma_n(x)}{\delta_n} + \sum_{k=0}^{n-1} \frac{\sigma_k(x \pm \delta_k) - \sigma_k(x)}{\delta_k}.$$

The function σ_n is monotone on either $[x - \delta_n, x]$ or $[x, x + \delta_n]$, since it is monotone on intervals of length 4^{-n} and the contiguous interval $[x - \delta_n, x, x + \delta_n]$ at x is of length 4^{-n}. The slope of σ_n is $\pm 3^n$. Thus, either

$$\left| \frac{\sigma_n(x + \delta_n) - \sigma_n(x)}{\delta_n} \right| = 3^n \quad \text{or} \quad \left| \frac{\sigma_n(x - \delta_n) - \sigma_n(x)}{\delta_n} \right| = 3^n.$$

The terms with $k < n$ are crudely estimated from the slope of σ_k being $\pm 3^k$:

$$\left| \frac{\sigma_k(x \pm \delta_k) - \sigma_k(x)}{\delta_k} \right| \le 3^k.$$

Thus

$$\left| \frac{\Delta f}{\Delta x} \right| \ge 3^n - (3^{n-1} + \cdots + 1) = 3^n - \frac{3^n - 1}{3 - 1} = \frac{1}{2}(3^n + 1),$$

which tends to ∞ as $\delta_n \to 0$, so $f'(x)$ does not exist. $\square$

By simply writing down a sawtooth series as above, Weierstrass showed that there exists a nowhere differentiable continuous function. Yet more amazing is the fact that *most* continuous functions (in a reasonable sense defined below) are nowhere differentiable. If you could pick a continuous function at random, it would be nowhere differentiable.

Recall that the set $D \subset M$ is dense in M if D meets every nonempty open subset W of M, $D \cap W \neq \emptyset$. The intersection of two dense sets need not be dense; it can be empty, as is the case with $\mathbb{Q}$ and $\mathbb{Q}^c$ in $\mathbb{R}$. On the other hand if U, V are open-dense sets in M then $U \cap V$ is open-dense in M. For if W is any nonempty open subset of M then $U \cap W$ is a nonempty open subset of M, and by denseness of V, we see that V meets $U \cap W$; i.e., $U \cap V \cap W$ is nonempty and $U \cap V$ meets W.

Moral Open dense sets do a good job of being dense.

The countable intersection $G = \bigcap G_n$ of open-dense sets is called a **thick** (or **residual**[†]) subset of M, due to the following result, which we will apply in the complete metric space $C^0([a,b], \mathbb{R})$. Extending our vocabulary in a natural way we say that the complement of a thick set is **thin** (or **meager**). A subset H of M is thin if and only if it is a countable union of nowhere dense closed sets, $H = \bigcup H_n$. Clearly, thickness and thinness are topological properties. A thin set is the topological analog of a zero set (a set whose outer measure is zero).

32 Baire's Theorem *Every thick subset of a complete metric space M is dense in M. A nonempty, complete metric space is not thin. That is, if M is the union of countably many closed sets then at least one has nonempty interior.*

If all points in a thick subset of M satisfy some condition then the condition is said to be **generic**. We also say that "most" points of M obey the condition. As a consequence of Baire's theorem and the Weierstrass Approximation Theorem we will prove

33 Theorem *The generic $f \in C^0 = C^0([a,b], \mathbb{R})$ is differentiable at no point of $[a,b]$, nor does it even have a left or right derivative at any $x \in [a,b]$, nor is it monotone on any subinterval of $[a,b]$.*

Using Lebesgue's monotone differentiation theorem from Chapter 6 (monotonicity implies differentiability almost everywhere), one can see that the second assertion follows from the first, but below we give a direct proof.

[†] "Residual" is an unfortunate choice of words. It connotes smallness, when it should connote just the opposite.

Before getting into the proofs of Baire's theorem and Theorem 33, we further discuss thickness, thinness, and genericity. The empty set is always thin and the full space M is always thick in itself. A single open-dense subset is thick and a single closed nowhere dense subset is thin. $\mathbb{R} \setminus \mathbb{Z}$ is a thick subset of $\mathbb{R}$ and the Cantor set is a thin subset of $\mathbb{R}$. Likewise $\mathbb{R}$ is a thin subset of $\mathbb{R}^2$. The generic point of $\mathbb{R}$ does not lie in the Cantor set. The generic point of $\mathbb{R}^2$ does not lie on the x-axis. Although $\mathbb{R} \setminus \mathbb{Z}$ is a thick subset of $\mathbb{R}$ it is not a thick subset of $\mathbb{R}^2$. The set $\mathbb{Q}$ is a thin subset of $\mathbb{R}$. It is the countable union of its points, each of which is a closed nowhere dense set. $\mathbb{Q}^c$ is a thick subset of $\mathbb{R}$. The generic real number is irrational. In the same vein:

(a) The generic square matrix has determinant $\neq 0$.

(b) The generic linear transformation $\mathbb{R}^m \to \mathbb{R}^m$ is an isomorphism.

(c) The generic linear transformation $\mathbb{R}^m \to \mathbb{R}^{m-k}$ is onto.

(d) The generic linear transformation $\mathbb{R}^m \to \mathbb{R}^{m+k}$ is one-to-one.

(e) The generic pair of lines in $\mathbb{R}^3$ are skew (nonparallel and disjoint).

(f) The generic plane in $\mathbb{R}^3$ meets the three coordinate axes in three distinct points.

(g) The generic n^{th}-degree polynomial has n distinct roots.

In an incomplete metric space such as $\mathbb{Q}$, thickness and thinness have no bite – every subset of $\mathbb{Q}$, even the empty set, is thick in $\mathbb{Q}$.

Proof of Baire's Theorem If $M = \emptyset$, the proof is trivial, so we assume $M \neq \emptyset$. Let $G = \bigcap G_n$ be a thick subset of M, each G_n being open-dense in M. Let $p_0 \in M$ and $\epsilon > 0$ be given. Choose a sequence of points $p_n \in M$ and radii $r_n > 0$ such that $r_n < 1/n$ and

$$M_{2r_1}(p_1) \subset M_\epsilon(p_0)$$
$$M_{2r_2}(p_2) \subset M_{r_1}(p_1) \cap G_1$$
$$\cdots$$
$$M_{2r_n}(p_n) \subset M_{r_{n-1}}(p_{n-1}) \cap G_1 \cap \cdots \cap G_{n-1}.$$

See Figure 105. Then

$$M_\epsilon(p_0) \supset \overline{M}_{r_1}(p_1) \supset \overline{M}_{r_2}(p_2) \supset \cdots.$$

The diameters of these closed sets tend to 0 as $n \to \infty$. Thus (p_n) is a Cauchy sequence and it converges to some $p \in M$ by completeness. The point p belongs to each set $\overline{M}_{r_n}(p_n)$ and therefore it belongs to each G_n. Thus $p \in G \cap M_\epsilon(p_0)$ and G is dense in M.

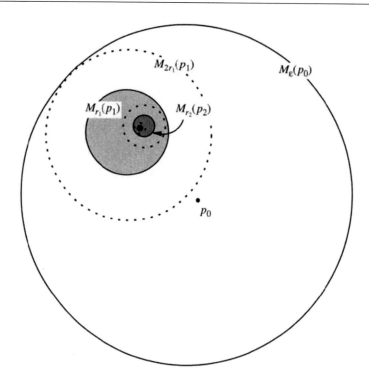

Figure 105 The closed neighborhoods $\overline{M}_{r_n}(p_n)$ nest down to a point.

To check that M is not thin, we take complements. Suppose that $M = \bigcup K_n$ and K_n is closed. If each K_n has empty interior then each $G_n = K_n^c$ is open-dense and

$$G = \bigcap G_n = \left(\bigcup K_n \right)^c = \emptyset,$$

a contradiction to density of G. □

34 Corollary *No subset of a complete nonempty metric space is both thick and thin.*

Proof If S is both a thick and thin subset of M then $M \setminus S$ is also both thick and thin. The intersection of two thick subsets of M is thick, so $\emptyset = S \cap (M \setminus S)$ is a thick subset of M. By Baire's Theorem, this empty set is dense in M, so M is empty, contrary to the hypothesis. □

Proof of Theorem 33 For $n \in \mathbb{N}$ define

$$R_n = \{f \in C^0 : \forall x \in [a,\ b - 1/n] \ \exists h > 0 \text{ such that } \left| \frac{\Delta f}{h} \right| > n\}$$

$$L_n = \{f \in C^0 : \forall x \in [a + 1/n,\ b] \ \exists h < 0 \text{ such that } \left| \frac{\Delta f}{h} \right| > n\}$$

$$G_n = \{f \in C^0 : f \text{ restricted to any interval of length } 1/n \text{ is nonmonotone}\},$$

where $\Delta f = f(x+h) - f(x)$. We claim that each of these sets is open-dense in C^0.

To check denseness it is enough to prove that the closures of R_n, L_n, and G_n contain the set $\mathcal{P}$ of polynomials. For by the Weierstrass Approximation Theorem $\mathcal{P}$ is dense in C^0. (A set whose closure contains a dense set is dense itself.)

Fix n, fix a $P \in \mathcal{P}$, and let $\epsilon > 0$ be given. Consider a sawtooth function σ which has period $< 1/n$, size $< \epsilon$, and

$$\min_x \{|\text{slope}_x(\sigma)|\} \;>\; n + \max_x \{|\text{slope}_x(P)|\}$$

Since the slopes of σ are far greater than those of P, the slopes of $f = P + \sigma$ alternate in sign with period $< 1/2n$. At any $x \in [a, \; b - 1/n]$, f has a rightward slope of either n or $-n$. Thus $f \in R_n$. Likewise $f \in L_n$ and $f \in G_n$, so the three sets are dense in C^0.

Next we prove R_n is open. Let $f \in R_n$ be given. For each $x \in [a, \; b - 1/n]$ there is an $h = h(x) > 0$ such that

$$\left| \frac{f(x+h) - f(x)}{h} \right| \;>\; n.$$

Since f is continuous, there is a neighborhood T_x of x in $[a,b]$ and a constant $\nu = \nu(x) > 0$ such that this same h yields

$$\left| \frac{f(t+h) - f(t)}{h} \right| \;>\; n + \nu$$

for all $t \in T_x$. Since $[a, \; b - 1/n]$ is compact, finitely many of these neighborhoods T_x cover it, say $T_{x_1}, \dots, T_{x_m}$. Continuity of f implies that for all $t \in \overline{T}_{x_i}$ we have

$$(25) \qquad \left| \frac{f(t+h_i) - f(t)}{h_i} \right| \;\geq\; n + \nu_i,$$

where $h_i = h(x_i)$ and $\nu_i = \nu(x_i)$. These m inequalities for points t in the m sets T_{x_i} remain nearly valid if f is replaced by a function g with $d(f,g)$ small enough. Then (25) becomes

$$(26) \qquad \left| \frac{g(t+h_i) - g(t)}{h_i} \right| \;>\; n,$$

which means that $g \in R_n$ and R_n is open in C^0. Similarly L_n is open in C^0.

Checking that G_n is open is easier. If (f_k) is a sequence of functions in G_n^c and $f_k \rightrightarrows f$ then we must show that $f \in G_n^c$. Each f_k is monotone on some interval I_k

of length $1/n$. There is a subsequence of these intervals that converges to a limit interval I. Its length is $1/n$ and by uniform convergence, f is monotone on I. Hence G_n^c is closed and G_n is open, which completes the proof that each set R_n, L_n, G_n is open-dense in C^0.

Finally, if f belongs to the thick set

$$\bigcap_{n=1}^{\infty} R_n \cap L_n \cap G_n$$

then for each $x \in [a, b]$ there are sequences $h_n^{\pm} \neq 0$ such that $h_n^- < 0 < h_n^+$ and

$$\left| \frac{f(x + h_n^-) - f(x)}{h_n^-} \right| > n \qquad \left| \frac{f(x + h_n^+) - f(x)}{h_n^+} \right| > n.$$

The numerator of these fractions is at most $2\|f\|$, so $h_n^{\pm} \to 0$ as $n \to \infty$. Thus f is not differentiable at x, nor does it even have a left or right derivative at x. Also, f is nonmonotone on every interval of length $1/n$. Since every interval J contains an interval of length $1/n$ when n is large enough, f is nonmonotone on J. $\qquad \square$

Further generic properties of continuous functions have been studied, and you might read about them in the books *A Primer of Real Functions* by Ralph Boas, *Differentiation of Real Functions* by Andrew Bruckner, or *A Second Course in Real Functions* by van Rooij and Schikhof.

8* Spaces of Unbounded Functions

When we contemplate equicontinuity, how important is it that the functions we deal with are bounded, or have domain $[a, b]$ and target $\mathbb{R}$? To some extent we can replace $[a, b]$ with a metric space X and $\mathbb{R}$ with a complete metric space Y. Let $\mathcal{F}$ denote the set of all functions $f : X \to Y$. Recall from Exercise 2.116 that the metric d_Y on Y gives rise to a bounded metric

$$\rho(y, y') = \frac{d_Y(y, y')}{1 + d_Y(y, y')},$$

where $y, y' \in Y$. Note that $\rho < 1$. Convergence and Cauchyness with respect to ρ and d_Y are equivalent. Thus completeness of Y with respect to d_Y implies completeness with respect to ρ. In the same way we give $\mathcal{F}$ the metric

$$d(f, g) = \sup_{x \in X} \frac{d_Y(fx, gx)}{1 + d_Y(fx, gx)}.$$

A function $f \in \mathcal{F}$ is **bounded** with respect to d_Y if and only if for any constant function c we have $\sup_x d_Y(f(x), c) < \infty$; i.e., $d(f, c) < 1$. Unbounded functions have $d(f, c) = 1$.

35 Theorem *In the space $\mathcal{F}$ equipped with the metric d,*

  (a) *Uniform convergence of (f_n) is equivalent to d-convergence.*

  (b) *Completeness of Y implies completeness of $\mathcal{F}$.*

  (c) *The set $\mathcal{F}_b$ of bounded functions is closed in $\mathcal{F}$.*

  (d) *The set $C^0(X, Y)$ of continuous functions is closed in $\mathcal{F}$.*

Proof (a) $f = \underset{n \to \infty}{\text{unif} \lim} f_n$ means that $d_Y(f_n(x), f(x)) \rightrightarrows 0$, which means that $d(f_n, f) \to 0$.

  (b) If (f_n) is Cauchy in $\mathcal{F}$ and Y is complete then, just as in Section 1, $f(x) = \underset{n \to \infty}{\lim} f_n(x)$ exists for each $x \in X$. Cauchyness with respect to the metric d implies uniform convergence and thus $d(f_n, f) \to 0$.

  (c) If $f_n \in \mathcal{F}_b$ and $d(f_n, f) \to 0$ then $\sup_x d_Y(f_n(x), f(x)) \to 0$. Since f_n is bounded, so is f.

  (d) The proof that C^0 is closed in $\mathcal{F}$ is the same as in Section 1. $\qquad \square$

The Arzelà-Ascoli theorem is trickier. A family $\mathcal{E} \subset \mathcal{F}$ is **uniformly equicontinuous** if for each $\epsilon > 0$ there is a $\delta > 0$ such that $f \in \mathcal{E}$ and $d_X(x, t) < \delta$ imply $d_Y(f(x), f(t)) < \epsilon$. If the δ depends on x but not on $f \in \mathcal{E}$ then $\mathcal{E}$ is **pointwise equicontinuous**.

36 Theorem *Pointwise equicontinuity implies uniform equicontinuity if X is compact.*

Proof Suppose not. Then there exists $\epsilon > 0$ such that for each $\delta = 1/n$ we have points $x_n, t_n \in X$ and functions $f_n \in \mathcal{E}$ with $d_X(x_n, t_n) < 1/n$ and $d_Y(f_n(x_n), f_n(t_n)) \geq \epsilon$. By compactness of X we may assume that $x_n \to x_0$. Then $t_n \to x_0$, which leads to a contradiction of pointwise equicontinuity at x_0. $\qquad \square$

37 Theorem *If the sequence of functions $f_n : X \to Y$ is uniformly equicontinuous, X is compact, and for each $x \in X$, the sequence $(f_n(x))$ lies in a compact subset of Y, then (f_n) has a uniformly convergent subsequence.*

Proof Being compact, X has a countable dense subset D. Then the proof of the Arzelà Ascoli Theorem in Section 3 becomes a proof of Theorem 37. $\qquad \square$

The space X is **σ-compact** if it is a countable union of compact sets, $X = \bigcup X_i$. For example $\mathbb{Z}, \mathbb{Q}, \mathbb{R}$ and $\mathbb{R}^m$ are σ-compact, while any uncountable set equipped with the discrete metric is not σ-compact.

38 Theorem *If X is σ-compact and if (f_n) is a sequence of pointwise equicontinuous functions such that for each $x \in X$, the sequence $(f_n(x))$ lies in a compact subset of Y, then (f_n) has a subsequence that converges uniformly to a limit on each compact subset of X.*

Proof Express X as $\bigcup X_i$ with X_i compact. By Theorem 36 $(f_n|_{X_i})$ is uniformly equicontinuous. By Theorem 37 there is a subsequence $f_{1,n}$ that converges uniformly on X_1, and it has a sub-subsequence $f_{2,n}$ that converges uniformly on X_2, and so on. A diagonal subsequence (g_m) converges uniformly on each X_i. Thus (g_m) converges pointwise. If $A \subset X$ is compact, then $(g_m|_A)$ is uniformly equicontinuous and pointwise convergent. By the proof of the Arzelà Ascoli propagation theorem, $(g_m|_A)$ converges uniformly. $\qquad\square$

39 Corollary *If (f_n) is a sequence of pointwise equicontinuous functions $\mathbb{R} \to \mathbb{R}$, and for some $x_0 \in \mathbb{R}, (f_n(x_0))$ is bounded then (f_n) has a subsequence that converges uniformly on every compact subset of $\mathbb{R}$.*

Proof Let $[a, b]$ be any interval containing x_0. By Theorem 36, the restrictions of f_n to $[a, b]$ are uniformly equicontinuous, and there is a $\delta > 0$ such that if $t, s \in [a, b]$ then $|t - s| < \delta$ implies that $|f_n(t) - f_n(s)| < 1$. Each point $x \in [a, b]$ can be reached in $\leq N$ steps of length $< \delta$, starting at x_0, if $N > (b - a)/\delta$. Thus $|f_n(x)| \leq |f_n(x_0)| + N$, and $(f_n(x))$ is bounded for each $x \in \mathbb{R}$. A bounded subset of $\mathbb{R}$ has compact closure and Theorem 38 gives the corollary. $\qquad\square$

Exercises

In these exercises, $C^0 = C^0([a, b], \mathbb{R})$ is the space of continuous real-valued functions defined on the closed interval $[a, b]$. It is equipped with the sup norm, $\|f\| = \sup\{|f(x)| : x \in [a, b]\}$.

1. Let M, N be metric spaces.
   (a) Formulate the concepts of pointwise convergence and uniform convergence for sequences of functions $f_n : M \to N$.
   (b) For which metric spaces are the concepts equivalent?
2. Suppose that $f_n \rightrightarrows f$ where f and f_n are functions from the metric space M to the metric space N. (Assume nothing about the metric spaces such as compactness, completeness, etc.) If each f_n is continuous prove that f is continuous. [Hint: Review the proof of Theorem 1.]
3. Let $f_n : [a, b] \to \mathbb{R}$ be a sequence of piecewise continuous functions, each of which is continuous at the point $x_0 \in [a, b]$. Assume that $f_n \rightrightarrows f$ as $n \to \infty$.
   (a) Prove that f is continuous at x_0. [Hint: Review the proof of Theorem 1.]
   (b) Prove or disprove that f is piecewise continuous.
4. (a) If $f_n : \mathbb{R} \to \mathbb{R}$ is uniformly continuous for each $n \in \mathbb{N}$ and if $f_n \rightrightarrows f$ as $n \to \infty$, prove or disprove that f is uniformly continuous.
   (b) What happens for functions from one metric space to another instead of $\mathbb{R}$ to $\mathbb{R}$?
5. Suppose that $f_n : [a, b] \to \mathbb{R}$ and $f_n \rightrightarrows f$ as $n \to \infty$. Which of the following discontinuity properties (see Exercise 3.36) of the functions f_n carries over to the limit function? (Prove or give a counterexample.)
   (a) No discontinuities.
   (b) At most ten discontinuities.
   (c) At least ten discontinuities.
   (d) Finitely many discontinuities.
   (e) Countably many discontinuities, all of jump type.
   (f) No jump discontinuities.
   (g) No oscillating discontinuities.
**6. (a) Prove that C^0 and $\mathbb{R}$ have equal cardinality. [Clearly there are at least as many functions as there are real numbers, for C^0 includes the constant functions. The issue is to show that there are no more continuous functions than there are real numbers.]
   (b) Is the same true if we replace $[a, b]$ with $\mathbb{R}$ or a separable metric space?
   (c) In the same vein, prove that the collection $\mathcal{T}$ of open subsets of $\mathbb{R}$ and $\mathbb{R}$ itself have equal cardinality.
   (d) What about more general metric spaces in place of $\mathbb{R}$?

7. Consider a sequence of functions f_n in C^0. The graph G_n of f_n is a compact subset of $\mathbb{R}^2$.
    (a) Prove that (f_n) converges uniformly as $n \to \infty$ if and only if the sequence (G_n) in $\mathcal{K}(\mathbb{R}^2)$ converges to the graph of a function $f \in C^0$. (The space $\mathcal{K}$ was discussed in Exercise 2.147.)
    (b) Formulate equicontinuity in terms of graphs.
8. Is the sequence of functions $f_n : \mathbb{R} \to \mathbb{R}$ defined by

$$f_n(x) = \cos(n + x) + \log(1 + \frac{1}{\sqrt{n+2}} \sin^2(n^n x))$$

equicontinuous? Prove or disprove.
9. If $f : \mathbb{R} \to \mathbb{R}$ is continuous and the sequence $f_n(x) = f(nx)$ is equicontinuous, what can be said about f?
10. Give an example to show that a sequence of functions may be uniformly continuous, pointwise equicontinuous, but not uniformly equicontinuous, when their domain M is noncompact.
11. If every sequence of pointwise equicontinuous functions $M \to \mathbb{R}$ is uniformly equicontinuous, does this imply that M is compact?
12. Prove that if $\mathcal{E} \subset C_b^0(M, N)$ is equicontinuous then so is its closure.
13. Suppose that (f_n) is a sequence of functions $\mathbb{R} \to \mathbb{R}$ and for each compact subset $K \subset \mathbb{R}$, the restricted sequence $(f_n|_K)$ is pointwise bounded and pointwise equicontinuous.
    (a) Does it follow that there is a subsequence of (f_n) that converges pointwise to a continuous limit function $\mathbb{R} \to \mathbb{R}$?
    (b) What about uniform convergence?
14. Recall from Exercise 2.78 that a metric space M is chain connected if for each $\epsilon > 0$ and each $p, q \in M$ there is a chain $p = p_0, \dots, p_n = q$ in M such that

$$d(p_{k-1}, p_k) < \epsilon \quad \text{for} \quad 1 \leq k \leq n.$$

A family $\mathcal{F}$ of functions $f : M \to \mathbb{R}$ is bounded at $p \in M$ if the set $\{f(p) : f \in \mathcal{F}\}$ is bounded in $\mathbb{R}$.

Show that M is chain connected if and only if pointwise boundedness of an equicontinuous family at one point of M implies pointwise boundedness at every point of M.
15. A continuous, strictly increasing function $\mu : (0, \infty) \to (0, \infty)$ is a **modulus of continuity** if $\mu(s) \to 0$ as $s \to 0$. A function $f : [a, b] \to \mathbb{R}$ has modulus of continuity μ if $|f(s) - f(t)| \leq \mu(|s - t|)$ for all $s, t \in [a, b]$.
    (a) Prove that a function is uniformly continuous if and only if it has a modulus of continuity.
    (b) Prove that a family of functions is equicontinuous if and only if its members have a common modulus of continuity.

16. Consider the modulus of continuity $\mu(s) = Ls$ where L is a positive constant.
    (a) What is the relation between C^μ and the set of Lipschitz functions with Lipschitz constant $\leq L$?
    (b) Replace $[a, b]$ with $\mathbb{R}$ and answer the same question.
    (c) Replace $[a, b]$ with $\mathbb{N}$ and answer the same question.
    (d) Formulate and prove a generalization of (a).
17. Consider a modulus of continuity $\mu(s) = Hs^\alpha$ where $0 < \alpha \leq 1$ and $0 < H < \infty$. A function with this modulus of continuity is said to be **α-Hölder**, with α-Hölder constant H. See also Exercise 3.2.
    (a) Prove that the set $C^\alpha(H)$ of all continuous functions defined on $[a, b]$ which are α-Hölder and have α-Hölder constant $\leq H$ is equicontinuous.
    (b) Replace $[a, b]$ with (a, b). Is the same thing true?
    (c) Replace $[a, b]$ with $\mathbb{R}$. Is it true?
    (d) What about $\mathbb{Q}$?
    (e) What about $\mathbb{N}$?
18. Suppose that (f_n) is an equicontinuous sequence in C^0 and $p \in [a, b]$ is given.
    (a) If $(f_n(p))$ is a bounded sequence of real numbers, prove that (f_n) is uniformly bounded.
    (b) Reformulate the Arzelà-Ascoli Theorem with the weaker boundedness hypothesis in (a).
    (c) Can $[a, b]$ be replaced with (a, b)?, $\mathbb{Q}$?, $\mathbb{R}$?, $\mathbb{N}$?
    (d) What is the correct generalization?
19. If M is compact and A is dense in M, prove that for each $\delta > 0$ there is a finite subset $\{a_1, \ldots, a_k\} \subset A$ which is **δ-dense** in M in the sense that each $x \in M$ lies within distance δ of at least one of the points $a_1, \ldots, a_k$.
*20. Given constants $\alpha, \beta > 0$ define

$$f_{\alpha,\beta}(x) = x^\alpha \sin(x^\beta)$$

for $x > 0$.
    (a) For which pairs α, β is $f_{\alpha,\beta}$ uniformly continuous?
    (b) For which sets of (α, β) in $(0, \infty)^2$ is the family equicontinuous?
    [Hint: Draw picture of the graphs when $\alpha \geq 2$ or $\beta \geq 2$. How about $\alpha > 1$ or $\beta > 1$?]
21. Suppose that $\mathcal{E} \subset C^0$ is equicontinuous and bounded.
    (a) Prove that $\sup\{f(x) : f \in \mathcal{E}\}$ is a continuous function of x.
    (b) Show that (a) fails without equicontinuity.
    (c) Show that this continuous-sup property does not imply equicontinuity.
    (d) Assume that the continuous-sup property is true for each subset $\mathcal{F} \subset \mathcal{E}$. Is $\mathcal{E}$ equicontinuous? Give a proof or counterexample.

22. Give an example of a sequence of smooth equicontinuous functions $f_n : [a, b] \to \mathbb{R}$ whose derivatives are not uniformly bounded.

23. Let M be a compact metric space, and let (i_n) be a sequence of isometries $i_n : M \to M$.
    (a) Prove that there exists a subsequence i_{n_k} that converges to an isometry i as $k \to \infty$.
    (b) Infer that the space of self-isometries of M is compact.
    (c) Does the inverse isometry $i_{n_k}^{-1}$ converge to i^{-1}? (Proof or counterexample.)
    (d) Infer that the group of orthogonal 3×3 matrices is compact. [Hint: Is it true that each orthogonal 3×3 matrix defines an isometry of the unit 2-sphere to itself?]
    (e) How about the group of $m \times m$ orthogonal matrices?

*24. Suppose that a sequence of continuous functions $f_n : [a, b] \to \mathbb{R}$ converges monotonically down to a continuous function f. (That is, for each $x \in [a, b]$ we have $f_1(x) \geq f_2(x) \geq f_3(x) \geq \ldots$ and $f_n(x) \to f(x)$ as $n \to \infty$.)
    (a) Prove that the convergence is uniform.
    (b) What if the sequence is increasing instead of decreasing?
    (c) What if you replace $[a, b]$ with $\mathbb{R}$?
    (d) What if you replace $[a, b]$ with a compact metric space or $\mathbb{R}^m$?

25. Suppose that $f : M \to M$ is a contraction, but M is not necessarily complete.
    (a) Prove that f is uniformly continuous.
    (b) Why does (a) imply that f extends uniquely to a continuous map $\widehat{f} : \widehat{M} \to \widehat{M}$, where $\widehat{M}$ is the completion of M?
    (c) Is $\widehat{f}$ a contraction?

26. Give an example of a contraction of an incomplete metric space that has no fixed-point.

27. Suppose that $f : M \to M$ and for all $x, y \in M$, if $x \neq y$ then $d(fx, fy) < d(x, y)$. Such an f is a **weak contraction**.
    (a) Is a weak contraction a contraction? (Proof or counterexample.)
    (b) If M is compact is a weak contraction a contraction? (Proof or counterexample.)
    (c) If M is compact, prove that a weak contraction has a unique fixed-point.

28. Suppose that $f : \mathbb{R} \to \mathbb{R}$ is differentiable and its derivative satisfies $|f'(x)| < 1$ for all $x \in \mathbb{R}$.
    (a) Is f a contraction?
    (b) A weak one?
    (c) Does it have a fixed-point?

29. Give an example to show that the fixed-point in Brouwer's Theorem need not be unique.

30. Give an example of a continuous map of a compact, nonempty, path-connected metric space into itself that has no fixed-point.

31. On page 233 it is shown that if $b_n \int_{-1}^{1}(1-t^2)^n\, dt = 1$ then for some constant c, and for all $n \in \mathbb{N}$, $b_n \le c\sqrt{n}$. What is the best (i.e., smallest) value of c that you can prove works? (A calculator might be useful here.)

32. Let M be a compact metric space, and let C^{Lip} be the set of continuous functions $f : M \to \mathbb{R}$ that obey a Lipschitz condition: For some L and all $p, q \in M$ we have

$$|fp - fq| \le Ld(p,q).$$

   *(a) Prove that C^{Lip} is dense in $C^0(M, \mathbb{R})$. [Hint: Stone-Weierstrass.]

   ***(b) If $M = [a, b]$ and $\mathbb{R}$ is replaced by some other complete, path-connected metric space, is the result true or false?

   ***(c) If M is a general compact metric space and Y is a complete metric space, is $C^{\text{Lip}}(M, Y)$ dense in $C^0(M, Y)$? (Would M equal to the Cantor set make a good test case?)

33. Consider the ODE $x' = x$ on $\mathbb{R}$. Show that its solution with initial condition x_0 is $t \mapsto e^t x_0$. Interpret $e^{t+s} = e^t e^s$ in terms of the flow property.

34. Consider the ODE $y' = 2\sqrt{|y|}$ where $y \in \mathbb{R}$.

   (a) Show that there are many solutions to this ODE, all with the same initial condition $y(0) = 0$. Not only does $y(t) = 0$ solve the ODE, but also $y(t) = t^2$ does for $t \ge 0$.

   (b) Find and graph other solutions such as $y(t) = 0$ for $t \le c$ and $y(t) = (t-c)^2$ for $t \ge c > 0$.

   (c) Does the existence of these nonunique solutions to the ODE contradict Picard's Theorem? Explain.

   *(d) Find all solutions with initial condition $y(0) = 0$.

35. Consider the ODE $x' = x^2$ on $\mathbb{R}$. Find the solution of the ODE with initial condition x_0. Are the solutions to this ODE defined for all time or do they escape to infinity in finite time?

36. Suppose that the ODE $x' = f(x)$ on $\mathbb{R}$ is bounded, $|f(x)| \le M$ for all x.

   (a) Prove that no solution of the ODE escapes to infinity in finite time.

   (b) Prove the same thing if f satisfies a Lipschitz condition, or more generally, if there are constants C, K such that $|f(x)| \le C|x| + K$ for all x.

   (c) Repeat (a) and (b) with $\mathbb{R}^m$ in place of $\mathbb{R}$.

   (d) Prove that if $f : \mathbb{R}^m \to \mathbb{R}^m$ is uniformly continuous then the condition stated in (b) is true. Infer that solutions of uniformly continuous ODEs defined on $\mathbb{R}^m$ do not escape to infinity in finite time.

37. (a) Prove **Borel's Lemma, which states that given any sequence whatsoever of real numbers (a_r), there is a smooth function $f : \mathbb{R} \to \mathbb{R}$ such that $f^{(r)}(0) = a_r$. [Hint: Try $f = \sum \beta_k(x) a_k x^k / k!$ where β_k is a well-chosen bump function.]

(b) Infer that there are many Taylor series with radius of convergence $R = 0$.

(c) Construct a smooth function whose Taylor series at every x has radius of convergence $R = 0$. [Hint: Try $\sum \beta_k(x)e(x + q_k)$ where $\{q_1, q_2, \ldots\} = \mathbb{Q}$.]

*38. Suppose that $T \subset (a, b)$ clusters at some point of (a, b) and that $f, g : (a, b) \to \mathbb{R}$ are analytic. Assume that for all $t \in T$ we have $f(t) = g(t)$.

(a) Prove that $f = g$ everywhere in (a, b).

(b) What if f and g are only C^∞?

(c) What if T is an infinite set but its only cluster points are a and b?

(d) Find a necessary and sufficient condition for a subset $Z \subset (a, b)$ to be the **zero locus of an analytic function f defined on (a, b), $Z = \{x \in (a, b) : f(x) = 0\}$. [Hint: Think Taylor. The result in (a) is known as the **Identity Theorem**. It states that if an equality between analytic functions is known to hold for points of T then it is an "identity," an equality that holds everywhere.]

39. Let M be any metric space with metric d. Fix a point $p \in M$ and for each $q \in M$ define the function $f_q(x) = d(q, x) - d(p, x)$.

(a) Prove that f_q is a bounded, continuous function of $x \in M$, and that the map $q \mapsto f_q$ sends M isometrically onto a subset M_0 of $C_b^0(M, \mathbb{R})$.

(b) Since $C_b^0(M, \mathbb{R})$ is complete, infer that an isometric copy of M is dense in a complete metric space, namely the closure of M_0, and hence that we have a second proof of the Completion Theorem 2.80.

40. As explained in Section 8, a metric space M is σ-compact if it is the countable union of compact subsets, $M = \bigcup M_i$.

(a) Why is it equivalent to require that M is the monotone union of compact subsets,

$$M = \hat{\bigcup} M_i$$

i.e., $M_1 \subset M_2 \subset \ldots$?

(b) Prove that a σ-compact metric space is separable.

(c) Prove that $\mathbb{Z}, \mathbb{Q}, \mathbb{R}, \mathbb{R}^m$ are σ-compact

*(d) Prove that C^0 is not σ-compact. [Hint: Think Baire.]

*(e) If $M = \hat{\bigcup} \text{int}(M_i)$ and each M_i is compact, M is $\boldsymbol{\sigma^*}$**-compact**. Prove that M is σ^*-compact if and only if it is separable and locally compact. Infer that $\mathbb{Z}, \mathbb{R}$, and $\mathbb{R}^m$ are σ^*-compact but $\mathbb{Q}$ is not.

(f) Assume that M is σ^*-compact, $M = \hat{\bigcup} \text{int}(M_i)$, with each M_i compact. Prove that this monotone union "engulfs" all compacts in M, in the sense that if $A \subset M$ is compact, then for some i, $A \subset M_i$.

(g) If $M = \hat{\bigcup} M_i$ and each M_i is compact show by example that this engulfing property may fail, even when M itself is compact.

**(h) Prove or disprove that a complete σ-compact metric space is σ^*-compact.

41. (a) Give an example of a function $f : [0,1] \times [0,1] \to \mathbb{R}$ such that for each
    fixed x, the function $y \mapsto f(x,y)$ is a continuous function of y, and for
    each fixed y, the function $x \mapsto f(x,y)$ is a continuous function of x, but f
    is not continuous.

 (b) Suppose in addition that the set of functions

$$\mathcal{E} = \{x \mapsto f(x,y) : y \in [0,1]\}$$

    is equicontinuous. Prove that f is continuous.

42. Prove that $\mathbb{R}$ cannot be expressed as the countable union of Cantor sets.

43. What is the joke in the following picture?

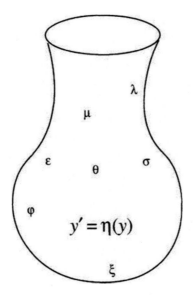

More Prelim Problems

1. Let f and f_n, $n \in \mathbb{N}$, be functions from $\mathbb{R}$ to $\mathbb{R}$. Assume that $f_n(x_n) \to f(x)$ as $n \to \infty$ and $x_n \to x$. Show that f is continuous. (Note: The functions f_n are not assumed to be continuous.)

2. Suppose that $f_n \in C^0$ and for each $x \in [a, b]$,

$$f_1(x) \geq f_2(x) \geq \ldots,$$

and $\lim_{n \to \infty} f_n(x) = 0$. Is the sequence equicontinuous? Give a proof or counterexample. [Hint: Does $f_n(x)$ converge uniformly to 0, or does it not?]

3. Let E be the set of all functions $u : [0, 1] \to \mathbb{R}$ such that $u(0) = 0$ and u satisfies a Lipschitz condition with Lipschitz constant 1. Define $\phi : E \to \mathbb{R}$ according to the formula

$$\phi(u) = \int_0^1 \left(u(x)^2 - u(x) \right) dx.$$

Prove that there exists a function $u \in E$ at which $\phi(u)$ attains an absolute maximum.

4. Let (g_n) be a sequence of twice-differentiable functions defined on $[0, 1]$, and assume that for all n, $g_n(0) = g_n'(0)$. Suppose also that for all $n \in \mathbb{N}$ and all $x \in [0, 1]$, $|g_n''(x)| \leq 1$. Prove that there is a subsequence of (g_n) converging uniformly on $[0, 1]$.

5. Let (a_n) be a sequence of nonzero real numbers. Prove that the sequence of functions

$$f_n(x) = \frac{1}{a_n} \sin(a_n x) + \cos(x + a_n)$$

has a subsequence converging to a continuous function.

6. Suppose that $f : \mathbb{R} \to \mathbb{R}$ is differentiable, $f(0) = 0$, and $f'(x) > f(x)$ for all $x \in \mathbb{R}$. Prove that $f(x) > 0$ for all $x > 0$.

7. Suppose that $f : [a, b] \to \mathbb{R}$ and the limits of $f(x)$ from the left and the right exist at all points of $[a, b]$. Prove that f is Riemann integrable.

8. Let $h : [0, 1) \to \mathbb{R}$ be a uniformly continuous function where $[0, 1)$ is the half-open interval. Prove that there is a unique continuous map $g : [0, 1] \to \mathbb{R}$ such that $g(x) = h(x)$ for all $x \in [0, 1)$.

9. Assume that $f : \mathbb{R} \to \mathbb{R}$ is uniformly continuous. Prove that there are constants A, B such that $|f(x)| \leq A + B|x|$ for all $x \in \mathbb{R}$.

10. Suppose that $f(x)$ is defined on $[-1, 1]$ and that its third derivative exists and is continuous. (That is, f is of class C^3.) Prove that the series

$$\sum_{n=0}^{\infty} \left(n(f(1/n) - f(-1/n)) - 2f'(0) \right)$$

converges.

11. Let $A \subset \mathbb{R}^m$ be compact, $x \in A$. Let (x_n) be a sequence in A such that every convergent subsequence of (x_n) converges to x.

    (a) Prove that the sequence (x_n) converges.

    (b) Give an example to show if A is not compact, the result in (a) is not necessarily true.

12. Let $f : [0,1] \to \mathbb{R}$ be continuously differentiable, with $f(0) = 0$. Prove that

$$\|f\|^2 \leq \int_0^1 (f'(x))^2 dx$$

    where $\|f\| = \sup\{|f(t)| : 0 \leq t \leq 1\}$.

13. Let $f_n : \mathbb{R} \to \mathbb{R}$ be differentiable functions, $n = 1, 2, \ldots$, with $f_n(0) = 0$ and $|f_n'(x)| \leq 2$ for all n, x. Suppose that

$$\lim_{n \to \infty} f_n(x) = g(x)$$

    for all x. Prove that g is continuous.

14. Let X be a nonempty connected set of real numbers. If every element of X is rational, prove that X has only one element.

15. Let $k \geq 0$ be an integer and define a sequence of maps $f_n : \mathbb{R} \to \mathbb{R}$ as

$$f_n(x) = \frac{x^k}{x^2 + n}$$

    $n = 1, 2, \ldots$. For which values of k does the sequence converge uniformly on $\mathbb{R}$? On every bounded subset of $\mathbb{R}$?

16. Let $f : [0,1] \to \mathbb{R}$ be Riemann integrable over $[b,1]$ for every b such that $0 < b \leq 1$.

    (a) If f is bounded, prove that f is Riemann integrable over $[0,1]$.

    (b) What if f is not bounded?

17. (a) Let S and T be connected subsets of the plane $\mathbb{R}^2$ having a point in common. Prove that $S \cup T$ is connected.

    (b) Let $\{S_\alpha\}$ be a family of connected subsets of $\mathbb{R}^2$ all containing the origin. Prove that $\bigcup S_\alpha$ is connected.

18. Let $f : \mathbb{R} \to \mathbb{R}$ be continuous. Suppose that $\mathbb{R}$ contains a countably infinite set S such that

$$\int_p^q f(x)\, dx = 0$$

    if p and q are not in S. Prove that f is identically zero.

19. Let $f : \mathbb{R} \to \mathbb{R}$ satisfy $f(x) \leq f(y)$ for $x \leq y$. Prove that the set where f is not continuous is finite or countably infinite.

20. Let (g_n) be a sequence of Riemann integrable functions from $[0,1]$ into $\mathbb{R}$ such that $|g_n(x)| \leq 1$ for all n, x. Define

$$G_n(x) = \int_0^x g_n(t)\, dt.$$

Prove that a subsequence of (G_n) converges uniformly.

21. Prove that every compact metric space has a countable dense subset.

22. Show that for any continuous function $f : [0,1] \to \mathbb{R}$ and for any $\epsilon > 0$ there is a function of the form

$$g(x) = \sum_{k=0}^{n} C_k x^k$$

for some $n \in \mathbb{N}$, and $|g(x) - f(x)| < \epsilon$ for all x in $[0,1]$.

23. Give an example of a function $f : \mathbb{R} \to \mathbb{R}$ having all three of the following properties:

    (a) $f(x) = 0$ for all $x < 0$ and $x > 2$.
    (b) $f'(1) = 1$.
    (c) f has derivatives of all orders.

24. (a) Give an example of a differentiable function $f : \mathbb{R} \to \mathbb{R}$ whose derivative is not continuous.
    (b) Let f be as in (a). If $f'(0) < 2 < f'(1)$ prove that $f'(x) = 2$ for some $x \in [0,1]$.

25. Let $U \subset \mathbb{R}^m$ be an open set. Suppose that the map $h : U \to \mathbb{R}^m$ is a homeomorphism from U onto $\mathbb{R}^m$ which is uniformly continuous. Prove that $U = \mathbb{R}^m$.

26. Let (f_n) be a sequence of continuous maps $[0,1] \to \mathbb{R}$ such that

$$\int_0^1 (f_n(y))^2\, dy \leq 5$$

for all n. Define $g_n : [0,1] \to \mathbb{R}$ by

$$g_n(x) = \int_0^1 \sqrt{x+y}\, f_n(y)\, dy$$

    (a) Find a constant $K \geq 0$ such that $|g_n(x)| \leq K$ for all n.
    (b) Prove that a subsequence of the sequence (g_n) converges uniformly.

27. Consider the following properties of a map $f : \mathbb{R}^m \to \mathbb{R}$.

    (a) f is continuous.
    (b) The graph of f is connected in $\mathbb{R}^m \times \mathbb{R}$.
    Prove or disprove the implications (a) $\Rightarrow$ (b), (b) $\Rightarrow$ (a).

28. Let (P_n) be a sequence of real polynomials of degree ≤ 10. Suppose that

$$\lim_{n \to \infty} P_n(x) = 0$$

for all $x \in [0,1]$. Prove that $P_n(x) \rightrightarrows 0$, $0 \le x \le 1$. What can you say about $P_n(x)$ for $4 \le x \le 5$?

29. Give an example of a subset of $\mathbb{R}$ having uncountably many connected components. Can such a subset be open? Closed? Does your answer change if $\mathbb{R}^2$ replaces $\mathbb{R}$?

30. For each $(a, b, c) \in \mathbb{R}^3$ consider the series

$$\sum_{n=3}^{\infty} \frac{a^n}{n^b (\log n)^c}.$$

Determine the values of a, b, and c for which the series converges absolutely, converges conditionally, diverges.

31. Let X be a compact metric space and $f : X \to X$ an isometry. (That is, $d(f(x), f(y)) = d(x, y)$ for all $x, y \in X$.) Prove that $f(X) = X$.

32. Prove or disprove: $\mathbb{Q}$ is the countable intersection of open subsets of $\mathbb{R}$.

33. Let $f : \mathbb{R} \to \mathbb{R}$ be continuous and

$$\int_{-\infty}^{\infty} |f(x)| \, dx < \infty.$$

Show that there is a sequence (x_n) in $\mathbb{R}$ such that $x_n \to \infty$, $x_n f(x_n) \to 0$, and $x_n f(-x_n) \to 0$ as $n \to \infty$.

34. Let $f : [0,1] \to \mathbb{R}$ be a continuous function. Evaluate the following limits (with proof):

(a) $\displaystyle \lim_{n\to\infty} \int_0^1 x^n f(x) \, dx$ (b) $\displaystyle \lim_{n\to\infty} n \int_0^1 x^n f(x) \, dx.$

35. Let K be an uncountable subset of $\mathbb{R}^m$. Prove that there is a sequence of distinct points in K which converges to some point of K.

36. Prove or give a counterexample: Every connected locally pathwise-connected set in $\mathbb{R}^m$ is pathwise-connected.

37. Let (f_n) be a sequence of continuous functions $[0,1] \to \mathbb{R}$ such that $f_n(x) \to 0$ for each $x \in [0,1]$. Suppose that

$$\left| \int_0^1 f_n(x) \, dx \right| \le K$$

for all n where K is a constant. Does $\int_0^1 f_n(x) \, dx$ converge to 0 as $n \to \infty$? Prove or give a counterexample.

38. Let E be a closed, bounded, and nonempty subset of $\mathbb{R}^m$ and let $f : E \to E$ be a function satisfying $|f(x) - f(y)| < |x - y|$ for all $x, y \in E$, $x \ne y$. Prove that there is one and only one point $x_0 \in E$ such that $f(x_0) = x_0$.

39. Let $f : [0, 2\pi] \to \mathbb{R}$ be a continuous function such that

$$\int_0^{2\pi} f(x) \sin(nx) \, dx = 0$$

for all integers $n \geq 1$. Prove that f is identically constant.

40. Let $f_1, f_2, \ldots$ be continuous real-valued functions on $[0, 1]$ such that for each $x \in [0, 1]$, $f_1(x) \geq f_2(x) \geq \ldots$. Assume that for each x, $f_n(x)$ converges to 0 as $n \to \infty$. Does f_n converge uniformly to 0? Give a proof or counterexample.

41. Let $f : [0, \infty) \to [0, \infty)$ be a monotonically decreasing function with

$$\int_0^\infty f(x) \, dx < \infty.$$

Prove that $\lim_{x \to \infty} x f(x) = 0$.

42. Suppose that $F : \mathbb{R}^m \to \mathbb{R}^m$ is continuous and satisfies

$$|F(x) - F(y)| \geq \lambda |x - y|$$

for all $x, y \in \mathbb{R}^m$ and some constant $\lambda > 0$. Prove that F is one-to-one, is onto, and has a continuous inverse.

43. Show that $[0, 1]$ cannot be written as a countably infinite union of disjoint closed subintervals.

44. Prove that a continuous function $f : \mathbb{R} \to \mathbb{R}$ which sends open sets to open sets must be monotonic.

45. Let $f : [0, \infty) \to \mathbb{R}$ be uniformly continuous and assume that

$$\lim_{b \to \infty} \int_0^b f(x) \, dx$$

exists (as a finite limit). Prove that $\lim_{x \to \infty} f(x) = 0$.

46. Prove or supply a counterexample: If f and g are continuously differentiable functions defined on the interval $0 < x < 1$ which satisfy the conditions

$$\lim_{x \to 0} f(x) = 0 = \lim_{x \to 0} g(x) \qquad \text{and} \qquad \lim_{x \to 0} \frac{f(x)}{g(x)} = c$$

and if g and g' never vanish, then $\lim_{x \to 0} \frac{f'(x)}{g'(x)} = c$. (This is a converse of L'Hôpital's rule.)

47. Prove or provide a counterexample: If the function f from $\mathbb{R}$ to $\mathbb{R}$ has both a left and a right limit at each point of $\mathbb{R}$, then the set of discontinuities is at most countable.

48. Prove or supply a counterexample: If f is a nondecreasing real-valued function on $[0,1]$ then there is a sequence f_n, $n = 1, 2, \ldots$, of continuous functions on $[0,1]$ such that for each x in $[0,1]$, $\lim_{n \to \infty} f_n(x) = f(x)$.

49. Show that if f is a homeomorphism of $[0,1]$ onto itself then there is a sequence of polynomials $P_n(x)$, $n = 1, 2, \ldots$, such that $P_n \to f$ uniformly on $[0,1]$ and each P_n is a homeomorphism of $[0,1]$ onto itself. [Hint: First assume that f is C^1.]

50. Let f be a C^2 function on the real line. Assume that f is bounded with bounded second derivative. Let $A = \sup_x |f(x)|$ and $B = \sup_x |f''(x)|$. Prove that

$$\sup_x |f'(x)| \le 2\sqrt{AB}.$$

51. Let f be continuous on $\mathbb{R}$ and let

$$f_n(x) = \frac{1}{n} \sum_{k=0}^{n-1} f\left(x + \frac{k}{n}\right).$$

Prove that $f_n(x)$ converges uniformly to a limit on every finite interval $[a, b]$.

52. Let f be a real-valued continuous function on the compact interval $[a, b]$. Given $\epsilon > 0$, show that there is a polynomial p such that

$$p(a) = f(a), \qquad p'(a) = 0, \quad \text{and} \quad |p(x) - f(x)| < \epsilon$$

for all $x \in [a, b]$.

53. A function $f : [0,1] \to \mathbb{R}$ is said to be **upper semicontinuous** if, given $x \in [0,1]$ and $\epsilon > 0$, there exists a $\delta > 0$ such that $|y - x| < \delta$ implies that $f(y) < f(x) + \epsilon$. Prove that an upper semicontinuous function on $[0,1]$ is bounded above and attains its maximum value at some point $p \in [0,1]$.

54. Let $f(x)$, $0 \le x \le 1$, be a continuous real function with continuous derivative $f'(x)$. Let M be the supremum of $|f'(x)|$, $0 \le x \le 1$. Prove the following: For $n = 1, 2, \ldots$,

$$\left| \frac{1}{n} \sum_{k=0}^{n-1} f\left(\frac{k}{n}\right) - \int_0^1 f(x)\, dx \right| \le \frac{M}{2n}.$$

55. Let K be a compact subset of $\mathbb{R}^m$ and let (B_j) be a sequence of open balls which cover K. Prove that there is an $\epsilon > 0$ such that each ϵ-ball centered at a point of K is contained in at least one of the balls B_j.

56. Let f be a continuous real-valued function on $[0, \infty)$ such that

$$\lim_{x \to \infty} \left(f(x) + \int_0^x f(t)\, dt \right)$$

exists (and is finite). Prove that $\lim_{x \to \infty} f(x) = 0$.

57. A standard theorem asserts that a continuous real-valued function on a compact set is bounded. Prove the converse: If K is a subset of $\mathbb{R}^m$ and if every continuous real-valued function defined on K is bounded, then K is compact.

58. Let $\mathcal{F}$ be a uniformly bounded equicontinuous family of real-valued functions defined on the metric space X. Prove that the function

$$g(x) = \sup\{f(x) : f \in \mathcal{F}\}$$

is continuous.

59. Suppose that (f_n) is a sequence of nondecreasing functions which map the unit interval into itself. Suppose that $\lim_{n\to\infty} f_n(x) = f(x)$ pointwise and that f is a continuous function. Prove that $f_n(x) \to f(x)$ uniformly as $n \to \infty$. Note that the functions f_n are not necessarily continuous.

60. Does there exist a continuous real-valued function $f(x)$, $0 \le x \le 1$, such that

$$\int_0^1 x f(x)\,dx = 1 \qquad \text{and} \qquad \int_0^1 x^n f(x)\,dx = 0$$

for all $n = 0, 2, 3, 4, 5, \ldots$? Give a proof or counterexample.

61. Let f be a continuous, strictly increasing function from $[0, \infty)$ onto $[0, \infty)$ and let $g = f^{-1}$ (the inverse, not the reciprocal). Prove that

$$\int_0^a f(x)\,dx + \int_0^b g(y)\,dy \ge ab$$

for all positive numbers a, b, and determine the condition for equality.

62. Let f be a function $[0, 1] \to \mathbb{R}$ whose graph $\{(x, f(x)) : x \in [0, 1]\}$ is a closed subset of the unit square. Prove that f is continuous.

63. Let (a_n) be a sequence of positive numbers such that $\sum a_n$ converges. Prove that there exists a sequence of numbers $c_n \to \infty$ as $n \to \infty$ such that $\sum c_n a_n$ converges.

64. Let $f(x, y)$ be a continuous real-valued function defined on the unit square $[0, 1] \times [0, 1]$. Prove that $g(x) = \max\{f(x, y) : y \in [0, 1]\}$ is continuous.

65. Let the function f from $[0, 1]$ to $[0, 1]$ have the following properties. It is of class C^1, $f(0) = 0 = f(1)$, and f' is nonincreasing (i.e., f is concave). Prove that the arclength of the graph of f does not exceed 3.

66. Let A be the set of all positive integers that do not contain the digit 9 in their decimal expansions. Prove that

$$\sum_{a \in A} \frac{1}{a} < \infty.$$

That is, A defines a convergent subseries of the harmonic series.

5

Multivariable Calculus

This chapter presents the natural geometric theory of calculus in n dimensions.

1 Linear Algebra

It will be taken for granted that you are familiar with the basic concepts of linear algebra – vector spaces, linear transformations, matrices, determinants, and dimension. In particular, you should be aware of the fact that an $m \times n$ matrix A with entries a_{ij} is more than just a static array of mn numbers. It is dynamic. It can act. It defines a **linear transformation** $T_A : \mathbb{R}^n \to \mathbb{R}^m$ that sends n-space to m-space according to the formula

$$T_A(v) = \sum_{i=1}^{m} \sum_{j=1}^{n} a_{ij} v_j e_i$$

where $v = \sum v_j e_j \in \mathbb{R}^n$ and $e_1, \ldots, e_n$ is the standard basis of $\mathbb{R}^n$. (Equally, $e_1, \ldots, e_m$ is the standard basis of $\mathbb{R}^m$.)

The set $\mathcal{M} = \mathcal{M}(m, n)$ of all $m \times n$ matrices with real entries a_{ij} is a vector space. Its vectors are matrices. You add two matrices by adding the corresponding entries, $A + B = C$ where $a_{ij} + b_{ij} = c_{ij}$. Similarly, if $\lambda \in \mathbb{R}$ is a scalar then λA is the matrix with entries λa_{ij}. The dimension of the vector space $\mathcal{M}$ is mn, as can be seen by expressing each A as $\sum a_{ij} E_{ij}$ where E_{ij} is the matrix whose entries are 0, except for the $(ij)^{\text{th}}$ entry which is 1. Thus, as vector spaces, $\mathcal{M} = \mathbb{R}^{mn}$.

© Springer International Publishing Switzerland 2015
C.C. Pugh, *Real Mathematical Analysis*, Undergraduate Texts
in Mathematics, DOI 10.1007/978-3-319-17771-7_5

The set $\mathcal{L} = \mathcal{L}(\mathbb{R}^n, \mathbb{R}^m)$ of linear transformations $T : \mathbb{R}^n \to \mathbb{R}^m$ is also a vector space. You combine linear transformations as functions, $U = T + S$ being defined by $U(v) = T(v) + S(v)$, and λT being defined by $(\lambda T)(v) = \lambda T(v)$. The vectors in $\mathcal{L}$ are linear transformations. The mapping $A \mapsto T_A$ is an isomorphism $\mathcal{T} : \mathcal{M} \to \mathcal{L}$. The matrix A is said to represent the linear transformation $T_A : \mathbb{R}^n \to \mathbb{R}^m$. As a rule of thumb, think with linear transformations and compute with matrices.

Corresponding to composition of linear transformations is the product of matrices. If A is an $m \times k$ matrix and B is a $k \times n$ matrix then the product matrix $P = AB$ is the $m \times n$ matrix whose $(ij)^{\text{th}}$ entry is

$$p_{ij} = a_{i1}b_{1j} + \cdots + a_{ik}b_{kj} = \sum_{r=1}^{k} a_{ir}b_{rj}.$$

1 Theorem $T_A \circ T_B = T_{AB}$.

Proof For each pair of basis vectors $e_r \in \mathbb{R}^k$ and $e_j \in \mathbb{R}^n$ we have

$$T_A(e_r) = \sum_{i=1}^{m} a_{ir}e_i \qquad T_B(e_j) = \sum_{r=1}^{k} b_{rj}e_r.$$

Thus for each basis vector e_j we have

$$
\begin{aligned}
(T_A \circ T_B)(e_j) &= T_A\left(\sum_{r=1}^{k} b_{rj}e_r\right) = \sum_{r=1}^{k} b_{rj} T_A(e_r) = \sum_{r=1}^{k} b_{rj} \sum_{i=1}^{m} a_{ir}e_i \\
&= \sum_{r=1}^{k}\sum_{i=1}^{m} b_{rj}a_{ir}e_i = \sum_{i=1}^{m}\sum_{r=1}^{k} a_{ir}b_{rj}e_i \\
&= \sum_{i=1}^{m} p_{ij}e_i = T_{AB}(e_j).
\end{aligned}
$$

Two linear transformations that are equal on a basis are equal. $\qquad\qquad\square$

Theorem 1 expresses the pleasing fact that matrix multiplication corresponds naturally to composition of linear transformations. See also Exercise 6.

As explained in Chapter 1, a norm on a vector space V is a function $|\ \ | : V \to \mathbb{R}$ that satisfies three properties:

(a) For all $v \in V$ we have $|v| \geq 0$; and $|v| = 0$ if and only if $v = 0$.
(b) $|\lambda v| = |\lambda|\,|v|$.

(c) $|v + w| \leq |v| + |w|$.

(Note the abuse of notation in (b); $|\lambda|$ is the magnitude of the scalar λ and $|v|$ is the norm of the vector v.) Norms are used to make vector estimates, and vector estimates underlie multivariable calculus.

A vector space with a norm is a **normed space**. Its norm gives rise to a metric as

$$d(v, v') = |v - v'|.$$

Thus a normed space is a special kind of metric space.

If V, W are normed spaces then the **operator norm** of a linear transformation $T : V \to W$ is

$$\|T\| = \sup \left\{ \frac{|Tv|_W}{|v|_V} : v \neq 0 \right\}.$$

The operator norm of T is the **maximum stretch** that T imparts to vectors in V. The subscript on the norm indicates the space in question, which for simplicity is often suppressed.[†]

The composition of linear transformations obeys the norm inequality

$$\|T \circ S\| \leq \|T\| \, \|S\|$$

where $S : U \to V$ and $T : V \to W$. Thinking in terms of stretch, the inequality is clear: S stretches a vector $u \in U$ by at most $\|S\|$, and T stretches $S(u)$ by at most $\|T\|$. The net effect on u is a stretch of at most $\|T\| \, \|S\|$.

2 Theorem *Let $T : V \to W$ be a linear transformation from one normed space to another. The following are equivalent:*

(a) $\|T\| < \infty$.
(b) T *is uniformly continuous.*
(c) T *is continuous.*
(d) T *is continuous at the origin.*

Proof Assume (a), $\|T\| < \infty$. For all $v, v' \in V$, linearity of T implies that

$$|Tv - Tv'| \leq \|T\| \, |v - v'| ,$$

which gives (b), uniform continuity. Clearly (b) implies (c) implies (d).

[†]If $\|T\|$ is finite then T is said to be a **bounded linear transformation**. Unfortunately, this terminology conflicts with T being bounded as a mapping from the metric space V to the metric space W. The only linear transformation that is bounded in the latter sense is the zero transformation.

Assume (d) and take $\epsilon = 1$. There is a $\delta > 0$ such that if $u \in V$ and $|u| < \delta$ then

$$|Tu| < 1.$$

For any nonzero $v \in V$, set $u = \lambda v$ where $\lambda = \delta/2|v|$. Then $|u| = \delta/2 < \delta$ and

$$\frac{|Tv|}{|v|} = \frac{|Tu|}{|u|} < \frac{1}{|u|} = \frac{2}{\delta}$$

which implies $\|T\| < 2/\delta$ and verifies (a).   □

3 Theorem *Every linear transformation $T : \mathbb{R}^n \to W$ is continuous and every isomorphism $T : \mathbb{R}^n \to W$ is a homeomorphism.*

Proof The norm on $\mathbb{R}^n$ is the Euclidean norm. If $v = (v_1, \ldots, v_n) \in \mathbb{R}^n$ then

$$|v| = \sqrt{v_1^2 + \ldots + v_n^2}.$$

Let $|\ \ |_W$ denote the norm on W and let $M = \max\{|T(e_1)|_W, \ldots, |T(e_n)|_W\}$. For $v = \sum v_j e_j \in \mathbb{R}^n$ we have $|v_j| \leq |v|$ and

$$|Tv|_W \ \leq \ \sum_{j=1}^{n} |T(v_j e_j)|_W \ = \ \sum_{j=1}^{n} |v_j||T(e_j)|_W \ \leq \ n|v|M$$

which implies that $\|T\| \leq nM < \infty$. Theorem 2 implies that T is continuous.

Assume that $T : \mathbb{R}^n \to W$ is an isomorphism. We have just shown that T is continuous, but what about T^{-1}? Continuity of T implies that the T-image of the unit sphere is compact. Injectivity implies that $O \notin T(S^{n-1})$. Since O and $T(S^{n-1})$ are disjoint compact sets in the metric space W, there is a constant $c > 0$ such that for all $u \in S^{n-1}$ we have $d_W(Tu, O) = |Tu| \geq c$. For each nonzero $v \in \mathbb{R}^n$ we write $v = \lambda u$ where $\lambda = |v|$ and $u = v/|v|$ is a unit vector. Linearity of T implies $Tv = \lambda Tu$ which gives $|Tv| \geq c|v|$, i.e.,

$$|v| \leq \frac{|Tv|}{c}.$$

For each $w \in W$ let $v = T^{-1}(w)$. Then $w = Tv$ and

$$\left|T^{-1}(w)\right| = |v| \leq \frac{|Tv|}{c} = \frac{1}{c}|w|$$

gives $\left\|T^{-1}\right\| \leq 1/c < \infty$, and by Theorem 2 we get continuity of T^{-1}. A bicontinuous bijection is a homeomorphism.   □

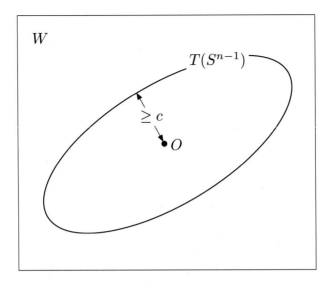

Figure 106 The minimum distance from $T(S^{n-1})$ to the origin is $\geq c$.

Geometrically speaking, the inequality $|Tv| \geq c\,|v|$ means that T shrinks each vector in $\mathbb{R}^n$ by a factor no smaller than c, so it follows that T^{-1} expands each vector in W by a factor no greater than $1/c$. The largest c with the property $|Tv| \geq c\,|v|$ for all v is the **conorm** of T. See Figure 106 and Exercise 4.

4 Corollary *In the world of finite-dimensional normed spaces, all linear transformations are continuous and all isomorphisms are homeomorphisms. In particular, if a finite-dimensional vector space is equipped with two different norms then the identity map is a homeomorphism between the two normed spaces. In particular $\mathcal{T} : \mathcal{M} \to \mathcal{L}$ is a homeomorphism.*

Proof Let V be an n-dimensional normed space and let $T : V \to W$ be a linear transformation. As you know from linear algebra, there is an isomorphism $H : \mathbb{R}^n \to V$. Theorem 3 implies that H is a homeomorphism. Therefore H^{-1} is a homeomorphism. Since $T \circ H$ is a linear transformation from $\mathbb{R}^n$ to W it is continuous. Thus

$$T = (T \circ H) \circ H^{-1}$$

is the composition of continuous maps so it is continuous.

Suppose that $T : V \to W$ is an isomorphism and V is finite-dimensional. Then W is finite-dimensional and $T^{-1} : W \to V$ is a linear transformation. Since every linear transformation from a finite-dimensional normed space to a normed space is continuous, T and T^{-1} are both continuous, so T is a homeomorphism.

Let a finite-dimensional vector space V be equipped with norms $|\ |_1$ and $|\ |_2$. Since the identity map is an isomorphism $V_1 \to V_2$ it is a homeomorphism. The same applies to the isomorphism $\mathcal{T}$ that assigns to a matrix A the corresponding linear transformation T_A.   □

2 Derivatives

A function of a real variable $y = f(x)$ has a derivative $f'(x)$ at x when

(1)
$$\lim_{h \to 0} \frac{f(x+h) - f(x)}{h} = f'(x).$$

If, however, x is a vector variable, (1) makes no sense. For what does it mean to divide by the vector increment h? Equivalent to (1) is the condition

$$f(x+h) = f(x) + f'(x)h + R(h) \quad \Rightarrow \quad \lim_{h \to 0} \frac{R(h)}{|h|} = 0,$$

which is easy to recast in vector terms.

Definition Let $f : U \to \mathbb{R}^m$ be given where U is an open subset of $\mathbb{R}^n$. The function f is **differentiable** at $p \in U$ with **derivative** $(Df)_p = T$ if $T : \mathbb{R}^n \to \mathbb{R}^m$ is a linear transformation and

(2)
$$f(p+v) = f(p) + T(v) + R(v) \quad \Rightarrow \quad \lim_{|v| \to 0} \frac{R(v)}{|v|} = 0.$$

We say that the Taylor remainder R is **sublinear** because it tends to 0 faster than $|v|$.

When $n = m = 1$, the multidimensional definition reduces to the standard one. This is because a linear transformation $\mathbb{R} \to \mathbb{R}$ is just multiplication by some real number, in this case multiplication by $f'(x)$.

Here is how to visualize Df. Take $m = n = 2$. The mapping $f : U \to \mathbb{R}^2$ distorts shapes nonlinearly; its derivative describes the linear part of the distortion. Circles are sent by f to wobbly ovals, but they become ellipses under $(Df)_p$. Lines are sent by f to curves, but they become straight lines under $(Df)_p$. See Figure 107 and also Appendix A.

This way of looking at differentiability is conceptually simple. Near p, f is the sum of three terms: A constant term $q = fp$, a linear term $(Df)_p v$, and a sublinear

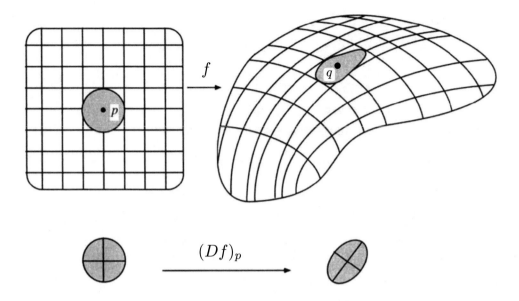

Figure 107 $(Df)_p$ is the linear part of f at p.

remainder term $R(v)$. Keep in mind what kind of an object the derivative is. It is not a number. It is not a vector. No, if it exists then $(Df)_p$ is a linear transformation from the domain space to the target space.

5 Theorem *If f is differentiable at p then it unambiguously determines $(Df)_p$ according to the limit formula, valid for all $u \in \mathbb{R}^n$,*

$$(3) \qquad\qquad (Df)_p(u) = \lim_{t \to 0} \frac{f(p + tu) - f(p)}{t}.$$

Proof Let T be a linear transformation that satisfies (2). Fix any $u \in \mathbb{R}^n$ and take $v = tu$. Then

$$\frac{f(p + tu) - f(p)}{t} = \frac{T(tu) + R(tu)}{t} = T(u) + \frac{R(tu)}{t\,|u|}|u|.$$

The last term converges to zero as $t \to 0$, which verifies (3). Limits, when they exist, are unambiguous and therefore if T' is a second linear transformation that satisfies (2) then $T(u) = T'(u)$ so $T = T'$. $\qquad\square$

6 Theorem *Differentiability implies continuity.*

Proof Differentiability at p implies that

$$|f(p + v) - f(p)| = |(Df)_p v + R(v)| \le \|(Df)_p\|\,|v| + |R(v)| \to 0$$

as $p + v \to p$. $\qquad\square$

Df is the **total derivative** or **Fréchet derivative**. In contrast, the ij^{th} **partial derivative** of f at p is the limit, if it exists,

$$\frac{\partial f_i(p)}{\partial x_j} = \lim_{t \to 0} \frac{f_i(p + te_j) - f_i(p)}{t}.$$

7 Corollary *If the total derivative exists then the partial derivatives exist and they are the entries of the matrix that represents the total derivative.*

Proof Substitute in (3) the vector $u = e_j$ and take the ith component of both sides of the resulting equation. □

As is shown in Exercise 15, the mere existence of partial derivatives does not imply differentiability. The simplest sufficient condition beyond the existence of the partials – and the simplest way to recognize differentiability – is given in the next theorem.

8 Theorem *If the partial derivatives of $f : U \to \mathbb{R}^m$ exist and are continuous then f is differentiable.*

Proof Let A be the matrix of partials at p, $A = [\partial f_i(p)/\partial x_j]$, and let $T : \mathbb{R}^n \to \mathbb{R}^m$ be the linear transformation that A represents. We claim that $(Df)_p = T$. We must show that the Taylor remainder

$$R(v) = f(p + v) - f(p) - Av$$

is sublinear. Draw a path $\sigma = [\sigma_1, \ldots, \sigma_n]$ from p to $q = p + v$ that consists of n segments parallel to the components of v. Thus $v = \sum v_j e_j$ and

$$\sigma_j(t) = p_{j-1} + tv_j e_j \qquad 0 \le t \le 1$$

is a segment from $p_{j-1} = p + \sum_{k<j} v_k e_k$ to $p_j = p_{j-1} + v_j e_j$. See Figure 108.

By the one-dimensional chain rule and mean value theorem applied to the differentiable real-valued function $g(t) = f_i \circ \sigma_j(t)$ of one variable, there exists $t_{ij} \in (0, 1)$ such that

$$f_i(p_j) - f_i(p_{j-1}) = g(1) - g(0) = g'(t_{ij}) = \frac{\partial f_i(p_{ij})}{\partial x_j} v_j,$$

where $p_{ij} = \sigma_j(t_{ij})$. Telescoping $f_i(p + v) - f_i(p)$ along σ gives

$$\begin{aligned}
R_i(v) &= f_i(p + v) - f_i(p) - (Av)_i \\
&= \sum_{j=1}^{n} \left(f_i(p_j) - f_i(p_{j-1}) - \frac{\partial f_i(p)}{\partial x_j} v_j \right) \\
&= \sum_{j=1}^{n} \left\{ \frac{\partial f_i(p_{ij})}{\partial x_j} - \frac{\partial f_i(p)}{\partial x_j} \right\} v_j.
\end{aligned}$$

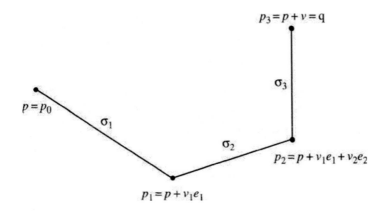

Figure 108 The segmented path σ from p to q

Continuity of the partials implies that the terms inside the curly brackets tend to 0 as $|v| \to 0$. Thus R is sublinear and f is differentiable at p.   □

Next we state and prove the basic rules of multivariable differentiation.

9 Theorem *Let f and g be differentiable. Then*

(a) $D(f + cg) = Df + cDg$.
(b) $D(constant) = 0$ and $D(T(x)) = T$.
(c) $D(g \circ f) = Dg \circ Df$.   *(**Chain Rule**)*
(d) $D(f \bullet g) = Df \bullet g + f \bullet Dg$.   *(**Leibniz Rule**)*

There is a fifth rule that concerns the derivative of the nonlinear inversion operator $\mathrm{Inv} : T \mapsto T^{-1}$. It is a glorified version of the formula

$$\frac{dx^{-1}}{dx} = -x^{-2},$$

and is discussed in Exercises 32 - 36.

Proof (a) Write the Taylor estimates for f and g and combine them to get the Taylor estimate for $f + cg$.

$$
\begin{aligned}
f(p + v) &= f(p) + (Df)_p(v) + R_f \\
g(p + v) &= g(p) + (Dg)_p(v) + R_g \\
(f + cg)(p + v) &= (f + cg)(p) + ((Df)_p + c(Dg)_p)(v) + R_f + cR_g.
\end{aligned}
$$

Since $R_f + cR_g$ is sublinear, $(Df)_p + c(Dg)_p$ is the derivative of $f + cg$ at p.

(b) If $f : \mathbb{R}^n \to \mathbb{R}^m$ is constant, $f(x) = c$ for all $x \in \mathbb{R}^n$, and if $O : \mathbb{R}^n \to \mathbb{R}^m$ denotes the zero transformation then the Taylor remainder $R(v) = f(p + v) - f(p) - O(v)$ is identically zero. Hence $D(constant)_p = O$.

$T : \mathbb{R}^n \to \mathbb{R}^m$ is a linear transformation. If $f(x) = T(x)$ for all x then substituting T itself in the Taylor expression gives the Taylor remainder $R(v) = f(p + v) - f(p) - T(v)$, which is identically zero. Hence $(DT)_p = T$.

Note that when $n = m = 1$, a linear function is of the form $f(x) = ax$, and the previous formula just states that $(ax)' = a$.

(c) Tacitly, we assume that the composite $g \circ f(x) = g(f(x))$ makes sense as x varies in a neighborhood of $p \in U$. The notation $Dg \circ Df$ refers to the composite of linear transformations and is written out as

$$D(g \circ f)_p = (Dg)_q \circ (Df)_p$$

where $q = f(p)$. The Chain Rule states that the derivative of a composite is the composite of the derivatives. Such a beautiful and natural formula *must* be true. See also Appendix A. Here is a proof.

It is convenient to write the remainder $R(v) = f(p + v) - f(p) - T(v)$ in a different form, defining the scalar function $\boldsymbol{\epsilon}(v)$ by

$$\boldsymbol{\epsilon}(v) = \begin{cases} \dfrac{|R(v)|}{|v|} & \text{if } v \neq 0 \\ 0 & \text{if } v = 0. \end{cases}$$

Sublinearity is equivalent to $\lim_{v \to 0} \boldsymbol{\epsilon}(v) = 0$. Think of $\boldsymbol{\epsilon}$ as an "error factor."

The Taylor expressions for f at p and g at $q = f(p)$ are

$$\begin{aligned} f(p + v) &= f(p) + Av + R_f \\ g(q + w) &= g(q) + Bw + R_g \end{aligned}$$

where $A = (Df)_p$ and $B = (Dg)_q$ as matrices. The composite is expressed as

$$g \circ f(p + v) = g(q + Av + R_f(v)) = g(q) + BAv + BR_f(v) + R_g(w)$$

where $w = Av + R_f(v)$. It remains to show that the remainder terms are sublinear with respect to v. First

$$|BR_f(v)| \leq \|B\| \, |R_f(v)|$$

is sublinear. Second,

$$|w| = |Av + R_f(v)| \leq \|A\| \, |v| + \epsilon_f(v)|v|.$$

Therefore,

$$|R_g(w)| \leq \epsilon_g(w) \, |w| \leq \epsilon_g(w)(\|A\| + \epsilon_f(v)) \, |v|.$$

Since $\epsilon_g(w) \to 0$ as $w \to 0$ and since $v \to 0$ implies that w does tend to 0, we see that $R_g(w)$ is sublinear with respect to v. It follows that $(D(g \circ f))_p = BA$ as claimed.

(d) To prove the Leibniz Product Rule, we must explain the notation $v \bullet w$. In $\mathbb{R}$ there is only one product, the usual multiplication of real numbers. In higher-dimensional vector spaces, however, there are many products and the general way to discuss products is in terms of bilinear maps.

A map $\beta : V \times W \to Z$ is **bilinear** if V, W, Z are vector spaces and for each fixed $v \in V$ the map $\beta(v, \, . \,) : W \to Z$ is linear, while for each fixed $w \in W$ the map $\beta(\, . \, , w) : V \to Z$ is linear. Examples are

(i) Ordinary real multiplication $(x, y) \mapsto xy$ is a bilinear map $\mathbb{R} \times \mathbb{R} \to \mathbb{R}$.
(ii) The dot product is a bilinear map $\mathbb{R}^n \times \mathbb{R}^n \to \mathbb{R}$.
(iii) The matrix product is a bilinear map $\mathcal{M}(m \times k) \times \mathcal{M}(k \times n) \to \mathcal{M}(m \times n)$.

The precise statement of (d) is that if $\beta : \mathbb{R}^k \times \mathbb{R}^\ell \to \mathbb{R}^m$ is bilinear while $f : U \to \mathbb{R}^k$ and $g : U \to \mathbb{R}^\ell$ are differentiable at p then the map $x \mapsto \beta(f(x), g(x))$ is differentiable at p and

$$(D\beta(f, g))_p(v) = \beta((Df)_p(v), g(p)) + \beta(f(p), (Dg)_p(v)).$$

Just as a linear transformation between finite-dimensional vector spaces has a finite operator norm, the same is true for bilinear maps:

$$\|\beta\| = \sup\{\frac{|\beta(v, w)|}{|v| \, |w|} : v, w \neq 0\} < \infty.$$

To check this we view β as a linear map $T_\beta : \mathbb{R}^k \to \mathcal{L}(\mathbb{R}^\ell, \mathbb{R}^m)$. According to Theorems 2 and 3, a linear transformation from one finite dimensional normed space to another is continuous and has finite operator norm. Thus the operator norm T_β is finite. That is,

$$\|T_\beta\| = \max \left\{ \frac{\|T_\beta(v)\|}{|v|} : v \neq 0 \right\} < \infty.$$

But $\|T_\beta(v)\| = \max\{|\beta(v, w)| \, / \, |w| : w \neq 0\}$, which implies that $\|\beta\| < \infty$.

Returning to the proof of the Leibniz Rule, we write out the Taylor estimates for f and g and plug them into β. If we use the notation $A = (Df)_p$ and $B = (Dg)_p$, then bilinearity implies

$$\beta(f(p+v),\ g(p+v)) = \beta(f(p) + Av + R_f,\ g(p) + Bv + R_g)$$
$$= \beta(f(p),\ g(p)) + \beta(Av,\ g(p)) + \beta(f(p),\ Bv)$$
$$+ \beta(f(p),\ R_g) + \beta(Av,\ Bv + R_g) + \beta(R_f,\ g(p) + Bv + R_g).$$

The last three terms are sublinear. For

$$
\begin{aligned}
|\beta(f(p), R_g)| &\leq \|\beta\|\,|f(p)|\,|R_g| \\
|\beta(Av,\ Bv + R_g)| &\leq \|\beta\|\,\|A\|\,|v|\,|Bv + R_g| \\
|\beta(R_f,\ g(p) + Bv + R_g)| &\leq \|\beta\|\,|R_f|\,|g(p) + Bv + R_g|
\end{aligned}
$$

Therefore $\beta(f,g)$ is differentiable and $D\beta(f,g) = \beta(Df, g) + \beta(f, Dg)$ as claimed. $\square$

Here are some applications of these differentiation rules:

10 Theorem *A function* $f : U \to \mathbb{R}^m$ *is differentiable at* $p \in U$ *if and only if each of its components* f_i *is differentiable at* p. *Furthermore, the derivative of its* i^{th} *component is the* i^{th} *component of the derivative.*

Proof Assume that f is differentiable at p and express the ith component of f as $f_i = \pi_i f$ where $\pi_i : \mathbb{R}^m \to \mathbb{R}$ is the projection that sends a vector $w = (w_1, \dots, w_m)$ to w_i. Since π_i is linear it is differentiable. By the Chain Rule, f_i is differentiable at p and

$$(Df_i)_p = (D\pi_i) \circ (Df)_p = \pi_i \circ (Df)_p.$$

The proof of the converse is equally natural. $\square$

Theorem 10 implies there is little loss of generality in assuming $m = 1$, i.e., that our functions are real-valued. Multidimensionality of the domain, not the target, is what distinguishes multivariable calculus from one-variable calculus.

11 Mean Value Theorem *If* $f : U \to \mathbb{R}^m$ *is differentiable on* U *and the segment* $[p, q]$ *is contained in* U *then*

$$|f(q) - f(p)| \leq M\,|q - p|$$

where $M = \sup\{\|(Df)_x\| : x \in U\}.$

Proof Fix any unit vector $u \in \mathbb{R}^m$. The function

$$g(t) = \langle u, f(p + t(q - p)) \rangle$$

is differentiable and we can calculate its derivative. By the one-dimensional Mean Value Theorem this gives some $\theta \in (0, 1)$ such that $g(1) - g(0) = g'(\theta)$. That is,

$$\langle u, f(q) - f(p) \rangle = g'(\theta) = \langle u, (Df)_{p+\theta(q-p)}(q - p) \rangle \leq M|q - p|.$$

A vector whose dot product with every unit vector is no larger than $M|q - p|$ has norm $\leq M|q - p|$. □

Remark The one-dimensional Mean Value Theorem is an equality

$$f(q) - f(p) = f'(\theta)(q - p)$$

and you might expect the same to be true for a vector-valued function if we replace $f'(\theta)$ by $(Df)_\theta$. Not so. See Exercise 17. The closest we can come to an equality form of the multidimensional Mean Value Theorem is the following.

12 C^1 Mean Value Theorem *If $f : U \to \mathbb{R}^m$ is of class C^1 (its derivative exists and is continuous) and if the segment $[p, q]$ is contained in U then*

$$\text{(4)} \qquad\qquad f(q) - f(p) = T(q - p)$$

*where T is the **average derivative** of f on the segment,*

$$T = \int_0^1 (Df)_{p+t(q-p)}\, dt.$$

Conversely, if there is a continuous family of linear maps $T_{pq} \in \mathcal{L}$ for which (4) holds then f is of class C^1 and $(Df)_p = T_{pp}$.

Proof The integrand takes values in the normed space $\mathcal{L}(\mathbb{R}^n, \mathbb{R}^m)$ and is a continuous function of t. The integral is the limit of Riemann sums

$$\sum_k (Df)_{p+t_k(q-p)}\, \Delta t_k,$$

which lie in $\mathcal{L}$. Since the integral is an element of $\mathcal{L}$ it has a right to act on the vector $q - p$. Alternatively, if you integrate each entry of the matrix that represents Df along the segment then the resulting matrix represents T. Fix an index i and apply the Fundamental Theorem of Calculus to the C^1 real-valued function of one variable

$$g(t) = f_i \circ \sigma(t)$$

where $\sigma(t) = p + t(q - p)$ parameterizes $[p, q]$. This gives

$$
\begin{aligned}
f_i(q) - f_i(p) \quad &= \quad g(1) - g(0) \;=\; \int_0^1 g'(t)\, dt \\
&= \quad \int_0^1 \sum_{j=1}^n \frac{\partial f_i(\sigma(t))}{\partial x_j}(q_j - p_j)\, dt \\
&= \quad \sum_{j=1}^n \int_0^1 \frac{\partial f_i(\sigma(t))}{\partial x_i}\, dt\,(q_j - p_j),
\end{aligned}
$$

which is the ith component of $T(q - p)$.

To check the converse, we assume that (4) holds for a continuous family of linear maps T_{pq}. Take $q = p + v$. The first-order Taylor remainder at p is

$$
R(v) \;=\; f(p + v) - f(p) - T_{pp}(v) \;=\; (T_{pq} - T_{pp})(v),
$$

which is sublinear with respect to v. Therefore $(Df)_p = T_{pp}$. $\square$

13 Corollary *Assume that U is connected. If $f : U \to \mathbb{R}^m$ is differentiable and for each point $x \in U$ we have $(Df)_x = 0$ then f is constant.*

Proof The enjoyable open and closed argument is left to you as Exercise 20. $\square$

We conclude this section with another useful rule – **differentiation past the integral**. See also Exercise 23.

14 Theorem *Assume that $f : [a, b] \times (c, d) \to \mathbb{R}$ is continuous and that $\partial f(x, y)/\partial y$ exists and is continuous. Then*

$$
F(y) \;=\; \int_a^b f(x, y)\, dx
$$

is of class C^1 and

(5)
$$
\frac{dF}{dy} \;=\; \int_a^b \frac{\partial f(x, y)}{\partial y}\, dx.
$$

Proof By the C^1 Mean Value Theorem, if h is small then

$$
\frac{F(y + h) - F(y)}{h} \;=\; \frac{1}{h}\int_a^b \left(\int_0^1 \frac{\partial f(x, y + th)}{\partial y}\, dt \right) h\, dx.
$$

The inner integral is the partial derivative of f with respect to y averaged along the segment from y to $y+h$. Continuity implies that this average converges to $\partial f(x, y)/\partial y$ as $h \to 0$, which verifies (5). Continuity of dF/dy follows from continuity of $\partial f/\partial y$. See Exercise 22. $\square$

3 Higher Derivatives

In this section we define higher-order multivariable derivatives. We do so in the same spirit as in the previous section – the second derivative will be the derivative of the first derivative, viewed naturally. Assume that $f : U \to \mathbb{R}^m$ is differentiable on U. The derivative $(Df)_x$ exists at each $x \in U$ and the map $x \mapsto (Df)_x$ defines a function

$$Df : U \to \mathcal{L}(\mathbb{R}^n, \mathbb{R}^m).$$

The derivative Df is the same sort of thing that f is, namely a function from an open subset of a vector space into another vector space. In the case of Df the target vector space is not $\mathbb{R}^m$ but rather the mn-dimensional space $\mathcal{L}$. If Df is differentiable at $p \in U$ then by definition

$$(D(Df))_p = (D^2 f)_p = \text{ the } \textbf{second derivative} \text{ of } f \text{ at } p$$

and f is **second-differentiable** at p. The second derivative at p is a linear map from $\mathbb{R}^n$ into $\mathcal{L}$. For each $v \in \mathbb{R}^n$, $(D^2 f)_p(v)$ belongs to $\mathcal{L}$ and therefore is a linear transformation $\mathbb{R}^n \to \mathbb{R}^m$ so $(D^2 f)_p(v)(w)$ is bilinear and we write it as

$$(D^2 f)_p(v, w).$$

(Recall that bilinearity is linearity in each variable separately.)

Third and higher derivatives are defined in the same way. If f is second-differentiable on U then $x \mapsto (D^2 f)_x$ defines a map

$$D^2 f : U \to \mathcal{L}^2$$

where $\mathcal{L}^2$ is the vector space of bilinear maps $\mathbb{R}^n \times \mathbb{R}^n \to \mathbb{R}^m$. If $D^2 f$ is differentiable at p then f is third-differentiable there, and its third derivative is the trilinear map $(D^3 f)_p = (D(D^2 f))_p$. And so on.

Just as for first derivatives, the relation between the second derivative and the second partial derivatives calls for thought. Express $f : U \to \mathbb{R}^m$ in component form as $f(x) = (f_1(x), \dots, f_m(x))$ where x varies in U.

15 Theorem *If $(D^2 f)_p$ exists then $(D^2 f_k)_p$ exists, the second partials at p exist, and*

$$(D^2 f_k)_p(e_i, e_j) = \frac{\partial^2 f_k(p)}{\partial x_i \partial x_j}.$$

Conversely, existence of the second partials implies existence of $(D^2 f)_p$, provided that the second partials exist at all points $x \in U$ near p and are continuous at p.

Proof Assume that $(D^2f)_p$ exists. Then $x \mapsto (Df)_x$ is differentiable at $x = p$ and the same is true of the matrix

$$M_x = \begin{bmatrix} \dfrac{\partial f_1}{\partial x_1} & \cdots & \dfrac{\partial f_1}{\partial x_n} \\ \cdots & \cdots & \cdots \\ \dfrac{\partial f_m}{\partial x_1} & \cdots & \dfrac{\partial f_m}{\partial x_n} \end{bmatrix}$$

that represents it; $x \mapsto M_x$ is differentiable at $x = p$. For according to Theorem 10, a vector function is differentiable if and only if its components are differentiable, and then the derivative of the kth component is the kth component of the derivative. A matrix is a special type of vector. Its components are its entries. Thus the entries of M_x are differentiable at $x = p$ and the second partials exist. Furthermore, the kth row of M_x is a differentiable vector function of x at $x = p$ and

$$(D(Df_k))_p(e_i)(e_j) = (D^2f_k)_p(e_i, e_j) = \lim_{t \to 0} \frac{(Df_k)_{p+te_i}(e_j) - (Df_k)_p(e_j)}{t}.$$

The first derivatives appearing in this fraction are the jth partials of f_k at $p + te_i$ and at p. Thus $\partial^2 f_k(p)/\partial x_i \partial x_j = (D^2f_k)_p(e_i, e_j)$ as claimed.

Conversely, assume that the second partials exist at all x near p and are continuous at p. Then the entries of M_x have partials that exist at all points q near p, and are continuous at p. Theorem 8 implies that $x \mapsto M_x$ is differentiable at $x = p$; i.e., f is second-differentiable at p. $\qquad \square$

The most important and surprising property of second derivatives is symmetry.

16 Theorem *If $(D^2f)_p$ exists then it is symmetric: For all $v, w \in \mathbb{R}^n$ we have*

$$(D^2f)_p(v, w) = (D^2f)_p(w, v).$$

Proof We will assume that f is real-valued (i.e., $m = 1$) because the symmetry assertion concerns the arguments of f rather than its values. For a variable $t \in [0, 1]$ we draw the parallelogram P determined by the vectors tv, tw and label the vertices with ± 1 as in Figure 109.

The quantity

$$\Delta = \Delta(t, v, w) = f(p + tv + tw) - f(p + tv) - f(p + tw) + f(p)$$

is the signed sum of f at the vertices of P. Clearly Δ is symmetric with respect to v, w,

$$\Delta(t, v, w) = \Delta(t, w, v).$$

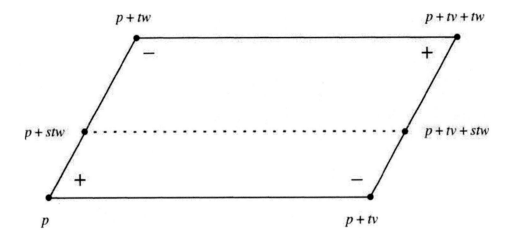

Figure 109 The parallelogram P has signed vertices.

We claim that

(6) $$(D^2 f)_p(v, w) = \lim_{t \to 0} \frac{\Delta(t, v, w)}{t^2},$$

from which symmetry of $D^2 f$ follows.

Fix t, v, w and write $\Delta = g(1) - g(0)$ where

$$g(s) = f(p + tv + stw) - f(p + stw).$$

Since f is differentiable, so is g. By the one-dimensional Mean Value Theorem there exists $\theta \in (0, 1)$ with $\Delta = g'(\theta)$. By the Chain Rule $g'(\theta)$ can be written in terms of Df and we get

$$\Delta = g'(\theta) = (Df)_{p+tv+\theta tw}(tw) - (Df)_{p+\theta tw}(tw).$$

Taylor's estimate applied to the differentiable function $u \mapsto (Df)_u$ at $u = p$ gives

$$(Df)_{p+x} = (Df)_p + (D^2 f)_p(x, .\,) + R(x, .\,)$$

where $R(x, .\,) \in \mathcal{L}(\mathbb{R}^n, \mathbb{R}^m)$ is sublinear with respect to x. Writing out this estimate for $(Df)_{p+x}$ first with $x = tv + \theta tw$ and then with $x = \theta tw$ gives

$$
\begin{aligned}
\frac{\Delta}{t^2} &= \frac{1}{t}\left\{\left[(Df)_p(w) + (D^2 f)_p(tv + \theta tw, w) + R(tv + \theta tw, w)\right]\right. \\
&\quad - \left.\left[(Df)_p(w) + (D^2 f)_p(\theta tw, w) + R(\theta tw, w)\right]\right\} \\
&= (D^2 f)_p(v, w) + \frac{R(tv + \theta tw, w)}{t} - \frac{R(\theta tw, w)}{t}
\end{aligned}
$$

Bilinearity was used to combine the two second derivative terms. Sublinearity of $R(x, w)$ with respect to x implies that the last two terms tend to 0 as $t \to 0$, which completes the proof of (6). Since $(D^2 f)_p$ is the limit of a symmetric (although nonlinear) function of v, w it too is symmetric.   □

Remark The fact that $D^2 f$ can be expressed directly as a limit of values of f is itself interesting. It should remind you of its one-dimensional counterpart,

$$f''(x) = \lim_{h \to 0} \frac{f(x+h) + f(x-h) - 2f(x)}{h^2}.$$

17 Corollary *Corresponding mixed second partials of a second-differentiable function are equal,*

$$\frac{\partial^2 f_k(p)}{\partial x_i \partial x_j} = \frac{\partial^2 f_k(p)}{\partial x_j \partial x_i}.$$

Proof The equalities

$$\frac{\partial^2 f_k(p)}{\partial x_i \partial x_j} = (D^2 f_k)_p(e_i, e_j) = (D^2 f_k)_p(e_j, e_i) = \frac{\partial^2 f_k(p)}{\partial x_j \partial x_i}$$

follow from Theorem 15 and the symmetry of $D^2 f$.   □

The mere existence of the second-order partials does not imply second order differentiability, nor does it imply equality of corresponding mixed second partials. See Exercise 24.

18 Corollary *The rth derivative, if it exists, is symmetric: Permutation of the vectors $v_1, \ldots, v_r$ does not affect the value of $(D^r f)_p(v_1, \ldots, v_r)$. Corresponding mixed higher-order partials are equal.*

Proof The induction argument is left to you as Exercise 29.   □

In my opinion Theorem 16 is quite natural even though its proof is tricky. It proceeds from a pointwise hypothesis to a pointwise conclusion – whenever the second derivative exists it is symmetric. No assumption is made about continuity of partials. It is possible that f is second-differentiable at p and nowhere else. See Exercise 25. All the same, it remains standard to prove equality of mixed partials under stronger hypotheses, namely, that $D^2 f$ is continuous. See Exercise 27.

We conclude this section with a brief discussion of the rules of higher-order differentiation. It is simple to check that the rth derivative of $f + cg$ is $D^r f + cD^r g$.

Also, if β is k-linear and $k < r$ then $f(x) = \beta(x, \ldots, x)$ has $D^r f = 0$. On the other hand, if $k = r$ then $(D^r f)_p = r! \operatorname{Symm}(\beta)$ where $\operatorname{Symm}(\beta)$ is the symmetrization of β. See Exercise 28.

The Chain Rule for r^{th} derivatives is a bit complicated. The difficulties arise from the fact that x appears in two places in the expression for the first-order Chain Rule, $(D(g \circ f))_x = (Dg)_{f(x)} \circ (Df)_x$, and so, differentiating this product produces

$$(D^2 g)_{f(x)} \circ (Df)_x^2 + (Dg)_{f(x)} \circ (D^2 f)_x.$$

(The meaning of $(Df)_x^2$ needs clarification.) Differentiating again produces four terms, two of which combine. The general formula is

$$(D^r (g \circ f))_x = \sum_{k=1}^{r} \sum_{\mu} (D^k g)_{f(x)} \circ (D^\mu f)_x$$

where the sum on μ is taken as μ runs through all partitions of $\{1, \ldots, r\}$ into k disjoint subsets. See Exercise 41.

The higher-order Leibniz rule is left for you as Exercise 42.

Smoothness Classes

A map $f : U \to \mathbb{R}^m$ is of **class C^r** if it is r^{th}-order differentiable at each $p \in U$ and its derivatives depend continuously on p. (Since differentiability implies continuity, all the derivatives of order less than r are automatically continuous. Only the rth derivative is in question.) If f is of class C^r for all r then it is **smooth** or of **class C^∞**. According to the differentiation rules, these smoothness classes are closed under the operations of linear combination, product, and composition. We discuss next how they are closed under limits.

Let (f_k) be a sequence of C^r functions $f_k : U \to \mathbb{R}^m$. The sequence is

(a) **Uniformly C^r convergent** if for some C^r function $f : U \to \mathbb{R}^m$ we have

$$f_k \rightrightarrows f \qquad Df_k \rightrightarrows Df \qquad \ldots \qquad D^r f_k \rightrightarrows D^r f$$

as $k \to \infty$.

(b) **Uniformly C^r Cauchy** if for each $\epsilon > 0$ there is an N such that for all $k, \ell \geq N$ and all $x \in U$ we have

$$|f_k(x) - f_\ell(x)| < \epsilon \quad \|(Df_k)_x - (Df_\ell)_x\| < \epsilon \ldots \|(D^r f_k)_x - (D^r f_\ell)_x\| < \epsilon.$$

19 Theorem *Uniform C^r convergence and Cauchyness are equivalent.*

Proof Convergence always implies the Cauchy condition. As for the converse, first assume that $r = 1$. We know that f_k converges uniformly to a continuous function f and the derivative sequence converges uniformly to a continuous limit

$$Df_k \rightrightarrows G.$$

We claim that $Df = G$. Fix $p \in U$ and consider points q in a small convex neighborhood of p. The C^1 Mean Value Theorem and uniform convergence imply that as $k \to \infty$ we have

$$f_k(q) - f_k(p) \quad = \quad \int_0^1 (Df_k)_{p+t(q-p)} \, dt \, (q - p)$$

$$\Downarrow \qquad\qquad\qquad\qquad \Downarrow$$

$$f(q) - f(p) \quad = \quad \int_0^1 G(p + t(q - p)) \, dt \, (q - p).$$

This integral of G is a continuous function of q that reduces to $G(p)$ when $p = q$. By the converse part of the C^1 Mean Value Theorem, f is differentiable and $Df = G$. Therefore f is C^1 and f_k converges C^1 uniformly to f as $k \to \infty$, completing the proof when $r = 1$.

Now suppose that $r \geq 2$. The maps $Df_k : U \to \mathcal{L}$ form a uniformly C^{r-1} Cauchy sequence. The limit, by induction, is C^{r-1} uniform; i.e., as $k \to \infty$ we have

$$D^s(Df_k) \rightrightarrows D^s G$$

for all $s \leq r - 1$. Hence f_k converges C^r uniformly to f as $k \to \infty$, completing the induction. □

The **C^r norm** of a C^r function $f : U \to \mathbb{R}^m$ is

$$\|f\|_r = \max\{\sup_{x \in U} |f(x)|, \dots, \sup_{x \in U} \|(D^r f)_x\|\}.$$

The set of functions with $\|f\|_r < \infty$ is denoted $C^r(U, \mathbb{R}^m)$.

20 Corollary $\|\ \ \|_r$ *makes $C^r(U, \mathbb{R}^m)$ a **Banach space** – a complete normed vector space.*

Proof The norm properties are easy to check; completeness follows from Theorem 19. □

21 $C^r M$-test *If $\sum M_k$ is a convergent series of constants and if $\|f_k\|_r \leq M_k$ for all k then the series of functions $\sum f_k$ converges in $C^r(U, \mathbb{R}^m)$ to a function f. Term-by-term differentiation of order $\leq r$ is valid, i.e., for all $s \leq r$ we have $D^s f = \sum_k D^s f_k$.*

Proof Obvious from the preceding corollary. □

4 Implicit and Inverse Functions

Let $f : U \to \mathbb{R}^m$ be given, where U is an open subset of $\mathbb{R}^n \times \mathbb{R}^m$. Fix attention on a point $(x_0, y_0) \in U$ and write $f(x_0, y_0) = z_0$. Our goal is to solve the equation

(7) $$f(x, y) = z_0$$

near (x_0, y_0). More precisely, we hope to show that the set of points (x, y) near (x_0, y_0) at which $f(x, y) = z_0$, the so-called z_0-locus of f, is the graph of a function $y = g(x)$. If so, g is the **implicit function** defined by (7). See Figure 110.

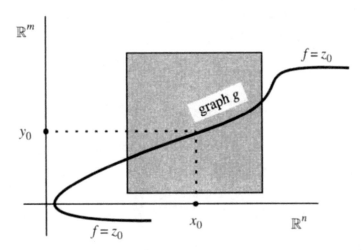

Figure 110 Near (x_0, y_0) the z_0-locus of f is the graph of a function
$$y = g(x).$$

Under various hypotheses we will show that g exists, is unique, and is differentiable. The main assumption, which we make throughout this section, is that

the $m \times m$ matrix $\quad B = \left[\dfrac{\partial f_i(x_0, y_0)}{\partial y_j}\right] \quad$ is invertible.

Equivalently the linear transformation that B represents is an isomorphism $\mathbb{R}^m \to \mathbb{R}^m$.

22 Implicit Function Theorem *If the function f above is C^r, $1 \leq r \leq \infty$, then near (x_0, y_0), the z_0-locus of f is the graph of a unique function $y = g(x)$. Besides, g is C^r.*

Proof Without loss of generality we suppose that (x_0, y_0) is the origin in $\mathbb{R}^n \times \mathbb{R}^m$ and $z_0 = 0$ in $\mathbb{R}^m$. The Taylor expression for f is

$$f(x, y) = Ax + By + R$$

where A is the $m \times n$ matrix

$$A = \left[\frac{\partial f_i(x_0, y_0)}{\partial x_j} \right]$$

and R is sublinear. Solving $f(x, y) = 0$ for $y = gx$ is equivalent to solving

(8) $$y = -B^{-1}(Ax + R(x, y)).$$

In the unlikely event that R does not depend on y, (8) is an explicit formula for gx and the implicit function is an explicit function. In general, the idea is that the remainder R depends so weakly on y that we can switch it to the left-hand side of (8), absorbing it in the y-term.

Solving (8) for y as a function of x is the same as finding a fixed-point of

$$K_x : y \mapsto -B^{-1}(Ax + R(x, y)),$$

so we hope to show that K_x contracts. The remainder R is a C^1 function, and $(DR)_{(0,0)} = 0$. Therefore if r is small and $|x|, |y| \leq r$ then

$$\|B^{-1}\| \left\| \frac{\partial R(x, y)}{\partial y} \right\| \leq \frac{1}{2}.$$

By the Mean Value Theorem this implies that

$$
\begin{aligned}
|K_x(y_1) - K_x(y_2)| &\leq \|B^{-1}\|\, |R(x,\ y_1) - R(x,\ y_2)| \\
&\leq \|B^{-1}\| \left\| \frac{\partial R}{\partial y} \right\| |y_1 - y_2| \leq \frac{1}{2}|y_1 - y_2|
\end{aligned}
$$

for $|x|, |y_1|, |y_2| \leq r$. Due to continuity at the origin, if $|x| \leq \tau \ll r$ then

$$|K_x(0)| \leq \frac{r}{2}.$$

Thus, for each $x \in X$, K_x contracts Y into itself where X is the τ-neighborhood of 0 in $\mathbb{R}^n$ and Y is the closure of the r-neighborhood of 0 in $\mathbb{R}^m$. See Figure 111.

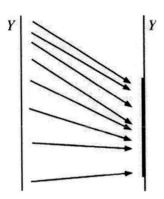

Figure 111 K_x contracts Y into itself.

By the Contraction Mapping Principle, K_x has a unique fixed point $g(x)$ in Y. This implies that near the origin, the zero locus of f is the graph of a function $y = g(x)$.

It remains to check that g is C^r. First we show that g obeys a Lipschitz condition at 0. We have

$$
\begin{aligned}
|gx| &= |K_x(gx) - K_x(0) + K_x(0)| \;\le\; \mathrm{Lip}(K_x)\,|gx - 0| + |K_x(0)| \\
&\le\; \frac{|gx|}{2} + |B^{-1}(Ax + R(x,0))| \;\le\; \frac{|gx|}{2} + 2L|x|
\end{aligned}
$$

where $L = \|B^{-1}\|\,(\|A\| + 1)$ and $|x|$ is small. Thus g satisfies the Lipschitz condition

$$
|gx| \;\le\; 4L|x|.
$$

In particular g is continuous at $x = 0$.

Note the trick here. The term $|gx|$ appears on both sides of the inequality but since its coefficient on the r.h.s. is smaller than that on the l.h.s., they combine to give a nontrivial inequality.

By the Chain Rule, the derivative of g at the origin, if it does exist, must satisfy $A + B(Dg)_0 = 0$, so we aim to show that $(Dg)_0 = -B^{-1}A$. Since gx is a fixed-point of K_x we have $gx = -B^{-1}A(x + R)$ and the Taylor estimate for g at the origin is

$$
\begin{aligned}
|g(x) - g(0) - (-B^{-1}Ax)| &= |B^{-1}R(x, gx)| \;\le\; \|B^{-1}\|\,|R(x, gx)| \\
&\le\; \|B^{-1}\|\,\mathbf{e}(x, gx)(|x| + |gx|) \\
&\le\; \|B^{-1}\|\,\mathbf{e}(x, gx)(1 + 4L)|x|
\end{aligned}
$$

where $\epsilon(x, y) \to 0$ as $(x, y) \to (0, 0)$. Since $gx \to 0$ as $x \to 0$, the error factor $\epsilon(x, gx)$ does tend to 0 as $x \to 0$, the remainder is sublinear with respect to x, and g is differentiable at 0 with $(Dg)_0 = -B^{-1}A$.

All facts proved at the origin hold equally at points (x, y) on the zero locus near the origin. For the origin is nothing special. Thus, g is differentiable at x and $(Dg)_x = -B_x^{-1} \circ A_x$ where

$$A_x = \frac{\partial f(x, gx)}{\partial x} = \left[\frac{\partial f_i}{\partial x_j}\right]_{(x, gx)} \qquad B_x = \frac{\partial f(x, gx)}{\partial y} = \left[\frac{\partial f_i}{\partial y_j}\right]_{(x, gx)}.$$

Since gx is continuous (being differentiable) and f is C^1, A_x and B_x are continuous functions of x. According to Cramer's Rule for finding the inverse of a matrix, the entries of B_x^{-1} are explicit, algebraic functions of the entries of B_x, and therefore they depend continuously on x. Therefore g is C^1.

To complete the proof that g is C^r we apply induction. For $2 \leq r < \infty$, assume the theorem is true for $r - 1$. When f is C^r this implies that g is C^{r-1}. Because they are composites of C^{r-1} functions, A_x and B_x are C^{r-1}. Because the entries of B_x^{-1} depend algebraically on the entries of B_x, B_x^{-1} is also C^{r-1}. Therefore $(Dg)_x$ is C^{r-1} and g is C^r. If f is C^∞, we have just shown that g is C^r for all finite r and thus g is C^∞. $\square$

Exercises 35 and 36 discuss the properties of matrix inversion avoiding Cramer's Rule and finite dimensionality.

Next we are going to deduce the Inverse Function Theorem from the Implicit Function Theorem. A fair question is: Since they turn out to be equivalent theorems, why not do it the other way around? Well, in my own experience the Implicit Function Theorem is more basic and flexible. I have at times needed forms of the Implicit Function Theorem with weaker differentiability hypotheses respecting x than y and they do not follow from the Inverse Function Theorem. For example, if we merely assume that $B = \partial f(x_0, y_0)/\partial y$ is invertible, that $\partial f(x, y)/\partial x$ is a continuous function of (x, y), and that f is continuous (or Lipschitz) then the local implicit function of f is continuous (or Lipschitz). It is not necessary to assume that f is of class C^1.

Just as a homeomorphism is a continuous bijection whose inverse is continuous, so a C^r **diffeomorphism** is a C^r bijection whose inverse is C^r, $1 \leq r \leq \infty$. The inverse being C^r is not automatic. The example to remember is $f(x) = x^3$. It is a smooth homeomorphism $\mathbb{R} \to \mathbb{R}$ but is not a diffeomorphism because its inverse fails to be differentiable at the origin. Since differentiability implies continuity, every diffeomorphism is a homeomorphism.

Diffeomorphisms are to C^r things as isomorphisms are to algebraic things. The sphere and ellipsoid are diffeomorphic under a diffeomorphism $\mathbb{R}^3 \to \mathbb{R}^3$ but the sphere and the surface of the cube are only homeomorphic, not diffeomorphic.

23 Inverse Function Theorem *If the derivative of f is invertible then f is a local diffeomorphism.*

Proof Invertibility of a matrix implies the matrix is square, so $m = n$. Then we have $f : U \to \mathbb{R}^m$, where U is an open subset of $\mathbb{R}^m$, and at some $p \in U$, $(Df)_p$ is assumed to be invertible. We assume f is C^r, $1 \leq r \leq \infty$, and set

$$F(x, y) = f(x) - y \qquad q = f(p)$$

for $(x, y) \in U \times \mathbb{R}^m$. Clearly F is C^r, $F(p, q) = 0$, and the derivative of F with respect to x at (p, q) is $(Df)_p$. We claim that f is a diffeomorphism from a neighborhood of p to a neighborhood of q.

Since $(Df)_p$ is an isomorphism we can apply the Implicit Function Theorem (with x and y interchanged!) to find neighborhoods U_p of p and V_q of q and a C^r implicit function $h : V_q \to U_p$ uniquely defined by the equation

$$F(hy, y) = f(hy) - y = 0.$$

This means that h is a "local right inverse" for f in the sense that $f \circ h = \mathrm{id}\,|_{V_q}$. Since $F(p, q) = 0$, uniqueness implies $p = hq$, and $(Df)_p \circ (Dh)_q = I$ implies $(Dh)_q$ is invertible.

We claim that h is also a "local left inverse" for f, and hence that f is a local diffeomorphism. We can apply the same analysis with h in place of f since it is C^r, it sends q to p, and its derivative at q is invertible. Consequently h has a unique local right inverse, say g. It satisfies $h \circ g = \mathrm{id}$ locally and we get

$$f = f \circ (h \circ g) = (f \circ h) \circ g = g.$$

Thus $h \circ f = h \circ g = \mathrm{id}$ shows that h is a local left inverse for f and we have $h = f^{-1}$ on a neighborhood of q. $\qquad\square$

5* The Rank Theorem

The **rank** of a linear transformation $T : \mathbb{R}^n \to \mathbb{R}^m$ is the dimension of its range. In terms of matrices, the rank is the size of the largest minor with nonzero determinant.

If T is onto then its rank is m. If it is one-to-one then its rank is n. A standard formula in linear algebra states that

$$\text{rank}\, T + \text{nullity}\, T = n$$

where nullity is the dimension of the kernel of T. A differentiable function $f : U \to \mathbb{R}^m$ has constant rank k if for all $p \in U$ the rank of $(Df)_p$ is k.

An important property of rank is that if T has rank k and $\|S - T\|$ is small then S has rank $\geq k$. The rank of T can increase under a small perturbation of T but it cannot decrease. Thus, if f is C^1 and $(Df)_p$ has rank k then automatically $(Df)_x$ has rank $\geq k$ for all x near p. See Exercise 43.

The Rank Theorem describes maps of constant rank. It says that locally they are just like linear projections. To formalize this we say that maps $f : A \to B$ and $g : C \to D$ are equivalent (for want of a better word) if there are bijections $\alpha : A \to C$ and $\beta : B \to D$ such that $g = \beta \circ f \circ \alpha^{-1}$. An elegant way to express this equation is a **commutative diagram**

$$
\begin{array}{ccc}
A & \xrightarrow{\ f\ } & B \\
\alpha \downarrow & & \downarrow \beta \\
C & \xrightarrow{\ g\ } & D.
\end{array}
$$

Commutativity means that for each $a \in A$ we have $\beta(f(a)) = g(\alpha(a))$. Following the maps around the rectangle clockwise from A to D gives the same result as following them around it counterclockwise. The α, β are "changes of variable." If f, g are C^r and α, β are C^r diffeomorphisms, $1 \leq r \leq \infty$, then f and g are said to be C^r **equivalent**, and we write $f \approx_r g$. As C^r maps, f and g are indistinguishable.

24 Lemma C^r *equivalence is an equivalence relation and it has no effect on rank.*

Proof Since diffeomorphisms form a group, $\approx_r$ is an equivalence relation. Also, if $g = \beta \circ f \circ \alpha^{-1}$ then the chain rule implies

$$Dg = D\beta \circ Df \circ D\alpha^{-1}.$$

Since $D\beta$ and $D\alpha^{-1}$ are isomorphisms, Df and Dg have equal rank. □

The linear projection $P : \mathbb{R}^n \to \mathbb{R}^m$

$$P(x_1, \ldots, x_n) = (x_1, \ldots, x_k, 0, \ldots, 0)$$

has rank k. It projects $\mathbb{R}^n$ onto the k-dimensional subspace $\mathbb{R}^k \times 0 \subset \mathbb{R}^m$. (We assume that $k \leq n, m$.) The $m \times n$ matrix of P is

$$\begin{bmatrix} I_{k \times k} & 0 \\ 0 & 0 \end{bmatrix}.$$

25 Rank Theorem *Locally, a C^r constant-rank-k map is C^r equivalent to a linear projection onto a k-dimensional subspace.*

As an example, think of the radial projection $\pi : \mathbb{R}^3 \setminus \{0\} \to S^2$, where $\pi(v) = v/|v|$. It has constant rank 2, and is locally indistinguishable from linear projection of $\mathbb{R}^3$ to the (x, y)-plane.

Proof Let $f : U \to \mathbb{R}^m$ have constant rank k and let $p \in U$ be given. We will show that on a neighborhood of p we have $f \approx_r P$.

Step 1. Define translations of $\mathbb{R}^n$ and $\mathbb{R}^m$ by

$$\tau : \mathbb{R}^n \to \mathbb{R}^n \qquad \tau' : \mathbb{R}^m \to \mathbb{R}^m$$
$$z \mapsto z + p \qquad z' \mapsto z' - fp.$$

The translations are diffeomorphisms of $\mathbb{R}^n$ and $\mathbb{R}^m$ and they show that f is C^r equivalent to $\tau' \circ f \circ \tau$, a C^r map that sends 0 to 0 and has constant rank k. Thus, it is no loss of generality to assume in the first place that p is the origin in $\mathbb{R}^n$ and fp is the origin in $\mathbb{R}^m$. We do so.

Step 2. Let $T : \mathbb{R}^n \to \mathbb{R}^n$ be an isomorphism that sends $0 \times \mathbb{R}^{n-k}$ onto the kernel of $(Df)_0$. Since the kernel has dimension $n - k$, there is such a T. Let $T' : \mathbb{R}^m \to \mathbb{R}^m$ be an isomorphism that sends the image of $(Df)_0$ onto $\mathbb{R}^k \times 0$. Since $(Df)_0$ has rank k, there is such a T'. Then $f \approx_r T' \circ f \circ T$. This map sends the origin in $\mathbb{R}^n$ to the origin in $\mathbb{R}^m$, while its derivative at the origin has kernel $0 \times \mathbb{R}^{n-k}$ and range $\mathbb{R}^k \times 0$. Thus it is no loss of generality to assume in the first place that f has these properties. We do so.

Step 3. Write

$$(x, y) \in \mathbb{R}^k \times \mathbb{R}^{n-k} \qquad f(x, y) = (f_X(x, y), f_Y(x, y)) \in \mathbb{R}^k \times \mathbb{R}^{m-k}.$$

We are going to find a $g \approx_r f$ such that

$$g(x, 0) = (x, 0).$$

The matrix of $(Df)_0$ is

$$\begin{bmatrix} A & 0 \\ 0 & 0 \end{bmatrix}$$

where A is $k \times k$ and invertible. By the Inverse Function Theorem the map

$$\sigma : x \mapsto f_X(x, 0)$$

is a diffeomorphism $\sigma : X \to X'$ where X and X' are small neighborhoods of the origin in $\mathbb{R}^k$ and f_X denotes the first k components of f. For $x' \in X'$, set

$$h(x') = f_Y(\sigma^{-1}(x'),\, 0).$$

This makes h a C^r map $X' \to \mathbb{R}^{m-k}$, and

$$h(\sigma(x)) = f_Y(x, 0)$$

where f_Y denotes the final $m - k$ components of f. The image of $X \times 0$ under f is the graph of h. For

$$
\begin{aligned}
f(X \times 0) &= \{f(x, 0) : x \in X\} = \{(f_X(x, 0),\, f_Y(x, 0)) : x \in X\} \\
&= \{(f_X(\sigma^{-1}(x'), 0),\, f_Y(\sigma^{-1}(x'), 0)) : x' \in X'\} \\
&= \{(x',\, h(x')) : x' \in X'\}.
\end{aligned}
$$

See Figure 112.

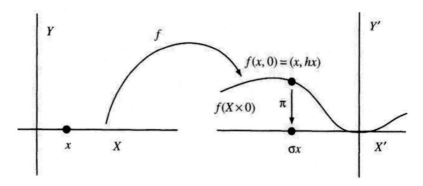

Figure 112 The f-image of $X \times 0$ is the graph of h.

If $(x', y') \in X' \times \mathbb{R}^{m-k}$ then we define

$$\psi(x', y') = (\sigma^{-1}(x'), \, y' - h(x')).$$

Since ψ is the composite of C^r diffeomorphisms,

$$(x', y') \mapsto (x', \, y' - h(x')) \mapsto (\sigma^{-1}(x'), \, y' - h(x')),$$

it too is a C^r diffeomorphism. (Alternatively, you could compute the derivative of ψ at the origin and apply the Inverse Function Theorem.) We observe that $g = \psi \circ f \approx_r f$ satisfies

$$
\begin{aligned}
g(x, 0) &= \psi \circ (f_X(x, 0), \, f_Y(x, 0)) \\
&= (\sigma^{-1} \circ f_X(x, 0), \, f_Y(x, 0) - h(f_X(x, 0))) = (x, 0).
\end{aligned}
$$

Thus it is no loss of generality to assume in the first place that $f(x, 0) = (x, 0)$. We do so. (This means that f sends the k-plane $\mathbb{R}^k \times 0 \subset \mathbb{R}^n$ into the k-plane $\mathbb{R}^k \times 0 \subset \mathbb{R}^m$.)

Step 4. Finally, we find a local diffeomorphism φ in the neighborhood of 0 in $\mathbb{R}^n$ so that $f \circ \varphi$ is the projection map $P(x, y) = (x, 0)$.

Define $F(\xi, x, y) = f_X(\xi, y) - x$. It is a map from $\mathbb{R}^k \times \mathbb{R}^k \times \mathbb{R}^{n-k}$ into $\mathbb{R}^k$. The equation

$$F(\xi, x, y) = 0$$

defines $\xi = \xi(x, y)$ implicitly in a neighborhood of the origin. For at the origin the derivative of F with respect to ξ is the invertible matrix $I_{k \times k}$. Thus ξ is a C^r map from $\mathbb{R}^n$ into $\mathbb{R}^k$ and $\xi(0, 0) = 0$. We claim that

$$\varphi(x, y) = (\xi(x, y), y)$$

is a local diffeomorphism of $\mathbb{R}^n$ and $G = f \circ \varphi$ is P.

The derivative of $\xi(x, y)$ with respect to x at the origin can be calculated from the Chain Rule (this was done in general for implicit functions) and since $F(\xi, x, y) \equiv 0$ we have

$$
0 \;=\; \frac{dF(\xi(x, y), x, y)}{dx} \;=\; \frac{\partial F}{\partial \xi} \frac{\partial \xi}{\partial x} + \frac{\partial F}{\partial x} \;=\; I_{k \times k} \frac{\partial \xi}{\partial x} - I_{k \times k}.
$$

That is, at the origin $\partial \xi / \partial x$ is the identity matrix. Thus,

$$(D\varphi)_0 = \begin{bmatrix} I_{k \times k} & * \\ 0 & I_{(n-k) \times (n-k)} \end{bmatrix}$$

which is invertible no matter what $*$ is. Clearly $\varphi(0) = 0$. By the Inverse Function Theorem, φ is a local C^r diffeomorphism on a neighborhood of the origin and G is C^r equivalent to f. By Lemma 24, G has constant rank k.

We have

$$
\begin{aligned}
G(x,y) &= f \circ \varphi(x,y) = f(\xi(x,y),\, y) \\
&= (f_X(\xi,y),\, f_Y(\xi,y)) = (x,\, G_Y(x,y)).
\end{aligned}
$$

Therefore $G_X(x,y) = x$ and

$$
DG = \begin{bmatrix} I_{k \times k} & 0 \\ * & \dfrac{\partial G_Y}{\partial y} \end{bmatrix}.
$$

At last we use the constant-rank hypothesis. (Until now, it has been enough that Df has rank $\geq k$.) The only way that a matrix of this form can have rank k is that

$$
\frac{\partial G_Y}{\partial y} \equiv 0.
$$

See Exercise 43. By Corollary 13 to the Mean Value Theorem this implies that in a neighborhood of the origin, G_Y is independent of y. Thus

$$
G_Y(x,y) = G_Y(x,0) = f_Y(\xi(x,0),0),
$$

which is 0 because (by Step 3) $f_Y = 0$ on $\mathbb{R}^k \times 0$. The upshot is that $G \approx_r f$ and $G(x,y) = (x,0)$; i.e., $G = P$. See also Exercise 31. By Lemma 24, steps 1-4 concatenate to give a C^r equivalence between the original constant-rank map f and the linear projection P. □

In the following three corollaries U is an open subset of $\mathbb{R}^n$.

26 Corollary *If $f : U \to \mathbb{R}^m$ has rank k at p then it is locally C^r equivalent to a map of the form $G(x,y) = (x, g(x,y))$ where $g : \mathbb{R}^n \to \mathbb{R}^{m-k}$ is C^r and $x \in \mathbb{R}^k$.*

Proof This was shown in the proof of the Rank Theorem before we used the assumption that f has constant-rank k. □

27 Corollary *If $f : U \to \mathbb{R}$ is C^r and $(Df)_p$ has rank 1 then in a neighborhood of p the level sets $\{x \in U : f(x) = c\}$ form a stack of C^r nonlinear discs of dimension $n-1$.*

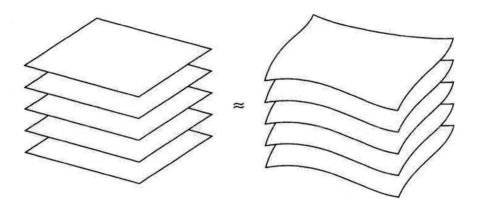

Figure 113 Near a rank-one point, the level sets of $f : U \to \mathbb{R}$ are
diffeomorphic to a stack of $(n-1)$-dimensional planes.

Proof Near p the rank can not decrease, so f has constant rank 1 near p. The level
sets of a projection $\mathbb{R}^n \to \mathbb{R}$ form a stack of $(n-1)$-dimensional planes and the level
sets of f are the images of these planes under the equivalence diffeomorphism in the
Rank Theorem. See Figure 113. □

28 Corollary *If $f : U \to \mathbb{R}^m$ has rank n at p then locally the image of U under f is
a diffeomorphic copy of the n-dimensional disc.*

Proof Near p the rank can not decrease, so f has constant rank n near p. The Rank
Theorem says that f is locally C^r equivalent to $x \mapsto (x, 0)$. (Since $k = n$, the y-
coordinates are absent.) Thus the local image of U is diffeomorphic to a neighborhood
of 0 in $\mathbb{R}^n \times 0$ which is an n-dimensional disc. □

The geometric meaning of the diffeomorphisms ψ and φ is illustrated in the Figures 114 and 115.

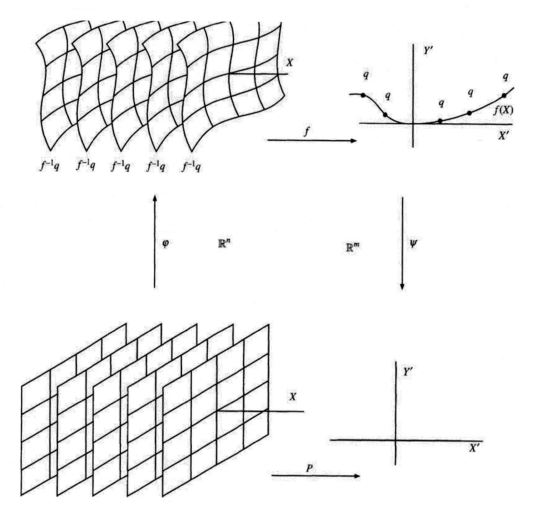

Figure 114 f has constant rank 1.

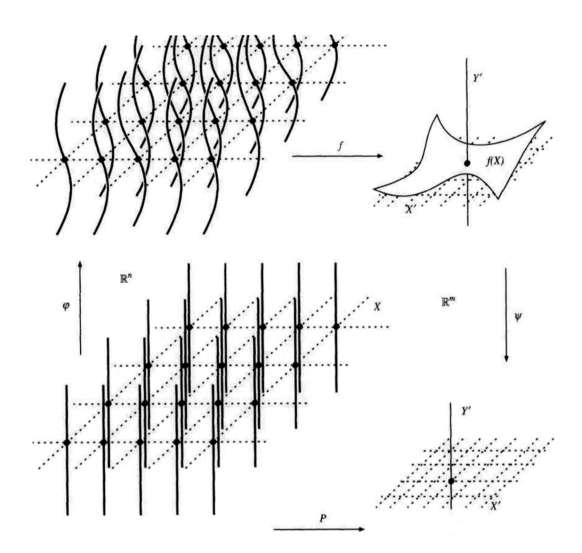

Figure 115 f has constant rank 2.

6* Lagrange Multipliers

In sophomore calculus you learn how to maximize a function $f(x, y, z)$ subject to a "constraint" or "side condition" $g(x, y, z) = constant$ by the Lagrange multiplier method. Namely, the maximum can occur only at a point p where the gradient of f is a scalar multiple of the gradient of g,

$$\operatorname{grad}_p f = \lambda \operatorname{grad}_p g.$$

The factor λ is the **Lagrange multiplier**. The goal of this section is a natural, mathematically complete explanation of the Lagrange multiplier method which amounts to gazing at the right picture.

First, the natural hypotheses are

(a) f and g are C^1 real-valued functions defined on some region $U \subset \mathbb{R}^3$.
(b) For some constant c, the set $S = g^{\text{pre}}(c)$ is compact, nonempty, and $\operatorname{grad}_q g \neq 0$ for all $q \in S$.

The conclusion is

(c) The restriction of f to the set S, $f|_S$, has a maximum, say M, and if $p \in S$ has $f(p) = M$ then there is a λ such that $\operatorname{grad}_p f = \lambda \operatorname{grad}_p g$.

The method is utilized as follows. You are given[†] f and g, and you are asked to find a point $p \in S$ at which $f|_S$ is maximum. Compactness implies that a maximum point exists. Your job is to find it. You first locate all points $q \in S$ at which the gradients of f and g are linearly dependent; i.e., one gradient is a scalar multiple of the other. They are "candidates" for the maximum point. You then evaluate f at each candidate and the one with the largest f-value is the maximum. Done.

Of course you can find the minimum the same way. It too will be among the candidates, and it will have the smallest f-value. In fact, the candidates are exactly the critical points of $f|_S$, the points $x \in S$ such that

$$\frac{fy - fx}{|y - x|} \to 0$$

as $y \in S$ tends to x.

[†]Sometimes you are merely given f and S. Then you must think up an appropriate g such that (b) is true.

Now we explain why the Lagrange multiplier method works. Recall that the **gradient** of a function $h(x, y, z)$ at $p \in U$ is the vector

$$\text{grad}_p h = \left(\frac{\partial h(p)}{\partial x}, \frac{\partial h(p)}{\partial y}, \frac{\partial h(p)}{\partial z} \right) \in \mathbb{R}^3.$$

Assume hypotheses (a), (b) and that $f|_S$ attains its maximum value M at $p \in S$. We must prove (c) – the gradient of f at p is a scalar multiple of the gradient of g at p. If $\text{grad}_p f = 0$ then $\text{grad}_p f = 0 \cdot \text{grad}_p g$, which verifies (c) degenerately. Thus it is fair to assume that $\text{grad}_p f \neq 0$.

By the Rank Theorem, in the neighborhood of a point at which the gradient of f is nonzero, the f-level surfaces are like a stack of pancakes. (The pancakes are infinitely thin and may be somewhat curved. Alternatively, you can picture the level surfaces as layers of an onion skin or as a pile of transparency foils.)

To arrive at a contradiction, assume that $\text{grad}_p f$ is not a scalar multiple of $\text{grad}_p g$. The angle between the gradients is nonzero. Gaze at the f-level surfaces $f = M \pm \epsilon$ for ϵ small. The way these f-level surfaces meet the g-level surface S is shown in Figure 116.

The surface S is a knife blade that slices through the f-pancakes. The knife blade is perpendicular to grad g, while the pancakes are perpendicular to grad f. There is a positive angle between these gradient vectors, so the knife is not tangent to the pancakes. Rather, S slices transversely through each f-level surface near p, and $S \cap \{f = M + \epsilon\}$ is a curve that passes near p. The value of f on this curve is $M + \epsilon$, which contradicts the assumption that $f|_S$ attains a maximum at p. Therefore $\text{grad}_p f$ is, after all, a scalar multiple of $\text{grad}_p g$ and the proof of (c) is complete.

There is a higher-dimensional version of the Lagrange multiplier method. A C^1 function $f : U \to \mathbb{R}$ is defined on an open set $U \subset \mathbb{R}^n$, and it is constrained to a compact "surface" $S \subset U$ defined by k simultaneous equations

$$g_1(x_1, \ldots, x_n) = c_1$$
$$\ldots$$
$$g_k(x_1, \ldots, x_n) = c_k.$$

We assume the functions g_i are C^1 and their gradients are linearly independent. The higher-dimensional Lagrange multiplier method asserts that if $f|_S$ achieves a maximum at p then $\text{grad}_p f$ is a linear combination of $\text{grad}_p g_1, \ldots, \text{grad}_p g_k$. In contrast to Protter and Morrey's presentation on pages 369-372 of their book, A

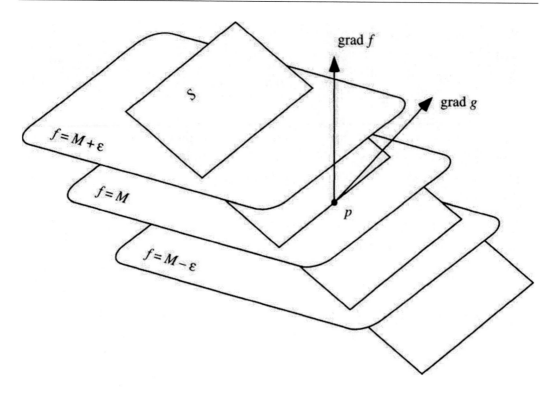

grad f

grad g

S

$f = M + \varepsilon$

$f = M$

$f = M - \varepsilon$

p

Figure 116 S cuts through all the f-level surfaces near p.

First Course in Real Analysis, the proof is utterly simple: It amounts to examining the situation in the right coordinate system at p.

It is no loss of generality to assume that p is the origin in $\mathbb{R}^n$ and that $c_1, \ldots, c_k$, $f(p)$ are zero. Also, we can assume that $\operatorname{grad}_p f \neq 0$, since otherwise it is already a trivial linear combination of the gradients of the g_i. Then choose vectors $w_{k+2}, \ldots, w_n$ so that

$$\operatorname{grad}_0 g_1, \ldots, \operatorname{grad}_0 g_k, \operatorname{grad}_0 f, w_{k+2}, \ldots, w_n$$

is a vector basis of $\mathbb{R}^n$. For $k + 2 \leq i \leq n$ define

$$h_i(x) = \langle w_i, x \rangle.$$

The map $x \mapsto F(x) = (g_1(x), \ldots, g_k(x), f(x), h_{k+2}(x), \ldots, h_n(x))$ is a local diffeomorphism of $\mathbb{R}^n$ to itself since the derivative of F at the origin is the $n \times n$ matrix of linearly independent column vectors

$$(DF)_0 = [\, \operatorname{grad}_0 g_1 \ \ldots \ \operatorname{grad}_0 g_k \ \operatorname{grad}_0 f \ w_{k+2} \ \ldots \ w_n \,].$$

Think of the functions $y_i = F_i(x)$ as new coordinates on a neighborhood of the origin in $\mathbb{R}^n$. With respect to these coordinates, the surface S is the coordinate plane

$0 \times \mathbb{R}^{n-k}$ on which the coordinates $y_1, \ldots, y_k$ are zero and f is the $(k+1)^{\text{st}}$ coordinate function y_{k+1}. This coordinate function obviously does not attain a maximum on the coordinate plane $0 \times \mathbb{R}^{n-k}$, so $f|_S$ attains no maximum at p.

7 Multiple Integrals

In this section we generalize to n variables the one-variable Riemann integration theory appearing in Chapter 3. For simplicity, we assume throughout that the function f we integrate is real-valued, as contrasted to vector-valued, and at first we assume that f is a function of only two variables.

Consider a rectangle $R = [a, b] \times [c, d]$ in $\mathbb{R}^2$. Partitions P and Q of $[a, b]$ and $[c, d]$

$$P : a = x_0 < x_1 < \ldots < x_m = b \qquad Q : c = y_0 < y_1 < \ldots < y_n = d$$

give rise to a "grid" $G = P \times Q$ of rectangles

$$R_{ij} = I_i \times J_j$$

where $I_i = [x_{i-1}, x_i]$ and $J_j = [y_{j-1}, y_j]$. Let $\Delta x_i = x_i - x_{i-1}$, $\Delta y_j = y_j - y_{j-1}$, and denote the area of R_{ij} as

$$|R_{ij}| = \Delta x_i \, \Delta y_j.$$

Let S be a choice of sample points $(s_{ij}, t_{ij}) \in R_{ij}$. See Figure 117.

Given $f : R \to \mathbb{R}$, the corresponding Riemann sum is

$$R(f, G, S) = \sum_{i=1}^{m} \sum_{j=1}^{n} f(s_{ij}, t_{ij}) |R_{ij}|.$$

If there is a number to which the Riemann sums converge as the mesh of the grid (the diameter of the largest rectangle) tends to zero then f is Riemann integrable and that number is the Riemann integral

$$\int_R f = \lim_{\text{mesh } G \to 0} R(f, G, S).$$

The lower and upper sums of a bounded function f with respect to the grid G are

$$L(f, G) = \sum m_{ij} |R_{ij}| \qquad U(f, G) = \sum M_{ij} |R_{ij}|$$

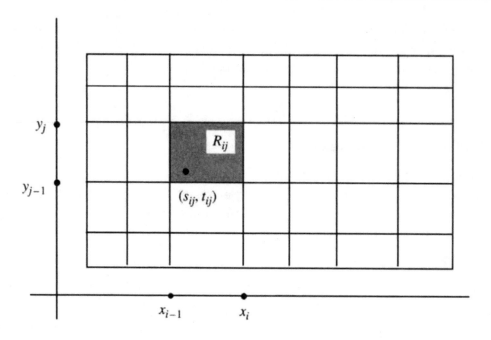

Figure 117 A grid and a sample point

where m_{ij} and M_{ij} are the infimum and supremum of $f(s,t)$ as (s,t) varies over R_{ij}. The lower integral is the supremum of the lower sums and the upper integral is the infimum of the upper sums.

The proofs of the following facts are conceptually identical to the one-dimensional versions explained in Chapter 3:

(a) If f is Riemann integrable then it is bounded.

(b) The set of Riemann integrable functions $R \to \mathbb{R}$ is a vector space $\mathcal{R} = \mathcal{R}(R)$ and integration is a linear map $\mathcal{R} \to \mathbb{R}$.

(c) The constant function $f = k$ is integrable and its integral is $k|R|$.

(d) If $f, g \in \mathcal{R}$ and $f \le g$ then

$$\int_R f \le \int_R g.$$

(e) Every lower sum is less than or equal to every upper sum, and consequently the lower integral is no greater than the upper integral,

$$\underline{\int}_R f \le \overline{\int}_R f.$$

(f) For a bounded function, Riemann integrability is equivalent to the equality of the lower and upper integrals, and integrability implies equality of the lower, upper, and Riemann integrals.

The Riemann-Lebesgue Theorem is another result that generalizes naturally to multiple integrals. It states that a bounded function is Riemann integrable if and only if its discontinuities form a zero set.

First of all, $Z \subset \mathbb{R}^2$ is a **zero set** if for each $\epsilon > 0$ there is a countable covering of Z by open rectangles S_ℓ whose total area is less than ϵ:

$$\sum_\ell |S_\ell| < \epsilon.$$

By the $\epsilon/2^\ell$ construction, a countable union of zero sets is a zero set.

As in dimension 1, we express the discontinuity set of our function $f : R \to \mathbb{R}$ as the union

$$D = \bigcup_{k \in \mathbb{N}} D_k,$$

where D_k is the set of points $z \in R$ at which the oscillation is $\geq 1/k$. (See Exercise 3.19.) That is,

$$\mathrm{osc}_z f = \lim_{r \to 0} \mathrm{diam}\left(f(R_r(z))\right) \geq 1/k$$

where $R_r(z)$ is the r-neighborhood of z in R. The set D_k is compact.

Assume that $f : R \to \mathbb{R}$ is Riemann integrable. It is bounded and its upper and lower integrals are equal. Fix $k \in \mathbb{N}$. Given $\epsilon > 0$, there exists $\delta > 0$ such that if G is a grid with mesh $< \delta$ then

$$U(f, G) - L(f, G) < \epsilon.$$

Fix such a grid G. Each R_{ij} in the grid that contains in its interior a point of D_k has $M_{ij} - m_{ij} \geq 1/k$, where m_{ij} and M_{ij} are the infimum and supremum of f on R_{ij}. The other points of D_k lie in the zero set of gridlines $x_i \times [c, d]$ and $[a, b] \times y_j$. Since $U - L < \epsilon$, the total area of these rectangles with oscillation $\geq 1/k$ does not exceed $k\epsilon$. Since k is fixed and ϵ is arbitrary, D_k is a zero set. Taking $k = 1, 2, \ldots$ shows that the discontinuity set $D = \bigcup D_k$ is a zero set.

Conversely, assume that f is bounded and D is a zero set. Fix any $k \in \mathbb{N}$. Each $z \in R \setminus D_k$ has a neighborhood $W = W_z$ such that

$$\sup\{f(w) : w \in W\} - \inf\{f(w) : w \in W\} < 1/k.$$

Since D_k is a zero set, it can be covered by countably many open rectangles S_ℓ of small total area, say

$$\sum |S_\ell| < \sigma.$$

Let $\mathcal{V}$ be the covering of R by the neighborhoods W with small oscillation, and the rectangles S_ℓ. Since R is compact, $\mathcal{V}$ has a positive Lebesgue number λ. Take a grid with mesh $< \lambda$. This breaks the sum

$$U - L = \sum (M_{ij} - m_{ij})|R_{ij}|$$

into two parts – the sum of those terms for which R_{ij} is contained in a neighborhood W with small oscillation, plus a sum of terms for which R_{ij} is contained in one of the rectangles S_ℓ. The latter sum is less than $2M\sigma$, while the former is less than $|R|/k$. Thus, when k is large and σ is small, $U - L$ is small, which implies Riemann integrability. To summarize,

The Riemann-Lebesgue Theorem remains valid

for functions of several variables.

Now we come to the first place that multiple integration has something new to say. Suppose that $f : R \to \mathbb{R}$ is bounded and define the lower and upper **slice integrals**

$$\underline{F}(y) = \int_{\underline{\;}a}^{b} f(x,y)\,dx \qquad \overline{F}(y) = \int_{a}^{\overline{\;}b} f(x,y)\,dx.$$

For each fixed $y \in [c,d]$, these are the lower and upper integrals of the single-variable function $f_y : [a,b] \to \mathbb{R}$ defined by $f_y(x) = f(x,y)$. They are the integrals of $f(x,y)$ on the horizontal line $y = \text{const}$. See Figure 118.

29 Fubini's Theorem *If f is Riemann integrable then so are $\underline{F}$ and $\overline{F}$. Moreover,*

$$\int_R f \;=\; \int_c^d \underline{F}\,dy \;=\; \int_c^d \overline{F}\,dy.$$

Since $\underline{F} \le \overline{F}$ and the integral of their difference is zero, it follows from the one-dimensional Riemann-Lebesgue Theorem that there exists a linear zero set $Y \subset [c,d]$ such that if $y \notin Y$ then $\underline{F}(y) = \overline{F}(y)$. That is, the integral of $f(x,y)$ with respect to x exists for almost all y and we get the more common way to write the Fubini formula

$$\iint_R f\,dxdy \;=\; \int_c^d \left[\int_a^b f(x,y)\,dx \right] dy.$$

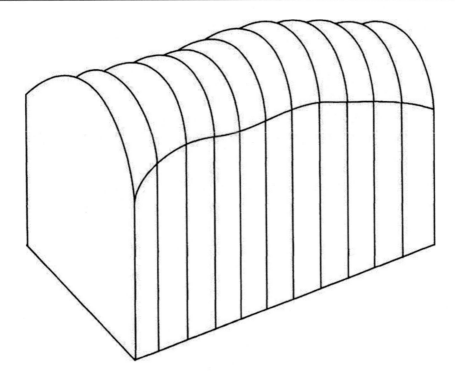

Figure 118 Fubini's Theorem is like sliced bread.

There is, however, an ambiguity in this formula. What is the value of the integrand $\int_a^b f(x, y)\, dx$ when $y \in Y$? For such a y, $\underline{F}(y) < \overline{F}(y)$ and the integral of $f(x, y)$ with respect to x does not exist. The answer is that we can choose any value between $\underline{F}(y)$ and $\overline{F}(y)$. The integral with respect to y will be unaffected. See also Exercise 47.

Proof of Fubini's Theorem If $G = P \times Q$ partitions $[a, b] \times [c, d]$ we claim that

$$(9) \qquad\qquad L(f, G) \leq L(\underline{F}, Q)$$

Fix any partition interval $J_j \subset [c, d]$. If $y \in J_j$ then

$$m_{ij} = \inf\{f(s, t) : (s, t) \in R_{ij}\} \leq \inf\{f(s, y) : s \in I_i\} = m_i(f_y).$$

Thus, for all $y \in J_j$ we have

$$\sum_{i=1}^m m_{ij}\Delta x_i \leq \sum_{i=1}^m m_i(f_y)\Delta x_i = L(f_y, P) \leq \underline{F}(y).$$

Since $m_j(\underline{F}) = \inf\{\underline{F}(y) : y \in J_j\}$, it follows that

$$\sum_{i=1}^m m_{ij}\Delta x_i \leq m_j(\underline{F}).$$

Therefore

$$L(f, G) = \sum_{j=1}^{n}\sum_{i=1}^{m} m_{ij}\Delta x_i \Delta y_j \leq \sum_{j=1}^{n} m_j(\underline{F})\Delta y_j = L(\underline{F}, Q)$$

which gives (9). Analogously, $U(\overline{F}, Q) \leq U(f, G)$. Thus

$$L(f, G) \leq L(\underline{F}, Q) \leq U(\overline{F}, Q) \leq U(f, G).$$

Since f is integrable, the outer terms of this inequality differ by arbitrarily little when the mesh of G is small. Taking infima and suprema over all grids $G = P \times Q$ gives

$$\int_R f = \sup_G L(f, G) \leq \sup_Q L(\underline{F}, Q) \leq \inf_Q U(\underline{F}, Q)$$

$$\leq \inf_G U(f, G) = \int_R f.$$

The resulting equality of these five quantities implies that $\underline{F}$ is integrable and its integral on $[c, d]$ equals that of f on R. The integral of $\overline{F}$ is handled the same way. $\square$

30 Corollary *If f is Riemann integrable then the order of integration – first x then y or vice versa – is irrelevant to the value of the iterated integral,*

$$\int_c^d \left[\int_a^b f(x, y)\, dx \right] dy = \int_a^b \left[\int_c^d f(x, y)\, dy \right] dx.$$

Proof Both iterated integrals equal the integral of f over R. $\square$

A geometric consequence of Fubini's Theorem concerns the calculation of the area of plane regions by a slice method. Corresponding slice methods are valid in higher dimensions too.

31 Cavalieri's Principle *The area of a region $S \subset R$ is the integral with respect to x of the length of its vertical slices,*

$$\text{area}(S) = \int_a^b \text{length}(S_x)\, dx,$$

provided that the boundary of S is a zero set.

Proof Deriving Cavalieri's Principle from Fubini's Theorem is mainly a matter of definition. For we define the length of a subset of $\mathbb{R}$ and the area of a subset of $\mathbb{R}^2$ to be the integrals of their characteristic functions. The requirement that ∂S is a zero set is made so that χ_S is Riemann integrable. It is met if S has a smooth, or piecewise smooth, boundary. See Appendix B for a delightful discussion of the historical origin of Cavalieri's Principle, and see Chapter 6 for the more general geometric definition of length and area in terms of outer measure. $\square$

The second new aspect of multiple integration concerns the change of variables formula. It is the higher-dimensional version of integration by substitution. We will suppose that $\varphi : U \to W$ is a C^1 diffeomorphism between open subsets of $\mathbb{R}^2$, that $R \subset U$, and that a Riemann integrable function $f : W \to \mathbb{R}$ is given. The **Jacobian** of φ at $z \in U$ is the determinant of the derivative,

$$\operatorname{Jac}_z \varphi = \det(D\varphi)_z.$$

32 Change of Variables Formula *Under the preceding assumptions we have*

$$\int_R f \circ \varphi \cdot |\operatorname{Jac}\varphi| = \int_{\varphi R} f.$$

See Figure 119.

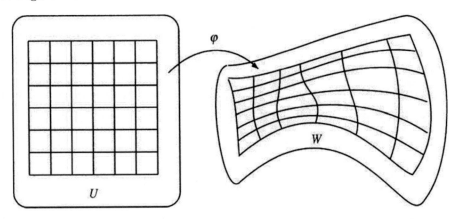

Figure 119 φ is a change of variables.

If S is a bounded subset of $\mathbb{R}^2$, its **area** (or **Jordan content**) is by definition the integral of its characteristic function χ_S, if the integral exists. When the integral does exist we say that S is **Riemann measurable**. See also Appendix D of Chapter 6. According to the Riemann-Lebesgue Theorem, S is Riemann measurable if and only if its boundary is a zero set. For χ_S is discontinuous at z if and only if z is a boundary point of S. See Exercise 44. The characteristic function of a rectangle R is Riemann integrable and its integral is $|R|$, so we are justified in using the same notation for area of a general set S, namely,

$$|S| = \text{area}(S) = \int \chi_S.$$

33 Proposition *If $T : \mathbb{R}^2 \to \mathbb{R}^2$ is an isomorphism then for every Riemann measurable set $S \subset \mathbb{R}^2$, $T(S)$ is Riemann measurable and*

$$|T(S)| = |\det T||S|.$$

Proposition 33 is a version of the Change of Variables Formula in which $\varphi = T$, $R = S$, and $f = 1$. It remains true in $\mathbb{R}^n$ and is called the **Volume Multiplier Formula**. It leads to a *definition* of the determinant of a linear transformation $\mathbb{R}^n \to \mathbb{R}^n$ as a volume multiplier.

Proof As is shown in linear algebra, the matrix A that represents T is a product of elementary matrices

$$A = E_1 \cdots E_k.$$

Each elementary 2×2 matrix is one of the following types:

$$\begin{bmatrix} \lambda & 0 \\ 0 & 1 \end{bmatrix} \qquad \begin{bmatrix} 1 & 0 \\ 0 & \lambda \end{bmatrix} \qquad \begin{bmatrix} 0 & 1 \\ 1 & 0 \end{bmatrix} \qquad \begin{bmatrix} 1 & \sigma \\ 0 & 1 \end{bmatrix}$$

where $\lambda > 0$. The first three matrices represent isomorphisms whose effect on I^2 is obvious: I^2 is converted to the rectangles $\lambda I \times I$, $I \times \lambda I$, and I^2. In each case the area agrees with the magnitude of the determinant. The fourth matrix is a **shear matrix**. Its isomorphism converts I^2 to the parallelogram

$$\Pi \;=\; \{(x,y) \in \mathbb{R}^2 : \sigma y \leq x \leq 1 + \sigma y \text{ and } 0 \leq y \leq 1\}.$$

Π is Riemann measurable since its boundary is a zero set. By Fubini's Theorem, we get

$$|\Pi| = \int \chi_\Pi = \int_0^1 \left[\int_{x=\sigma y}^{x=1+\sigma y} 1 \, dx \right] dy = 1 = \det E.$$

Exactly the same thinking shows that for any rectangle R, not merely the unit square,

(10) $$|E(R)| = |\det E||R|.$$

We claim that (10) implies that for any Riemann measurable set S, $E(S)$ is Riemann measurable and

(11) $$|E(S)| = |\det E||S|.$$

Let $\epsilon > 0$ be given. Choose a grid G on $R \supset S$ with mesh so small that the rectangles R of G satisfy

$$(12) \qquad |S| - \epsilon \;\leq\; \sum_{R \subset S} |R| \;\leq\; \sum_{R \cap S \neq \emptyset} |R| \;\leq\; |S| + \epsilon.$$

The interiors of the inner rectangles – those with $R \subset S$ – are disjoint, and therefore for each $z \in \mathbb{R}^2$ we have

$$\sum_{R \subset S} \chi_{\mathrm{int}\,R}(z) \leq \chi_S(z).$$

The same is true after we apply E, namely

$$\sum_{R \subset S} \chi_{\mathrm{int}(E(R))}(z) \leq \chi_{E(S)}(z).$$

Linearity and monotonicity of the integral, and Riemann measurability of the sets $E(R)$ imply that

$$(13) \qquad \sum_{R \subset S} |E(R)| = \sum_{R \subset S} \int \chi_{\mathrm{int}(E(R))} = \sum_{R \subset S} \underline{\int} \chi_{\mathrm{int}(E(R))} \leq \underline{\int} \chi_{E(S)}.$$

Similarly,

$$\chi_{E(S)}(z) \leq \sum_{R \cap S \neq \emptyset} \chi_{E(R)}(z)$$

which implies that

$$(14) \qquad \overline{\int} \chi_{E(S)} \leq \sum_{R \cap S \neq \emptyset} \overline{\int} \chi_{E(R)} = \sum_{R \cap S \neq \emptyset} \int \chi_{E(R)} = \sum_{R \cap S \neq \emptyset} |E(R)|.$$

By (10) and (12), (13) and (14) become

$$|\det E|(|S| - \epsilon) \;\leq\; |\det E| \sum_{R \subset S} |R|$$

$$\leq\; \underline{\int} \chi_{E(S)} \;\leq\; \overline{\int} \chi_{E(S)} \;\leq\; |\det E| \sum_{R \cap S \neq \emptyset} |R|$$

$$\leq\; |\det E|(|S| + \epsilon).$$

Since these upper and lower integrals do not depend on ϵ and ϵ is arbitrarily small, they equal the common value $|\det E|\,|S|$, which completes the proof of (11).

The determinant of a matrix product is the product of the determinants. Since the matrix of T is the product of elementary matrices, $E_1 \cdots E_k$, (11) implies that if S is Riemann measurable then so is $T(S)$ and

$$\begin{aligned} |T(S)| &= |E_1 \cdots E_k(S)| \\ &= |\det E_1| \cdots |\det E_k||S| = |\det T||S|. \qquad \square \end{aligned}$$

We isolate two facts in preparation for the proof of the Change of Variables Formula. We use the maximum coordinate norm on $\mathbb{R}^2$ and the associated operator norm on its linear transformations. With respect to this norm on $\mathbb{R}^2$, the closed r-neighborhood of p is a square $S_r(p)$ of side length $2r$ and center p. We write S_r when $p = (0, 0)$.

34 Lemma *Suppose that $\psi : U \to \mathbb{R}^2$ is C^1, $0 \in U$, $\psi(0) = 0$, and for all $u \in U$ we have $\|(D\psi)_u - Id\| \leq \sigma$. If $S_r \subset U$ then*

$$S_{r(1-\sigma)} \subset \psi(S_r) \subset S_{r(1+\sigma)}.$$

Proof We have $\psi(u) = u + \rho(u)$ where the remainder ρ is C^1, $\rho(0) = 0$, and $\|D\rho\| \leq \sigma$. If $|u| \leq r$ this implies $|\psi(u)| \leq r(1+\sigma)$, so $\psi(S_r) \subset S_{r(1+\sigma)}$ as claimed. The other half of the assertion is a quantitative version of the Inverse Function Theorem.

For each $v \in S_{r(1-\sigma)}$ we claim there is a unique $u \in S_r$ such that $v = \psi(u)$, and hence $S_{r(1-\sigma)} \subset \psi(S_r)$. Define $K_v : S_r \to \mathbb{R}^2$ by

$$K_v(u) = v - \rho(u).$$

It contracts S_r into itself. For $|v| \leq r(1 - \sigma)$ and $|u| \leq r$ imply

$$|K_v(u)| \leq r(1 - \sigma) + \sigma r = r,$$

while $|v| \leq r(1 - \sigma)$ and $|u|, |u'| \leq r$ imply

$$\left| K_v(u) - K_v(u') \right| = \left| \rho(u') - \rho(u) \right| \leq \sigma \left| u - u' \right|$$

by the Mean Value Theorem. The contraction K_v has a unique fixed point u. It satisfies $u = K_v(u) = v - \rho(u)$, so $\psi(u) = u + \rho(u) = v$. □

35 Lemma *The Lipschitz image of a zero set is a zero set.*

Proof Suppose that Z is a zero set and $h : Z \to \mathbb{R}^2$ satisfies a Lipschitz condition

$$|h(z) - h(z')| \leq L|z - z'|.$$

Given $\epsilon > 0$, there is a countable covering of Z by squares S_k such that

$$\sum_k |S_k| < \epsilon.$$

See Exercise 45. Each set $S_k \cap Z$ has diameter $\leq \operatorname{diam} S_k$ and therefore $h(Z \cap S_k)$ has diameter $\leq L \operatorname{diam} S_k$. As such it is contained in a square S'_k of edge length $L \operatorname{diam} S_k$. The squares S'_k cover $h(Z)$ and

$$\sum_k |S'_k| \leq L^2 \sum_k (\operatorname{diam} S_k)^2 = 2L^2 \sum_k |S_k| \leq 2L^2 \epsilon.$$

Therefore $h(Z)$ is a zero set. $\qquad\qquad\qquad\qquad\qquad\qquad\qquad\qquad\qquad$ $\square$

Proof of the Change of Variables Formula Recall what our the assumptions are: $\varphi : U \to W$ is a C^1 diffeomorphism, $f : W \to \mathbb{R}$ is Riemann integrable, R is a rectangle in U, and it is asserted that

(15)
$$\int_R f \circ \varphi \cdot |\operatorname{Jac} \varphi| = \int_{\varphi R} f.$$

Let D' be the set of discontinuity points of f. It is a zero set. Then

$$D = \varphi^{-1}(D')$$

is the set of discontinuity points of $f \circ \varphi$. The C^1 Mean Value Theorem implies that φ^{-1} is Lipschitz, Lemma 35 implies that D is a zero set, and the Riemann-Lebesgue Theorem implies that $f \circ \varphi$ is Riemann integrable. Since $|\operatorname{Jac} \varphi|$ is continuous, it is Riemann integrable and so is the product $f \circ \varphi \cdot |\operatorname{Jac} \varphi|$. In short, the l.h.s. of (15) makes sense.

Since φ is a diffeomorphism, it is a homeomorphism and it carries the boundary of R to the boundary of φR. The former boundary is a zero set and by Lemma 35 so is the latter. Thus $\chi_{\varphi R}$ is Riemann integrable. Choose a rectangle R' that contains φR. Then the r.h.s. of (15) becomes

$$\int_{\varphi R} f = \int_{R'} f \cdot \chi_{\varphi R},$$

which also makes sense. It remains to show that the two sides of (15) not only make sense but are equal.

The idea is simple: Use a grid G to break R into squares R_{ij} of radius r, prove an approximate result on each R_{ij}, add the results, and take a limit as r tends to zero. The details are not so simple.

Let z_{ij} be the center point of R_{ij} and call

$$A_{ij} = (D\varphi)_{z_{ij}} \qquad \varphi(z_{ij}) = w_{ij} \qquad \varphi(R_{ij}) = W_{ij}.$$

The Taylor approximation to φ on R_{ij} is the affine map

$$\phi_{ij}(z) = w_{ij} + A_{ij}(z - z_{ij}).$$

The composite $\psi = \phi_{ij}^{-1} \circ \varphi$ sends z_{ij} to itself and its derivative at z_{ij} is the identity transformation. Uniform continuity of $(D\varphi)_z$ on R implies that for all R_{ij} and for all $z \in R_{ij}$ we have $\|(D\psi)_z - \mathrm{id}\| < \sigma$, and $\sigma = \sigma(r) \to 0$ as $r \to 0$. Lemma 34 implies that $\psi(R_{ij})$ is sandwiched between squares of radius $r(1 \pm \sigma)$ centered at z_{ij}. Applying ϕ_{ij} to this square sandwich gives the parallelogram sandwich

$$(16) \qquad \phi_{ij}((1 - \sigma)R_{ij}) \subset \varphi(R_{ij}) = W_{ij} \subset \phi_{ij}((1 + \sigma)R_{ij}).$$

where $(1 \pm \sigma)R_{ij}$ refers to the $(1 \pm \sigma)$-dilation of R_{ij} centered at z_{ij}. See Figure 120.

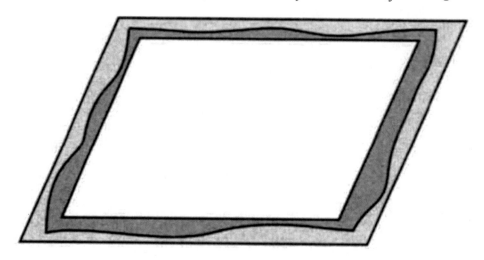

Figure 120 How we magnify the picture and sandwich a nonlinear
parallelogram between two linear ones

By Proposition 33 this gives the area estimate

$$(17) \qquad (1 - \sigma)^2 J_{ij} |R_{ij}| \leq |W_{ij}| \leq (1 + \sigma)^2 J_{ij} |R_{ij}|$$

where $J_{ij} = |\operatorname{Jac}_{z_{ij}} \varphi|$.

Let m_{ij} and M_{ij} be the infimum and supremum of $f \circ \varphi$ on R_{ij}. Then, for all $w \in \varphi R = W$ we have

$$\sum m_{ij} \chi_{\operatorname{int} W_{ij}}(w) \leq f(w) \leq \sum M_{ij} \chi_{W_{ij}}(w)$$

which integrates to

$$(18) \qquad \sum m_{ij} |W_{ij}| \leq \int_{\varphi R} f \leq \sum M_{ij} |W_{ij}|.$$

Using (17), we replace $|W_{ij}|$ in (18), getting

$$(1-\sigma)^2 \sum m_{ij} J_{ij} |R_{ij}| \leq \int_{\varphi R} f \leq (1+\sigma)^2 \sum M_{ij} J_{ij} |R_{ij}|.$$

Uniform continuity of the Jacobian implies that there exists $\tau = \tau(r)$ such that for all $z, z' \in R_{ij}$ we have

$$1 - \tau \leq \frac{\mathrm{Jac}_z \, \varphi}{\mathrm{Jac}_{z'} \, \varphi} \leq 1 + \tau$$

and $\tau(r) \to 0$ as $r \to 0$. This gives

$$(19) \qquad (1-\tau)L \leq \sum m_{ij} J_{ij} |R_{ij}| \leq \sum M_{ij} J_{ij} |R_{ij}| \leq (1+\tau)U$$

where L and U are the lower and upper sums of $f \circ \varphi(z) \, |\mathrm{Jac}_z \, \varphi|$ with respect to the grid G. Thus, the integral of f on φR is sandwiched between $(1-\tau)(1-\sigma)^2 L$ and $(1+\tau)(1+\sigma)^2 U$. As the grid mesh r tends to zero, the sandwich shrinks to the integrals $\int_R f \circ \varphi(z) \, |\mathrm{Jac}_z \, \varphi|$ and $\int_{\varphi R} f$, which completes the proof of their equality. $\square$

Finally, here is a sketch of the n-dimensional theory. Instead of a two-dimensional rectangle we have a box

$$R = [a_1, b_1] \times \cdots \times [a_n, b_n].$$

Riemann sums of a function $f : R \to \mathbb{R}$ are defined as before: Take a grid G of small boxes R_ℓ in R, take a sample point s_ℓ in each, and set

$$R(f, G, S) = \sum f(s_\ell) |R_\ell|$$

where $|R_\ell|$ is the product of the edge lengths of the small box R_ℓ and S is the set of sample points. If the Riemann sums converge to a limit it is the integral. The general theory, including the Riemann-Lebesgue Theorem, is the same as in dimension 2.

Fubini's Theorem is proved by induction on n, and has the same meaning: Integration on a box can be done slice by slice, and the order in which the iterated integration is performed has no effect on the answer.

The Change of Variables Formula has the same statement, only now the Jacobian is the determinant of an $n \times n$ matrix. In place of area we have volume, the n-dimensional volume of a set $S \subset \mathbb{R}^n$ being the integral of its characteristic function.

The volume-multiplier formula, Proposition 33, has essentially the same proof but the elementary matrix notation is messier. (It helps to realize that the following types of elementary row operations suffice for row reduction: Transposition of two adjacent rows, multiplication of the first row by λ, and addition of the second row to the first.) The proof of the Change of Variables Formula itself differs only in that $(1 \pm \sigma)^2$ becomes $(1 \pm \sigma)^n$.

In Chapter 6 we give a different proof of the Volume Multiplier Formula, namely Theorem 6.15. It is valid in n-dimensions, not just in the plane, and it relies on polar form instead of elementary matrices. Also, it applies to sets that are Lebesgue measurable, not just Riemann measurable. See page 397.

8 Differential Forms

The Riemann integral notation

$$\sum_{i=1}^{n} f(t_i)\Delta x_i \quad \approx \quad \int_a^b f(x)\,dx$$

may lead one to imagine the integral as an "infinite sum of infinitely small quantities $f(x)dx$." Although this idea itself seems to lead nowhere, it points to a good question – how do you give an independent meaning to the symbol $f\,dx$? The answer: differential forms. Not only does the theory of differential forms supply coherent, independent meanings for $f\,dx$, dx, dy, df, $dxdy$, and even for d and x separately, but it also unifies vector calculus results. A single result, the General Stokes Formula for differential forms

$$\int_M d\omega = \int_{\partial M} \omega,$$

encapsulates all integral theorems about divergence, gradient, and curl.

The presentation of differential forms in this section appears in the natural generality of n dimensions, and as a consequence it is unavoidably fraught with complicated index notation – armies of i's, j's, double subscripts, multi-indices, and so on. Your endurance may be tried.

First, consider a function $y = F(x)$. Normally, you think of F as the function, x as the input variable, and y as the output variable. But you can also take a *dual* approach and think of x as the function, F as the input variable, and y as the output variable. After all, why not? It's a kind of mathematical yin/yang.

Now consider a path integral the way it is defined in calculus,

$$\int_C f\,dx + g\,dy = \int_0^1 f(x(t), y(t))\frac{dx(t)}{dt}\,dt + \int_0^1 g(x(t), y(t))\frac{dy(t)}{dt}\,dt.$$

f and g are smooth real-valued functions of (x, y) and C is a smooth path parameterized by $(x(t), y(t))$ as t varies on $[0, 1]$. Normally you think of the integral as a number that depends on the functions f and g. Taking the dual approach you can think of it as a number that depends on the path C. This will be our point of view. It parallels that found in Rudin's *Principles of Mathematical Analysis*.

Definition A **differential 1-form** is a function that sends paths to real numbers and which can be expressed as a path integral in the previous notation. The **name** of this particular differential 1-form is $f\,dx + g\,dy$

In a way, this definition begs the question. For it simply says that the standard calculus formula for path integrals should be read in a new way – as a function of the integration domain. Doing so, however, is illuminating, for it leads you to ask: Just what property of C does the differential 1-form $f\,dx + g\,dy$ measure?

First take the case that $f(x, y) = 1$ and $g(x, y) = 0$. Then the path integral is

$$\int_C dx = \int_0^1 \frac{dx(t)}{dt}\,dt = x(1) - x(0)$$

which is the "net x-variation" of the path C. In functional notation it becomes $dx : C \mapsto x(1) - x(0)$, which means that dx assigns to each path C its net x-variation. Similarly dy assigns to each path its net y-variation. The word "net" is important. Negative x-variation cancels positive x-variation, and negative y-variation cancels positive y-variation. In the world of forms, orientation matters.

What about $f\,dx$? The function f "weights" x-variation. If the path C passes through a region in which f is large, its x-variation is magnified accordingly, and the integral $\int_C f\,dx$ reflects the net f-weighted x-variation of C. In functional notation

$$f\,dx : C \mapsto \text{net } f\text{-weighted } x\text{-variation of } C.$$

Similarly, $g\,dy$ assigns to a path its net g-weighted y-variation, and the 1-form $f\,dx + g\,dy$ assigns to C the sum of the two variations.

Figure 121 suggests why $\int_C y\,dx$ is positive and $\int_{C'} y\,dx$ is negative: The weight factor is positive on C and negative on C'. On the other hand, if the weight factor is the constant c then both integrals are $c(q - p)$.

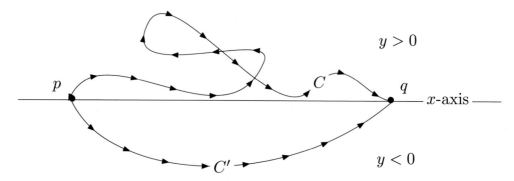

Figure 121 C and C' are paths from p to q where p and q lie on the x-axis. The integrals $\int_C y\,dx$ and $\int_{C'} y\,dx$ express the net y-weighted x-variation along C and C'.

Terminology A **functional** on a set X is a function from X to $\mathbb{R}$.

Differential 1-forms are functionals on the set of paths. Some functionals on the set of paths are differential forms but others are not. For instance, assigning to each path its arclength is a functional that is not a form. For if C is a path parameterized by $(x(t), y(t))$ then $(x^*(t), y^*(t)) = (x(1-t), y(1-t))$ parameterizes C in the reverse direction. Arclength is unaffected but the value of every 1-form on the path changes sign. Hence, arclength is not a 1-form. A more trivial example is the functional that assigns to each path the number 1.

Definition A **k-cell** in $\mathbb{R}^n$ is a smooth map[†] $\varphi : I^k \to \mathbb{R}^n$ where I^k is the unit k-cube. If $k = 1$ then φ is a path. The set of k-cells is $C_k(\mathbb{R}^n)$. The set of functionals on $C_k(\mathbb{R}^n)$ is $C^k(\mathbb{R}^n)$.

A k-cell φ need not be a diffeomorphism to its image. φ can be noninjective and its derivative can have zero determinant at many points. For this reason cells are often called "singular cells." Singularities are permitted. For example, if $0 \le t \le 1$ then $\varphi : t \mapsto ((2t-1)^3, (2t-1)^2)$ is a smooth 1-cell in the plane, despite the fact that its image has a cusp at the origin. See Figure 122.

[†]Normally, a smooth map should be defined on an open set. Here and below we mean there is a $\widehat{\varphi} : U \to \mathbb{R}^n$ such that $I^k \subset U \subset \mathbb{R}^k$, U is open, $\widehat{\varphi}(x) = \varphi(x)$ for all $x \in I^k$, and $\widehat{\varphi}$ is smooth.

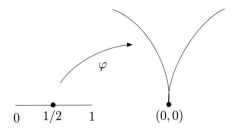

Figure 122 This smooth 1-cell is a path with a cusp. It is part of the graph of $y = x^{2/3}$.

This flexibility is a good thing. It lets the closed disc and many other planar regions be (the images of) 2-cells. See page 354, Figure 130, and Exercise 70.

Integrating a k-form over a k-cell φ with $k \geq 2$ requires Jacobian determinants. To simplify notation we write $I = (i_1, \ldots, i_k)$; it is always a k-tuple of distinct integers in $\{1, \ldots, n\}$. The **Jacobian** of φ_I at $u \in I^k$ is the $k \times k$ determinant

$$\frac{\partial \varphi_I}{\partial u} = \det \begin{bmatrix} \dfrac{\partial \varphi_{i_1}}{\partial u_1} & \cdots & \dfrac{\partial \varphi_{i_1}}{\partial u_k} \\ \vdots & \vdots & \vdots \\ \dfrac{\partial \varphi_{i_k}}{\partial u_1} & \cdots & \dfrac{\partial \varphi_{i_k}}{\partial u_k} \end{bmatrix}.$$

(We assume the integers in I are distinct, since otherwise the determinant is zero.) If $k = 1$ and $I = (i)$ then $\partial \varphi_I / \partial u$ is just $\partial \varphi_i / \partial u$, while if $k = 2$ and $I = (1, 3)$ then $\partial \varphi_I / \partial u$ is the 2×2 determinant

$$\frac{\partial \varphi_I}{\partial u} = \frac{\partial(\varphi_1, \varphi_3)}{\partial(u_1, u_2)} = \det \begin{bmatrix} \dfrac{\partial \varphi_1}{\partial u_1} & \dfrac{\partial \varphi_1}{\partial u_2} \\ \dfrac{\partial \varphi_3}{\partial u_1} & \dfrac{\partial \varphi_3}{\partial u_2} \end{bmatrix}.$$

Notation We denote $x \in I^k$ as $x = (x_1, \ldots, x_k)$ and $y \in \mathbb{R}^n$ as $y = (y_1, \ldots, y_n)$. Thus $y = \varphi(x)$ and $\varphi_i(x_1, \ldots, x_k) = y_i$ is the ith component of a k-cell $\varphi : I^k \to \mathbb{R}^n$ at x. Alternately, to match calculus notation, if $k = 1$ we write t instead of x_1; if $k = 2$ we write (s, t) instead of (x_1, x_2); if $n = 3$ we write (x, y, z) instead of (y_1, y_2, y_3). For example $dxdy$ is a 2-form in $\mathbb{R}^3$. It is the same as $dx_1 dx_2$ but $dxdy$ is a more familiar name for it. A planar path φ is $\varphi(t) = (\varphi_1(t), \varphi_2(t)) = (x(t), y(t))$. Finally, $u = (u_1, \ldots, u_k)$ is a dummy Riemann integration variable.

Definition The I**-shadow area** of φ is the functional on $C_k(\mathbb{R}^n)$, the set of k-cells,

$$dy_I : \varphi \mapsto \int_{I^k} \frac{\partial \varphi_I}{\partial u} \, du$$

where $I = (i_1, \ldots, i_k)$, $\varphi_I = (\varphi_{i_1}, \ldots, \varphi_{i_k})$, and the integral notation is shorthand,

$$\int_{I^k} \frac{\partial \varphi_I}{\partial u} \, du \; = \; \int_0^1 \cdots \int_0^1 \frac{\partial(\varphi_{i_1}, \ldots, \varphi_{i_k})}{\partial(u_1, \ldots, u_k)} \, du_1 \ldots du_k.$$

For 1-forms the definition is nothing new. The integral of dx on the path $\varphi(t) = (x(t), y(t))$ is the integral of the 1×1 Jacobian $dx(t)/dt$, namely

$$\int_0^1 \frac{dx(t)}{dt} \, dt = x(1) - x(0)$$

which is the net x-variation of φ. In the shadow terminology it is the "x-area" of φ.

Just as for paths, shadow area can be positive or negative. It is the **signed area** of the shadow of φ on the I-plane, i.e., the signed area of its projection $\pi_I(\varphi(I^k))$. After all, the Jacobian can be negative and it only involves the I-components of φ. No components φ_j with $j \notin I$ appear in $\partial \varphi_I / \partial u$. See Figure 123.

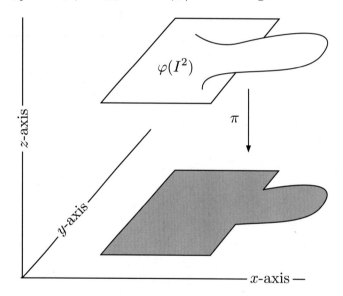

Figure 123 A pseudopod emerging from a rectangle. It is a 2-cell φ in $\mathbb{R}^3$ that casts a shadow in the xy-plane.

If f is a smooth function on $\mathbb{R}^n$ then $f dy_I$ is the functional

$$f \, dy_I : \varphi \mapsto \int_{I^k} f(\varphi(u)) \frac{\partial \varphi_I}{\partial u} \, du.$$

The function f "weights" I-area.

Definition The functional dy_I is a **basic k-form** and $f\,dy_I$ is a **simple k-form**, while a sum of simple k-forms is a **general k-form**,

$$\omega \;=\; \sum_I f_I\,dy_I \;:\; \varphi \;\mapsto\; \sum_I (f_I\,dy_I)(\varphi).$$

The careful reader will detect some abuse of notation. Here, the k-tuple $I = (i_1,\ldots,i_k)$ is used to index a collection of scalar coefficient functions $\{f_I\}$, whereas I is also used to reduce an n-vector $(y_1,\ldots,y_n)$ to a k-vector $y_I = (y_{i_1},\ldots,y_{i_k})$, to name the corresponding k-plane in $\mathbb{R}^n$, and to denote the basic k-form dy_I. Besides this, I is the unit interval and the identity matrix. Please persevere.

To underline the fact that a form is an integral we write

$$\omega(\varphi) = \int_\varphi \omega.$$

Notation $C_k(\mathbb{R}^n)$ is the set of all k-cells in $\mathbb{R}^n$, $C^k(\mathbb{R}^n)$ is the set of all functionals on $C_k(\mathbb{R}^n)$, and $\Omega^k(\mathbb{R}^n)$ is the set of k-forms on $\mathbb{R}^n$. Thus $\Omega^k \subset C^k$.

Because a determinant changes sign under a row transposition, k-forms satisfy the **signed commutativity** property: If π permutes I to πI then

$$dy_{\pi I} = \mathrm{sgn}(\pi)dy_I$$

where $\mathrm{sgn}(\pi)$ is the sign of the permutation π. In particular, $dy_{(1,2)} = -dy_{(2,1)}$ signifies that xy-area is the negative of yx-area, that is $dxdy = -dydx$, a formula that is certainly familiar from Sophomore Calculus. Because a determinant is zero if it has a repeated row, $dy_I = 0$ if I has a repeated entry. In particular $dxdx$ is the zero functional on $C_2(\mathbb{R}^2)$.

Upshot The integral of the basic 2-form $dxdy$ over a 2-cell φ in $\mathbb{R}^3$ is the net area of its shadow on the xy-plane. ("Net" means negative area cancels positive area.) The same holds for the other coordinate planes and in higher dimensions – *net shadow area equals the integral of the basic form.*

Example Consider a 2-cell $\varphi : I^2 \to \mathbb{R}^3$. What is its xy-area? By definition it is the integral of the Jacobian $\partial(\varphi_1,\varphi_2)/\partial(s,t)$ over the unit square in (s,t)-space. Suppose that φ is given by the formula

$$\varphi(s,t) = \begin{cases} (s, t(1 - ms), t) & \text{if } 0 \le s \le 1/2 \\ (s, t(1 - m + ms), t) & \text{if } 1/2 \le s \le 1. \end{cases}$$

φ is only piecewise smooth but never mind. If the slope m is 4 then the signed xy-area of φ is zero. If $m > 4$ it is negative.

I^2 has four edges. φ sends the bottom edge to itself by the identity map, it sends the top edge to the piecewise linear V-shaped path in the plane $z = 1$ from $(0, 1, 1)$ to $(1/2, 1 - m/2, 1)$ to $(1, 1, 1)$. Finally φ sends the left and right edges to lines of slope 1 that join $(0, 0, 0)$ to $(0, 1, 1)$ and $(1, 0, 0)$ to $(1, 1, 1)$. Figure 124 shows the projection of the cell on the xy-plane.

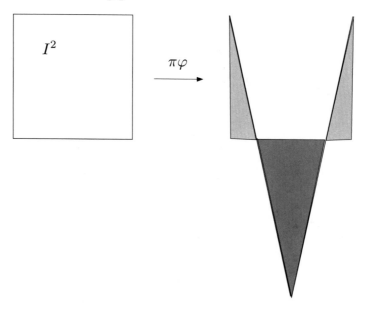

Figure 124 $\pi\varphi$ fixes all points of the square's lower edge, left edge, and right edge. It sends the upper edge to the V-shaped path from $(0, 1)$ to $(1, 1)$. For fixed s, $\pi\varphi(s, t)$ is affine in t. Positive shadow area is lightly shaded and negative shadow area heavily shaded. The total signed xy-area of φ is negative when $m > 4$. When $m \geq 2$ the cell φ resembles a ship's prow or an upside down cow catcher.

Form Naturality

It is a common error to confuse a cell, which a smooth mapping, with its image, which is point set – but the error is fairly harmless.

36 Theorem *Integrating a k-form over k-cells that differ by a reparameterization of I^k produces the same answer up to a factor of ± 1, and this factor of ± 1 is determined by whether the reparameterization preserves or reverses orientation.*

Proof If T is an orientation-preserving diffeomorphism of I^k to itself then the Jacobian $\partial T/\partial u$ is positive. The product determinant formula and the change of variables formula for multiple integrals applied to the simple form $\omega = f\, dy_I$ give

$$
\begin{aligned}
\int_{\varphi \circ T} \omega &= \int_{I^k} f(\varphi \circ T(u)) \frac{\partial(\varphi \circ T)_I}{\partial u}\, du \\
&= \int_{I^k} f(\varphi \circ T(u)) \left(\frac{\partial \varphi_I}{\partial v}\right)_{v=T(u)} \frac{\partial T}{\partial u}\, du \\
&= \int_{I^k} f(\varphi(v)) \frac{\partial \varphi_I}{\partial v}\, dv = \int_\varphi \omega.
\end{aligned}
$$

Taking sums shows that $\int_{\varphi \circ T} \omega = \int_\varphi \omega$ continues to hold for general forms $\omega \in \Omega^k$. If T reverses orientation, its Jacobian is negative. In the change of variables formula appears the absolute value of the Jacobian, which causes $\int_{\varphi \circ T} \omega$ to change sign. $\square$

A particular case of the previous theorem concerns line integrals in the plane. The integral of a 1-form over a curve C does not depend on how C is parameterized. If we first parameterize C using a parameter $t \in [0,1]$ and then reparameterize it by arclength $s \in [0,L]$ where L is the length of C and the orientation of C remains the same then integrals of 1-forms are unaffected. If $\omega = f\, dx + g\, dy$ then

$$
\begin{aligned}
\int_0^1 f(x(t), y(t)) \frac{dx(t)}{dt}\, dt &= \int_0^L f(x(s), y(s)) \frac{dx(s)}{ds}\, ds \\
\int_0^1 g(x(t), y(t)) \frac{dy(t)}{dt}\, dt &= \int_0^L g(x(s), y(s)) \frac{dy(s)}{ds}\, ds.
\end{aligned}
$$

Form Names

A k-tuple $A = (i_1, \ldots, i_k)$ **ascends** if $i_1 < \cdots < i_k$.

37 Proposition *Each k-form ω has a unique expression as a sum of simple k-forms with ascending k-tuple indices,*

$$
\omega = \sum f_A dy_A.
$$

Moreover, the coefficient $f_A(y)$ in this "ascending presentation" of ω is determined by the value of ω on small k-cells at y.

Proof For each I there is a unique permutation π such that $A = \pi I$ ascends. Signed commutativity lets us regroup and combine a sum of simple forms into terms in which the indices ascend, which gives existence of the ascending presentation $\omega = \sum f_A dy_A$.

Fix an ascending k-tuple $A = (i_1, \ldots, i_k)$ and fix a point $y \in \mathbb{R}^n$. Let $L : \mathbb{R}^k \to \mathbb{R}^n$ be the linear map sending u to $u_1 e_{i_1} + \cdots + u_k e_{i_k}$. For $r > 0$ the **inclusion cell** is $\iota : I^k \to \mathbb{R}^n$ where

$$\iota = \iota_{r,y} : u \mapsto y + rL(u).$$

ι sends I^k to a cube of side length r in the y_A-plane at y. As $r \to 0$, the cube shrinks to y. If I ascends then the Jacobian of ι is

$$\frac{\partial \iota_I}{\partial u} = \begin{cases} r^k & \text{if } I = A \\ 0 & \text{if } I \neq A. \end{cases}$$

Thus, if $I \neq A$ then $f_I dy_I(\iota) = 0$ and

$$\omega(\iota) = f_A dy_A(\iota) = r^k \int_{I^k} f_A(\iota(u)) \, du.$$

Continuity of f_A implies that

(20)
$$f_A(y) = \lim_{r \to 0} \frac{1}{r^k} \omega(\iota),$$

which is how the value of ω on small k-cells at y determines the coefficient $f_A(y)$. $\square$

38 Corollary *If $k > n$ then $\Omega^k(\mathbb{R}^n) = 0$.*

Proof There are no ascending k-tuples of integers in $\{1, \ldots, n\}$. $\square$

Moral A form may have many names, but it has a unique ascending name. Therefore if definitions or properties of a form are to be discussed in terms of a form's name then the use of ascending names avoids ambiguity.

Wedge Products

Let α be a k-form and β be an ℓ-form. Write them in their ascending presentations, $\alpha = \sum_A a_A dy_A$ and $\beta = \sum_B b_B dy_B$. Their **wedge product** is the $(k + \ell)$-form

$$\alpha \wedge \beta = \sum_{A,B} a_A b_B dy_{AB}$$

where $A = (i_1, \ldots, i_k)$, $B = (j_1, \ldots, j_\ell)$, $AB = (i_1, \ldots, i_k, j_1, \ldots, j_\ell)$, and the sum is taken over all ascending A, B. The use of ascending presentations avoids name ambiguity although Theorem 39 makes the ambiguity moot. A particular case of the definition is

$$dy_1 \wedge dy_2 = dy_{(1,2)}.$$

39 Theorem *The wedge product* $\wedge : \Omega^k \times \Omega^\ell \to \Omega^{k+l}$ *satisfies four natural conditions:*

(a) *distributivity:* $(\alpha + \beta) \wedge \gamma = \alpha \wedge \gamma + \beta \wedge \gamma$ *and* $\gamma \wedge (\alpha + \beta) = \gamma \wedge \alpha + \gamma \wedge \beta$.

(b) *insensitivity to presentations:* $\alpha \wedge \beta = \sum_{I,J} a_I b_J dy_{IJ}$ *for general presentations* $\alpha = \sum a_I dy_I$ *and* $\beta = \sum b_J dy_J$.

(c) *associativity:* $\alpha \wedge (\beta \wedge \gamma) = (\alpha \wedge \beta) \wedge \gamma$.

(d) *signed commutativity:* $\beta \wedge \alpha = (-1)^{k\ell} \alpha \wedge \beta$ *when* α *is a k-form and* β *is an* ℓ*-form. In particular* $dx \wedge dy = -dy \wedge dx$.

40 Lemma *The wedge product of basic forms satisfies*

$$dy_I \wedge dy_J = dy_{IJ}.$$

Proof #1 See Exercise 55.   □

Proof #2 If I and J ascend then the lemma merely repeats the definition of the wedge product. Otherwise, let π and ρ be permutations that make πI and ρJ ascend. Call σ the permutation of IJ that is π on the first k terms and ρ on the last ℓ. The sign of σ is $\text{sgn}(\pi)\,\text{sgn}(\rho)$ and

$$dy_I \wedge dy_J = \text{sgn}(\pi)\,\text{sgn}(\rho)\, dy_{\pi I} \wedge dy_{\rho J} = \text{sgn}(\sigma)\, dy_{\sigma(IJ)} = dy_{IJ}.$$ □

Proof of Theorem 39 (a) To check distributivity, suppose that $\alpha = \sum a_I dy_I$ and $\beta = \sum b_I dy_I$ are k-forms, while $\gamma = \sum c_J dy_J$ is an ℓ-form and all sums are ascending presentations. Then

$$\sum (a_I + b_I) dy_I$$

is the ascending presentation of $\alpha + \beta$ (this is the only trick in the proof) and

$$(\alpha + \beta) \wedge \gamma = \sum_{I,J} (a_I + b_I) c_J dy_{IJ} = \sum_{I,J} a_I c_J dy_{IJ} + \sum_{I,J} b_I c_J dy_{IJ},$$

which is $\alpha \wedge \gamma + \beta \wedge \gamma$, and verifies distributivity on the left. Distributivity on the right is checked in the same way.

(b) Let $\sum a_I dy_I$ and $\sum b_J dy_J$ be general nonascending presentations of α and β. By distributivity and Lemma 40 we have

$$\left(\sum_I a_I dy_I \right) \wedge \left(\sum_J b_J dy_J \right) = \sum_{I,J} a_I b_J dy_I \wedge dy_J = \sum_{I,J} a_I b_J dy_{IJ}$$

(c) By (b), to check associativity we need not use ascending presentations. Thus if $\alpha = \sum a_I dy_I$, $\beta = \sum b_J \, dy_J$, and $\gamma = \sum c_K \, dy_K$ then

$$\alpha \wedge (\beta \wedge \gamma) = \left(\sum_I a_I \, dy_I \right) \wedge \left(\sum_{J,K} b_J c_K \, dy_{JK} \right) = \sum_{I,J,K} a_I b_J c_K \, dy_{IJK},$$

which equals $(\alpha \wedge \beta) \wedge \gamma$.

(d) Associativity implies that it makes sense to write dy_I and dy_J as products $dy_{i_1} \wedge \cdots \wedge dy_{i_k}$ and $dy_{j_1} \wedge \cdots \wedge dy_{j_\ell}$. Thus,

$$dy_I \wedge dy_J \;=\; dy_{i_1} \wedge \cdots \wedge dy_{i_k} \wedge dy_{j_1} \wedge \cdots \wedge dy_{j_\ell}.$$

It takes $k\ell$ pair-transpositions to push each dy_i past each dy_j, which implies

$$dy_J \wedge dy_I \;=\; (-1)^{k\ell} dy_I \wedge dy_J.$$

Distributivity completes the proof of signed commutativity for general α and β. $\square$

The Exterior Derivative

Differentiating a form is subtle. The idea, as with all derivatives, is to imagine how the form changes under small variations of the point at which it is evaluated.

A 0-form is a smooth function $f(x)$. Its exterior derivative is by definition the functional on paths $\varphi : [0,1] \to \mathbb{R}^n$,

$$df : \varphi \mapsto f(\varphi(1)) - f(\varphi(0)).$$

41 Proposition *df is a 1-form; when $n = 2$ it is expressed as*

$$df = \frac{\partial f}{\partial x} \, dx + \frac{\partial f}{\partial y} \, dy.$$

In particular, $d(x) = dx$.

Proof When no abuse of notation occurs we use calculus shorthand and write $f_x = \partial f / \partial x$, $f_y = \partial f / \partial y$. Applied to φ, the form $\omega = f_x dx + f_y dy$ produces the number

$$\omega(\varphi) = \int_0^1 \left(f_x(\varphi(t)) \frac{dx(t)}{dt} + f_y(\varphi(t)) \frac{dy(t)}{dt} \right) dt.$$

By the Chain Rule the integrand is the derivative of $f \circ \varphi(t)$, so the Fundamental Theorem of Calculus implies that $\omega(\varphi) = f(\varphi(1)) - f(\varphi(0))$. Therefore $df = \omega$ as claimed. $\square$

Remark Just as with the 1-form dx, the 1-form df measures the net f-variation of a path from p to q. It is the difference $fq - fp$.

Definition Fix $k \geq 1$. Let $\sum f_A dy_A$ be the ascending presentation of a k-form ω. The **exterior derivative** of ω is the $(k+1)$-form

$$d\omega = \sum_A df_A \wedge dy_A.$$

The sum is taken over all ascending k-tuples A. *The derivative of $\omega = f dy_A$ amounts to how the coefficient f changes.* If f is constant then $d\omega = 0$.

Use of the ascending presentation makes the definition unambiguous although Theorem 42 makes this moot. Since df_A is a 1-form and dy_A is k-form, $d\omega$ is indeed a $(k+1)$-form. For example, we get

$$d(f dx + g dy) = (g_x - f_y) dx \wedge dy.$$

42 Theorem *Exterior differentiation $d : \Omega^k \to \Omega^{k+1}$ satisfies four natural conditions.*

(a) *It is linear: $d(\alpha + c\beta) = d\alpha + cd\beta$.*
(b) *It is insensitive to presentation: If $\sum f_I dy_I$ is a general presentation of ω then*
$$d\omega = \sum df_I \wedge dy_I.$$
(c) *It obeys a product rule: If α is a k-form and β is an ℓ-form then*

$$d(\alpha \wedge \beta) = d\alpha \wedge \beta + (-1)^k \alpha \wedge d\beta.$$

(d) *$d^2 = 0$. That is, $d(d\omega) = 0$ for all $\omega \in \Omega^k$.*

Proof (a) Linearity is easy and is left for the reader as Exercise 57.

(b) Let π make πI ascend. Linearity of d and associativity of $\wedge$ give

$$d(f_I dy_I) = \text{sgn}(\pi) \, d(f_I dy_{\pi I}) = \text{sgn}(\pi) \, d(f_I) \wedge dy_{\pi I} = d(f_I) \wedge dy_I.$$

Linearity of d promotes the result from simple forms to general ones.

(c) The ordinary Leibniz product rule for differentiating functions of two variables gives

$$\begin{aligned} d(fg) &= \frac{\partial fg}{\partial x} dx + \frac{\partial fg}{\partial y} dy \\ &= f_x g \, dx + f_y g \, dy + f g_x \, dx + f g_y \, dy \end{aligned}$$

which is $g\,df + f\,dg$, and verifies (c) for 0-forms in $\mathbb{R}^2$. The higher-dimensional case is similar. Next we consider simple forms $\alpha = f\,dy_I$ and $\beta = g\,dy_J$. Then

$$
\begin{aligned}
d(\alpha \wedge \beta) &= d(fg\,dy_{IJ}) = (g\,df + f\,dg) \wedge dy_{IJ} \\
&= (df \wedge dy_I) \wedge (g\,dy_J) + (-1)^k (f\,dy_I) \wedge (dg \wedge dy_J) \\
&= d\alpha \wedge \beta + (-1)^k \alpha \wedge d\beta.
\end{aligned}
$$

Distributivity completes the proof for general α and β.

The proof of (d) is fun. We check it first for the special 0-form x. By Proposition 41 the exterior derivative of x is dx and in turn the exterior derivative of dx is zero. For $dx = 1dx$, $d1 = 0$, and by definition, $d(1dx) = d(1) \wedge dx = 0$. For the same reason, $d(dy_I) = 0$.

Next we consider a smooth function $f : \mathbb{R}^2 \to \mathbb{R}$ and prove that $d^2 f = 0$. Since $d^2 x = d^2 y = 0$ we have

$$
\begin{aligned}
d^2 f &= d(f_x\,dx + f_y\,dy) &= d(f_x) \wedge dx + d(f_y) \wedge dy \\
&= (f_{xx}\,dx + f_{xy}\,dy) \wedge dx &+ (f_{yx}\,dx + f_{yy}\,dy) \wedge dy \\
&= f_{xx}\,dx \wedge dx &+ (f_{yx} - f_{xy})dx \wedge dy + f_{yy}\,dy \wedge dy &= 0
\end{aligned}
$$

since $dx \wedge dx = dy \wedge dy = 0$ and smoothness of f implies $f_{xy} = f_{yx}$.

The fact that $d^2 = 0$ for functions easily gives the same result for forms. The higher-dimensional case is similar.   □

Pushforward and Pullback

According to Theorem 36 forms behave naturally under cell reparameterization. What about changes of the target space instead of changes of the domain space? Let $T : \mathbb{R}^n \to \mathbb{R}^m$ be a smooth transformation. It induces a natural transformation on k-cells, $T_* : \varphi \mapsto T \circ \varphi$, called the **pushforward** of T. The k-cell φ in $\mathbb{R}^n$ gets pushed forward to become a k-cell $T_*\varphi$ on $\mathbb{R}^m$. Dual to the pushforward is the **pullback**. If α is a k-form on $\mathbb{R}^m$ then its pullback to $\mathbb{R}^n$ is the k-form $T^*\alpha$ that sends each k-cell φ to $\alpha(T \circ \varphi)$. This is expressed as the **duality equation**

$$
T^*\alpha(\varphi) = \alpha(T_*\varphi).
$$

The pushforward T_* goes the same direction as T, from $\mathbb{R}^n$ to $\mathbb{R}^m$, while the pullback T^* goes the opposite way.

It is clear that $T^*\alpha$ is a functional on the space of cells, $C_k(\mathbb{R}^n)$, and in the next theorem we show that it is actually a k-form. Thus $T^*: \Omega^k(\mathbb{R}^m) \to \Omega^k(\mathbb{R}^n)$.

Linearity of T^* is straightforward: $T^*(\alpha + \lambda\beta) = T^*\alpha + \lambda T^*\beta$, since for each φ we have

$$(T^*(\alpha + \lambda\beta))(\varphi) = (\alpha + \lambda\beta)(T \circ \varphi) = \alpha(T_*\varphi) + \lambda\beta(T_*\varphi) = T^*\alpha(\varphi) + \lambda T^*\beta(\varphi).$$

Figure 125 is what to remember.

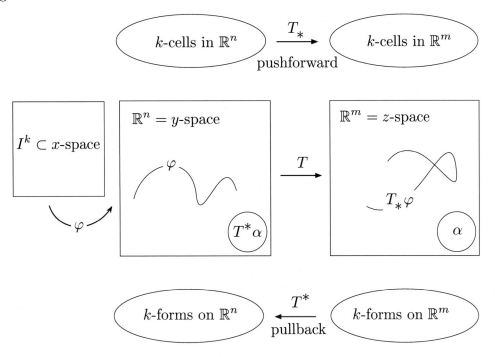

Figure 125 k-cells in $\mathbb{R}^n$ get pushed forward into $\mathbb{R}^m$ while k-forms on $\mathbb{R}^m$ get pulled back to $\mathbb{R}^n$.

43 Theorem *Pullbacks of forms obey the following four natural conditions.*

(a) *The pullback of a form is a form. In particular, $T^*(dz_I) = dT_I = dT_{i_1} \wedge \cdots \wedge dT_{i_k}$ where $T_I = (T_{i_1}, \ldots, T_{i_k})$ is the I-component of T.*

(b) *The pullback preserves wedge products: $T^*(\alpha \wedge \beta) = T^*\alpha \wedge T^*\beta$.*

(c) *The pullback commutes with the exterior derivative: $dT^* = T^*d$.*

(d) *The pullback commutes with the integral: $\int_{T \circ \varphi} \alpha = \int_\varphi T^*\alpha$.*

Proof (a) We rely on a nontrivial result in linear algebra, the **Cauchy-Binet Formula**, which concerns the determinant of a product matrix $AB = C$, where A is $k \times n$ and B is $n \times k$. See Appendix E.

In terms of Jacobians, the Cauchy-Binet Formula asserts that if the maps $\varphi : \mathbb{R}^k \to \mathbb{R}^n$ and $\psi : \mathbb{R}^n \to \mathbb{R}^k$ are smooth then the composite $\phi = \psi \circ \varphi : \mathbb{R}^k \to \mathbb{R}^k$ satisfies

$$\frac{\partial \phi}{\partial x} = \sum_A \frac{\partial \psi}{\partial y_A} \frac{\partial \varphi_A}{\partial x}.$$

The Jacobian $\partial \psi / \partial y_A$ is evaluated at $y = \varphi(x)$ and A ranges through all ascending k-tuples in $\{1, \ldots, n\}$. Applying this formula to the composition $T_I \circ \varphi$ gives

$$\text{Jac}_x(T_I \circ \varphi) = \sum_A \left(\frac{\partial T_I}{\partial y_A} \right)_{y=\varphi(x)} \left(\frac{\partial \varphi_A}{\partial x} \right)$$

where I is a k-tuple in $\{1, \ldots, m\}$ and $T_I = (T_{i_1}, \ldots, T_{i_k})$ is the I-component of T.

Let's start with a simple form $f \, dz_I$ on $\mathbb{R}^m$ and its pullback evaluated on φ, $T^*(f \, dz_I)(\varphi) = f \, dz_I(T \circ \varphi)$. By the definition of how $f \, dz_I$ acts on the k-cell $T \circ \varphi$ and the Cauchy-Binet Formula, we have

$$T^*(f \, dz_I)(\varphi) = \int_{I^k} f \circ T \circ \varphi(u) \left(\frac{\partial (T \circ \varphi)_I}{\partial u} \right) du$$

$$= \int_{I^k} f \circ T \circ \varphi(u) \left(\frac{\partial (T_I \circ \varphi)}{\partial u} \right) du$$

$$= \int_{I^k} f \circ T \circ \varphi(u) \sum_A \left(\frac{\partial T_I}{\partial y_A} \right)_{y=\varphi(u)} \left(\frac{\partial \varphi_A}{\partial u} \right) du,$$

which is exactly the same answer we would get by letting the k-form

$$\sum_A \left(f \circ T(y) \left(\frac{\partial T_I}{\partial y_A} \right)_y \right) dy_A$$

act on φ. Therefore the pullback of a simple form is a form:

(21) $$T^*(f \, dz_I) = \sum_A (T^*f) \frac{\partial T_I}{\partial y_A} dy_A$$

Linearity of T^* promotes (21) from simple forms to general forms, and completes the proof that the pullback of a form is a form.

It remains to check that $T^*(dz_I) = dT_I$. If $I = (i_1, \ldots, i_k)$ then distributivity of the wedge product and the definition of the exterior derivative of a function imply that

$$dT_I = dT_{i_1} \wedge \cdots \wedge dT_{i_k} = \left(\sum_{s_1=1}^n \frac{\partial T_{i_1}}{\partial y_{s_1}} dy_{s_1} \right) \wedge \cdots \wedge \left(\sum_{s_k=1}^n \frac{\partial T_{i_k}}{\partial y_{s_k}} dy_{s_k} \right)$$

$$= \sum_{s_1,\ldots,s_k=1}^n \frac{\partial T_{i_1}}{\partial y_{s_1}} \cdots \frac{\partial T_{i_k}}{\partial y_{s_k}} dy_{s_1} \wedge \cdots \wedge dy_{s_k}$$

The indices $i_1, \ldots, i_k$ are fixed. All terms with repeated dummy indices $s_1, \ldots, s_k$ are zero, so the sum is really taken as $(s_1, \ldots, s_k)$ varies in the set of k-tuples with no repeated entry, and then we know that $(s_1, \ldots, s_k)$ can be expressed uniquely as $(s_1, \ldots, s_k) = \pi A$ for an ascending $A = (j_1, \ldots, j_k)$ and a permutation π. Also, $dy_{s_1} \wedge \cdots \wedge dy_{s_k} = \text{sgn}(\pi)\, dy_A$. This gives

$$dT_I = \sum_A \left(\sum_\pi \text{sgn}(\pi) \frac{\partial T_{i_1}}{\partial y_{\pi(j_1)}} \cdots \frac{\partial T_{i_k}}{\partial y_{\pi(j_k)}} \right) dy_A = \sum_A \left(\frac{\partial T_I}{\partial y_A} \right) dy_A$$

and hence $T^*(dz_I) = dT_I$. Here we used the description of the determinant from Appendix E.

(b) For 0-forms it is clear that the pullback of a product is the product of the pullbacks, $T^*(fg) = T^*f\, T^*g$. Suppose that α is a simple k-form and β is a simple ℓ-form. Then $\alpha = f\, dz_I$, $\beta = g\, dz_J$, and $\alpha \wedge \beta = fg\, dz_{IJ}$. By (a) we get

$$T^*(\alpha \wedge \beta) = T^*(fg) dT_{IJ} = T^*f\, T^*g\, dT_I \wedge dT_J = T^*\alpha \wedge T^*\beta.$$

Wedge distributivity and pullback linearity complete the proof of (b).

(c) If α is a form of degree 0, $\alpha = f \in \Omega^0(\mathbb{R}^m)$, then

$$
\begin{aligned}
T^*(df) &= T^* \left(\sum_{i=1}^m \frac{\partial f}{\partial z_i} dz_i \right) \\
&= \sum_{i=1}^m T^* \left(\frac{\partial f}{\partial z_i} \right) T^*(dz_i) \\
&= \sum_{i=1}^m \left(\frac{\partial f}{\partial z_i} \right)_{z=T(y)} dT_i \\
&= \sum_{j=1}^n \left(\sum_{i=1}^m \left(\frac{\partial f}{\partial z_i} \right)_{z=T(y)} \left(\frac{\partial T_i}{\partial y_j} \right) \right) dy_j,
\end{aligned}
$$

which is merely the Chain Rule expression for $d(f \circ T) = d(T^*f)$,

$$d(f \circ T) = \sum_{j=1}^n \left(\frac{\partial f(T(y))}{\partial y_j} \right) dy_j.$$

Thus, $T^*d\alpha = dT^*\alpha$ for 0-forms.

Next consider a simple k-form $\alpha = f\,dz_I$ with $k \geq 1$. Using (a), the degree-zero case, and the wedge differentiation formula, we get

$$
\begin{aligned}
d(T^*\alpha) &= d(T^*f\,dT_I) \\
&= d(T^*f) \wedge dT_I + (-1)^0 T^*f \wedge d(dT_I) \\
&= T^*(df) \wedge dT_I = T^*(df \wedge dy_I) = T^*(d\alpha).
\end{aligned}
$$

Linearity promotes this to general k-forms and completes the proof of (c).

(d) is merely a restatement of the duality equation $T^*\alpha(\varphi) = \alpha(T_*\varphi)$ since

$$
\int_\varphi T^*\alpha = T^*\alpha(\varphi) = \alpha(T_*\varphi) = \int_{T_*\varphi} \alpha = \int_{T\circ\varphi} \alpha,
$$

which completes the proof. $\square$

9 The General Stokes Formula

In this section we establish the general Stokes formula as

$$
\int_\varphi d\omega = \int_{\partial\varphi} \omega,
$$

where $\omega \in \Omega^k(\mathbb{R}^n)$ and $\varphi \in C_{k+1}(\mathbb{R}^n)$. Then, as special cases, we reel off the standard formulas of vector calculus. Finally, we discuss antidifferentiation of forms and briefly introduce de Rham cohomology.

First we verify Stokes' formula on a cube, and then get the general case by means of the pullback.

Definition A k-**chain** is a formal linear combination[†] of k-cells,

$$
\Phi = \sum_{j=1}^N a_j \varphi_j,
$$

where $a_1, \ldots, a_N$ are real constants. The integral of a k-form ω over Φ is

$$
\int_\Phi \omega = \sum_{j=1}^N a_j \int_{\varphi_j} \omega.
$$

[†]To be more precise, but no more informative, we form an infinite-dimensional vector space V using an uncountable basis consisting of all all k-cells in $\mathbb{R}^n$. Then $\Phi = \sum_{j=1}^N a_j \varphi_j$ is a vector in V.

Definition The **boundary** of a $(k+1)$-cell φ is the k-chain

$$\partial\varphi \;=\; \sum_{j=1}^{k+1}(-1)^{j+1}(\varphi\circ\iota^{j,1} - \varphi\circ\iota^{j,0})$$

where $\iota : I^{k+1} \to \mathbb{R}^{k+1}$ is the identity-inclusion and its rear face and front face are

$$\iota^{j,0} \;:\; (u_1,\ldots,u_k) \;\mapsto\; (u_1,\ldots,u_{j-1},0,u_j,\ldots,u_k)$$
$$\iota^{j,1} \;:\; (u_1,\ldots,u_k) \;\mapsto\; (u_1,\ldots,u_{j-1},1,u_j,\ldots,u_k).$$

They are k-cells. See Figure 126. The jth **dipole** of φ is the two term k-chain

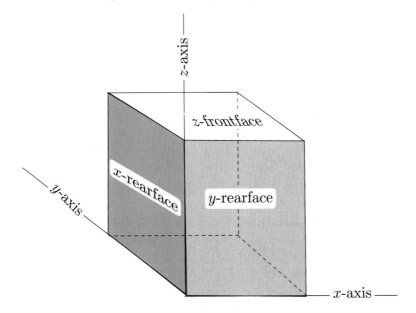

Figure 126 The rear inclusions $\iota^{1,0}$ and $\iota^{2,0}$ are the x-rearface and the y-rearface. The front inclusion $\iota^{3,1}$ is the z-frontface, the top of the cube.

$\delta^j\varphi = \varphi\circ\iota^{j,1} - \varphi\circ\iota^{j,0}$, so $\partial\varphi$ is the alternating sum of dipoles

$$\partial\varphi \;=\; \sum_{j=1}^{k+1}(-1)^{j+1}\delta^j\varphi.$$

Since $\delta^j\iota = \iota^{j,1} - \iota^{j,0}$ this means $\delta^j\varphi$ is the pushforward of $\delta^j\iota$, $\delta^j\varphi = \varphi_*(\delta^j\iota)$.

44 Stokes' Formula for a Cube *Assume that* $k+1 = n$. *If* $\omega \in \Omega^k(\mathbb{R}^n)$ *and* $\iota : I^n \to \mathbb{R}^n$ *is the identity-inclusion n-cell in $\mathbb{R}^n$ then*

$$\int_\iota d\omega \;=\; \int_{\partial\iota} \omega.$$

Proof Write ω as

$$\omega = \sum_{i=1}^{n} f_i(x) dx_1 \wedge \cdots \wedge \widehat{dx_i} \wedge \cdots \wedge dx_n,$$

where the hat above the term dx_i is standard notation to indicate that dx_i is deleted. The exterior derivative of ω is

$$\begin{aligned} d\omega &= \sum_{i=1}^{n} df_i \wedge dx_1 \wedge \cdots \wedge \widehat{dx_i} \wedge \cdots \wedge dx_n \\ &= \sum_{i=1}^{n} (-1)^{i-1} \frac{\partial f_i}{\partial x_i} dx_1 \wedge \cdots \wedge dx_n \\ &= \sum_{i=1}^{n} (-1)^{i+1} \frac{\partial f_i}{\partial x_i} dx_1 \wedge \cdots \wedge dx_n \end{aligned}$$

which implies that $\int_\iota d\omega$ is the alternating sum of Riemann integrals over I^n,

$$\int_\iota d\omega = \sum_{i=1}^{n} (-1)^{i+1} \int_{I^n} \frac{\partial f_i}{\partial x_i} \, dx_1 \ldots dx_n.$$

Deleting the jth component of the rear j^{th} face $\iota^{j,0}(u)$ gives the k-tuple $(u_1, \ldots, u_k)$, while deleting any other component gives a k-tuple with a component that remains constant as u varies. The same is true of the jth front face. Thus the Jacobians are

$$\frac{\partial(\iota^{j,0})_I}{\partial u} = \frac{\partial(\iota^{j,1})_I}{\partial u} = \begin{cases} 1 & \text{if } I = (1, \ldots, \widehat{j}, \ldots, n) \\ 0 & \text{otherwise,} \end{cases}$$

and so the jth dipole integral of ω is zero except when $i = j$, and in that case

$$\int_{\delta^j \iota} \omega = \int_0^1 \cdots \int_0^1 (f_j(u_1, \ldots, u_{j-1}, 1, u_j, \ldots, u_k) \\ - f_j(u_1, \ldots, u_{j-1}, 0, u_j, \ldots, u_k)) \, du_1 \ldots du_k.$$

By the Fundamental Theorem of Calculus we can substitute the integral of a derivative for the f_j difference; and by Fubini's Theorem the order of integration in ordinary multiple integration is irrelevant. This expresses $\int_{\delta^j \iota} \omega$ as a Riemann integral over I^n,

$$\int_{\delta^j \iota} \omega = \int_0^1 \cdots \int_0^1 \frac{\partial f_j}{\partial x_j} dx_1 \ldots dx_n,$$

so the alternating dipole sum $\sum (-1)^{j+1} \int_{\delta^j \iota} \omega$ equals $\int_\iota d\omega$. $\qquad \square$

45 Stokes' Formula for a General Cell *If ω is an $(n-1)$-form in $\mathbb{R}^m$ and φ is an n-cell in $\mathbb{R}^m$ then*

$$\int_\varphi d\omega = \int_{\partial\varphi} \omega.$$

Proof Viewing φ in the rôle of a smooth transformation, it follows from Theorem 43(c,d) that

$$\int_\varphi d\omega = \int_{\varphi \circ \iota} d\omega = \int_\iota \varphi^* d\omega = \int_\iota d\varphi^* \omega = \int_{\partial\iota} \varphi^* \omega = \int_{\varphi_* \partial\iota} \omega = \int_{\partial\varphi} \omega.$$

where $\iota : I^n \to \mathbb{R}^n$ is the identity-inclusion. For $\varphi_*(\partial\iota) = \partial\varphi$. $\square$

Stokes' Formula on Manifolds

If $M \subset \mathbb{R}^n$ divides into m-cells diffeomorphic to I^m and its boundary divides into $(m-1)$-cells diffeomorphic to I^{m-1} as shown in Figure 127, then there is a version of Stokes' Formula for the m-dimensional **manifold** M. Namely, if ω is an $(m-1)$-form then

$$\int_M d\omega = \int_{\partial M} \omega.$$

It is required that the boundary $(m-1)$-cells that are interior to M cancel each other out. This prohibits M being the Möbius band or another nonorientable set. The m-cells are said to "tile" M.

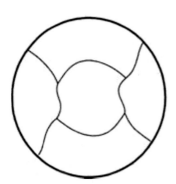
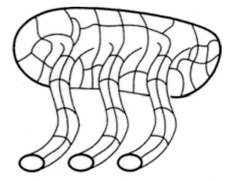

Figure 127 Two dimensional manifolds in $\mathbb{R}^2$ and $\mathbb{R}^3$. The boundary of M, drawn darker, may have several connected components.

Vector Calculus

The Fundamental Theorem of Calculus can be viewed as a special case of Stokes' Formula

$$\int_M d\omega = \int_{\partial M} \omega$$

by taking $M = [a, b] \subset \mathbb{R}^1$ and $\omega = f$. The integral of ω over the 0-chain $\partial M = b - a$ is $f(b) - f(a)$, while the integral of $d\omega$ over M is $\int_a^b f'(x)\, dx$. Likewise, if $f : \mathbb{R}^2 \to \mathbb{R}$ is smooth then the integral of the 1-form $df = f_x\, dx + f_y\, dy$ is "path independent" in the sense that if φ, ψ are paths from p to q then

$$\int_\varphi df = \int_\psi df.$$

After all, paths are 1-cells and both integrals equal $f(q) - f(p)$. The same holds in $\mathbb{R}^3$ and $\mathbb{R}^n$.

Second, **Green's Formula** in the plane,

$$\iint_D (g_x - f_y)dxdy \;\; = \;\; \int_C f dx + g dy,$$

is also a special case when we take $\omega = f\, dx + g\, dy$. Here, the region D is bounded by the curve C. It is a manifold of 2-cells in the plane.

Third, the **Gauss Divergence Theorem**

$$\iiint_D \operatorname{div} F \;\; = \;\; \iint_S \operatorname{flux} F,$$

is a consequence of Stokes' Formula. Here, $F = (f, g, h)$ is a smooth vector field defined on $U \subset \mathbb{R}^3$. (The notation indicates that f is the x-component of F, g is its y-component, and h is its z-component.) The **divergence** of F is the scalar function

$$\operatorname{div} F = f_x + g_y + h_z.$$

If φ is a 2-cell in U then the integral

$$\int_\varphi f\, dy \wedge dz \;\; + \;\; g\, dz \wedge dx \;\; + \;\; h\, dx \wedge dy$$

is the **flux** of F across φ. Let S be a compact manifold of 2-cells. The total flux across S is the sum of the flux across its 2-cells. If S bounds a region $D \subset U$ then the Gauss Divergence Theorem is just Stokes' Formula with

$$\omega \;\; = \;\; f\, dy \wedge dz \;\; + \;\; g\, dz \wedge dx \;\; + \;\; h\, dx \wedge dy.$$

For $d\omega = \text{div}\, F\, dx \wedge dy \wedge dz$.

Finally, the **curl** of a vector field $F = (f, g, h)$ is the vector field

$$(h_y - g_z,\ f_z - h_x,\ g_x - f_y).$$

Applying Stokes' Formula to the form $\omega = f\, dx + g\, dy + h\, dz$ gives

$$\int_S (h_y - g_z)\, dy \wedge dz\ +\ (f_z - h_x)\, dz \wedge dx\ +\ (g_x - f_y)\, dx \wedge dy$$

$$= \int_C f\, dx + g\, dy + h\, dz$$

where S is a surface bounded by the closed curve C. The first integral is the total curl across S, while the second is the circulation of F at the boundary. Their equality is **Stokes' Curl Theorem**. See Corollaries 50 and 51 for further vector calculus results.

Closed Forms and Exact Forms

A form is **closed** if its exterior derivative is zero. It is **exact** if it is the exterior derivative of some other form. Since $d^2 = 0$, every exact form is closed:

$$\omega = d\alpha\ \Rightarrow\ d\omega = d(d\alpha) = 0.$$

When is the converse true? That is, when can we antidifferentiate a closed form ω and find α such that $\omega = d\alpha$? If the forms are defined on $\mathbb{R}^n$ then the answer "always" is the Poincaré Lemma. See below. But if the forms are defined on some subset U of $\mathbb{R}^n$, and if they do not extend to smooth forms defined on all of $\mathbb{R}^n$, then the answer depends on the topology of U.

There is one case that should be familiar from calculus: Every closed 1-form $\omega = f\, dx + g\, dy$ on $\mathbb{R}^2$ is exact. See Exercise 58. With more work the result holds for every $U \subset \mathbb{R}^n$ that is **simply connected** in the sense that each closed curve in U can be continuously shrunk to a point in U without leaving U.

If $U \subset \mathbb{R}^2$ is not simply connected then there are 1-forms on it that are closed but not exact. The standard example is

$$\omega = \frac{-y}{r^2}dx + \frac{x}{r^2}dy$$

where $r^2 = x^2 + y^2$. Its domain of definition is the "punctured plane" $\mathbb{R}^2 \setminus \{O\}$. See Exercise 65.

In $\mathbb{R}^3$ it is instructive to consider the 2-form

$$\omega = \frac{x}{r^3} dy \wedge dz + \frac{y}{r^3} dz \wedge dx + \frac{z}{r^3} dx \wedge dy.$$

ω is defined on U, which is $\mathbb{R}^3$ minus the origin. U is a spherical shell with inner radius 0 and outer radius ∞. The form ω is closed but not exact despite the fact that U is simply connected. See Exercise 59.

46 Poincaré Lemma *If ω is a closed k-form on $\mathbb{R}^n$ then it is exact.*

Proof In fact a better result is true. There are "integration operators"

$$L_k : \Omega^k(\mathbb{R}^n) \to \Omega^{k-1}(\mathbb{R}^n)$$

with the property that $Ld + dL =$ identity. That is, for all $\omega \in \Omega^k(\mathbb{R}^n)$ we have

$$(L_{k+1}d + dL_k)(\omega) = \omega.$$

From the existence of these integration operators, the Poincaré Lemma is immediate. For if $d\omega = 0$ then we have

$$\omega = L(d\omega) + dL(\omega) = dL(\omega),$$

which shows that ω is exact with antiderivative $\alpha = L(\omega)$.

The construction of L is tricky. First we consider a k-form β, not on $\mathbb{R}^n$, but on $\mathbb{R}^{n+1}$. It can be expressed uniquely as

$$(22) \qquad \beta = \sum_I f_I \, dx_I + \sum_J g_J \, dt \wedge dx_J$$

where $f_I = f_I(x,t)$, $g_J = g_J(x,t)$, and $(x,t) \in \mathbb{R}^{n+1} = \mathbb{R}^n \times \mathbb{R}$. The first sum is taken over all ascending k-tuples I in $\{1, \ldots, n\}$, and the second over all ascending $(k-1)$-tuples J in $\{1, \ldots, n\}$. The exterior derivative of β is

$$(23) \quad d\beta = \sum_{I,\ell} \frac{\partial f_I}{\partial x_\ell} dx_\ell \wedge dx_I + \sum_I \frac{\partial f_I}{\partial t} dt \wedge dx_I + \sum_{J,\ell} \frac{\partial g_J}{\partial x_\ell} dx_\ell \wedge dt \wedge dx_J$$

where $\ell = 1, \ldots, n$.

Then we define operators

$$N : \Omega^k(\mathbb{R}^{n+1}) \to \Omega^{k-1}(\mathbb{R}^n)$$

by setting

$$N(\beta) = \sum_J \left(\int_0^1 g_J(x,t)\, dt \right) dx_J.$$

The operator N only looks at terms of the form in which dt appears. It ignores the others. We claim that for all $\beta \in \Omega^k(\mathbb{R}^{n+1})$ we have

(24) $$(dN + Nd)(\beta) = \sum_I (f_I(x,1) - f_I(x,0))\, dx_I$$

where the coefficients f_I take their meaning from (22). By Theorem 14 it is legal to differentiate past the integral sign. From (23) and the definition of N we get

$$N(d\beta) = \sum_I \left(\int_0^1 \frac{\partial f_I}{\partial t}\, dt \right) dx_I - \sum_{J,\ell} \left(\int_0^1 \frac{\partial g_J}{\partial x_\ell}\, dt \right) dx_\ell \wedge dx_J$$

$$dN(\beta) = \sum_{J,\ell} \left(\int_0^1 \frac{\partial g_J}{\partial x_\ell}\, dt \right) dx_\ell \wedge dx_J.$$

For the coefficients in $N(\beta)$ are independent of t. Therefore

$$(dN + Nd)(\beta) = \sum_I \left(\int_0^1 \frac{\partial f_I}{\partial t}\, dt \right) dx_I = \sum_I (f_I(x,1) - f_I(x,0)) dx_I,$$

as claimed in (24).

Then we define a **cone map** $\rho : \mathbb{R}^{n+1} \to \mathbb{R}^n$ by

$$\rho(x,t) = tx,$$

and set $L = N \circ \rho^*$. See Figure 128. Commutativity of pullback and d gives

(25) $$Ld + dL = N\rho^* d + dN\rho^* = (Nd + dN)\rho^*,$$

so it behooves us to work out $\rho^*(\omega)$. First suppose that ω is simple, say $\omega = h\, dx_I \in \Omega^k(\mathbb{R}^n)$. Since $\rho(x,t) = (tx_1, \ldots, tx_n)$ we have

$$\begin{aligned}
\rho^*(h\, dx_I) &= (\rho^* h)(\rho^*(dx_I)) = h(tx) d\rho_I \\
&= h(tx)(d(tx_{i_1}) \wedge \cdots \wedge d(tx_{i_k})) \\
&= h(tx)((t\, dx_{i_1} + x_{i_1}\, dt) \wedge \cdots \wedge (t\, dx_{i_k} + x_{i_k} dt)) \\
&= h(tx)(t^k dx_I) + \text{ terms that include } dt
\end{aligned}$$

where $I = \{i_1, \ldots, i_k\}$. From (24) we conclude that

$$(Nd + dN) \circ \rho^*(h dx_I) = (h(1x)1^k - h(0x)0^k) dx_I = h dx_I,$$

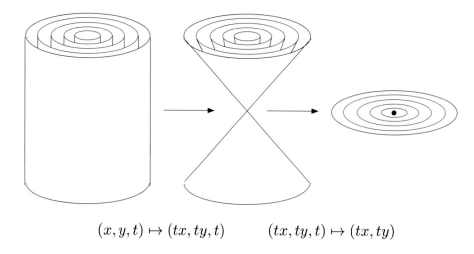

$$(x, y, t) \mapsto (tx, ty, t) \qquad (tx, ty, t) \mapsto (tx, ty)$$

Figure 128 When $n + 1 = 3$ the cone map sends vertical cylinders to vertical cones, which are then projected to the plane.

and from (25) we get

(26) $$(Ld + dL)(hdx_I) = hdx_I.$$

The linearity of L and d promote (26) to general k-forms,

$$(Ld + dL)\omega = \omega,$$

and as remarked at the outset, the existence of such an L implies that closed forms on $\mathbb{R}^n$ are exact. $\qquad \square$

47 Corollary *If U is diffeomorphic to $\mathbb{R}^n$ then all closed forms on U are exact.*

Proof Let $T : U \to \mathbb{R}^n$ be a diffeomorphism and assume that ω is a closed k-form on U. Set $\alpha = (T^{-1})^*\omega$. Since pullback commutes with d we see that α is a closed k-form on $\mathbb{R}^n$. By the Poincaré Lemma there is a $(k-1)$-form μ on $\mathbb{R}^n$ with $\alpha = d\mu$. Then

$$dT^*\mu = T^*d\mu = T^*\alpha = T^* \circ (T^{-1})^*\omega = (T^{-1} \circ T)^*\omega = \text{id}^*\omega = \omega$$

which shows that ω is exact with antiderivative $T^*\mu$. $\qquad \square$

48 Corollary *Locally, closed forms defined on open subsets of $\mathbb{R}^n$ are exact.*

Proof Locally an open subset of $\mathbb{R}^n$ is diffeomorphic to $\mathbb{R}^n$. $\qquad \square$

49 Corollary *If $U \subset \mathbb{R}^n$ is open and starlike (in particular, if U is convex) then closed forms on U are exact.*

Proof A **starlike** set $U \subset \mathbb{R}^n$ contains a point p such that the line segment from each $q \in U$ to p lies in U. Every starlike open set in $\mathbb{R}^n$ is diffeomorphic to $\mathbb{R}^n$. See Exercise 52. □

50 Corollary *A smooth vector field F on $\mathbb{R}^3$ (or on an open set diffeomorphic to $\mathbb{R}^3$) is the gradient of a scalar function if and only if its curl is everywhere zero.*

Proof If $F = \operatorname{grad}\phi$ then

$$F = (\phi_x, \phi_y, \phi_z) \quad \Rightarrow \quad \operatorname{curl}F = (\phi_{zy} - \phi_{yz},\ \phi_{xz} - \phi_{zx},\ \phi_{yx} - \phi_{xy}) = 0.$$

On the other hand, if $F = (f, g, h)$ then

$$\operatorname{curl}F = 0 \quad \Rightarrow \quad \omega = f\,dx + g\,dy + h\,dz$$

is closed and therefore exact. A function ϕ with $d\phi = \omega$ has gradient F. □

51 Corollary *A smooth vector field on $\mathbb{R}^3$ (or on an open set diffeomorphic to $\mathbb{R}^3$) has everywhere zero divergence if and only if it is the curl of some other vector field.*

Proof If $F = (f, g, h)$ and $G = \operatorname{curl}F$ then

$$G = (h_y - g_z,\ f_z - h_x,\ g_x - f_y)$$

so the divergence of G is zero. On the other hand, if the divergence of $G = (A, B, C)$ is zero then the form

$$\omega = A\,dy \wedge dz\ +\ B\,dz \wedge dx\ +\ C\,dx \wedge dy$$

is closed and therefore exact. If the form $\alpha = f\,dx + g\,dy + h\,dz$ has $d\alpha = \omega$ then $F = (f, g, h)$ has $\operatorname{curl}F = G$. □

Cohomology

The set of exact k-forms on U is usually denoted $B^k(U)$, while the set of closed k-forms is denoted $Z^k(U)$. ("B" is for boundary and "Z" is for cycle.) Both are vector subspaces of $\Omega^k(U)$ and

$$B^k(U) \subset Z^k(U).$$

The quotient vector space

$$H^k(U) = Z^k(U)/B^k(U)$$

is the kth **de Rham cohomology group** of U. Its members are the "cohomology classes" of U. As was discussed above, if U is simply connected then $H^1(U) = 0$. Also, $H^2(U) \neq 0$ when U is the three-dimensional spherical shell. If U is starlike then $H^k(U) = 0$ for all $k > 0$, and $H^0(U) = \mathbb{R}$. Cohomology necessarily reflects the *global* topology of U. For locally, closed forms are exact. The relation between the cohomology of U and its topology is the subject of algebraic topology, the basic idea being that the more complicated the set U (think of Swiss cheese), the more complicated is its cohomology, and vice versa. The book *From Calculus to Cohomology* by Madsen and Tomehave provides a beautiful exposition of the subject.

Differential Forms Viewed Pointwise

The preceding part of this chapter presents differential forms as "abstract integrands" – things which it makes sense to write after an integral sign. But they are not defined as functions that have values point by point. Rather they are special functionals on the space of cells. This is all well and good since it provides a clean path to the main result about forms, the Stokes Formula.

A different path to Stokes involves multilinear functionals. You have already seen bilinear functionals like the dot product. It is a map $\beta : \mathbb{R}^n \times \mathbb{R}^n \to \mathbb{R}$ with various properties, the first being that for each $v \in \mathbb{R}^n$ the maps

$$w \mapsto \beta(v, w) \quad \text{and} \quad w \mapsto \beta(w, v)$$

are linear. We say β is linear in each vector variable separately. A map $\beta : \mathbb{R}^n \times \cdots \times \mathbb{R}^n \to \mathbb{R}$ which is linear in each vector variable separately is a **k-multilinear functional**. (Its domain is the Cartesian product of k copies $\mathbb{R}^n$.) It is **alternating** if for each permutation π of $\{1, \ldots, k\}$ we have

$$\beta(v_1, \ldots, v_k) = \text{sgn}(\pi)\beta(v_{\pi(1)}, \ldots, v_{\pi(k)}).$$

The set of alternating k-linear forms is a vector space $\mathcal{A}^k$, and one can view $\omega \in \Omega^k(\mathbb{R}^n)$ at a point p as a member $\omega_p \in \mathcal{A}^k$. It is a certain type of tensor that we integrate over a cell as p varies in the cell; the vectors on which ω_p is evaluated are tangent to the cell at p. You can read about this approach to differential forms in Michael Spivak's book *Calculus on Manifolds*.

10* The Brouwer Fixed-Point Theorem

Let $B = B^n$ be the closed unit n-ball,

$$B = \{x \in \mathbb{R}^n : |x| \leq 1\}.$$

The following is one of the deep results in topology and analysis:

52 Brouwer Fixed-Point Theorem *If $F : B \to B$ is continuous then it has a fixed-point, a point $p \in B$ such that $F(p) = p$.*

Proof The proof is relatively short and depends on Stokes' Theorem. Note that Brouwer's Theorem is trivial when $n = 0$, for B^0 is a point and is the fixed-point of F. Also, if $n = 1$ then, as observed on page 242, the result is a consequence of the Intermediate Value Theorem on $B^1 = [-1, 1]$. For the continuous function $F(x) - x$ is nonnegative at $x = -1$ and nonpositive at $x = +1$, so at some $p \in [-1, 1]$ we have $F(p) - p = 0$; i.e., $F(p) = p$.

The strategy of the proof in higher dimensions is to suppose that there does exist a continuous $F : B \to B$ which fails to have a fixed-point, and from this supposition to derive a contradiction, namely that the volume of B is zero. The first step in the proof is standard.

Step 1. The existence of a continuous $F : B \to B$ without a fixed-point implies the existence of a smooth **retraction** T of a neighborhood U of B to ∂B. The map T sends U to ∂B and fixes every point of ∂B.

If F has no fixed-point as x varies in B, then compactness of B implies there is some $\mu > 0$ such that for all $x \in B$ we have

$$|F(x) - x| > \mu.$$

The Stone-Weierstrass Theorem then produces a multivariable polynomial $\widetilde{F} : \mathbb{R}^n \to \mathbb{R}^n$ that $\mu/2$-approximates F on B. The map

$$G(x) = \frac{1}{1 + \mu/2} \widetilde{F}(x)$$

is smooth and sends B into the interior of B. It μ-approximates F on B, so it too has no fixed-point. The restriction of G to a small neighborhood U of B also sends U into B and has no fixed-point.

Figure 129 shows how to construct the retraction T from the map G. Since G is smooth, so is T.

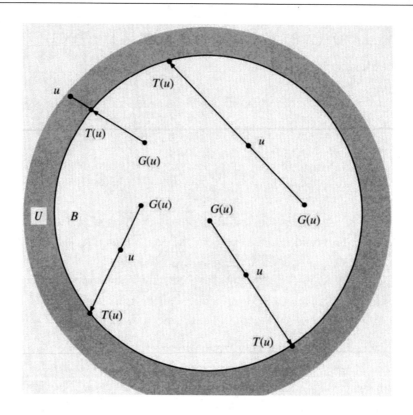

Figure 129 T retracts U onto ∂B. The point $u \in U$ is sent by T to the unique point $u' = T(u)$ at which the segment $[u, G(u)]$, extended through u, crosses the sphere ∂B.

<u>Step 2.</u> T^* kills all n-forms. If there is a point $p \in U$ such that $(DT)_p$ is invertible then the Inverse Function Theorem implies TU contains an open n-dimensional ball at fp. Since no such ball is contained in $\partial B = TU$, DT is nowhere invertible, its Jacobian determinant $\partial T/\partial u$ is everywhere zero, and $T^* : \Omega^n(U) \to \Omega^n(U)$ is the zero map.

<u>Step 3.</u> There is a map $\varphi : I^n \to B$ that exhibits B as an n-cell such that

(a) φ is smooth.

(b) $\varphi(I^n) = B$ and $\varphi(\partial I^n) = \partial B$.

(c) $\displaystyle\int_{I^n} \frac{\partial \varphi}{\partial u}\, du > 0$.

To construct φ, start with a smooth function $\sigma : \mathbb{R} \to \mathbb{R}$ such that $\sigma(r) = 0$ for $r \leq 1/2$, $\sigma'(r) > 0$ for $1/2 < r < 1$, and $\sigma(r) = 1$ for $r \geq 1$. Then define $\psi : \mathbb{R}^n \to \mathbb{R}^n$

by

$$\psi(v) = \begin{cases} v + \sigma(|v|)\left(\dfrac{v}{|v|} - v\right) & \text{if } v \neq 0 \\ 0 & \text{if } v = 0. \end{cases}$$

See Figure 130 and Exercise 53. Since $\sigma(|v|) = 0$ when $|v| \leq 1/2$, ψ is smooth.

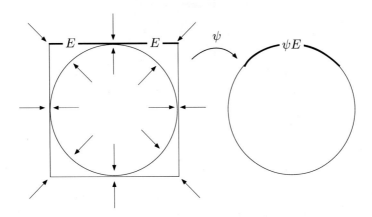

Figure 130 The map ψ crushes all of $\mathbb{R}^n$ onto the closed unit ball B^n. It is a diffeomorphism of the interior of B^n to itself, and fixes each point of $\partial B^n = S^{n-1}$. Its derivative has rank $n-1$ at each point of $\mathbb{R}^n \setminus \text{int } B^n$. Restricted to each $(n-1)$-dimensional face E of the cube $[-1,1]^n$, ψ is a diffeomorphism from the interior of E to one of the $2n$ open cubical polar caps on S^{n-1}. See also Figure 131 and Exercise 52.

The map ψ carries the sphere S_r of radius r to the sphere S_ρ of radius ρ where $\rho(r) = r + \sigma(r)(1-r)$, sending each radial line into itself. Fix the n-cell

$$\varphi = \psi \circ \kappa$$

where κ scales I^n to $[-1,1]^n$ by the affine map $\kappa : u \mapsto v = (2u_1 - 1, \ldots, 2u_n - 1)$. Then

(i) φ is smooth since ψ and κ are smooth.

(ii) φ sends ∂I^n to ∂B since ψ sends $\partial([-1,1]^n)$ to ∂B.

(iii) It is left as Exercise 70 to show that the Jacobian of ψ is $\rho'(r)\rho(r)^{n-1}/r^{n-1}$ when $r = |v|$. Thus, the Jacobian $\partial\varphi/\partial u$ is always nonnegative, and is identically equal to 2^n on the ball of radius $1/4$ at the center of I^n, so its integral on I^n is positive.

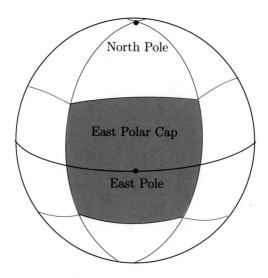

Figure 131 There are six polar caps at the six poles of the 2-sphere.

<u>Step 4.</u> Consider an $(n-1)$-form α. If $\nu : I^{n-1} \to \mathbb{R}^n$ is an $(n-1)$-cell whose image lies in ∂B then

$$\int_\nu \alpha = \int_{T \circ \nu} \alpha = \int_\nu T^*\alpha$$

since T is the identity map on ∂B. The $(n-1)$-dimensional faces of $\varphi : I^n \to B$ lie in ∂B. Thus

(27) $$\int_{\partial \varphi} \alpha = \int_{\partial \varphi} T^*\alpha.$$

<u>Step 5.</u> Now we get the contradiction. Consider the specific $(n-1)$-form

$$\alpha = x_1 \, dx_2 \wedge \cdots \wedge dx_n.$$

Note that $d\alpha = dx_1 \wedge \cdots \wedge dx_n$ is n-dimensional volume and

$$\int_\varphi d\alpha = \int_{I^n} \frac{\partial \varphi}{\partial u} \, du > 0.$$

In fact the integral is the volume of B. However, we also have

$$
\begin{aligned}
\int_\varphi d\alpha &= \int_{\partial\varphi} \alpha & \text{by Stokes' Theorem on a cell} \\
&= \int_{\partial\varphi} T^*\alpha & \text{by Equation (27)} \\
&= \int_\varphi dT^*\alpha & \text{by Stokes' Theorem on a cell} \\
&= \int_\varphi T^* d\alpha & \text{by (c) in Theorem 43} \\
&= 0 & \text{by Step 2.}
\end{aligned}
$$

This is a contradiction – an integral can not simultaneously be zero and positive. The assumption that there exists a continuous $F : B \to B$ with no fixed-point has led to a contradiction. Therefore it is untenable and every F does have a fixed-point. $\quad\square$

Appendix A　　Perorations of Dieudonné

In his classic book, *Foundations of Analysis*, Jean Dieudonné of the French Bourbaki school writes

"The subject matter of this Chapter [Chapter VIII on differential calculus] is nothing else but the elementary theorems of Calculus, which however are presented in a way which will probably be new to most students. That presentation which throughout adheres strictly to our general 'geometric' outlook on Analysis, aims at keeping as close as possible to the fundamental idea of Calculus, namely the local approximation of functions by linear functions. In the classical teaching of Calculus, this idea is immediately obscured by the accidental fact that, on a one-dimensional vector space, there is a one-to-one correspondence between linear forms and numbers, and therefore the derivative at a point is defined as a number instead of a linear form. This slavish subservience to the shibboleth of numerical interpretation at any cost becomes much worse when dealing with functions of several variables: One thus arrives, for instance, at the classical formula"... "giving the partial derivatives of a composite function, which has lost any trace of intuitive meaning, whereas the natural statement of the theorem is of course that the (total) derivative of a composite function is the composite of their derivatives"..., "a very sensible formulation when one thinks in terms of linear approximation."

"This 'intrinsic' formulation of Calculus, due to its greater 'abstraction', and in particular to the fact that again and again, one has to leave the initial spaces and climb higher and higher to new 'function spaces' (especially when dealing with the theory of higher derivatives), certainly requires some mental effort, contrasting with the comfortable routine of the classical formulas. But we believe the result is well worth the labor, as it will prepare the student to the still more general idea of Calculus on a differentiable manifold; the reader who wants to have a glimpse of that theory and of the questions to which it leads can look into the books of Chevalley and de Rham. Of course, he will observe in these applications, all the vector spaces which intervene have finite dimension; if that gives him an additional feeling of security, he may of course add that assumption to all the theorems of this chapter. But he will inevitably realize that this does not make the proofs shorter or simpler by a single line; in other words the hypothesis of finite dimension is entirely irrelevant to the material developed below; we have therefore thought it best to dispense with it altogether, although the applications of Calculus which deal with the finite-dimensional case still by far exceed the others in number and importance."

I share most of Dieudonné's opinions expressed here. And where else will you read the phrase "slavish subservience to the shibboleth of numerical interpretation at any cost"?

Appendix B   The History of Cavalieri's Principle

The following is from Marsden and Weinstein's *Calculus*.

The idea behind the slice method goes back, beyond the invention of calculus, to Francesco Bonaventura Cavalieri (1598-1647), a student of Galileo and then professor at the University of Bologna. An accurate report of the events leading to Cavalieri's discovery is not available, so we have taken the liberty of inventing one.

Cavalieri's delicatessen usually produced bologna in cylindrical form, so that the volume would be computed as π. radius^2. length. One day the casings were a bit weak, and the bologna came out with odd bulges. The scale was not working that day, either, so the only way to compute the price of the bologna was in terms of its volume.

Cavalieri took his best knife and sliced the bologna into n very thin slices, each of thickness x, and measured the radii, r_1, r_2, $\ldots$, r_n of the slices (fortunately they were all round). He then estimated the volume to be $\sum_{i=1}^{n} \pi r_i^2 x$, the sum of the volumes of the slices.

Cavalieri was moonlighting from his regular job as a professor at the University of Bologna. That afternoon he went back to his desk and began the book *Geometria indivisibilium continuorum nova quandum ratione promota* (Geometry shows the continuous indivisibility between new rations and getting promoted), in which he stated what is now known as Cavalieri's principle: If two solids are sliced by a family of parallel planes in such a way that corresponding sections have equal areas, then the two solids have the same volume.

The book was such a success that Cavalieri sold his delicatessen and retired to a life of occasional teaching and eternal glory.

Appendix C   A Short Excursion into the Complex Field

The field $\mathbb{C}$ of complex numbers corresponds bijectively with $\mathbb{R}^2$. The complex number $z = x + iy \in \mathbb{C}$ corresponds to $(x, y) \in \mathbb{R}^2$. A function $T : \mathbb{C} \to \mathbb{C}$ is complex linear if for all $\lambda, z, w \in \mathbb{C}$ we have

$$T(z + w) = T(z) + T(w) \quad \text{and} \quad T(\lambda z) = \lambda T(z).$$

Since $\mathbb{C}$ is a one-dimensional complex vector space the value $\mu = T(1)$ determines T, namely, $T(z) = \mu z$ for all z. If $z = x + iy$ and $\mu = \alpha + i\beta$ then $\mu z = (\alpha x - \beta y) + i(\beta x + \alpha y)$. In $\mathbb{R}^2$ terms $T : (x, y) \mapsto ((\alpha x - \beta y), (\beta x + \alpha y))$ which shows that T is a linear transformation $\mathbb{R}^2 \to \mathbb{R}^2$ whose matrix is

$$\begin{bmatrix} \alpha & -\beta \\ \beta & \alpha \end{bmatrix}.$$

The form of this matrix is special. For instance it could never be $\begin{bmatrix} 2 & 1 \\ 1 & 1 \end{bmatrix}$.

A complex function of a complex variable $f(z)$ has a **complex derivative** $f'(z)$ if the complex ratio $(f(z+h) - f(z))/h$ tends to $f'(z)$ as the complex number h tends to zero. Equivalently,

$$\frac{f(z+h) - f(z) - f'(z)h}{h} \to 0$$

as $h \to 0$. Write $f(z) = u(x, y) + iv(x, y)$ where $z = x + iy$, and u, v are real-valued functions of two real variables. Define $F : \mathbb{R}^2 \to \mathbb{R}^2$ by $F(x, y) = (u(x, y), \; v(x, y))$. Then F is $\mathbb{R}$-differentiable with derivative matrix

$$DF = \begin{bmatrix} \dfrac{\partial u}{\partial x} & \dfrac{\partial u}{\partial y} \\[2mm] \dfrac{\partial v}{\partial x} & \dfrac{\partial v}{\partial y} \end{bmatrix}.$$

Since this derivative matrix is the $\mathbb{R}^2$ expression for multiplication by the complex number $f'(z)$, it must have the $\begin{bmatrix} \alpha & -\beta \\ \beta & \alpha \end{bmatrix}$ form. This demonstrates a basic fact about complex differentiable functions – their real and imaginary parts, u and v, satisfy the

53 Cauchy-Riemann Equations

$$\frac{\partial u}{\partial x} = \frac{\partial v}{\partial y} \qquad and \qquad \frac{\partial u}{\partial y} = -\frac{\partial v}{\partial x}.$$

Appendix D   Polar Form

The shape of the image of a unit ball under a linear transformation T is not an issue that is used directly in anything we do in Chapter 5 but it certainly underlies the geometric outlook on linear algebra.

Question. What shape is the $(n-1)$-sphere S^{n-1}?

Answer. Round.

Question. What shape is $T(S^{n-1})$?

Answer. Ellipsoidal. See also Exercise 39.

Let $z = x + iy$ be a nonzero complex number. Its polar form is $z = re^{i\theta}$ where $r > 0$ and $0 \le \theta < 2\pi$, and $x = r\cos\theta$, $y = r\sin\theta$. Multiplication by z breaks up into multiplication by r, which is just dilation, and multiplication by $e^{i\theta}$, which is rotation of the plane by angle θ. As a matrix the rotation is

$$\begin{bmatrix} \cos\theta & -\sin\theta \\ \sin\theta & \cos\theta \end{bmatrix}.$$

The polar coordinates of (x, y) are (r, θ).

Analogously, consider an isomorphism $T : \mathbb{R}^n \to \mathbb{R}^n$. Its **polar form** is

$$T = OP$$

where O and P are isomorphisms $\mathbb{R}^n \to \mathbb{R}^n$ such that

(a) O is like $e^{i\theta}$; it is an orthogonal isomorphism.
(b) P is like r; it is positive definite symmetric (PDS) isomorphism.

Orthogonality of O means that for all $v, w \in \mathbb{R}^n$ we have

$$\langle Ov, Ow \rangle = \langle v, w \rangle,$$

while P being PDS means that for all nonzero vectors $v, w \in \mathbb{R}^n$ we have

$$\langle Pv, v \rangle > 0 \text{ and } \langle Pv, w \rangle = \langle v, Pw \rangle.$$

The notation $\langle v, w \rangle$ indicates the usual dot product on $\mathbb{R}^n$.

The polar form $T = OP$ reveals everything geometric about T. The geometric effect of O is nothing. It is an isometry and changes no distances or shapes. It is rigid. The effect of a PDS operator P is easy to describe. In linear algebra it is shown that there exists a basis $\mathcal{B} = \{u_1, \ldots, u_n\}$ of orthonormal vectors (the vectors are of unit length and are mutually perpendicular) and with respect to this basis we have

$$P = \begin{bmatrix} \lambda_1 & 0 & \cdots & & & \\ 0 & \lambda_2 & 0 & & \cdots & \\ & & \cdots & & & \\ & \cdots & 0 & \lambda_{n-1} & 0 & \\ & \cdots & & & 0 & \lambda_n \end{bmatrix}$$

The diagonal entries λ_i are positive. P stretches each u_i by the factor λ_i. Thus P stretches the unit sphere to an n-dimensional ellipsoid. The u_i are its axes. The norm of P and hence of T is the largest λ_i, while the conorm is the smallest λ_i. The ratio of the largest to the smallest, the **condition number**, is the eccentricity of the ellipsoid.

Upshot Except for the harmless orthogonal factor O, an isomorphism is no more geometrically complicated than a diagonal matrix with positive entries.

54 Polar Form Theorem *Each isomorphism $T : \mathbb{R}^n \to \mathbb{R}^n$ factors as $T = OP$, where O is orthogonal and P is PDS.*

Proof Recall that the transpose of $T : \mathbb{R}^n \to \mathbb{R}^n$ is the unique isomorphism T^t satisfying the equation

$$\langle Tv, w \rangle = \langle v, T^t w \rangle$$

for all $v, w \in \mathbb{R}^n$. Thus the condition $\langle Pv, w \rangle = \langle v, Pw \rangle$ in the definition of PDS means exactly that $P^t = P$.

Let T be a given isomorphism $T : \mathbb{R}^n \to \mathbb{R}^n$. We must find its factors O and P. We just write them down as follows. Consider the composite $T^t \circ T$. It is PDS because

$$(T^t T)^t = (T^t)(T^t)^t = T^t T \quad \text{and} \quad \langle T^t Tv, v \rangle = \langle Tv, Tv \rangle > 0.$$

Every PDS transformation has a unique PDS square root, just as does every positive real number r. (To see this, take the diagonal matrix with entries $\sqrt{\lambda_i}$ in place of λ_i.) Thus $T^t T$ has a PDS square root and this is the factor P that we seek,

$$P^2 = T^t T.$$

By P^2 we mean the composite $P \circ P$. In order for the formula $T = OP$ to hold with this choice of P we must have $O = TP^{-1}$. To finish the proof we merely must check that TP^{-1} actually is orthogonal. Magically,

$$
\begin{aligned}
\langle Ov, Ow \rangle &= \langle TP^{-1}v, TP^{-1}w \rangle = \langle P^{-1}v, T^t TP^{-1}w \rangle \\
&= \langle P^{-1}v, Pw \rangle = \langle P^t P^{-1}v, w \rangle = \langle PP^{-1}v, w \rangle \\
&= \langle v, w \rangle
\end{aligned}
$$

which implies that O is orthogonal. $\qquad \square$

55 Corollary *Under any invertible $T : \mathbb{R}^n \to \mathbb{R}^n$ the unit ball is sent to an ellipsoid.*

Proof Write T in polar form $T = OP$. The image of the unit ball under P is an ellipsoid. The orthogonal factor O merely rotates the ellipsoid. $\qquad \square$

Appendix E   Determinants

A permutation of a set S is a bijection $\pi : S \to S$. That is, π is one-to-one and onto. We assume the set S is finite, $S = \{1, \ldots, k\}$. The **sign** of π is

$$\mathrm{sgn}(\pi) = (-1)^r$$

where r is the number of reversals – i.e., the number of pairs i, j such that

$$i < j \ \text{ and } \pi(i) > \pi(j).$$

56 Proposition *Every permutation is the composite of pair transpositions; the sign of a composite permutation is the product of the signs of its factors; and the sign of a pair transposition is -1.*

The proof of this combinatorial proposition is left to the reader. Although the factorization of a permutation π into pair transpositions is not unique, the number of factors, say t, satisfies $(-1)^t = \mathrm{sgn}(\pi)$.

Definition The **determinant** of a $k \times k$ matrix A is the sum

$$\det A = \sum_{\pi} \mathrm{sgn}(\pi) a_{1\pi(1)} a_{2\pi(2)} \cdots a_{k\pi(k)}$$

where π ranges through all permutations of $\{1, \ldots, k\}$.

Equivalent definitions appear in standard linear algebra courses. One of the key facts about determinants is the product rule: For two $k \times k$ matrices we have

$$\det AB = \det A \det B.$$

It extends to nonsquare matrices as follows.

57 Cauchy-Binet Formula *Assume that $k \leq n$. If A is a $k \times n$ matrix and B is an $n \times k$ matrix, then the determinant of the product $k \times k$ matrix $AB = C$ is given by the formula*

$$\det C = \sum_{J} \det A^J \det B_J,$$

where J ranges through the set of ascending k-tuples in $\{1, \ldots, n\}$, A^J is the $k \times k$ minor of A whose column indices j belong to J, while B_J is the $k \times k$ minor of B whose row indices i belong to J. See Figure 132.

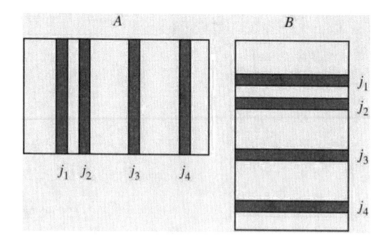

Figure 132 The paired 4×4 minors of A and B are determined by the 4-tuple $J = (j_1, j_2, j_3, j_4)$.

Proof Note that special cases of the Cauchy-Binet Formula occur when $k = 1$ or $k = n$. When $k = 1$, C is the 1×1 matrix that is the dot product of an A-row vector of length n times a B-column vector of height n. The 1-tuples J in $\{1, \ldots, n\}$ are just single integers, $J = (1), \ldots, J = (n)$, and the product formula is immediate. In the second case, $k = n$, we have the usual product determinant formula because there is only one ascending k-tuple in $\{1, \ldots, k\}$, namely $J = (1, \ldots, k)$.

To handle the general case, define the sum

$$S(A, B) = \sum_J \det A^J \det B_J$$

as above. Consider an elementary $n \times n$ matrix E. We claim that

$$S(A, B) = S(AE, E^{-1}B).$$

Since there are only two types of elementary matrices, this is not too hard a calculation, and is left to the reader. Then we perform a sequence of elementary column operations on A to put it in lower triangular form

$$A' = AE_1 \ldots E_r = \begin{bmatrix} \alpha_{11} & 0 & \cdots & \cdots & 0 & \cdots & 0 \\ \alpha_{21} & \alpha_{22} & \cdots & \cdots & 0 & \cdots & 0 \\ \vdots & \vdots & \ddots & & \vdots & & \vdots \\ \alpha_{k1} & \alpha_{k2} & \cdots & \alpha_{kk} & 0 & \cdots & 0 \end{bmatrix}.$$

About $B' = E_r^{-1} \ldots E_1^{-1} B$ we observe only that

$$AB = A'B' = A'^{J_0} B'_{J_0}$$

where $J_0 = (1, \ldots, k)$. Since elementary column operations do not affect S we have

$$S(A, B) = S(AE_1, E_1^{-1}B) = S(AE_1E_2, E_2^{-1}E_1^{-1}B) = \ldots = S(A', B').$$

All terms in the sum that defines $S(A', B')$ are zero except the J_0^{th}, and thus

$$\det(AB) = \det A'^{J_0} \det B'_{J_0} = S(A', B') = S(A, B)$$

as claimed. $\square$

Exercises

1. Let $T : V \to W$ be a linear transformation, and let $p \in V$ be given. Prove that the following are equivalent.
   (a) T is continuous at the origin.
   (b) T is continuous at p.
   (c) T is continuous at at least one point of V.

2. Let $\mathcal{L}$ be the vector space of continuous linear transformations from a normed space V to a normed space W. Show that the operator norm makes $\mathcal{L}$ a normed space.

3. Let $T : V \to W$ be a linear transformation between normed spaces. Show that

$$
\begin{aligned}
\|T\| &= \sup\{|Tv| : |v| < 1\} \\
&= \sup\{|Tv| : |v| \leq 1\} \\
&= \sup\{|Tv| : |v| = 1\} \\
&= \inf\{M : v \in V \Rightarrow |Tv| \leq M|v|\}.
\end{aligned}
$$

4. The **conorm** of a linear transformation $T : \mathbb{R}^n \to \mathbb{R}^m$ is

$$
m(T) = \inf \left\{ \frac{|Tv|}{|v|} : v \neq 0 \right\}.
$$

   It is the **minimum stretch** that T imparts to vectors in $\mathbb{R}^n$. Let U be the unit ball in $\mathbb{R}^n$.
   (a) Show that the norm and conorm of T are the radii of the smallest ball that contains TU and, when $n = m$, the largest ball contained in TU.
   (b) If T is an isomorphism prove that $m(T) = \|T^{-1}\|^{-1}$.
   (c) If $m = n$, $T = I + S$, and $\|S\| < 1$ prove that $m(T) > 0$. [Hint: The inequality $|u + v| \geq |u| - |v|$ is useful because it implies $|Tu| \geq |u| - |Su|$.] How can you infer that T is an isomorphism?
   (d) If the norm and conorm of T are equal, what can you say about T?

5. Formulate and prove the fact that function composition is associative. Why can you infer that matrix multiplication is associative?

6. Let $\mathcal{M}_n$ and $\mathcal{L}_n$ be the vector spaces of $n \times n$ matrices and linear transformations $\mathbb{R}^n \to \mathbb{R}^n$. Let $\mathcal{T} : \mathcal{M}_n \to \mathcal{L}_n$ send A to T_A as on page 277.
   (a) Look up the definition of "ring" in your algebra book.
   (b) Show that $\mathcal{M}_n$ and $\mathcal{L}_n$ are rings with respect to matrix multiplication and composition.
   (c) Show that $\mathcal{T} : \mathcal{M}_n \to \mathcal{L}_n$ is a ring isomorphism.

7. Two norms $|\ |_1$ and $|\ |_2$ on a vector space are **comparable**[†] if there are

---

[†]From an analyst's point of view, the choice between comparable norms has little importance. At worst it affects a few constants that turn up in estimates.

positive constants c and C such that for all nonzero vectors in V we have

$$c \leq \frac{|v|_1}{|v|_2} \leq C.$$

(a) Prove that comparability is an equivalence relation on norms.
(b) Prove that any two norms on a finite-dimensional vector space are comparable. [Hint: Use Theorem 3.]
(c) Consider the norms

$$|f|_{L^1} = \int_0^1 |f(t)|\, dt \quad \text{and} \quad |f|_{C^0} = \max\{|f(t)| : t \in [0,1]\},$$

defined on the infinite-dimensional vector space C^0 of continuous functions $f : [0,1] \to \mathbb{R}$. Show that the norms are not comparable by finding functions $f \in C^0$ whose integral norm is small but whose C^0 norm is 1.

*8. Let $|\ \ | = |\ \ |_{C^0}$ be the supremum norm on C^0 as in the previous exercise. Define an integral transformation $T : C^0 \to C^0$ by

$$T : f \mapsto \int_0^x f(t)\, dt.$$

(a) Show that T is linear, continuous, and find its norm.
(b) Let $f_n(t) = \cos(nt)$, $n = 1, 2, \ldots$. What is $T(f_n)$?
(c) Is the set of functions $K = \{f_n : n \in \mathbb{N}\}$ closed? Bounded? Compact?
(d) Is $T(K)$ compact? How about its closure?

9. Give an example of two 2×2 matrices such that the operator norm of the product is less than the product of the operator norms.

10. In the proof of Theorem 3 we used the fact that with respect to the Euclidean norm, the length of a vector is at least as large as the length of any of its components. Show by example that this is false for some norms in $\mathbb{R}^2$. [Hint: Consider the matrix

$$A = \begin{bmatrix} 3 & -2 \\ -2 & 2 \end{bmatrix}.$$

Use A to define an inner product $\langle v, w \rangle_A = \sum v_i a_{ij} w_j$ on $\mathbb{R}^2$, and use the inner product to define a norm

$$|v|_A = \sqrt{\langle v, v \rangle_A}.$$

(What properties must A have for the sum to define an inner product? Does A have these properties?) With respect to this norm, what are the lengths of e_1, e_2, and $v = e_1 + e_2$?]

11. Consider the shear matrix
$$S = \begin{bmatrix} 1 & s \\ 0 & 1 \end{bmatrix}$$
and the linear transformation $S : \mathbb{R}^2 \to \mathbb{R}^2$ it represents. Calculate the norm and conorm of S. [Hint: Using polar form, it suffices to calculate the norm and conorm of the positive definite symmetric part of S. Recall from linear algebra that the eigenvalues of the square of a matrix A are the squares of the eigenvalues of A.]

12. What is the one-line proof that if V is a finite-dimensional normed space then its unit sphere $\{v : |v| = 1\}$ is compact?

13. The set of invertible $n \times n$ matrices is open in $\mathcal{M}_n$. Is it dense?

14. An $n \times n$ matrix is **diagonalizable** if there is a change of basis in which it becomes diagonal.
    (a) Is the set of diagonalizable matrices open in $\mathcal{M}_n$?
    (b) Closed?
    (c) Dense?

15. Show that both partial derivatives of the function
$$f(x,y) = \begin{cases} \dfrac{xy}{x^2 + y^2} & \text{if } (x,y) \neq (0,0) \\ 0 & \text{if } (x,y) = (0,0) \end{cases}$$
exist at the origin but the function is not differentiable there.

16. Let $f : \mathbb{R}^2 \to \mathbb{R}^3$ and $g : \mathbb{R}^3 \to \mathbb{R}$ be defined by $f = (x,y,z)$ and $g = w$ where
$$w = w(x,y,z) = xy + yz + zx$$
$$x = x(s,t) = st \quad y = y(s,t) = s\cos t \quad z = z(s,t) = s\sin t.$$

    (a) Find the matrices that represent the linear transformations $(Df)_p$ and $(Dg)_q$ where $p = (s_0, t_0) = (1,0)$ and $q = f(p)$.
    (b) Use the Chain Rule to calculate the 1×2 matrix $[\partial w/\partial s,\ \partial w/\partial t]$ that represents $(D(g \circ f))_p$.
    (c) Plug the functions $x = x(s,t)$, $y = y(s,t)$, and $z = z(s,t)$ directly into $w = w(x,y,z)$, and recalculate $[\partial w/\partial s,\ \partial w/\partial t]$, verifying the answer given in (b).
    (d) Examine the statements of the multivariable chain rules that appear in your old calculus book and observe that they are nothing more than the components of various product matrices.

17. Let $f : U \to \mathbb{R}^m$ be differentiable, $[p,q] \subset U \subset \mathbb{R}^n$, and ask whether the direct generalization of the one-dimensional Mean Value Theorem is true: Does there exist a point $\theta \in [p,q]$ such that

    (28) $$f(q) - f(p) = (Df)_\theta(q - p)?$$

(a) Take $n = 1$, $m = 2$, and examine the function

$$f(t) = (\cos t, \sin t)$$

for $\pi \leq t \leq 2\pi$. Take $p = \pi$ and $q = 2\pi$. Show that there is no $\theta \in [p, q]$ which satisfies (28).

**(b) Assume that the set of derivatives

$$\{(Df)_x \in \mathcal{L}(\mathbb{R}^n, \mathbb{R}^m) : x \in [p, q]\}$$

is convex. Prove there exists $\theta \in [p, q]$ which satisfies (28). [Hint: Google "support plane."]

(c) How does (b) imply the one-dimensional Mean Value Theorem?

18. The **directional derivative** of $f : U \to \mathbb{R}^m$ at $p \in U$ in the direction u is the limit, if it exists,

$$\nabla_p f(u) = \lim_{t \to 0} \frac{f(p + tu) - f(p)}{t}.$$

(Often one requires that $|u| = 1$.)

(a) If f is differentiable at p, why is it obvious that the directional derivative exists in each direction u?

(b) Show that the function $f : \mathbb{R}^2 \to \mathbb{R}$ defined by

$$f(x, y) = \begin{cases} \dfrac{x^3 y}{x^4 + y^2} & \text{if } (x, y) \neq (0, 0) \\ 0 & \text{if } (x, y) = (0, 0) \end{cases}$$

has $\nabla_{(0,0)} f(u) = 0$ for all u but is not differentiable at $(0, 0)$.

*19. Using the functions in Exercises 15 and 18, show that the composite of functions whose partial derivatives exist may fail to have partial derivatives, and the composite of functions whose directional derivatives exist may fail to have directional derivatives. (That is, the classes of these functions are not closed under composition, which is further reason to define multidimensional differentiability in terms of Taylor approximation, and not in terms of partial or directional derivatives.)

20. Assume that U is a connected open subset of $\mathbb{R}^n$ and $f : U \to \mathbb{R}^m$ is differentiable everywhere on U. If $(Df)_p = 0$ for all $p \in U$, show that f is constant.

21. For U as above, assume that f is second-differentiable everywhere and $(D^2 f)_p = 0$ for all p. What can you say about f? Generalize to higher-order differentiability.

22. If Y is a metric space and $f : [a, b] \times Y \to \mathbb{R}$ is continuous, show that

$$F(y) = \int_a^b f(x, y) \, dx$$

is continuous.

23. Assume that $f : [a, b] \times Y \to \mathbb{R}^m$ is continuous, Y is an open subset of $\mathbb{R}^n$, the partial derivatives $\partial f_i(x, y)/\partial y_j$ exist, and they are continuous. Let $D_y f$ be the linear transformation $\mathbb{R}^n \to \mathbb{R}^m$ which is represented by the $m \times n$ matrix of partials.
    (a) Show that
$$F(y) = \int_a^b f(x, y)\, dx$$
    is of class C^1 and
$$(DF)_y = \int_a^b (D_y f)\, dx.$$
    This generalizes Theorem 14 to higher dimensions.
    (b) Generalize (a) to higher-order differentiability.

24. Show that all second partial derivatives of the function $f : \mathbb{R}^2 \to \mathbb{R}$ defined by
$$f(x, y) = \begin{cases} \dfrac{xy(x^2 - y^2)}{x^2 + y^2} & \text{if } (x, y) \neq (0, 0) \\ 0 & \text{if } (x, y) = (0,\ 0) \end{cases}$$

    exist everywhere, but the mixed second partials are unequal at the origin, $\partial^2 f(0, 0)/\partial x \partial y \neq \partial^2 f(0, 0)/\partial y \partial x$.

*25. Construct an example of a C^1 function $f : \mathbb{R} \to \mathbb{R}$ that is second-differentiable only at the origin. (Infer that this phenomenon occurs also in higher dimensions.)

26. Suppose that $u \mapsto \beta_u$ is a continuous function from $U \subset \mathbb{R}^n$ into $\mathcal{L}(\mathbb{R}^m, \mathbb{R}^m)$.
    (a) If for all $u \in U$, β_u is symmetric, prove that its average over each $W \subset U$ is symmetric.
    (b) Conversely, prove that if the average over all small two-dimensional parallelograms in U is symmetric then β_u is symmetric for all $u \in U$. (That is, if for some $p \in U$, β_p is not symmetric, prove that its average over some small two-dimensional parallelogram at p is also not symmetric.)
    (c) Generalize (a) and (b) by replacing $\mathcal{L}$ with a finite-dimensional space E, and the subset of symmetric bilinear maps with a linear subspace of E: The average values of a continuous function always lie in the subspace if and only if the values do.

*27. Assume that $f : U \to \mathbb{R}^m$ is of class C^2 and show that $D^2 f$ is symmetric by the following integral method. With reference to the signed sum Δ of f at the vertices of the parallelogram P in Figure 109, use the C^1 Mean Value Theorem to show that
$$\Delta = \left(\int_0^1 \int_0^1 (D^2 f)_{p+sv+tw}\, ds dt \right)(v, w).$$

Infer symmetry of $(D^2 f)_p$ from symmetry of Δ and Exercise 26.

28. Let $\beta : \mathbb{R}^n \times \cdots \times \mathbb{R}^n \to \mathbb{R}^m$ be r-linear. Define its "symmetrization" as

$$\text{symm}(\beta)(v_1, \ldots, v_r) = \frac{1}{r!} \sum_{\pi} \beta(v_{\pi(1)}, \ldots, v_{\pi(r)}),$$

where π ranges through the set of permutations of $\{1, \ldots, r\}$.

(a) Prove that $\text{symm}(\beta)$ is symmetric.

(b) If β is symmetric prove that $\text{symm}(\beta) = \beta$.

(c) Is the converse to (b) true?

(d) Prove that $\alpha = \beta - \text{symm}(\beta)$ is antisymmetric in the sense that if π is any permutation of $\{1, \ldots, r\}$ then

$$\alpha(v_{\pi(1)}, \ldots, v_{\pi(r)}) = \text{sgn}(\pi)\alpha(v_1, \ldots, v_r).$$

Infer that $\mathcal{L}^r = \mathcal{L}_s^r \oplus \mathcal{L}_a^r$ where $\mathcal{L}_s^r$ and $\mathcal{L}_a^r$ are the subspaces of symmetric and antisymmetric r-linear transformations.

(e) Let $\beta \in \mathcal{L}^2(\mathbb{R}^2, \mathbb{R})$ be defined by

$$\beta((x, y), (x', y')) = xy'.$$

Express β as the sum of a symmetric and an antisymmetric bilinear transformation.

*29. Prove Corollary 18 that r^{th}-order differentiability implies symmetry of $D^r f$, $r \geq 3$, in one of two ways.

(a) Use induction to show that $(D^r f)_p(v_1, \ldots, v_r)$ is symmetric with respect to permutations of $v_1, \ldots, v_{r-1}$ and of $v_2, \ldots, v_r$. Then take advantage of the fact that r is strictly greater than 2.

(b) Define the signed sum Δ of f at the vertices of the paralleletope P spanned by $v_1, \ldots, v_r$, and show that it is the average of $D^r f$. Then proceed as in Exercise 27.

30. Consider the equation

(29) $xe^y + ye^x = 0.$

(a) Observe that there is no way to write down an explicit solution $y = y(x)$ of (29) in a neighborhood of the point $(x_0, y_0) = (0, 0)$.

(b) Why, nevertheless, does there exist a C^∞ solution $y = y(x)$ of (29) near $(0, 0)$?

(c) What is its derivative at $x = 0$?

(d) What is its second derivative at $x = 0$?

(e) What does this tell you about the graph of the solution?

(f) Do you see the point of the Implicit Function Theorem better?

**31. Consider a function $f : U \to \mathbb{R}$ such that

    (i) U is a connected open subset of $\mathbb{R}^2$.

    (ii) f is C^1.

    (iii) For each $(x, y) \in U$ we have

$$\frac{\partial f(x, y)}{\partial y} = 0.$$

    (a) If U is a disc show that f is independent of y.

    (b) Construct such an f of class C^∞ which does depend on y.

    (c) Show that the f in (b) can not be analytic.

    (d) Why does your example in (b) not invalidate the proof of the Rank Theorem on page 306?

32. Let G denote the set of invertible $n \times n$ matrices.

    (a) Prove that G is an open subset of $\mathcal{M}(n \times n)$.

    (b) Prove that G is a group. (It is called the **general linear group**.)

    (c) Prove that the inversion operator Inv : $A \mapsto A^{-1}$ is a homeomorphism of G onto G.

    (d) Prove that Inv is a diffeomorphism and show that its derivative at A is the linear transformation $\mathcal{M} \to \mathcal{M}$,

$$X \mapsto -A^{-1} \circ X \circ A^{-1}.$$

    (e) Relate this formula to the ordinary derivative of $1/x$ at $x = a$.

33. Observe that $Y = \text{Inv} \, X$ solves the implicit function problem

$$F(X, Y) - I = 0,$$

where $F(X, Y) = X \circ Y$. Assume it is known that Inv is smooth and use the Chain Rule to derive from this equation the formula for the derivative of Inv.

34. Use Gaussian elimination to prove that the entries of the matrix A^{-1} depend smoothly (in fact analytically) on the entries of A.

*35. Give a proof that the inversion operator Inv is analytic (i.e., is defined locally by a convergent power series) as follows:

    (a) If $T \in \mathcal{L}(\mathbb{R}^n, \mathbb{R}^n)$ and $\|T\| < 1$ show that the series of linear transformations

$$I + T + T^2 + \ldots + T^k + \ldots$$

converges to a linear transformation S, and

$$S \circ (I - T) = I = (I - T) \circ S,$$

where I is the identity transformation.

    (b) Infer from (a) that inversion is analytic at I.

(c) In general, if $T_0 \in G$ and $\|T\| < 1/\left\|T_0^{-1}\right\|$ show that

$$\mathrm{Inv}(T_0 - T) = \mathrm{Inv}(I - T_0^{-1} \circ T) \circ T_0^{-1},$$

and infer that Inv is analytic at T_0.

(d) Infer from the general fact that analyticity implies smoothness that inversion is smooth.

(This proof avoids Cramer's Rule and makes little use of finite-dimensionality.)

*36. Give a proof of smoothness of Inv by the following bootstrap method.

(a) Using Exercise 4c on page 366 and the identity

$$X^{-1} - Y^{-1} = X^{-1} \circ (Y - X) \circ Y^{-1},$$

give a simple proof that Inv is continuous.

(b) Infer that $Y = \mathrm{Inv}(X)$ is a continuous solution of the C^∞ implicit function problem

$$F(X,Y) - I = 0,$$

where $F(X,Y) = X \circ Y$ as in Exercise 33. Since the proof of the C^1 Implicit Function Theorem relies only on continuity of Inv, it is not circular reasoning to conclude that Inv is C^1.

(c) Assume simultaneously that the C^r Implicit Function Theorem has been proved and that Inv is known to be C^{r-1}. Prove that Inv is C^r and that the C^{r+1} Implicit Function Theorem is true.

(d) Conclude logically that Inv is smooth and the C^∞ Implicit Function Theorem is true.

(This proof avoids Cramer's Rule and also makes little use of finite dimensionality.)

*37. Use polar decomposition to give an alternate proof of the volume-multiplier formula.

**38. Consider the set S of all 2×2 matrices $X \in \mathcal{M} = \mathcal{M}_2$ that have rank 1.

(a) Show that in a neighborhood of the matrix

$$X_0 = \begin{bmatrix} 1 & 0 \\ 0 & 0 \end{bmatrix}$$

S is diffeomorphic to a two-dimensional disc.

(b) Is this true (locally) for all matrices $X \in S$?

(c) Describe S globally. (How many connected components does it have? Is it closed in $\mathcal{M}$? If not, what are its limit points and how does S approach them? What is the intersection of S with the unit sphere in $\mathcal{M}$?, etc.)

39. Draw pictures of all the possible shapes of $T(S^2)$ where $T : \mathbb{R}^3 \to \mathbb{R}^3$ is a linear transformation and S^2 is the 2-sphere. (Don't forget the cases in which T has rank < 3.)

40. Let M and N be $n \times n$ real matrices.
    (a) If N represents a positive definite symmetric transformation $\mathbb{R}^n \to \mathbb{R}^n$, prove that there exist orthogonal matrices O_1 and O_2, and a diagonal matrix D with positive entries such that $N = O_1 D O_2$.
    (b) If M is invertible prove that there exist orthogonal matrices O_3 and O_4, and a diagonal matrix D with positive entries such that $M = O_3 D O_4$.
    (c) What can you prove if M is not invertible?

**41. Suppose that f and g are r^{th}-order differentiable and that the composite $h = g \circ f$ makes sense. A partition divides a set into nonempty disjoint subsets. Prove the Higher Order Chain Rule,

$$(D^r h)_p = \sum_{k=1}^{r} \sum_{\mu \in P(k,r)} (D^k g)_q \circ (D^\mu f)_p$$

where μ partitions $\{1, \ldots, r\}$ into k subsets, and $q = f(p)$. In terms of r-linear transformations, this notation means

$$(D^r h)_p(v_1, \ldots, v_r)$$
$$= \sum_{k=1}^{r} \sum_{\mu} (D^k g)_q((D^{|\mu_1|} f)_p(v_{\mu_1}), \ \ldots \ , (D^{|\mu_k|} f)_p(v_{\mu_k}))$$

where $|\mu_i| = \#\mu_i$ and v_{μ_i} is the $|\mu_i|$-tuple of vectors v_j with $j \in \mu_i$. (Symmetry implies that the order of the vectors v_j in the $|\mu_i|$-tuple v_{μ_i} and the order in which the partition blocks $\mu_1, \ldots, \mu_k$ occur are irrelevant.)

**42. Suppose that β is bilinear and $\beta(f, g)$ makes sense. If f and g are r^{th}-order differentiable at p, find the Higher-Order Leibniz Formula for $D^r(\beta(f, g))_p$. [Hint: First derive the formula in dimension 1.]

43. Suppose that $T : \mathbb{R}^n \to \mathbb{R}^m$ has rank k.
    (a) Show there exists a $\delta > 0$ such that if $S : \mathbb{R}^n \to \mathbb{R}^m$ and $\|S - T\| < \delta$ then S has rank $\geq k$.
    (b) Give a specific example in which the rank of S can be greater than the rank of T, no matter how small δ is.
    (c) Give examples of linear transformations of rank k for each k where $0 \leq k \leq \min\{n, m\}$.

44. Let $S \subset M$ be given.
    (a) Define the characteristic function $\chi_S : M \to \mathbb{R}$.
    (b) If M is a metric space, show that $\chi_S(x)$ is discontinuous at x if and only if x is a boundary point of S.

45. On page 315 there is a definition of $Z \subset \mathbb{R}^2$ being a zero set that involves open rectangles.

(a) Show that the definition is unaffected if we require that the rectangles covering Z are open squares.

(b) What if we permit the squares or rectangles to be nonopen?

(c) What if we use discs or other shapes instead of squares and rectangles?

*46. Assume that $S \subset \mathbb{R}^2$ is bounded.

(a) Prove that if S is Riemann measurable then so are its interior and closure.

(b) Suppose that the interior and closure of S are Riemann measurable and $|\operatorname{int}(S)| = |\overline{S}| < \infty$. Prove that S is Riemann measurable.

(c) Show that some open bounded subsets of $\mathbb{R}^2$ are not Riemann measurable. See Appendix E in Chapter 6.

*47. The ambiguity in Fubini's formula $\int_R f = \int_c^d \left(\int_a^b f(x,y)\,dx \right) dy$ revolves around writing $\int_a^b f(x,y)\,dx$ at points y where $f(x,y)$ fails to be integrable as a function of x. What we actually showed in the proof of Fubini's Theorem on pages 316-318 is that the lower and upper integrals of f with respect to x are integrable with respect to y, equal for almost every y, and their integrals with respect to y equal $\int_R f$. (The lower and upper integrals with respect to x are $\underline{F}(y)$ and $\overline{F}(y)$. They *always* exist.) Let $E \subset [c,d]$ denote the "equality set" $\{y : \underline{F}(y) = \overline{F}(y)\}$, and let $E(y)$ denote the "equality value" $\underline{F}(y) = \overline{F}(y)$ on E. The complement of E is a zero set. To remove the ambiguity one must "define" the integral of $f(x,y)$ with respect to x at points $y \in E^c$ – that's exactly where the integral doesn't exist! One way is to take the average, $F(y) = (\underline{F}(y) + \overline{F}(y))/2$. Then F is integrable, defined everywhere, agrees with $E(y)$ at all $y \in E$, and its integral equals $\int_R f$. A more general way is to choose *any* Riemann integrable function $F(y)$ which equals $E(y)$ on E. Then $F(y) = \int_a^b f(x,y)\,dx$ "by definition." Justify this approach by proving the following.

(a) Riemann integrable functions $g, h : [a,b] \to \mathbb{R}$ that are almost everywhere equal have equal integrals. [Hint: This is like Corollary 31 in Chapter 3. For any partition P of $[a,b]$, why is there a choice of sample points T such that the Riemann sums $R(g, P, T)$ and $R(h, P, T)$ are equal, and why does this imply that $\int g = \int h$?]

(b) Under the hypotheses above, if $F(y)$ is any function such that $\underline{F}(y) \le F(y) \le \overline{F}(y)$ for all $y \in [c,d]$, show that $F(y)$ is Riemann integrable and its integral agrees with those of $\underline{F}(y)$ and $\overline{F}(y)$.

(c) If $F(y)$ is any Riemann integrable function which equals $E(y)$ on E, show that the integral of F equals those of $\underline{F}$ and $\overline{F}$.

(d) Find a counterexample to (c) if we remove the hypothesis that F is Riemann integrable.

48. Set $f(x,y) = 1 - 1/q$ if $x, y \in \mathbb{Q} \cap [0,1]$ and $y = p/q$ in lowest terms. Set $f = 1$ otherwise. Prove that f is Riemann integrable on $R = [0,1]^2$, calculate $\underline{F}(y)$ and $\overline{F}(y)$, and prove that $\int_0^1 \underline{F}(y)\,dy = \int_0^1 \overline{F}(y)\,dy = \int_R f = 1$.

49. Using the Fundamental Theorem of Calculus, give a direct proof of Green's Formulas

$$-\iint_R f_y \, dxdy = \int_{\partial R} f \, dx \quad \text{and} \quad \iint_R g_x \, dxdy = \int_{\partial R} g \, dy$$

where R is a square in the plane and $f, g : \mathbb{R}^2 \to \mathbb{R}$ are smooth. (Assume that the edges of the square are parallel to the coordinate axes.)

50. Draw a **staircase** curve S_n that approximates the diagonal

$$\Delta = \{(x, y) \in \mathbb{R}^2 : 0 \le x = y \le 1\}$$

to within a tolerance $1/n$. See Figure 133. Suppose that $f, g : \mathbb{R}^2 \to \mathbb{R}$ are

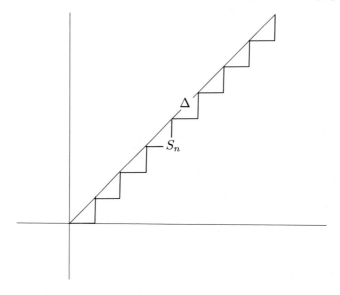

Figure 133 The staircase curve approximating the diagonal consists of both treads and risers.

smooth.

(a) Why does the length of S_n not converge to the length of Δ as $n \to \infty$?

(b) Despite (a), prove that

$$\int_{S_n} f \, dx \to \int_{\Delta} f \, dx \quad \text{and} \quad \int_{S_n} g \, dy \to \int_{\Delta} g \, dy$$

as $n \to \infty$.

(c) Repeat (b) with Δ replaced by the graph of a smooth function $h : [a, b] \to \mathbb{R}$.

(d) If C is a smooth simple closed curve in the plane, show that it is the union of finitely many arcs C_ℓ, each of which is the graph of a smooth function $y = h(x)$ or $x = h(y)$, and the arcs C_ℓ meet only at common endpoints.

(e) Infer that if (S_n) is a sequence of staircase curves that converges to C then

$$\int_{S_n} f\,dx + g\,dy \;\; \to \;\; \int_{C} f\,dx + g\,dy.$$

(f) Use (e) and Exercise 49 to give a proof of Green's Formulas on a general region $D \subset \mathbb{R}^2$ bounded by a smooth simple closed curve C, that relies on approximating[†] C, say from the inside, by staircase curves S_n which bound regions R_n composed of many small squares. (You may imagine that $R_1 \subset R_2 \subset \ldots$ and that $R_n \to D$.)

51. A region R in the plane is of type 1 if there are smooth functions $g_1 : [a, b] \to \mathbb{R}$, $g_2 : [a, b] \to \mathbb{R}$ such that $g_1(x) \le g_2(x)$ and

$$R = \{(x, y) : a \le x \le b \text{ and } g_1(x) \le y \le g_2(x)\}.$$

R is of type 2 if the roles of x and y can be reversed, and it is a **simple region** if it is of both type 1 and type 2.

   (a) Give an example of a region that is type 1 but not type 2.
   (b) Give an example of a region that is neither type 1 nor type 2.
   (c) Is every simple region starlike? Convex?
   (d) If a convex region is bounded by a smooth simple closed curve, is it simple?
   (e) Give an example of a region that divides into three simple subregions but not into two.
   *(f) If a region is bounded by a smooth simple closed curve C then it need not divide into a finite number of simple subregions. Find an example.
   (g) Infer that the standard proof of Green's Formulas for simple regions (as, for example, in J. Stewart's *Calculus*) does not immediately carry over to the general planar region R with smooth boundary; i.e., cutting R into simple regions can fail.
   ***(h) Is there a planar region bounded by a smooth simple closed curve such that for every linear coordinate system (i.e., a new pair of axes), the region does not divide into finitely many simple subregions? In other words, is Stewart's proof of Green's Theorem doomed?
   *(i) Show that if the curve C in (f) is analytic, then no such example exists. [Hint: C is analytic if it is locally the graph of a function defined by a convergent power series. A nonconstant analytic function has the property that for each x, there is some derivative of f which is nonzero, $f^{(r)}(x) \ne 0$.]

**52. Show that every starlike open subset of the plane is diffeomorphic to the plane. (The same is true in $\mathbb{R}^n$.)

[†]This staircase approximation proof generalizes to regions that are bounded by fractal, nondifferentiable curves such as the von Koch snowflake. As Jenny Harrison has shown, it also generalizes to higher dimensions.

**53. The 2-cell $\varphi : I^n \to B^n$ constructed in Step 3 of the proof of Brouwer's Theorem is smooth but not one-to-one. For it crushes the corners of I^n into ∂B.

    (a) Construct a homeomorphism $h : I^2 \to B^2$ where I^2 is the closed unit square and B^2 is the closed unit disc.

    (b) In addition make h in (a) be of class C^1 (on the closed square) and be a diffeomorphism from the interior of I^2 onto the interior of B^2. (The derivative of a diffeomorphism is everywhere nonsingular.)

    (c) Why can h not be a diffeomorphism from I^2 onto B^2?

    (d) Improve class C^1 in (b) to class C^∞.

54. If $K, L \subset \mathbb{R}^n$ and if there is a homeomorphism $h : K \to L$ that extends to $H : U \to V$ such that $U, V \subset \mathbb{R}^n$ are open, H is a homeomorphism, and H, H^{-1} are of class C^r with $1 \le r \le \infty$ then we say that K and L are **ambiently C^r-diffeomorphic.

    (a) In the plane, prove that the closed unit square is ambiently diffeomorphic to a general rectangle and to a general parallelogram.

    (b) If K, L are ambiently diffeomorphic polygons in the plane, prove that K and L have the same number of vertices. (Do not count vertices at which the interior angle is 180 degrees.)

    (c) Prove that the closed square and closed disc are not ambiently diffeomorphic.

    (d) If K is a convex polygon that is ambiently diffeomorphic to a polygon L, prove that L is convex.

    (e) Is the converse to (b) true or false? What about in the convex case?

    (f) The closed disc is tiled by five ambiently diffeomorphic copies of the unit square as shown in Figure 134. Prove that it cannot be tiled by fewer.

    (g) Generalize to dimension $n \ge 3$ and show that the n-ball can be tiled by $2n + 1$ diffeomorphs of the n-cube. Can it be done with fewer?

    (h) Show that a triangle can be tiled by three diffeomorphs of the square. Infer that any surface that can be tiled by diffeomorphs of the triangle can also be tiled by diffeomorphs of the square. What happens in higher dimensions?

55. Choose at random I, J, two non-ascending triples of integers between 1 and 9. Using only the wedge definition on page 334, check that $dx_I \wedge dx_J = dx_{IJ}$.

56. True or false? For every k-form α we have $\alpha \wedge \alpha = 0$.

57. Show that $d : \Omega^k \to \Omega^{k+1}$ is a linear vector space homomorphism.

58. Using Stokes' Formula (but not the Poincaré Lemma and its consequences), prove that closed 1-forms are exact (i.e., $d\omega = 0 \Rightarrow \omega = dh$ for some h) when defined on $\mathbb{R}^2$ or on any convex open subset of $\mathbb{R}^2$ as follows.

    (a) If $\varphi, \psi : [0, 1] \to U$ are paths from p to q, define

$$\sigma(s, t) = (1 - s)\varphi(t) + s\psi(t)$$

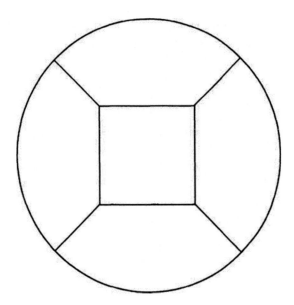

Figure 134 Five diffeomorphs of the square tile the disc.

for $0 \leq s, t \leq 1$ and observe it is a smooth 2-cell.

(b) If $\omega = f\,dx + g\,dy$ is a closed 1-form, how does Stokes' Formula imply $\int_{\varphi} \omega = \int_{\psi} \omega$, and what does this mean about path independence?

(c) Show that if p is held fixed then

$$h(q) = \int_{\varphi} \omega$$

is smooth and $dh = \omega$.

(d) What if U is nonconvex but diffeomorphic to $\mathbb{R}^2$?

(e) What about higher dimensions?

*59. For $0 < a < b$ the **spherical shell** is the set

$$U = \{(x, y, x) \in \mathbb{R}^3 : a^2 < x^2 + y^2 + z^2 < b^2\}.$$

It is the open region between spheres of radius a and b. If C is any closed curve in U (i.e., the image of a continuous map $\gamma : S^1 \to U$), show that C can be shrunk to a point without leaving U. That is, U is simply connected. [Hint: Why is there a point of U not in C, and how does this help? Gazing at Figure 135 may be a good idea.]

*60. Prove that the closure of the spherical shell is simply connected.

61. Assume $d^2 f = 0$ for all smooth functions f, and prove that $d^2\omega = 0$ for all smooth k-forms ω.

62. Does there exist a continuous mapping from the circle to itself that has no fixed-point? What about the 2-torus? The 2-sphere?

63. Show that a smooth map $T : U \to V$ induces a linear map of cohomology groups $H^k(V) \to H^k(U)$ defined by

$$T^* : [\omega] \mapsto [T^*\omega].$$

Here, $[\omega]$ denotes the equivalence class of $\omega \in Z^k(V)$ in $H^k(V)$. The question amounts to showing that the pullback of a closed form ω is closed and that its cohomology class depends only on the cohomology class of ω.[†]

64. Prove that diffeomorphic open sets have isomorphic cohomology groups.

65. Show that the 1-form defined on $\mathbb{R}^2 \setminus \{(0,0)\}$ by

$$\omega = \frac{-y}{r^2} \, dx + \frac{x}{r^2} \, dy$$

is closed but not exact. Why do you think that this 1-form is often referred to as $d\theta$ and why is the name problematic?

66. Let $H \subset \mathbb{R}^3$ be the helicoid $\{(x, y, z) : x^2 + y^2 \neq 0$ and $z = \arctan y/x\}$ and let $\pi : H \to \mathbb{R}^2 \setminus \{(0,0)\}$ be the projection $(x, y, z) \mapsto (x, y)$.
    (a) For $\omega = (x \, dy - y \, dx)/r^2$ as in Exercise 65, why is $\pi^*\omega$ a closed 1-form on H?
    (b) Is it exact? That is, does there exist a smooth function $f : H \to \mathbb{R}$ such that $df = \omega$?
    (c) Is there more than one?
    (d) Is there more than one such that $f(1, 0, 0) = 0$?

67. Show that the 2-form defined on the spherical shell by

$$\omega = \frac{x}{r^3} \, dy \wedge dz \; + \; \frac{y}{r^3} \, dz \wedge dx \; + \; \frac{z}{r^3} \, dx \wedge dy$$

is closed but not exact.

68. True or false: If ω is closed then $f\omega$ is closed.
    True or false: If ω is exact then $f\omega$ is exact.

69. Is the wedge product of closed forms closed? Of exact forms exact? What about the product of a closed form and an exact form? Does this give a ring structure to the cohomology classes?

[†]A fancier way to present the proof of the Brouwer Fixed Point Theorem goes like this: As always, the question reduces to showing that there is no smooth retraction T of the n-ball to its boundary. Such a T would give a cohomology map $T^* : H^k(\partial B) \to H^k(B)$ where the cohomology groups of ∂B are those of its spherical shell neighborhood. The map T^* is seen to be a cohomology group isomorphism because $T \circ$ inclusion$_{\partial B}$ = inclusion$_{\partial B}$ and inclusion$_{\partial B}{}^*$ = identity. But when $k = n - 1 \geq 1$ the cohomology groups are nonisomorphic; they are computed to be $H^{n-1}(\partial B) = \mathbb{R}$ and $H^{n-1}(B) = 0$.

70. Prove that the n-cell $\psi : [-1, 1]^n \to B^n$ in the proof of the Brouwer Fixed-Point Theorem has Jacobian $\rho'(r)\rho(r)^{n-1} / r^{n-1}$ for $r = |v|$ as claimed on page 355.

71. The **Hairy Ball Theorem states that any continuous vector field X in $\mathbb{R}^3$ that is everywhere tangent to the 2-sphere S is zero at some point of S. Here is an outline of a proof for you to fill in. (If you imagine the vector field as hair combed on a sphere, there must be a cowlick somewhere.)

(a) Show that the Hairy Ball Theorem is equivalent to a fixed-point assertion: Every continuous map of S to itself that is sufficiently close to the identity map $S \to S$ has a fixed-point. (This is not needed below but it is interesting.)

(b) If a continuous vector field on S has no zero on or inside a small simple closed curve $C \subset S$, show that the net angular turning of X along C as judged by an observer who takes a tour of C in the counterclockwise direction is -2π. (The observer walks along C in the counterclockwise direction when S is viewed from the outside, and he measures the angle that X makes with respect to his own tangent vector as he walks along C. By convention, clockwise angular variation is negative.) Show also that the net turning is $+2\pi$ if the observer walks along C in the clockwise direction.

(c) If C_t is a continuous family of simple closed curves on S, $a \leq t \leq b$, and if X never equals zero at points of C_t, show that the net angular turning of X along C_t is independent of t. (This is a case of a previous exercise stating that a continuous integer-valued function of t is constant.)

(d) Imagine the following continuous family of simple closed curves C_t. For $t = 0$, C_0 is the Arctic Circle. For $0 \leq t \leq 1/2$, the latitude of C_t decreases while its circumference increases as it oozes downward, becomes the Equator, and then grows smaller until it becomes the Antarctic Circle when $t = 1/2$. For $1/2 \leq t \leq 1$, C_t maintains its size and shape, but its new center, the South Pole, slides up the Greenwich Meridian until at $t = 1$, C_t regains its original arctic position. See Figure 135. Its orientation has reversed. Orient the Arctic Circle C_0 positively and choose an orientation on each C_t that depends continuously on t. To reach a contradiction, suppose that X has no zero on S.

   (i) Why is the total angular turning of X along C_0 equal to -2π?.

   (ii) Why is it $+2\pi$ on C_1?

   (iii) Why is this a contradiction to (c) unless X has a zero somewhere?

   (iv) Conclude that you have proved the Hairy Ball Theorem.

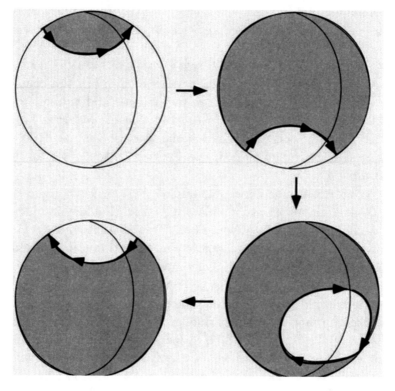

Figure 135 A deformation of the Arctic Circle that reverses its orientation.

6

Lebesgue Theory

This chapter presents a geometric theory of Lebesgue measure and integration. In calculus you certainly learned that the integral is the area under the curve. With a good definition of area that is the point of view I advance here. Deriving the basic theory of Lebesgue integration then becomes a matter of inspecting the right picture. See Appendix E for the geometric relation between Riemann integration and Lebesgue integration.

Throughout the chapter definitions and theorems are stated in $\mathbb{R}^n$ but proved in $\mathbb{R}^2$. Multidimensionality can complicate a proof's notation but never its logic.

1 Outer Measure

How should you measure the length of a subset of the line? If the set to be measured is simple, so is the answer. The length of the interval (a, b) is $b - a$. But what is the length of the set of rational numbers? of the Cantor set? As is often the case in analysis we proceed by inequalities and limits. In fact one might distinguish the fields of algebra and analysis solely according to their use of equalities versus inequalities.

Definition The **length** of an interval $I = (a, b)$ is $b - a$. It is denoted $|I|$. The **Lebesgue outer measure** of a set $A \subset \mathbb{R}$ is

$$m^*A = \inf \left\{ \sum_k |I_k| : \{I_k\} \text{ is a covering of } A \text{ by open intervals} \right\}.$$

© Springer International Publishing Switzerland 2015
C.C. Pugh, *Real Mathematical Analysis*, Undergraduate Texts
in Mathematics, DOI 10.1007/978-3-319-17771-7_6

Tacitly we assume that the covering is countable; the series $\sum_k |I_k|$ is its **total length**. (Recall that "countable" means either finite or denumerable.) The outer measure of A is the infimum of the total lengths of all possible coverings $\{I_k\}$ of A by open intervals. If every series $\sum_k |I_k|$ diverges then by definition $m^*A = \infty$.

Outer measure is defined for every $A \subset \mathbb{R}$. It measures A from the outside as do calipers. A dual approach measures A from the inside. It is called **inner measure**, is denoted $m_* A$, and is discussed in Section 4.

Three properties of outer measure (the "axioms of outer measure") are easy to check.

1 Theorem   (a) *The outer measure of the empty set is 0, $m^* \emptyset = 0$.*
   (b) *If $A \subset B$ then $m^*A \leq m^*B$.*
   (c) *If $A = \bigcup_{n=1}^{\infty} A_n$ then $m^*A \leq \sum_{n=1}^{\infty} m^*A_n$.*

Proof (b) and (c) are called **monotonicity** and **countable subadditivity**.

   (a) This is obvious. Every interval covers the empty set.
   (b) This is obvious. Every covering of B is also a covering of A.
   (c) This uses the $\epsilon/2^n$ trick. Given $\epsilon > 0$ there exists for each n a covering $\{I_{k,n} : k \in \mathbb{N}\}$ of A_n such that

$$\sum_{k=1}^{\infty} |I_{k,n}| \leq m^*A_n + \frac{\epsilon}{2^n}.$$

The collection $\{I_{k,n} : k, n \in \mathbb{N}\}$ covers A and

$$\sum_{k,n} |I_{k,n}| = \sum_{n=1}^{\infty} \sum_{k=1}^{\infty} |I_{k,n}| \leq \sum_{n=1}^{\infty} (m^*A_n + \frac{\epsilon}{2^n}) = \sum_{n=1}^{\infty} m^*A_n + \epsilon.$$

Thus the infimum of the total lengths of coverings of A by open intervals is $\leq \sum_n m^*A_n + \epsilon$, and since $\epsilon > 0$ is arbitrary the infimum is $\leq \sum_n m^*A_n$, which is what (c) asserts.   □

Next, suppose you have a set A in the plane and you want to measure its area. Here is the natural way to do it.

Definition The **area** of a rectangle $R = (a, b) \times (c, d)$ is $|R| = (b - a) \cdot (d - c)$ and the (planar) outer measure of $A \subset \mathbb{R}^2$ is the infimum of the total area of countable coverings of A by open rectangles R_k

$$m^*A \;=\; \inf \left\{ \sum_k |R_k| \,:\, \{R_k\} \text{ covers } A \right\}.$$

See Figure 136.

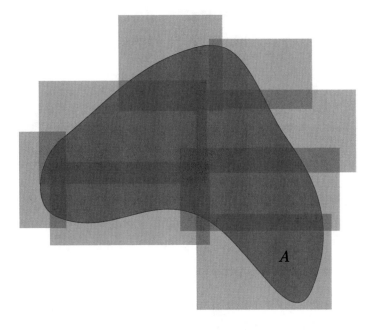

Figure 136 Rectangles that cover A

Because it is so natural, the preceding definition makes perfect sense in higher dimensions too.

Definition An open box $B \subset \mathbb{R}^n$ is the Cartesian product n open intervals, $B = \prod_k I_k$. Its n-dimensional volume $|B|$ is the product of their lengths. The n-dimensional outer measure of $A \subset \mathbb{R}^n$ is the infimum of the total volume of countable coverings of A by open boxes B_k

$$m^*A \;=\; \inf \left\{ \sum_k |B_k| \,:\, \{B_k\} \text{ covers } A \right\}.$$

If need be, we decorate $|\quad|$ and m^* with subscripts "1", "2", or "n" to distinguish the linear, planar, and n-dimensional quantities. As in the linear case we write $|R|$ and $|B|$ only for *open* rectangles and boxes. The outer measure axioms – monotonicity, countable subadditivity, and the outer measure of the empty set being zero – are true for planar outer measure too. See also Exercise 2.

Definition If $Z \subset \mathbb{R}^n$ has outer measure zero then it is a **zero set**.

2 Proposition *Every subset of a zero set is a zero set. The countable union of zero sets is a zero set. Each plane $P_i(a) = \{(x_1, \ldots, x_n) \in \mathbb{R}^n : x_i = a\}$ is a zero set in $\mathbb{R}^n$.*

Proof Monotonicity implies $m^*(Z') \leq m^* Z = 0$ whenever Z' is a subset of a zero set Z. If $m^*(Z_k) = 0$ for all $k \in \mathbb{N}$ and $Z = \bigcup Z_k$ then by Theorem 1(c) we have

$$m^* Z \leq \sum_k m^* Z_k = 0.$$

We assume $n = 2$. The "plane" $P_i(a)$ is the line $\{x = a\}$ when $i = 1$ or $\{y = a\}$ when $i = 2$. Given $\epsilon > 0$ we can cover the line $P_1(a)$ with rectangles $R_k = I_k \times J_k$ where

$$I_k = (a - \epsilon/k2^{k+2},\ a + \epsilon/k2^{k+2}) \qquad J_k = (-k, k).$$

The total area of these rectangles is ϵ so $P_1(a)$ is a zero set. $\square$

The next theorem states a property of outer measure that seems obvious.

3 Theorem *The linear outer measure of $[a, b]$ is $b - a$; the planar outer measure $[a, b] \times [c, d]$ is $(b - a) \cdot (d - c)$; the n-dimensional outer measure of a closed box is the product of its edge lengths.*

Proof for $n = 1$ For each $\epsilon > 0$ the open interval $(a - \epsilon,\ b + \epsilon)$ covers $[a, b]$. Thus $m^*([a, b]) \leq (b + \epsilon) - (a - \epsilon) = b - a + 2\epsilon$. By the ϵ-principle we get $m^*([a, b]) \leq b - a$.

To get the reverse inequality we must show that if $\{I_i\}$ is a countable open covering of $[a, b]$ then $\sum |I_i| \geq b - a$. Since $[a, b]$ is compact it suffices to prove this for finite open coverings $\{I_1, \ldots, I_N\}$. Let $I_i = (a_i, b_i)$. If $N = 1$ then $(a_1, b_1) \supset [a, b]$ implies $a_1 < a \leq b < b_1$ so $b - a < |I_1|$. That's the base case of the induction.

Assume that for each open covering of a compact interval $[c, d]$ by N open intervals $\{J_j\}$ we have $d - c < \sum_{j=1}^{N} |J_j|$, and let $\{I_i\}$ be a covering of $[a, b]$ by $N + 1$ open intervals $I_i = (a_i, b_i)$. We claim that $\sum_{i=1}^{N+1} |I_i| > b - a$. One of the intervals contains a, say it is $I_1 = (a_1, b_1)$. If $b_1 \geq b$ then $I_1 \supset [a, b)$ and again $a_1 < a \leq b \leq b_1$ implies that $\sum_{i=1}^{N+1} |I_i| \geq |I_1| = b_1 - a_1 > b - a$. On the other hand, if $b_1 < b$ then

$$[a, b] = [a, b_1) \cup [b_1, b]$$

and $|I_1| > b_1 - a$. The compact interval $[b_1, b]$ is covered by $I_2, \ldots, I_{N+1}$. By induction we have $\sum_{i=2}^{N+1} |I_i| > b - b_1$. Thus

$$\sum_{i=1}^{N+1} |I_i| = |I_1| + \sum_{i=2}^{N+1} |I_i| > (b_1 - a) + (b - b_1) = b - a$$

which completes the induction and the proof. $\qquad\qquad\qquad\qquad\qquad\square$

The preceding inductive proof does not carry over to rectangles. For a rectangle has no left to right order. However, the following grid proof works for all n.

Grid proof for $n = 2$ Let $R = [a, b] \times [c, d] \subset \mathbb{R}^2$. It is simple to see that $m^*R \leq (b - a) \cdot (d - c)$. To check the reverse inequality, consider any countable covering of R by open rectangles R_k. We must show that $\sum |R_k| \geq (b - a) \cdot (d - c)$. Since R is compact the covering has a positive Lebesgue number λ. Partition $[a, b]$ and $[c, d]$ into intervals $I_i = (a_i, b_i)$ and $J_j = (c_j, d_j)$ such that each rectangle $S_{ij} = I_i \times J_j$ has diameter $< \lambda$. Then $\sum |S_{ij}| = |R|$. Also, the union of the rectangles S_{ij} contained in any R_k forms a smaller rectangle $R_k' \subset R_k$. (Strictly speaking R_k' includes edges of the S_{ij} but they form a zero set.) See Figure 137.

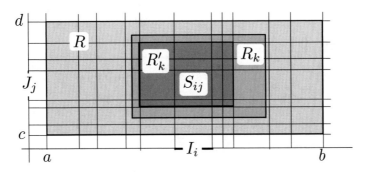

Figure 137 R_k contains R_k', heavily shaded. It consists of sixteen grid rectangles S_{ij}. Their total area is $\leq |R_k|$.

Thus

$$\sum_{S_{ij} \subset R_k} |S_{ij}| \leq |R_k|.$$

It follows that

$$(b - a) \cdot (d - c) = |R| = \sum |S_{ij}| \leq \sum_k \sum_{S_{ij} \subset R_k} |S_{ij}| = \sum_k |R_k'| \leq \sum_k |R_k|,$$

which completes the proof. $\qquad\qquad\qquad\qquad\qquad\qquad\qquad\qquad\square$

4 Corollary *The formulas $m^*I = b - a$, $m^*R = (b - a) \cdot (d - c)$, and $m^*B = \prod_k m^*(I_k)$ hold also for intervals, rectangles, and boxes that are open or partly open. In particular, $m^*I = |I|$, $m^*R = |R|$, and $m^*B = |B|$ for open intervals, rectangles, and boxes.*

Proof Let I be any interval with endpoints $a < b$ and let $\epsilon > 0$ be given. (We assume $\epsilon < (b - a)/2$ without loss of generality.) The closed intervals $J = [a + \epsilon, \ b - \epsilon]$ and $J' = [a - \epsilon, \ b + \epsilon]$ sandwich I as $J \subset I \subset J'$. By Theorem 3 we have $m^*J = b - a - 2\epsilon$ and $m^*J' = b - a + 2\epsilon$. Thus

$$
\begin{array}{ccccc}
m^*J & \leq & m^*I & \leq & m^*J' \\
\| & & & & \| \\
b - a - 2\epsilon & \leq & b - a & \leq & b - a + 2\epsilon.
\end{array}
$$

Then m^*I and $b - a$ are sandwiched between $b - a - 2\epsilon$ and $b - a + 2\epsilon$ for all $\epsilon > 0$, which implies the two constants m^*I and $b - a$ are equal. The sandwich method works equally well for rectangles and boxes. $\qquad\square$

2 Measurability

If A and B are subsets of disjoint intervals in $\mathbb{R}$ it is easy to show that

$$
m^*(A \sqcup B) = m^*A + m^*B.
$$

But what if A and B are merely disjoint? Is the formula still true? The answer is "yes" if the sets have an additional property called measurability, and "no" in general as is shown in Appendix D. Measurability is the rule and nonmeasurability the exception. The sets you meet in analysis – open sets, closed sets, their unions, differences, etc. – all are measurable. See Section 4.

Definition A set $E \subset \mathbb{R}$ is **(Lebesgue) measurable** if the division $E|E^c$ of $\mathbb{R}$ is so "clean" that for each "test set" $X \subset \mathbb{R}$ we have

(1) $$ m^*X = m^*(X \cap E) + m^*(X \cap E^c). $$

The definition of measurability in higher dimensions is analogous. A set $E \subset \mathbb{R}^n$ is measurable if $E|E^c$ divides each $X \subset \mathbb{R}^n$ so cleanly that (1) is true for n-dimensional outer measure.

We denote by $\mathcal{M} = \mathcal{M}(\mathbb{R}^n)$ the collection of all Lebesgue measurable subsets of $\mathbb{R}^n$. If E is measurable its **Lebesgue measure** is m^*E, which we write as mE, dropping the asterisk to emphasize the measurability of E.

Which sets are measurable? It is obvious that the empty set is measurable. It is also obvious that if a set is measurable then so is its complement, since $E|E^c$ and $E^c|E$ divide a test set X in the same way.

In the rest of this section we analyze measurability in the abstract. For the basic facts about measurability have nothing to do with $\mathbb{R}$ or $\mathbb{R}^n$. They hold for any "abstract outer measure."

Definition Let M be any set. The collection of all subsets of M is denoted as 2^M. An **abstract outer measure** on M is a function $\omega : 2^M \to [0, \infty]$ that satisfies the three axioms of outer measure: $\omega(\emptyset) = 0$, ω is monotone, and ω is countably subadditive. A set $E \subset M$ is **measurable** with respect to ω if $E|E^c$ is so clean that for each test set $X \subset M$ we have

$$\omega X = \omega(X \cap E) + \omega(X \cap E^c).$$

Example Given any set M there are two trivial outer measures on M. Counting outer measure assigns to a finite set $S \subset M$ its cardinality and assigns ∞ to every infinite set. The zero/infinity measure assigns outer measure zero to the empty set and ∞ to every other set. All sets are measurable with respect to these outer measures. See Exercise 10.

Example A less trivial outer measure weights Lebesgue outer measure. One sets $\omega I = e^{-c^2} |I|$, where c is the midpoint of the interval I, and then defines the outer measure of $A \subset \mathbb{R}$ to be the infimum of the total ω-area of countable interval coverings of A. Other weighting functions can be used.

5 Theorem *The collection $\mathcal{M}$ of measurable sets with respect to any outer measure on any set M is a σ-algebra and the outer measure restricted to this σ-algebra is countably additive. All zero sets are measurable and have no effect on measurability. In particular Lebesgue measure has these properties.*

A **σ-algebra** is a collection of sets that includes the empty set, is closed under complement, and is closed under countable union. **Countable additivity** of ω means that if $E_1, E_2, \ldots$ are measurable with respect to ω then

$$E = \bigsqcup_i E_i \quad \Rightarrow \quad \omega E = \sum_i \omega E_i.$$

Proof Let $\mathcal{M}$ denote the collection of measurable sets with respect to the outer measure ω on M. First we deal with zero sets, sets for which $\omega Z = 0$. By monotonicity, if Z is a zero set and X is a test set then

$$\omega X \;\le\; \omega(X \cap Z) + \omega(X \cap Z^c) \;=\; 0 + \omega(X \cap Z^c) \;\le\; \omega X$$

implies Z is measurable. Likewise, if $E|E^c$ divides X cleanly then the same is true for $(E \cup Z)|(E \cup Z)^c$ and $(E \setminus Z)|(E \setminus Z)^c$. That is, Z has no effect on measurability.

To check that $\mathcal{M}$ is a σ-algebra we must show that it contains the empty set, is closed under complements, and is closed under countable union. By the definition of outer measure the empty set is a zero set so it is measurable, $\emptyset \in \mathcal{M}$. Also, since $E|E^c$ divides a test set X in the same way that $E^c|E$ does, $\mathcal{M}$ is closed under complements. To check that $\mathcal{M}$ is closed under countable union takes four preliminary steps:

(a) $\mathcal{M}$ is closed under differences.

(b) $\mathcal{M}$ is closed under finite union.

(c) ω is finitely additive on $\mathcal{M}$.

(d) ω satisfies a special countable addition formula.

(a) For measurable sets E_1, E_2, and a test set X, draw the Venn diagram in Figure 138 where X is represented as a disc. To check measurability of $E_1 \setminus E_2$ we must verify the equation

$$2 + 134 \;=\; 1234$$

where $2 = \omega(X \cap (E_1 \setminus E_2))$, $134 = \omega(X \cap (E_1 \setminus E_2)^c)$, $1234 = \omega X$, etc. Since E_1 divides any set cleanly, $134 = 1 + 34$, and since E_2 divides any set cleanly, $34 = 3 + 4$. Thus

$$2 + 134 \;=\; 2 + 1 + 3 + 4 \;=\; 1 + 2 + 3 + 4.$$

For the same reason $1234 = 12 + 34 = 1 + 2 + 3 + 4$ which completes the proof of (a).

(b) Suppose that E_1, E_2 are measurable and $E = E_1 \cup E_2$. Since $E^c = E_1^c \setminus E_2$, (a) implies that $E^c \in \mathcal{M}$ and thus $E \in \mathcal{M}$. For more than two sets, induction shows that if $E_1, \ldots, E_n \in \mathcal{M}$ then $E_1 \cup \ldots \cup E_n \in \mathcal{M}$.

(c) If $E_1, E_2 \in \mathcal{M}$ are disjoint then E_1 divides $E = E_1 \sqcup E_2$ cleanly, so

$$\omega E \;=\; \omega(E \cap E_1) + \omega(E \cap E_1^c) \;=\; \omega E_1 + \omega E_2,$$

which is additivity for pairs of measurable sets. For more than two measurable sets, induction implies that ω is **finitely additive** on $\mathcal{M}$; i.e., if $E_1, \ldots, E_n \in \mathcal{M}$ then

$$E \;=\; \coprod_{i=1}^{n} E_i \quad \Rightarrow \quad \omega E \;=\; \sum_{i=1}^{n} \omega E_i.$$

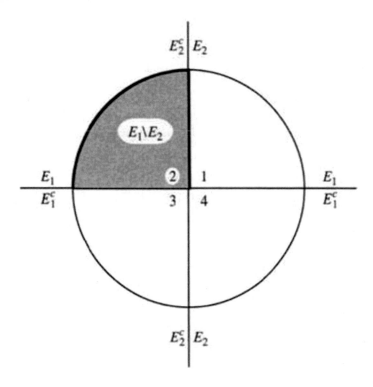

Figure 138 The picture that proves $\mathcal{M}$ is closed under differences.

In particular, zero sets have no effect on measure: $m(E \cup Z) = mE = m(E \setminus Z)$.

(d) Given a test set $X \subset M$ and a countable disjoint union of measurable sets $E = \bigsqcup E_i$ we claim that

$$(2) \qquad \omega X = \sum_{i=1}^{\infty} \omega(X \cap E_i) + \omega(X \cap E^c).$$

(Taking $X = E$, (2) implies countable additivity of measurable sets, $\omega E = \sum \omega(E_i)$, but does not imply measurability of E.) (b) implies that the finite union $F = \bigsqcup_{i=1}^{k} E_i$ is measurable so it divides X cleanly: $\omega X = \omega(X \cap F) + \omega(X \cap F^c)$. Since $F \subset E$, we have $F^c \supset E^c$, which gives

$$\omega X \geq \omega(X \cap F) + \omega(X \cap E^c).$$

The individual clean divisions by the E_i imply $\omega(X \cap F) = \sum_{i=1}^{k} \omega(X \cap E_i)$. Thus

$$\omega X \geq \sum_{i=1}^{k} \omega(X \cap E_i) + \omega(X \cap E^c).$$

Since the constant ωX dominates every partial sum, it dominates the whole series:

$$\omega X \;\geq\; \sum_{i=0}^{\infty} \omega(X \cap E_i) + \omega(X \cap E^c).$$

The reverse inequality is always true by subadditivity of ω, so we get equality, which is the special additivity formula (2).

It remains to show that the countable union of measurable sets $E = \bigcup E_i$ is measurable. Since E can be re-expressed as a countable disjoint union of measurable sets $E_i' = E_i \setminus (E_1 \cup \cdots \cup E_{i-1})$, it's enough to check that $E = \bigsqcup E_i$ is measurable. Applying (2) to the test set $Y = X \cap E$ gives

$$\omega(X \cap E) \;=\; \sum \omega(X \cap E \cap E_i) + \omega(X \cap E \cap E^c) \;=\; \sum \omega(X \cap E_i).$$

Substituting this into (2) itself gives measurability of E and completes the proof that $\mathcal{M}$ is a σ-algebra. $\square$

From countable additivity we deduce a very useful fact about measures. It applies to any outer measure ω, in particular to Lebesgue outer measure.

6 Measure Continuity Theorem *If $\{E_k\}$ and $\{F_k\}$ are sequences of measurable sets then*

upward measure continuity	$E_k \uparrow E \;\Rightarrow\; \omega E_k \uparrow \omega E$
downward measure continuity	$F_k \downarrow F \text{ and } \omega F_1 < \infty \;\Rightarrow\; \omega F_k \downarrow \omega F.$

Proof The notation $E_k \uparrow E$ means that $E_1 \subset E_2 \subset \ldots$ and $E = \bigcup E_k$. Write E disjointly as $E = \bigsqcup E_k'$ where $E_k' = E_k \setminus (E_1 \cup \ldots \cup E_{k-1})$. Countable additivity for measurable sets gives

$$\omega E \;=\; \sum_{n=1}^{\infty} \omega E_n'.$$

Also, the kth partial sum of the series equals ωE_k, so ωE_k converges upward to ωE. The notation $F_k \downarrow F$ means that $F_1 \supset F_2 \supset \ldots$ and $F = \bigcap F_k$. Write F_1 disjointly as

$$F_1 \;=\; \left(\bigsqcup_{k=1}^{\infty} F_k' \right) \sqcup F$$

where $F_k' = F_k \setminus F_{k+1}$. Then $F_k = \bigsqcup_{n \geq k} F_n' \sqcup F$. The countable additivity formula for measurable sets

$$\omega F_1 = \omega F + \sum_{n=1}^{\infty} \omega F_n'$$

implies the series converges to a finite limit, so its tails converge to zero:

$$\omega F_k = \sum_{n=k}^{\infty} \omega F_n' + \omega F$$

converges downward to ωF as $k \to \infty$. $\qquad\qquad\qquad\qquad\qquad\square$

3 Meseomorphism

A **measure space** is a triple $(M, \mathcal{M}, \mu)$ where M is a set, $\mathcal{M}$ is a σ-algebra of subsets of M, and μ is a measure on $\mathcal{M}$. That is, $\mu : \mathcal{M} \to [0, \infty]$ has the three properties

(a) $\mu(\emptyset) = 0$

(b) μ is monotone: $A \subset B$ implies $\mu A \leq \mu B$.

(c) μ is countably additive on $\mathcal{M}$: $E = \bigsqcup E_i$ implies $\mu E = \sum \mu(E_i)$.

As proved in the previous section, every abstract outer measure ω on a set M restricts to a measure on the σ-algebra of ω-measurable sets $\mathcal{M} = \mathcal{M}(\omega)$. In particular, this is true for outer Lebesgue measure on $\mathbb{R}^n$, the measure space being the triple $(\mathbb{R}^n, \mathcal{M}, m)$ where $\mathcal{M}$ is the σ-algebra of Lebesgue measurable subsets of $\mathbb{R}^n$.

An isomorphism preserves algebraic structure. A homeomorphism preserves topological structure. A diffeomorphism preserves smooth structure. A meseomorphism preserves measure structure. More precisely, if $(M, \mathcal{M}, \mu)$ and $(M', \mathcal{M}', \mu')$ are measure spaces then a mapping $T : M \to M'$ is a

> **mesemorphism** if T sends each $E \in \mathcal{M}$ to $TE \in \mathcal{M}'$.
> **meseomorphism** if T is a bijection, and both T and T^{-1} are mesemorphisms.
> **mesisometry** if T is a meseomorphism and $\mu'(TE) = \mu E$ for each $E \in \mathcal{M}$. (Other terms for mesisometry are **measure preserving transformation** and **isomorphism of measure spaces**.)

7 Theorem *If a bijection increases outer measure by at most a factor t and its inverse increases outer measure by at most a factor $1/t$ then it is a meseomorphism. If $t = 1$ then it is a mesisometry.*

Proof Let $T : M \to M'$ be the bijection where M and M' are equipped with outer measures ω and ω'. For each $X \subset M$ we have

$$\omega X = \omega(T^{-1} \circ T(X)) \leq t^{-1} \omega'(TX) \leq t^{-1} t \omega X = \omega X.$$

Thus, if $X \subset M$ and $X' \subset M'$ then $\omega'(TX) = t \omega X$ and $\omega T^{-1}(X') = t^{-1} \omega'(X')$.

If $E \subset M$ is measurable and X' be a test set in M' then $X = T^{-1}(X')$ is a test set in M. Since T multiplies outer measure by t and E is measurable the clean division

$$
\begin{aligned}
\omega'(X') = t\,\omega X &= t\left(\omega(X \cap E) + \omega(X \cap E^c)\right) \\
&= t\left(t^{-1}\omega'(T(X \cap E)) + t^{-1}\omega'(T(X \cap E^c))\right) \\
&= \omega'(X' \cap TE) + \omega'(X' \cap T(E^c))
\end{aligned}
$$

shows TE is measurable. Likewise for T^{-1}, so T is a meseomorphism. If $t = 1$ then T preserves measure: it is a mesisometry. $\qquad\square$

These concepts may help you organize your thoughts while reading the next sections, but I think you will find Exercise 19 quite remarkable. One dimensional and n-dimensional Lebesgue measure are mesisometric. A meseomorphism is no respecter of topology. The more pertinent questions are – what do measurable subsets of $\mathbb{R}^n$ look like and how do they behave with respect to transformations of $\mathbb{R}^n$, not what does the measure algebra look like. In the next section we will show that measurable sets are not much different from open sets, and measurability is invariant under many (but not all) continuous transformations of $\mathbb{R}^n$.

4 Regularity

How is topology related to measure: which subsets of Euclidean space are measurable?

8 Theorem *Open sets and closed sets in $\mathbb{R}^n$ are Lebesgue measurable.*

9 Proposition *The half-spaces $[a, \infty) \times \mathbb{R}^{n-1}$ and $(a, \infty) \times \mathbb{R}^{n-1}$ are measurable in $\mathbb{R}^n$. So are all open boxes.*

Proof Without loss of generality we assume $n = 2$. Let $H = [a, \infty) \times \mathbb{R}$. We must show that H divides each test set X cleanly: $m^*X = m^*(X \cap H) + m^*(X \cap H^c)$. Since $a \times \mathbb{R}$ is a zero set in $\mathbb{R}^2$ and zero sets have no effect on outer measure (Theorem 5) we may assume that $X \cap (a \times \mathbb{R}) = \emptyset$. Set

$$
X^- = \{(x, y) \in X : x < a\} \qquad X^+ = \{(x, y) \in X : x > a\}.
$$

Then $X = X^- \sqcup X^+$. Given $\epsilon > 0$ there is a countable covering $\mathcal{R}$ by rectangles R with $\sum_{\mathcal{R}} |R| \leq m^*X + \epsilon$. Let $\mathcal{R}^\pm$ be the collection of rectangles $R^\pm = \{(x, y) \in R :$

$R \in \mathcal{R}$ and $\pm (x - a) > 0\}$. Then $\mathcal{R}^{\pm}$ covers $X^{\pm}$ and

$$m^*X \leq m^*(X \cap H) + m^*(X \cap H^c)$$
$$\leq \sum_{\mathcal{R}+} |R^+| + \sum_{\mathcal{R}-} |R^-| = \sum_{\mathcal{R}} |R| \leq m^*X + \epsilon.$$

Since $\epsilon > 0$ is arbitrary this gives measurability of $H = [a, \infty) \times \mathbb{R}$. Since the line $x = a$ is a planar zero set $(a, \infty) \times \mathbb{R}$ is also measurable. The vertical strip $(a, b) \times \mathbb{R}$ is measurable since it is the intersection

$$(a, \infty) \times \mathbb{R} \cap (-\infty, b) \times \mathbb{R}$$

and $(-\infty, b) \times \mathbb{R} = ([b, \infty) \times \mathbb{R})^c$. Interchanging the coordinates shows that the horizontal strip $\mathbb{R} \times (c, d)$ is also measurable. The rectangle $R = (a, b) \times (c, d)$ is the intersection of the strips and is therefore measurable. $\qquad\square$

Proof of Theorem 8 Let U be an open subset of $\mathbb{R}^n$. It is the countable union of open boxes.[†] Since $\mathcal{M}(\mathbb{R}^n)$ is a σ-algebra and a σ-algebra is closed with respect to countable unions, U is measurable. Since a σ-algebra is closed with respect to complements, every closed set in $\mathbb{R}^n$ is also measurable. $\qquad\square$

10 Corollary *The Lebesgue measure of a closed or partially box is the volume of its interior. The boundary of a box is a zero set.*

Proof This is just Proposition 2, Theorem 3, and Proposition 9. $\qquad\square$

Sets that are slightly more general than open sets and closed sets arise naturally. A countable intersection of open sets is called a G_δ-set and a countable union of closed sets is an F_σ-set. ("δ" stands for the German word *durchschnitt* and "σ" stands for "sum.") By De Morgan's laws, the complement of a G_δ-set is an F_σ-set and conversely. Clearly a homeomorphism sends G_δ-sets to G_δ-sets and F_σ-sets to F_σ-sets. Since the σ-algebra of measurable sets contains the open sets and the closed sets it also contains the G_δ-sets and the F_σ-sets.

11 Theorem *Lebesgue measure is **regular** in the sense that each measurable set E can be sandwiched between an F_σ-set and a G_δ-set, $F \subset E \subset G$, such that $G \setminus F$ is a zero set. Conversely, if there is such an $F \subset E \subset G$ then E is measurable.*

[†]This is a fact about separable metric spaces. Apply Exercises 127 and 131 from Chapter 2 to $\mathbb{R}^n$ equipped with the maximum coordinate metric, $d(x, y) = \max_i |x_i - y_i|$.

Proof We take $E \subset \mathbb{R}^2$. We assume first that E is bounded and choose a large rectangle R that contains E. We write $E^c = R \setminus E$. Measurability implies

$$mR = mE + m(E^c).$$

There are decreasing sequences of open sets U_n and V_n such that $U_n \supset E$, $V_n \supset E^c$, $m(U_n) \to mE$, and $m(V_n) \to m(E^c)$ as $n \to \infty$. Measurability of E implies $m(U_n \setminus E) \to 0$ and $m(V_n \setminus E^c) \to 0$. The complements $K_n = R \setminus V_n$ form an increasing sequence of closed subsets of E and

$$mK_n = mR - mV_n \quad \to \quad mR - m(E^c) = mE.$$

Thus $F = \bigcup K_n$ is an F_σ-set contained in E with $mF = mE$. Similarly, $G = \bigcap U_n$ is a G_δ-set that contains E and has $mG = mE$. Because all the measures are finite, the equality $mF = mE = mG$ implies that $m(G \setminus F) = 0$.

Conversely, if F is an F_σ-set, G is a G_δ-set, $F \subset E \subset G$, and $m(G \setminus F) = 0$ then E is measurable since $E = F \cup Z$, where $Z = E \cap (G \setminus F)$ is a zero set.

The unbounded case is left as Exercise 6. $\square$

12 Corollary *A bounded subset $E \subset \mathbb{R}^n$ is measurable if and only if it has a **regularity sandwich** $F \subset E \subset G$ such that F is an F_σ-set, G is a G_δ-set, and $mF = mG$.*

Proof If E is measurable, bounded or not, then Theorem 11 implies there is a regularity sandwich with $mF = mE = mG$. Conversely, if there is a regularity sandwich with $mF = mG$ then boundedness of E implies $mF < \infty$. Measurability of F and G imply $m(G \setminus F) = mG - mF = 0$ and Theorem 11 then implies E is measurable. $\square$

13 Corollary *Modulo zero sets, Lebesgue measurable sets are F_σ-sets and G_δ-sets.*

Proof $E = F \cup Z = G \setminus Z'$ for the zero sets $Z = E \setminus F$ and $Z' = G \setminus E$. $\square$

14 Corollary *A Lipeomorphism $h : \mathbb{R}^n \to \mathbb{R}^n$ is a meseomorphism.*

Proof Homeomorphisms send G_δ-sets to G_δ-sets and F_σ-sets to F_σ-sets. By Lemma 35 in Chapter 5, since h is Lipschitz it sends zero sets to zero sets. Thus h sends regularity sandwiches to regularity sandwiches:

$$F \subset E \subset G \quad \longrightarrow \quad hF \subset hE \subset hG,$$

which implies h sends measurable sets to measurable sets. The same holds for h^{-1}.$\square$

Affine Motions

An affine motion of $\mathbb{R}^n$ is an invertible linear transformation T followed by a translation. Translation does not affect Lebesgue measure, so we just deal with the linear map T.

15 Theorem *An affine motion* $T : \mathbb{R}^n \to \mathbb{R}^n$ *is a meseomorphism. It multiplies measure by* $|\det T|$.

16 Lemma *Every open set in n-space is a countable disjoint union of open cubes plus a zero set.*

Proof Take $n = 2$ and let $U \subset \mathbb{R}^2$ be open. Accept all the open unit dyadic squares that lie in U, and reject the rest. Bisect every rejected square into four equal subsquares. Accept the interiors of all these subsquares that lie in U, and reject the rest. Proceed inductively, bisecting the rejected squares, accepting the interiors of the resulting subsquares that lie in U, and rejecting the rest. In this way U is shown to be the countable union of disjoint, accepted, open dyadic squares, together with the points rejected at every step in the construction. See Figure 139. Rejected points

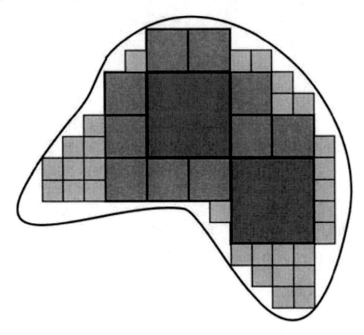

Figure 139 An open set is a countable union of dyadic cubes.

of U lie on horizontal or vertical dyadic lines. There are countably many such lines, each is a zero set, and so the rejected points in U form a zero set. □

17 Lemma *Every open set is a countable disjoint union of balls plus a zero set.*

Proof Again, take $n = 2$, and let $U \subset \mathbb{R}^2$ be an open set. By Lemma 16, U is a countable disjoint union of open squares plus a zero set. It therefore suffices to check that every open square S is a countable disjoint union of discs plus a zero set. As shown in Figure 140, S contains an open disc Δ whose area is greater than half the area of the square, $m(\Delta) > m(S)/2$.

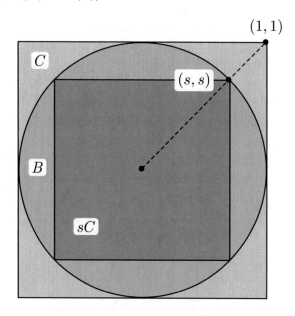

Figure 140 B is the largest ball in C and sC is the largest cube in B.

The difference $U_1 = S \setminus \overline{\Delta}$ is an open subset of S with $m(U_1) < m(S)/2$. By Lemma 16, U_1 is the disjoint countable union of small open squares S_i plus a zero set. Each S_i contains a small open disc Δ_i whose area is greater than half the area of S_i. The total area of just finitely many of the discs Δ_i is greater than half the total area of all the squares S_i. Thus, for some k,

$$U_2 = S \setminus \left(\overline{\Delta} \cup \bigcup_{i=1}^{k} \overline{\Delta}_i \right)$$

is an open subset of U_1 and $m(U_2) < m(U_1)/2 < m(S)/4$. See Figure 141.

Repeating this process gives a nested sequence of open sets $S \supset U_1 \supset U_2 \supset U_3 \supset \dots$. Each difference $U_{\ell-1} \setminus U_\ell$ contains finitely many disjoint open discs whose total area is greater than half $m(U_\ell)$. Call these discs "accepted at stage ℓ." Thus

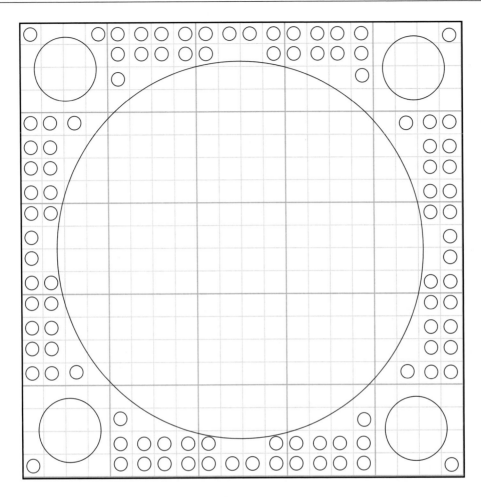

Figure 141 Each disc occupies greater than half the area of its square.

$m(U_\ell) \leq m(S)/2^\ell \to 0$ as $\ell \to \infty$, and by countable additivity mU is the total area of all the accepted discs.

In the n-dimensional case, $n \geq 3$, "half" is replaced by $1/n^{n/2}$. The case $n = 1$ is trivial: cubes *are* balls. $\qquad \square$

Proof of Theorem 15 T is Lipschitz with Lipschitz constant $\|T\|$. Its inverse has Lipschitz constant $\|T^{-1}\|$, so T is a Lipeomorphism. By Corollary 14, T is a meseomorphism, and it remains to check $m(TE) = |\det T| \, mE$.

First, we assume that T is given by a diagonal matrix D with diagonal entries $\lambda_1, \ldots, \lambda_n$. Its determinant is $d = \lambda_1 \cdots \lambda_n$. Clearly

$$|TR| \;=\; |d| \, |R|$$

for all open rectangles R. Let $E \subset \mathbb{R}^n$ be measurable and let $\epsilon > 0$ be given. There exists a covering of E by open rectangles R_k with total volume

$$\sum |R_k| \leq mE + \epsilon.$$

Then $\{T(R_k)\}$ is a covering of TE by open rectangles with total volume $|d| \sum |R_k|$, and it follows that $m(TE) \leq |d|\, mE + \epsilon$. Since ϵ is arbitrary, $m(TE) \leq |d|\, mE$. The same reasoning applies to T^{-1}, and $\det T^{-1} = d^{-1}$, so $mE \leq |d^{-1}|\, m(TE)$. Together, these two inequalities become $m(TE) \leq |d|\, mE \leq m(TE)$, completing the proof of the theorem when T is diagonal.

Next, we assume T is orthogonal and R is an open cube. Lemmas 16 and 17 imply that R is a countable disjoint union of balls B_i plus a zero set Z. The orthogonal transformation T sends a ball B of radius r and center c to a ball TB of radius r and center $T(c)$. For orthogonal transformations are isometries. That is, TB is a translate of B, so $m(TB) = mB$. Since TZ is a zero set, $m(TR) = mR$ for all open cubes R. As in the diagonal case this implies $m(TE) = mE$. Since the determinant of an orthogonal transformation is ± 1, this completes the proof of the theorem when T is orthogonal.

Finally, we use Polar Form (Appendix D in Chapter 5) to write

$$T = O_1 D O_2$$

where O_1 and O_2 are orthogonal and D is diagonal. Since $|\det T| = |\det D|$, the proof is complete. $\qquad\square$

18 Corollary *Rigid motions of $\mathbb{R}^n$ preserve Lebesgue measure. They are meseometries.*

Proof A rigid motion is an orthogonal transformation followed by a translation. $\quad\square$

Inner Measure, Hulls, and Kernels

Consider any bounded $A \subset \mathbb{R}^n$, measurable or not. m^*A is the infimum of the measure of open sets that contain A. The infimum is achieved by a G_δ-set that contains A. We call it a **hull** of A and denote it as H_A. It is unique up to a zero set. Dually, the inner measure of A is the supremum of the measure of closed sets it contains. The supremum is achieved by an F_σ-set contained in A. We call it a

kernel of A and denote it as K_A. It is unique up to a zero set.[†] We denote the inner measure of A as m_*A. It equals $m(K_A)$. Clearly $m_*A \leq m^*A$ and m_* measures A from the inside. Also, m_* is monotone: $A \subset B$ implies $m_*A \leq m_*B$. The hull and kernel are measure theoretic analogs of closure and interior in topology.

Definition The **measure theoretic boundary** of A is $\partial_m(A) = H_A \setminus K_A$.

Remark Theorem 11 implies that a bounded subset of $\mathbb{R}^n$ is measurable if and only if its inner and outer measures are equal, i.e., $m(\partial_m(A)) = 0$. *Lebesgue took this as his definition of measurability.* He said a bounded set is measurable if its inner and outer measures are equal, and an unbounded set is measurable if it is a countable union of bounded measurable sets. In contrast, the current definition which uses cleanness and test sets is due to Carathéodory. It is easier to use (there are fewer complements to consider), unboundedness has no effect on it, and it generalizes more easily to abstract measure spaces.

19 Theorem *If $A \subset B \subset \mathbb{R}^n$ and B is a box then A is measurable if and only if it divides B cleanly.*

Remark The theorem is also valid for a bounded measurable set B instead of a box, but it's most useful for boxes. It means you don't need to check clean division of all test sets, just clean division of one big box.

20 Lemma *If A is contained in a box B then $mB = m_*A + m^*(B \setminus A)$.*

Proof If $K \subset A$ is closed then $B \setminus K$ is open and contains $B \setminus A$. Measurability implies

$$mB = mK + m(B \setminus K).$$

Maximizing mK minimizes $m(B \setminus K)$ and vice versa. □

Proof of Theorem 19 Lemma 20 implies

$$m_*A + m^*(B \setminus A) = mB.$$

If A divides B cleanly then

$$m^*A + m^*(B \setminus A) = mB.$$

Finiteness of these four quantities permits subtraction, so $m_*A = m^*A$ and A is measurable. The converse is obvious: a measurable set divides *every* test set cleanly.□

[†]If A is unbounded we need to take a little more care. It is not enough to achieve the infimum or supremum if they are ∞. Rather, we demand that H_A is minimal in the sense that if $H \supset A$ and is measurable then $H_A \setminus H$ is a zero set. Similarly, we demand maximality of K_A in the sense that if $K \subset A$ and is measurable then $K \setminus K_A$ is a zero set. See Exercise 6.

5 Products and Slices

Regularity of Lebesgue measure has a number of uses such as in Exercises 21, 22, 23, 69, and 73. Here are some more.

21 Measurable Product Theorem *If* $A \subset \mathbb{R}^n$ *and* $B \subset \mathbb{R}^k$ *are measurable then* $A \times B$ *is measurable and*

$$m(A \times B) = mA \cdot mB.$$

By convention $0 \cdot \infty = 0 = \infty \cdot 0$.

22 Lemma *If* A *and* B *are boxes then* $A \times B$ *is measurable and* $m(A \times B) = mA \cdot mB$.

Proof $A \times B$ is a box and the product formula follows from Corollary 10.   □

23 Lemma *If* A *or* B *is a zero set then so is* $A \times B$.

Proof We assume $A, B \subset \mathbb{R}$ and $mA = 0$. If $\epsilon > 0$ and $\ell \in \mathbb{N}$ are given then we cover A with open intervals I_i whose total length is so small that the total area of the rectangles $I_i \times [-\ell, \ell]$ is $< \epsilon/2^\ell$. The union of all these rectangles covers $A \times \mathbb{R}$ and has measure $< \epsilon$. Thus $A \times \mathbb{R}$ is a zero set and so is its subset $A \times B$.   □

24 Lemma *If* U *and* V *are open then* $U \times V$ *is measurable and* $m(U \times V) = mU \cdot mV$.

Proof We assume $U, V \subset \mathbb{R}$. Since $U \times V$ is open it is measurable. Lemma 16 implies that $U = \bigsqcup_i I_i \cup Z_U$ and $V = \bigsqcup_j J_j \cup Z_V$, where I_i and J_j are open intervals while Z_U and Z_V are zero sets. Then

$$U \times V = \bigsqcup_{i,j} I_i \times J_j \ \cup \ Z$$

where $Z = (Z_U \times V) \cup (U \times Z_V)$ is a zero set by Lemma 23. Since

$$\left(\sum_i m(I_i) \right) \left(\sum_j m(J_j) \right) = \sum_{i,j} m(I_i) m(J_j) = \sum_{i,j} m(I_i \times J_j)$$

we conclude that $m(U \times V) = mU \cdot mV$.   □

Proof of the Measurable Product Theorem We assume $A, B \subset I$ are measurable where I is the unit interval. We claim that the hull of a product is the product of the hulls and the kernel of a product is the product of the kernels. Since hulls are

G_δ-sets their product is a G_δ-set and is therefore measurable. Similarly, the product of kernels is measurable. Clearly

$$K_A \times K_B \ \subset \ A \times B \ \subset \ H_A \times H_B$$

and $(H_A \times H_B) \setminus (K_A \times K_B) = (H_A \times (H_B \setminus K_B)) \cup ((H_A \setminus K_A) \times H_B)$. Measurability of A and B implies $m(H_A \setminus K_A) = m(H_B \setminus K_B) = 0$ so Lemma 23 gives

$$m(K_A \times K_B) \ = \ m(H_A \times H_B).$$

Since $A \times B$ is sandwiched between two measurable sets of the same finite measure, it is measurable and its measure equals their common value. That is,

(3) $$m(K_A \times K_B) \ = \ m(A \times B) \ = \ m(H_A \times H_B).$$

Let U_n and V_n be sequences of open sets in I converging down to H_A and H_B. Then $U_n \times V_n$ is a sequence of open sets in I^2 converging down to $H_A \times H_B$. Downward measure continuity implies $m(U_n \times V_n) \to m(H_A \times H_B)$. By Lemma 24 we have $m(U_n \times V_n) = m(U_n) \cdot m(V_n)$. Since $m(U_n) \to mA$ and $m(V_n) \to mB$ we conclude from (3) that $m(A \times B) = mA \cdot mB$. $\square$

Recall from Chapter 5 that the **slice** of $E \subset \mathbb{R}^n \times \mathbb{R}^k$ at $x \in \mathbb{R}^n$ is the set

$$E_x \ = \ \{y \in \mathbb{R}^k : (x,y) \in E\}.$$

Among other things, the next theorem lets us generalize the Measurable Product Theorem to nonmeasurable sets. See Exercise 73.

25 Zero Slice Theorem *If $E \subset \mathbb{R}^n \times \mathbb{R}^k$ is measurable then E is a zero set if and only if almost every slice of E is a (slice) zero set.*

26 Lemma *If $W \subset I^n$ is open and $X_\alpha = X_\alpha(W) = \{x : m(W_x) > \alpha\}$ then*

$$mW \ \geq \ m(X_\alpha(W)) \cdot \alpha.$$

Remark Think of the slices W_x with $x \in X_\alpha$ as "heavy." The lemma means there can't be too many (in the measure theoretic sense) heavy slices.

Proof The slices of an open set are open and are therefore measurable. Fix any $x \in X_\alpha$. There is a compact set $K(x) \subset W_x$ such that measure $> \alpha$. For $m(W_x) = m_*(W_x)$. Since $x \times K(x)$ is a compact subset of W, there is a product open set

$W(x) = U(x) \times V(x)$ with $W \supset W(x) \supset x \times K(x)$. Each $x' \in U(x)$ has $W_{x'} \supset V(x)$, which implies $U(x) \subset X_\alpha$, so $X_\alpha = \bigcup_{x \in X_\alpha} U(x)$ is open.

Let $K \subset X_\alpha$ be any compact set. The covering $\{U(x) : x \in X_\alpha\}$ of K reduces to a finite subcovering $\{U_i\}$, $1 \le i \le N$. As before, setting $U_i' = U_i \setminus (U_1 \cup \ldots U_{i-1})$ gives a disjoint covering of K by measurable subsets of X_α. Thus

$$mW \ge m(\bigsqcup_{i=1}^{N} U_i' \times V_i) = \sum_{i=1}^{N} m(U_i') \cdot m(V_i) > \sum_{i=1}^{N} m(U_i') \cdot \alpha \ge mK \cdot \alpha.$$

Taking the supremum of mK over all compact $K \subset X_\alpha$ gives $mW \ge m_*(X_\alpha) \cdot \alpha$. Since X_α is open, $m_*(X_\alpha) = m(X_\alpha)$ and we get the asserted inequality. $\qquad \square$

Proof of the Zero Slice Theorem As above, it is no great loss of generality to assume $n = k = 1$ and E is contained in the unit square.

We first assume that $mE = 0$ and prove that almost every slice of E is a zero set in the slice measure. Given $\epsilon > 0$ there is an open set W such that $I^2 \supset W \supset E$ and $mW < \epsilon$. Then $W_x \supset E_x$ for all x and $X_\alpha(W) \supset X_\alpha(E)$ where

$$X_\alpha(W) = \{x : m(W_x) > 0\} \qquad X_\alpha(E) = \{x : m^*(E_x) > \alpha\}$$

Note that we use outer measure on E_x. For although $mE = 0$, some of its slices can be nonmeasurable with respect to the slice measure. From Lemma 26 we have $\epsilon \ge mW \ge m(X_\alpha(W)) \cdot \alpha \ge m^*(X_\alpha(E)) \cdot \alpha$. Holding α fixed and infimizing over $\epsilon > 0$ gives $m^*(X_\alpha(E)) = 0$. Since $X_0(E)$ is the countable union of the zero sets $X_\alpha(E)$ with $1/\alpha \in \mathbb{N}$, we have $m(X_0(E)) = 0$, which means exactly that almost every slice of E is a zero set.

Conversely, we suppose E is measurable and $m(E_x) = 0$ for almost every x, and seek to prove that $mE = 0$. Let $Z = \{x : E_x \text{ is not a zero set}\}$. Z is a zero set. The slices E_x for which E_x is not a zero set are contained in $Z \times \mathbb{R}$ which, as proved above, is a zero set in $\mathbb{R}^2$. Then $E \setminus (Z \times \mathbb{R})$ is measurable, has the same measure as E, and so it is no loss of generality to assume that *every* slice E_x is a zero set.

It suffices to show that the inner measure of E is zero. For measurability implies $m_*E = m^*E$. (Here is where we use measurability of E.) We claim that every compact $K \subset E$ is a zero set. Let $\epsilon > 0$ be given. The slice K_x is compact and it has slice measure zero. Therefore it has an open neighborhood $V(x)$ such that $m(V(x)) < \epsilon$. Compactness of K implies that for all x' near x we have $K_{x'} \subset V(x)$. For otherwise there is a sequence (x_n, y_n) in K with $(x_n, y_n) \to (x, y)$ and $y \notin K_x$.

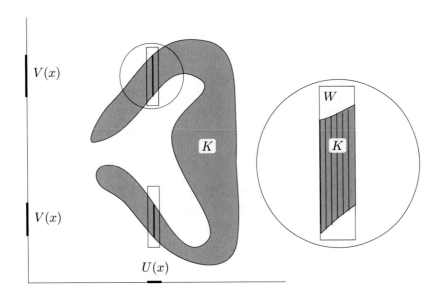

Figure 142 The open set $V(x)$ contains the slice K_x and has small measure. If x' lies in a small enough neighborhood $U(x)$ of x then the set $x' \times K_{x'}$ lies in $W(x) = U(x) \times V(x)$. These sets $x' \times K_{x'}$ are shown in the enlarged picture as vertical segments in K.

Closedness of K implies $(x, y) \in K$, so $y \in K_x$, a contradiction. Hence if $U(x)$ is small then for all $x' \in U(x)$ we have $x' \times K_{x'} \subset W(x) = U(x) \times V(x)$. See Figure 142.

The product neighborhoods $U(x) \times V(x)$ form an open covering of the compact set K, and therefore the covering reduces to a finite subcovering, say $\{U_i \times V_i\}$ with $1 \le i \le N$. The sets $U_i' = U_i \setminus (U_1 \cup \ldots U_{i-1})$ are disjoint and the sets $U_i' \times V_i$ cover K disjointly. Theorem 21 implies that

$$ mK \ \le \ \sum_{i=1}^{N} m(U_i' \times V_i) \ = \ \sum_{i=1}^{N} m(U_i') \cdot m(V_i) \ < \ \sum_{i=1}^{N} m(U_i') \cdot \epsilon \ \le \ \epsilon, $$

so $mK = 0$ and $m_* E = mE = 0$. $\qquad\qquad\qquad\qquad\qquad\qquad\qquad\qquad\square$

Remark Measurability of E is a necessary condition in Theorem 25. See Exercise 25.

6 Lebesgue Integrals

Following J.C. Burkill, we justify the maxim that the integral of a function is the area under its graph. Let $f : \mathbb{R} \to [0, \infty)$ be given.[†]

Definition The **undergraph** of f is

$$\mathcal{U}f = \{(x, y) \in \mathbb{R} \times [0, \infty) : 0 \leq y < f(x)\}.$$

The function f is **(Lebesgue) measurable** if $\mathcal{U}f$ is measurable with respect to planar Lebesgue measure, and if it is then the **Lebesgue integral** of f is the measure of the undergraph

$$\int f = m(\mathcal{U}f).$$

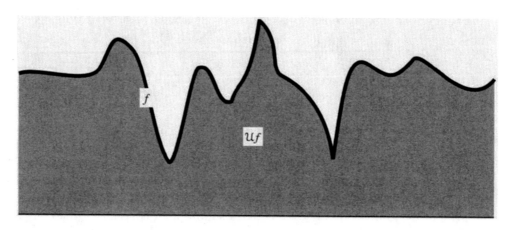

Figure 143 The geometric definition of the integral is the measure of the undergraph.

See Figure 143.

Burkill refers to the undergraph as the **ordinate set** of f. The notation for the Lebesgue integral intentionally omits the usual "dx" and the limits of integration to remind you that it is not merely the ordinary Riemann integral $\int_a^b f(x) \, dx$ or the improper Riemann integral $\int_{-\infty}^{\infty} f(x) \, dx$.

Since a measurable set can have infinite measure we permit $\int f = \infty$.

[†] In this section we deal with functions of one variable. The multivariable case in which $f : \mathbb{R}^n \to \mathbb{R}$ offers no new ideas, only new notation.

Definition The function $f : \mathbb{R} \to [0, \infty)$ is **Lebesgue integrable** if (it is measurable and) its integral is finite.[†] The set of integrable functions is denoted by L^1, $\mathcal{L}^1$, or $\mathcal{L}$.

The three basic convergence theorems for Lebesgue integrals are the Monotone Convergence Theorem, the Dominated Convergence Theorem, and Fatou's Lemma. Their proofs are easy if you look at the right undergraph pictures. We write $f_n \to f$ a.e. to indicate that $\lim\limits_{n\to\infty} f_n(x) = f(x)$ for **almost every** x, i.e., for all x not belonging to some zero set.[‡] (See Chapter 3 for previous use of the phrase "almost every" in connection with Riemann integrability.) However, we often abuse the notation by dropping the "a.e." for clarity. This is rarely a problem since Lebesgue theory systematically neglects zero sets; as Theorem 5 states, zero sets have no effect on measurability or measure, and thus no effect on integrals.[§]

27 Monotone Convergence Theorem *Assume that (f_n) is a sequence of measurable functions $f_n : \mathbb{R} \to [0, \infty)$ and $f_n \uparrow f$ a.e. as $n \to \infty$. Then*

$$\int f_n \uparrow \int f.$$

Proof Obvious from Figure 144. □

Definition The **completed undergraph** of $f : \mathbb{R} \to [0, \infty)$ is

$$\widehat{\mathcal{U}}f = \{(x, y) \in \mathbb{R} \times [0, \infty) : 0 \le y \le f(x)\}.$$

It is the undergraph plus the graph.

28 Proposition $\widehat{\mathcal{U}}f$ *is measurable if and only if $\mathcal{U}f$ is measurable, and if measurable then their measures are equal.*

Proof For $n \in \mathbb{N}$ let $T_{\pm n} : \mathbb{R}^2 \to \mathbb{R}^2$ send (x, y) to $(x, (1 \pm 1/n)y)$. The matrix that represents $T_{\pm n}$ is

$$\begin{bmatrix} 1 & 0 \\ 0 & 1 \pm 1/n \end{bmatrix}.$$

[†]Thus the integral of a measurable nonnegative function exists even if the function is not integrable. To avoid this abuse of language the word "summable" is sometimes used in place of "integrable" to indicate that $\int f < \infty$.

[‡]You may also come across the abbreviation "p.p." for the French *presque partout.*

[§]As informal, clunky notation you can decorate the standard symbols "$\to$", "$=$", "$\forall$", etc. with small zeros indicating "up to a zero set." Thus $f_n \overset{\circ}{\to} f$ would indicate a.e. convergence, $A \overset{\circ}{=} B$ would indicate set equality except for a zero set, $\overset{\circ}{\forall}$ would indicate for almost every, and so on. We will use the notation sparingly.

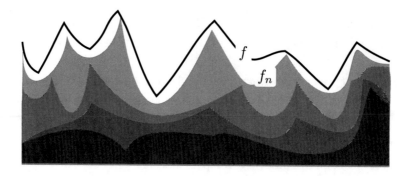

Figure 144 $f_n \uparrow f$ implies $\mathcal{U}f_n \uparrow \mathcal{U}f$. Upward measure continuity
(Theorem 6) then implies $\int f_n = m(\mathcal{U}f_n) \uparrow m(\mathcal{U}f) = \int f$.

By Theorem 15 $T_{\pm n}$ is a meseomorphism and $m(T_n(\mathcal{U}f)) = (1 + 1/n)m(\mathcal{U}f)$. The
intersection $\bigcap T_n(\mathcal{U}f)$ is $\widehat{\mathcal{U}}f$ except for points $(x, 0)$ of the x-axis at which $f(x) = 0$.
The x-axis is a planar zero set and has no effect on measurability. Therefore $\widehat{\mathcal{U}}f$ is
measurable and $m(\widehat{\mathcal{U}}f) = m(\mathcal{U}f)$.

Similarly, $\mathcal{U}f$ is the union of the sets $T_{-n}(\widehat{\mathcal{U}}f)$ except for points on the x-axis
and so measurability of $\widehat{\mathcal{U}}f$ implies measurability of $\mathcal{U}f$. Upward measure continuity
implies that

$$m(\mathcal{U}f) = \lim_{n \to \infty} (1 - 1/n)m(\widehat{\mathcal{U}}f) = m(\widehat{\mathcal{U}}f)$$

which completes the proof.   □

29 Corollary *If (f_n) is a sequence of integrable functions that converges monotonically downward to a limit function f almost everywhere then*

$$\int f_n \downarrow \int f.$$

Proof Since $m(\widehat{\mathcal{U}}(f_n)) = \int f_n$ is finite, downward measure continuity is valid. Proposition 28 then implies

$$\int f_n = m(\mathcal{U}(f_n)) = m(\widehat{\mathcal{U}}(f_n)) \downarrow m(\widehat{\mathcal{U}}f) = m(\mathcal{U}f) = \int f$$

as $n \to \infty$.   □

Definition If $f_n : X \to [0, \infty)$ is a sequence of functions then the lower and upper
envelope sequences are

$$\underline{f}_n(x) = \inf\{f_k(x) : k \geq n\} \qquad \overline{f}_n(x) = \sup\{f_k(x) : k \geq n\}.$$

We permit $\overline{f}_n(x) = \infty$.

30 Proposition $\mathcal{U}(\overline{f}_n) = \bigcup_{k \geq n} \mathcal{U}(f_k)$ *and* $\widehat{\mathcal{U}}(\underline{f}_n) = \bigcap_{k \geq n} \widehat{\mathcal{U}}(f_k).$

Proof We have

$$
\begin{aligned}
(x, y) \in \mathcal{U}(\overline{f}_n) &\iff y < \sup\{f_k(x) : k \geq n\} \\
&\iff \exists \ell \geq n \text{ such that } y < f_\ell(x) \\
&\iff \exists \ell \geq n \text{ such that } (x, y) \in \mathcal{U}(f_\ell) \\
&\iff (x, y) \in \bigcup_{k \geq n} \mathcal{U}(f_k).
\end{aligned}
$$

The other equality is checked the same way. $\qquad\square$

31 Dominated Convergence Theorem *If $f_n : \mathbb{R} \to [0, \infty)$ is a sequence of measurable functions such that $f_n \to f$ a.e. and if there exists a function $g : \mathbb{R} \to [0, \infty)$ whose integral is finite and which is an upper bound for all the functions f_n then f is integrable and $\int f_n \to \int f$ as $n \to \infty$.*

Proof Obvious from Figure 145. $\qquad\square$

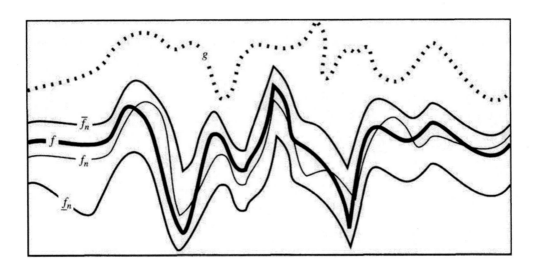

Figure 145 Dominated convergence. Proposition 30 implies the envelope functions are measurable. Due to the dominator g they are integrable. The Monotone Convergence Theorem and Corollary 29 imply their integrals converge to $\int f$. Since $\mathcal{U}(\underline{f}_n) \subset \mathcal{U}(f_n) \subset \widehat{\mathcal{U}}(\overline{f}_n)$ the integral of f_n also converges to $\int f$.

Remark If a dominator g with finite integral fails to exist then the assertion fails. For example, the sequence of steeple functions shown in Figure 89 on page 214, have integral n and converge at all x to the zero function as $n \to \infty$. See Exercise 33.

32 Corollary *The pointwise limit of measurable functions is measurable.*

Proof $\mathcal{U}(\underline{f}_n)$ is measurable and converges upward to $\mathcal{U}f$. $\qquad\square$

33 Fatou's Lemma *If $f_n : \mathbb{R} \to [0, \infty)$ is a sequence of measurable functions then*

$$\int \liminf f_n \ \leq \ \liminf \int f_n.$$

Proof The assertion is really more about lim infs than integrals. The lim inf of the sequence (f_n) is $f = \lim_{n \to \infty} \underline{f}_n$, where $\underline{f}_n$ is the lower envelope function. Since $\underline{f}_n \uparrow f$, the Monotone Convergence Theorem implies $\int \underline{f}_n \uparrow \int f$, and since $\underline{f}_n \leq f_n$ we have $\int f \leq \liminf \int f_n$. $\qquad\square$

Remark The inequality in Fatou's Lemma can be strict as is shown by the steeple functions. See Exercise 33.

Having established the three basic convergence theorems for Lebesgue integrals using mainly pictures of undergraphs, we collect some integration facts of a more mundane character.

34 Theorem *Let $f, g : \mathbb{R} \to [0, \infty)$ be measurable functions.*

(a) If $f \leq g$ then $\int f \leq \int g$.
(b) If $\mathbb{R} = \bigsqcup_{k=1}^{\infty} X_k$ and each X_k is measurable then

$$\int f = \sum_{k=1}^{\infty} \int_{X_k} f.$$

(c) If $X \subset \mathbb{R}$ is measurable then $mX = \int \chi_X$.
(d) If $mX = 0$ then $\int_X f = 0$.
(e) If $f(x) = g(x)$ almost everywhere then $\int f = \int g$.
(f) If $c \geq 0$ then $\int cf = c \int f$.
(g) The integral of f is zero if and only if $f(x) = 0$ for almost every x.
(h) $\int f + g = \int f + \int g$.

Proof Assertions (a) – (g) are obvious from what we know about measure.

(a) $f \leq g$ implies $\mathcal{U}f \subset \mathcal{U}g$ implies $m(\mathcal{U}f) \leq m(\mathcal{U}g)$.

(b) The product $X_k \times \mathbb{R}$ is measurable and its intersection with $\mathcal{U}f$ is $\mathcal{U}f|_{X_k}$. Thus $\mathcal{U}f = \bigsqcup_{k=1}^{\infty} \mathcal{U}f|_{X_k}$ and countable additivity of planar measure gives the result.

(c) The planar measure of the product $\mathcal{U}(\chi_X) = X \times [0,1)$ is mX.

(d) $\mathcal{U}f$ is contained in the product $X \times \mathbb{R}$ of zero planar measure.

(e) Almost everywhere equality of f and g means there is a zero set $Z \subset \mathbb{R}$ such that if $x \notin Z$ then $f(x) = g(x)$. Apply (b), (d) to $\mathbb{R} = Z \sqcup (\mathbb{R} \setminus Z)$.

(f) According to Theorem 15 scaling the y-axis by the factor c scales planar measure correspondingly.

(g) The Zero Slice Theorem (Theorem 25) asserts that $\mathcal{U}f$ is a zero set if and only if almost every vertical slice is a slice zero set. The vertical slices are the segments $[0, fx)$.

(h) This requires a new concept and a corresponding picture. See Theorem 35, Corollary 36, and Figure 146. $\square$

Definition If $f : \mathbb{R} \to \mathbb{R}$ then $\boldsymbol{f}$**-translation** $T_f : \mathbb{R}^2 \to \mathbb{R}^2$ sends the point (x, y) to the point $(x, y + f(x))$.

T_f slides points along the vertical lines $x \times \mathbb{R}$ and

$$T_f \circ T_g \;=\; T_{f+g} \;=\; T_g \circ T_f$$

so T_f is a bijection whose inverse is T_{-f}.

35 Theorem *If* $f : \mathbb{R} \to [0, \infty)$ *is integrable then* T_f *preserves planar Lebesgue measure; i.e., it is a mesisometry.*

Proof We must show that T_f bijects the class $\mathcal{M}$ of Lebesgue measurable subsets of $\mathbb{R}^2$ to itself and $m(T_f E) = mE$ for all $E \in \mathcal{M}$.

Consider Figure 146. It demonstrates that for any two nonnegative functions on $\mathbb{R}$ we have two ways to express $\mathcal{U}(f + g)$, namely

$$\mathcal{U}f \sqcup T_f(\mathcal{U}g) \;=\; \mathcal{U}(f + g) \;=\; T_g(\mathcal{U}f) \sqcup \mathcal{U}g.$$

First we consider the function

$$g(x) \;=\; \begin{cases} h & \text{if } x \in I \\ 0 & \text{otherwise} \end{cases}$$

where I is an interval in $\mathbb{R}$ and h is a positive constant. See Figure 147. The un-

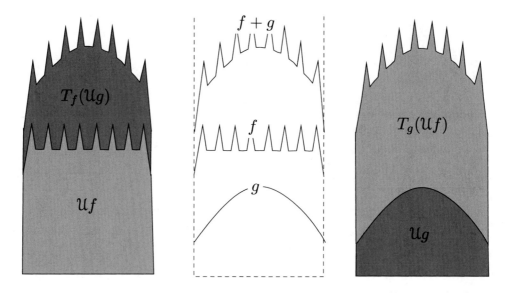

Figure 146 The undergraph of a sum

dergraph of g is the rectangle $R = I \times [0, h)$. The T_f-image of R is the same as the Tf_I-image, where $f_I(x) = f(x) \cdot \chi_I(x)$. Thus we can assume that $f(x) = 0$ for $x \notin I$. The map T_g is vertical translation by the constant h and since Lebesgue measure is translation invariant we get measurability of $T_g(\mathcal{U}f)$. Then $\mathcal{U}f \sqcup T_f R = T_g(\mathcal{U}f) \sqcup R$ implies $T_f R$ is measurable and

$$m(\mathcal{U}f) + m(T_f R) \;=\; m(T_g(\mathcal{U}f)) + mR.$$

Since $m(\mathcal{U}f) < \infty$, subtraction is legal and we get $m(T_f R) = mR$. If we translate R vertically by k then we have a rectangle $T_k R = I \times [k, h+k)$ and $T_f(T_k R) = T_k \circ T_f R$ implies that T_f sends each rectangle $I \times [c, d)$ to a measurable set of the same measure.

We claim that T_f never increases outer measure. If $S \subset \mathbb{R}^2$ and $\epsilon > 0$ is given then we cover S with countably many rectangles R_i such that

$$\sum m(R_i) \;\leq\; m^* S + \epsilon.$$

Then $T_f S$ is covered by countably many measurable sets $T_f(R_i)$ with total measure $\leq m^* S + \epsilon$. From countable subadditivity and the ϵ-Principle we deduce $m^*(T_f S) \leq m^* S$. The same is true for T_{-f} since

$$T_{-f} = \psi \circ T_f \circ \psi$$

where $\psi : \mathbb{R}^2 \to \mathbb{R}^2$ is the mesisometry sending (x, y) to $(x, -y)$. Neither T_f nor its inverse increase outer measure, so Theorem 7 implies T_f is a mesisometry. $\qquad\square$

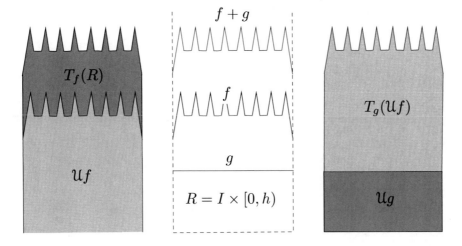

Figure 147 T_f translates R upward by f and T_g translates $\mathcal{U}f$ upward by h.

36 Corollary *If $f : \mathbb{R} \to [0, \infty)$ and $g : \mathbb{R} \to [0, \infty)$ are integrable then*

$$\int f + g = \int f + \int g.$$

Proof Since $\mathcal{U}(f + g) = \mathcal{U}f \sqcup T_f(\mathcal{U}g)$ and T_f is a mesisometry we see that $f + g$ is measurable and $m(\mathcal{U}(f + g)) = m(\mathcal{U}f) + m(\mathcal{U}g)$. That is, the integral of the sum is the sum of the integrals. □

Remark The standard proof of linearity of the Lebesgue integral is outlined in Exercise 47. It is no easier than this undergraph proof, and undergraphs at least give you a picture as guidance.

37 Corollary *If $f_k : \mathbb{R} \to [0, \infty)$ is a sequence of integrable functions then*

$$\sum_{k=1}^{\infty} \int f_k = \int \sum_{k=1}^{\infty} f_k.$$

Proof Let $F_n(x) = \sum_{k=1}^{n} f_k(x)$ be the nth partial sum and $F(x) = \sum_{k=1}^{\infty} f_k(x)$. Then $F_n(x) \uparrow F(x)$ as $n \to \infty$. The Monotone Convergence Theorem implies $\int F_n \to \int F$. Corollary 36 implies $\sum_{k=1}^{n} \int f_k = \int \sum_{k=1}^{n} f_k$ and the assertion follows. □

Until now we have assumed the integrand f is nonnegative. If f takes both positive and negative values we define

$$f_+(x) = \begin{cases} f(x) & \text{if } f(x) \geq 0 \\ 0 & \text{if } f(x) < 0 \end{cases} \qquad f_-(x) = \begin{cases} -f(x) & \text{if } f(x) < 0 \\ 0 & \text{if } f(x) \geq 0. \end{cases}$$

Then $f_{\pm} \geq 0$ and $f = f_{+} - f_{-}$. See Exercise 28. If $f_{\pm}$ are integrable we say that f is integrable and define its integral as

$$\int f = \int f_{+} - \int f_{-}.$$

38 Proposition *The set of measurable functions $f : \mathbb{R} \to \mathbb{R}$ is a vector space, the set of integrable functions is a subspace, and the integral is a linear map from the latter into $\mathbb{R}$.*

The proof is left to the reader as Exercise 32.

7 Italian Measure Theory

In Chapter 5 the slice method is developed in terms of Riemann integrals. Here we generalize to Lebesgue integrals. If $E \subset \mathbb{R}^k \times \mathbb{R}^n$ and $x \in \mathbb{R}^k$ then the **x-slice** through a point $x \in \mathbb{R}^k$ is

$$E_x = \{y \in \mathbb{R}^n : (x,y) \in E\}.$$

The **y-slice** is $E^y = \{x : (x,y) \in E\}$. Similarly, the **$x$-slice** and **$y$-slice** of a function $f : E \to \mathbb{R}$ are $f_x : y \mapsto f(x,y)$ and $f^y : x \mapsto f(x,y)$.

Remark In this section we frequently write dx and dy to indicate which variable is the integration variable.

39 Cavalieri's Principle *If E is measurable then almost every slice E_x of E is measurable, the function $x \mapsto m(E_x)$ is measurable, and its integral is*

$$(4) \qquad\qquad mE = \int m(E_x)\, dx.$$

(Note that mE refers to $(k + n)$-dimensional measure while $m(E_x)$ refers to n-dimensional measure.)

See Figure 148.

Proof We take $k = 1 = n$. The proof of the Zero Slice Theorem (Theorem 26) contains the hard work; if E is a zero set then it asserts that almost every slice E_x is a zero set, and since the integral of a function that vanishes almost everywhere is zero we get (4) for zero sets.

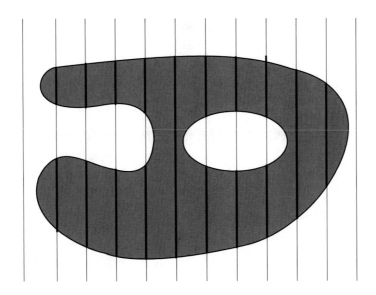

Figure 148 Slicing a planar set

(4) is obvious for boxes, and hence it holds also for open sets. After all, an open set is the disjoint union of boxes and a zero set, and slicing preserves disjointness.

The Dominated Convergence Theorem promotes (4) from open sets to bounded G_δ-sets.

(4) holds for bounded measurable sets since each is a bounded G_δ-set minus a zero set. The general measurable set E is a disjoint union of bounded measurable sets, $E = \bigsqcup E_i$, so countable additivity gives (4) for E. $\qquad\qquad\square$

The proof of Cavalieri's Principle in higher dimensions differs only notationally from the proof in $\mathbb{R}^2$. See also Appendix B of Chapter 5 and Exercise 44.

40 Corollary *The y-slices of an undergraph decrease monotonically as y increases, and the following formulas hold:*

$$(\mathcal{U}f)^a = \bigcup_{y>a} (\mathcal{U}f)^y \qquad (\widehat{\mathcal{U}}f)^a = \bigcap_{y<a} (\widehat{\mathcal{U}}f)^y.$$

Every horizontal slice of a measurable undergraph is measurable.

Proof Monotonicity and the formulas follow from

$$(\mathcal{U}f)^a = \{x : a < fx\} = \{x : \exists y > a \text{ such that } y < fx\}$$
$$(\widehat{\mathcal{U}}f)^a = \{x : a \le fx\} = \{x : \forall y < a \text{ we have } y \le fx\}.$$

We fix an arbitrary a and ask: Are the slices $(\mathcal{U}f)^a$ and $(\widehat{\mathcal{U}}f)^a$ measurable? Cavalieri's Principle implies that *almost* every horizontal slice of a measurable undergraph is measurable. Thus, there exist $y_n \downarrow a$ such that $(\mathcal{U}f)^{y_n}$ is measurable. By monotonicity, $(\mathcal{U}f)^a = \bigcup_n (\mathcal{U}f)^{y_n}$ gives measurability of $(\mathcal{U}f)^a$. Similarly for the completed undergraph.   □

41 Corollary *Undergraph measurability is equivalent to the more common definition using preimages.*

Proof We say that $f : \mathbb{R} \to [0, \infty)$ is **preimage measurable** if for each $a \in [0, \infty)$ the preimage $f^{\mathrm{pre}}[a, \infty) = \{x : fx \geq a\}$ is a measurable subset of the line. (See also Appendix A.) Since

$$f^{\mathrm{pre}}[a, \infty) \;=\; \{x : a \leq fx\} \;=\; (\widehat{\mathcal{U}}f)^a$$

by Corollary 40, we see that undergraph measurability implies preimage measurability. The converse follows from the equation

$$\mathcal{U}f \;=\; \bigcup_{0 \leq a \in \mathbb{Q}} f^{\mathrm{pre}}[a, \infty) \times [0, a).$$
□

As a consequence of Cavalieri's Principle in 3-space we get the integral theorems of Fubini and Tonelli. It is standard practice to refer to the integral of a function f on $\mathbb{R}^2$ as a double integral and to write it as

$$\int f = \iint f(x, y)\, dx dy.$$

It is also standard to write the iterated integral as

$$\int \left[\int f_x(y)\, dy \right] dx \;=\; \int \left[\int f(x, y)\, dy \right] dx.$$

42 Fubini-Tonelli Theorem *If $f : \mathbb{R}^2 \to [0, \infty)$ is measurable then almost every slice $f_x(y)$ is a measurable function of y, the function $x \mapsto \int f_x(y)\, dy$ is measurable, and the double integral equals the iterated integral,*

$$\iint f(x, y)\, dx dy \;=\; \int \left[\int f(x, y)\, dy \right] dx.$$

Proof The result follows from the simple observation that the slice of the undergraph is the undergraph of the slice,

(5) $(\mathcal{U}f)_x \;=\; \mathcal{U}f_x.$

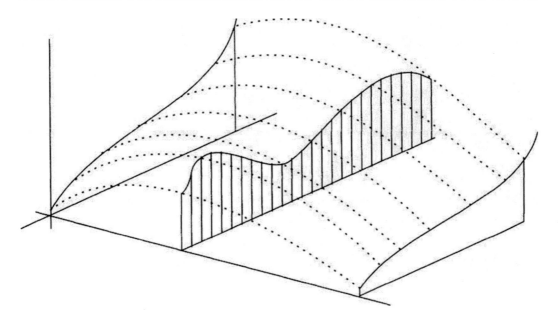

Figure 149 Slicing the undergraph

See Figure 149. For (5) implies that $m_2((\mathcal{U}f)_x) = m_2(\mathcal{U}f_x) = \int f(x,y)\,dy$, and then Cavalieri gives

$$\iint f(x,y)\,dxdy \;=\; m_3(\mathcal{U}f) \;=\; \int [m_2\,((\mathcal{U}f)_x)]\,dx$$
$$=\; \int \left[\int f(x,y)\,dy\right]dx. \qquad \square$$

43 Corollary *When $f : \mathbb{R}^2 \to [0,\infty)$ is measurable the order of integration in the iterated integrals is irrelevant,*

$$\int \left[\int f(x,y)\,dy\right]dx \;=\; \iint f(x,y)\,dxdy \;=\; \int \left[\int f(x,y)\,dx\right]dy.$$

(In particular if one of the three integrals is finite then so are the other two and all three are equal.)

Proof The difference between "x" and "y" is only notational. In contrast to the integration of differential forms, the orientation of the plane or 3-space plays no role in Lebesgue integration so the Fubini-Tonelli Theorem applies equally to x-slicing and y-slicing, which implies that both iterated integrals equal the double integral. $\square$

The multidimensional version of Cavalieri's Principle yields similar multi-integral results. See Exercise 54.

When f takes on both signs a little care must be taken to avoid subtracting ∞ from ∞.

44 Theorem *If $f : \mathbb{R}^2 \to \mathbb{R}$ is integrable (the double integral of f exists and is finite) then the iterated integrals exist and equal the double integral.*

Proof Split f into its positive and negative parts, $f = f_+ - f_-$, and apply the Fubini-Tonelli Theorem to each separately. Since the integrals are finite, subtraction is legal and the theorem follows for f. □

See Exercise 53 for an example in which trouble arises if you forget to assume that the double integral is finite.

8 Vitali Coverings and Density Points

The fact that every open covering of a closed and bounded subset of Euclidean space reduces to a finite subcovering is certainly an important component of basic analysis. In this section we present another covering theorem, this time the accent being on disjointness of the sets in the subcovering rather than on finiteness. The result is used to differentiate Lebesgue integrals.

Definition A covering $\mathcal{V}$ of a set A in a metric space M is a **Vitali covering** if for each point $p \in A$ and each $r > 0$ there is $V \in \mathcal{V}$ such that $p \in V \subset M_r p$ and V is not merely the singleton set $\{p\}$.

For example, if $A = [a, b]$, $M = \mathbb{R}$, and $\mathcal{V}$ consists of all intervals $[\alpha, \beta]$ with $\alpha \leq \beta$ and $\alpha, \beta \in \mathbb{Q}$ then $\mathcal{V}$ is a Vitali covering of A.

45 Vitali Covering Lemma *A Vitali covering of a bounded set $A \subset \mathbb{R}^n$ by closed balls reduces to an efficient disjoint subcovering of almost all of A.*

More precisely, given $\epsilon > 0$, $\mathcal{V}$ reduces to a countable subcollection $\{V_k\}$ such that

(a) The V_k are disjoint.
(b) $mU \leq m^*A + \epsilon$, where $U = \bigsqcup_{k=1}^{\infty} V_k$.
(c) $A \setminus U$ is a zero set.

Condition (b) is what we mean by $\{V_k\}$ being an "efficient" covering – the extra points covered form an ϵ-set. The sets $U_N = V_1 \sqcup \cdots \sqcup V_N$ "nearly" cover A in the sense that given $\epsilon > 0$, if N is large then U_N contains A except for an ϵ-set. After all, $U = \bigcup U_N$ contains A except for a zero set. See also Appendix E.

Boundedness of A is an unnecessary hypothesis. Also, the assumption that the sets $V \in \mathcal{V}$ are closed balls can be weakened somewhat. We discuss these improvements after the proof of the result as stated.

Proof of the Vitali Covering Lemma Given $\epsilon > 0$, there is a bounded open set $W \supset A$ such that $mW \leq m^*A + \epsilon$. Define

$$\mathcal{V}_1 = \{V \in \mathcal{V} : V \subset W\} \quad \text{and} \quad d_1 = \sup\{\text{diam } V : V \in \mathcal{V}_1\}.$$

$\mathcal{V}_1$ is still a Vitali covering of A. Since W is bounded d_1 is finite. Choose $V_1 \in \mathcal{V}_1$ with diam $V_1 \geq d_1/2$ and define

$$\mathcal{V}_2 = \{V \in \mathcal{V}_1 : V \cap V_1 = \emptyset\} \quad \text{and} \quad d_2 = \sup\{\text{diam } V : V \in \mathcal{V}_2\}.$$

Choose $V_2 \in \mathcal{V}_2$ with diam $V_2 \geq d_2/2$. In general,

$$\mathcal{V}_k = \{V \in \mathcal{V}_{k-1} : V \cap U_{k-1} = \emptyset\}$$
$$d_k = \sup\{\text{diam } V : V \in \mathcal{V}_k\}$$
$$V_k \in \mathcal{V}_k \text{ has } \text{diam } V_k \geq \frac{d_k}{2}$$

where $U_{k-1} = V_1 \sqcup \ldots \sqcup V_{k-1}$. This means that V_k has roughly maximal diameter among the $V \in \mathcal{V}$ that do not meet U_{k-1}. By construction, the balls V_k are disjoint and since they lie in W we have $m(\bigsqcup V_k) \leq mW \leq m^*A + \epsilon$, verifying (a) and (b). It remains to check (c).

If at any stage in the construction $\mathcal{V}_k = \emptyset$ then we have covered A with finitely many sets V_k, so (c) becomes trivial. We therefore assume that $V_1, V_2, \ldots$ form an infinite sequence. Additivity implies that $m(\bigsqcup V_k) = \sum mV_k$. Since each V_k is contained in W the series converges. This implies that diam $V_k \to 0$ as $k \to \infty$; i.e.,

(6) $$d_k \to 0 \text{ as } k \to \infty.$$

For each $N \in \mathbb{N}$ we claim that

(7) $$\bigcup_{k=N}^{\infty} 5V_k \quad \supset \quad A \setminus U_{N-1}$$

where $5V_k$ denotes the ball V_k dilated from its center by the factor 5. (These dilated balls need not belong to $\mathcal{V}$.)

Take any $a \in A \setminus U_{N-1}$. Since U_{N-1} is compact and $\mathcal{V}_1$ is Vitali, there is a ball $B \in \mathcal{V}_1$ such that $a \in B$ and $B \cap U_{N-1} = \emptyset$. That is, $B \in \mathcal{V}_N$. Assume that (7) fails. Then, for all $k \geq N$ we have

$$a \notin 5V_k.$$

Therefore $B \not\subset 5V_N$. Figure 150 shows that due to the choice of V_N with roughly maximal diameter, the fact that $5V_N$ fails to contain B implies that V_N is disjoint from B, so $B \in \mathcal{V}_{N+1}$. This continues for all $k > N$; namely for all $k > N$ we have $B \in \mathcal{V}_k$.

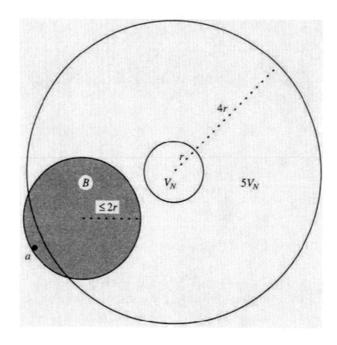

Figure 150 The unchosen ball B

Aha!

B was available for choice as the next V_k, $k > N$, but it was never chosen. Therefore the chosen V_k has a diameter at least half as large as that of B. The latter diameter is fixed, but (6) states that the former diameter tends to 0 as $k \to \infty$, a contradiction. Thus (7) is true.

It is easy to see that (7) implies (c). For let $\delta > 0$ be given. Choose N so large that

$$\sum_{k=N}^{\infty} m(V_k) < \frac{\delta}{5^n}$$

where $n = \dim \mathbb{R}^n$. Since the series $\sum m(V_k)$ converges this is possible. By (7) and the scaling law $m(tE) = t^n mE$ for n-dimensional measure we have

$$m^*(A \setminus U_{N-1}) \; \leq \; \sum_{k=N}^{\infty} m(5V_k) \; = \; 5^n \sum_{k=N}^{\infty} m(V_k) < \delta.$$

Since δ is arbitrary, $A \setminus U = \bigcap_k (A \setminus U_k)$ is a zero set.   □

Remark A similar strategy of covering reduction appears in the proof in Chapter 2 that sequential compactness implies covering compactness. Formally, the proof is expressed in terms of the Lebesgue number of the covering but the intuition is this: Given an open covering $\mathcal{U}$ of a sequentially compact set K, you choose a subcovering by first taking a $U_1 \in \mathcal{U}$ that covers about as much of K as possible, then taking $U_2 \in \mathcal{U}$ that covers about as much of the remainder of K as possible, and so on. If finitely many of these sets U_n fail to cover K then you take a sequence $x_n \in K \setminus (U_1 \cup \cdots \cup U_{n-1})$ and prove that it has no subsequence which converges in K. (The contradiction shows that in fact finitely many of the U_n you chose actually did cover K.) In short, when reducing a covering it is a good idea to *choose the biggest sets first*. This is exactly the Vitali outlook.

Removing the assumption that A is bounded presents no problem. Express $\mathbb{R}^n$ as $\bigsqcup D_i \cup Z$, where the D_i are the open unit dyadic cubes and Z is the zero set of hyperplanes having at least one integer coordinate. If $A \subset \mathbb{R}^n$ is unbounded then $A = \bigsqcup A_i \cup (A \cap Z)$, where $A_i = A \cap D_i$. Given a Vitali covering $\mathcal{V}$ of A by closed balls, we set

$$\mathcal{V}_i \; = \; \{V \in \mathcal{V} : V \subset D_i\}.$$

It is a Vitali covering of the bounded set A_i and therefore reduces to a disjoint $(\epsilon/2^i)$-efficient covering $\{V_{i,k} : k \in \mathbb{N}\}$ of almost all of A_i. Thus $\mathcal{V}$ reduces to a disjoint ϵ-efficient covering $\{V_{i,k} : i, k \in \mathbb{N}\}$ of almost all of A.

A further generalization involves the shapes of the sets $V \in \mathcal{V}$. If $| \; |_*$ is any norm on $\mathbb{R}^n$ then its closed ball of radius r at p is

$$B_*(r, p) = \{x \in \mathbb{R}^n : |x|_* \leq r\}.$$

The preceding proof of the Vitali Covering Lemma goes through word for word when we substitute balls with respect to the norm $| \; |_*$ for Euclidean balls. Even the factor 5 remains the same. If $| \; |_*$ is the taxicab norm then this gives the following result. See also Exercise 61.

46 Vitali Covering Lemma for Cubes *A Vitali covering of $A \subset \mathbb{R}^n$ by closed cubes[†] reduces to an efficient disjoint subcovering of almost all of A.*

Density Points

Let $E \subset \mathbb{R}^n$ be measurable. For $p \in \mathbb{R}^n$, define the **density** of E at p as

$$\delta(p, E) = \lim_{Q \downarrow p} \frac{m(E \cap Q)}{mQ},$$

if the limit exists, m being Lebesgue measure on $\mathbb{R}^n$. The notation $Q \downarrow p$ indicates that Q is a cube which contains p and shrinks down to p. It need not be centered at p. Clearly $0 \leq \delta \leq 1$. Points with $\delta = 1$ are called **density points** of E. The fraction that we're taking the limit of is the "relative measure" or **concentration** of E in Q. I like to write the concentration of E in Q as in chemistry,

$$\frac{m(E \cap Q)}{mQ} = [E : Q].$$

Existence of $\delta(p, E)$ means that for each $\epsilon > 0$ there exists an $\ell > 0$ such that if Q is any cube of edgelength $< \ell$ that contains p then the concentration of E in Q differs from $\delta(p, E)$ by $< \epsilon$.

Remark Demanding that that the cubes be centered at p produces the concept of **balanced density**. Balls or certain other shapes can be used instead of cubes. See Exercise 58, Exercise 61, the end of the preceding section, and Figure 151.

47 Lebesgue Density Theorem *If E is measurable then almost every $p \in E$ is a density point of E.*

Interior points of E are obviously density points of E, although sets like the irrationals or a fat Cantor set have empty interior, while still having plenty of density points.

Proof of the Lebesgue Density Theorem Without loss of generality we assume E is bounded. Take any a, $0 \leq a < 1$, and consider

$$E_a = \{p \in E : \underline{\delta}(E, p) < a\}$$

[†]The cubes are Cartesian products $I_1 \times \cdots \times I_n$, where the I_i are closed intervals, all of the same length.

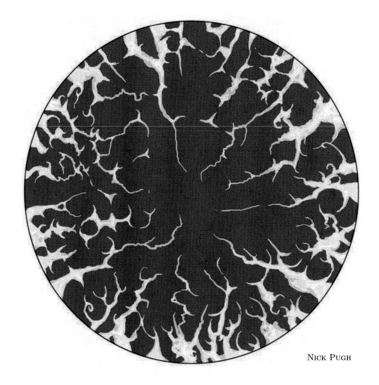

NICK PUGH

Figure 151 An artist's rendering of a density point

where $\underline{\delta}$ is the lower density, $\liminf_{Q \downarrow p}[E : Q]$. We will show that E_a has outer measure zero.

By assumption, at every $p \in E_a$ there are arbitrarily small cubes in which the concentration of E is $< a$. These cubes form a Vitali covering of E_a and by the Vitali Covering Lemma we can select a subcollection $Q_1, Q_2, \ldots$ such that the Q_k are disjoint, cover almost all of E_a, and nearly give the outer measure of E_a in the sense that

$$\sum_k m(Q_k) \;<\; m^*(E_a) + \epsilon.$$

(E_a turns out to be measurable but the Vitali Covering Lemma does not require us to know this in advance.) We get

$$m^*(E_a) = \sum_k m^*(E_a \cap Q_k)$$
$$\leq \sum_k m(E \cap Q_k) \;<\; a \sum_k m(Q_k) \;\leq\; a(m^*(E_a) + \epsilon)$$

which implies that $m^*(E_a) \leq a\epsilon/(1-a)$. Since $\epsilon > 0$ is arbitrary we have $m^*(E_a) = 0$.

The E_a are monotone increasing zero sets as $a \uparrow 1$. Letting $a = 1 - 1/\ell$ with $\ell = 1, 2, \ldots$, we see that the union of all the E_a with $a < 1$ is also a zero set, say Z. Points $p \in E \setminus Z$ have the property that as $Q \downarrow p$, the lim inf of the concentration of E in Q is $\geq a$ for all $a < 1$. Since the concentration is always ≤ 1 this means the limit of the concentration exists and equals 1 for all $p \in E \setminus Z$; i.e., almost every point of E is a density point of E. $\qquad\square$

48 Corollary *If E is measurable then $E \overset{\circ}{=} \mathrm{dp}(E)$, i.e., for almost every p we have*

$$\chi_E(p) = \lim_{Q \downarrow p}[E : Q].$$

Proof For almost every $p \in E$ we have $\lim_{Q \downarrow p}[E : Q] = 1$ and for almost every $q \in E^c$ we have $\lim_{Q \downarrow q}[E^c : Q] = 1$. Measurability of E implies $[E : Q] + [E^c : Q] = 1$, which completes the proof. $\qquad\square$

A consequence of the Lebesgue Density Theorem is that measurable sets are not "diffuse" – a measurable subset of $\mathbb{R}$ can not meet every interval (a, b) in a set of measure $c \cdot (b - a)$ where c is a constant, $0 < c < 1$. Instead, a measurable set must be "concentrated" or "clumpy." See Exercise 56. Also, looking at the complement E^c of E, we see that almost every point $x \in E^c$ has $\delta(E, x) = 0$. Thus, almost every point of E is a density point of E and almost every point of E^c is not.

As you might expect, Cavalieri's Principle meshes well with density points.

49 Theorem *For a measurable set $E \subset \mathbb{R}^k \times \mathbb{R}^\ell$, we have $(\mathrm{dp}(E))_x \overset{\circ}{=} \mathrm{dp}(E_x)$ for almost all $x \in \mathbb{R}^k$.*

Proof The Zero Slice Theorem 25 implies there is a zero set $Z_1 \subset \mathbb{R}^k$ such that for all $x \notin Z_1$, E_x is m_ℓ-measurable. The Lebesgue Density Theorem 47 implies $\mathrm{dp}(E) = E$ up to a zero set: $m_n(E \,\Delta\, \mathrm{dp}(E)) = 0$. Theorem 25 implies there is a second zero set $Z_2 \subset \mathbb{R}^k$ such that for all $x \notin Z_2$, $m_\ell((E \,\Delta\, \mathrm{dp}(E))_x) = 0$. Since $(E \,\Delta\, \mathrm{dp}(E))_x = E_x \,\Delta\, (\mathrm{dp}(E))_x$ for all x, we conclude that for all $x \notin Z = Z_1 \cup Z_2$ we have $(\mathrm{dp}(E))_x = \mathrm{dp}(E_x)$ up to a slice zero set, which is what the theorem asserts. $\square$

Recall the measure theoretic boundary of a set $A \subset \mathbb{R}^n$, $\partial_m(A) = H_A \setminus K_A$, as discussed on page 401. In the same vein we can define $\partial_d(A) = \mathrm{dp}^*(A) \setminus (\mathrm{dp}^*(A^c))^c$ where dp^* indicates the set of outer density points

$$\mathrm{dp}^*(A) = \{p : \lim_{Q \downarrow p} \frac{m^*(A \cap Q)}{mQ} = 1\}.$$

50 Theorem *Up to a zero set, $\partial_m(A) = \partial_d(A)$.*

Proof Taking enough complements, it suffices to check that $H_A = \mathrm{dp}^*(A)$ up to a zero set. Since $A \subset H_A$, it is clear that

$$\mathrm{dp}^*(A) \subset \mathrm{dp}^*(H_A) = \mathrm{dp}(H_A).$$

On the other hand, if Q is a cube at p then $m^*(A \cap Q) \geq m(H_A \cap Q)$. Otherwise $m^*(A \cap Q) < m(H_A \cap Q)$ and there is an open set $U \subset Q$ such that $U \supset (A \cap Q)$ and $mU < m(H_A \cap Q)$. See Figure 152. But then $G = (H_A \setminus Q) \cup \partial Q \cup U$ is a G_δ-set that

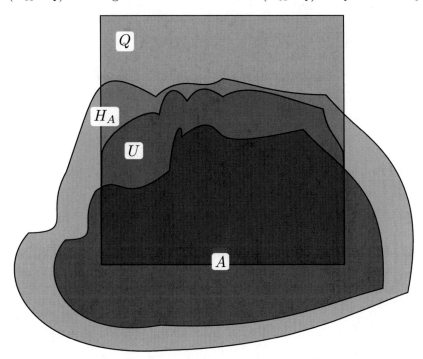

Figure 152 The sets H_A and $\mathrm{dp}^*(A)$ differ by a zero set. They differ from A by a zero set if and only if A is measurable.

contains A and is measure theoretically smaller than H_A, a contradiction. Thus

$$\frac{m^*(A \cap Q)}{mQ} \geq \frac{m(H_A \cap Q)}{mQ},$$

which implies that $\mathrm{dp}^*(A) \supset \mathrm{dp}(H_A)$, and we get equality: $\mathrm{dp}^*(A) = \mathrm{dp}(H_A)$. The Lebesgue Density Theorem 47 implies $\mathrm{dp}(H_A) \stackrel{\circ}{=} H_A$, which completes the proof. $\square$

Remark H_A is only defined up to a zero set, while $\mathrm{dp}^*(A)$ is uniquely defined. However, $H_A \supset A$ while $\mathrm{dp}^*(A)$ need not contain A. Neither is a perfect analog of the closure of A.

9 Calculus à la Lebesgue

In this section we write the integral in dimension 1 as $\int_A f(t)\,dt$ or as $\int_\alpha^\beta f(t)\,dt$ when $A = (\alpha, \beta)$.

Definition The **average** of a locally integrable function $f : \mathbb{R}^n \to \mathbb{R}$ over a measurable set $A \subset \mathbb{R}^n$ with finite positive measure is

$$\fint_A f \;=\; \frac{1}{mA} \int_A f \;=\; [f : A].$$

The notation $[f : A]$ suggests that we are looking at the concentration of f on A. It corresponds to the fact that $[E : A] = m(E \cap A)/mA = [\chi_E : A]$. By "locally integrable" we mean "integrable on a small enough neighborhood of each point in $\mathbb{R}^n$." Of course we assume that A lies in such a neighborhood. The **density** of f at p is

$$\delta(p, f) \;=\; \lim_{Q \downarrow p}[f : Q]$$

if the limit exists. Otherwise we use the lower and upper densities

$$\underline{\delta}(p, f) \;=\; \liminf_{Q \downarrow p}[f : Q] \qquad \overline{\delta}(p, f) \;=\; \limsup_{Q \downarrow p}[f : Q].$$

As above, $Q \downarrow p$ means that Q is a cube which contains p and shrinks down to p.

The following result is also called **Lebesgue's Fundamental Theorem of Calculus**.

51 Average Value Theorem *If $f : \mathbb{R}^n \to \mathbb{R}$ is locally integrable then for almost every $p \in \mathbb{R}^n$ the density of f at p exists and*

$$f(p) \;=\; \delta(p, f) \;=\; \lim_{Q \downarrow p} \fint_Q f.$$

Remark A more advanced way of understanding this theorem is to consider a new measure $\mu(E) = \int_E f$. The assertion is that the new measure has a Vitali property.

52 Lemma *If $g : \mathbb{R}^n \to [0, \infty)$ is integrable then for every small $\alpha > 0$ the set $X(\alpha, g) = \{p : \overline{\delta}(p, g) > \alpha\}$ has outer measure*

$$m^*(X(\alpha, g)) \;\le\; \frac{1}{\alpha} \int g.$$

Proof The set $X(\alpha, g)$ is covered by arbitrarily small cubes on which the average value of g exceeds α. By Vitali's Covering Lemma we have disjoint cubes Q_i with

$$\bigsqcup Q_i \supset X(\alpha, g)$$

up to a zero set, where the average of g on Q_i is $> \alpha$. Hence $\alpha \cdot m(Q_i) \leq \int_{Q_i} g$ and

$$\alpha \cdot m^*(X(\alpha, g)) \ \leq \ \sum \alpha \cdot m(Q_i) \ \leq \ \sum \int_{Q_i} g \ \leq \ \int g.$$

Dividing the first and last terms by α gives the assertion. $\square$

Proof of the Average Value Theorem Since the theorem is local, we can assume f is integrable on some cube X and zero off X. Similarly, we can assume $f \geq 0$.

Fix $\alpha > 0$ and partition $[0, \infty)$ by intervals $I_k = [k\alpha, (k+1)\alpha)$, $k = 0, 1, \ldots$. The sets $X_k = f^{\mathrm{pre}}(I_k)$ are measurable, $X = \bigsqcup_k X_k$, and since they are measurable, $X_k \overset{\circ}{=} \mathrm{dp}(X_k)$. Fix k, set $A = \bigcup_{j<k} X_j$, $B = X_k$, $C = \bigcup_{j>k} X_j$, and write

$$[f : Q] = \fint_Q f = \frac{1}{mQ} \left(\int_{A \cap Q} f + \int_{B \cap Q} f + \int_{C \cap Q} f \right).$$

We claim that for almost every $p \in B$, the interval $[k\alpha, (k+2)\alpha]$ contains $f(p)$, $\underline{\delta}(p, f)$, and $\overline{\delta}(p, f)$, which is nearly what the theorem asserts.

The first two terms behave very nicely since $[A : Q] \to 0$ and $[B : Q] \to 1$ as $Q \downarrow p \in B \cap \mathrm{dp}(B)$. In fact, the first term tends to zero since f is bounded (by $k\alpha$) on A, while the second has liminf and limsup between $k\alpha$ and $(k+1)\alpha$. Unboundedness of f on C creates the only problem.

Define f_n by chopping off the top of the graph of f,

$$f_n(x) = \begin{cases} f(x) & \text{if } f(x) \leq n \\ n & \text{if } f(x) \geq n. \end{cases}$$

Then $f = f_n + g_n$ where f_n is bounded (by n) and $f \geq g_1 \geq g_2 \geq \ldots$. Since the functions g_n tend pointwise to zero and are dominated by f, $\int g_n \to 0$ as $n \to \infty$.

Since α is fixed, Lemma 52 implies that the outer measure of

$$X(\alpha, g_n) = \{x : \overline{\delta}(x, g_n) > \alpha\}$$

tends to zero as $n \to \infty$. These sets nest downward as $n \to \infty$, so their intersection $X(\alpha) = \bigcap X(\alpha, g_n)$ is a zero set. Taking complements, we get a set of full measure $G(\alpha) = \bigcup G(\alpha, g_n) \subset X$ where $G(\alpha, g_n) = (X(\alpha, g_n))^c$ and $G(\alpha) = (X(\alpha))^c$.

We will verify our claim by showing that for each $p \in B \cap \mathrm{dp}(B) \cap G(\alpha)$, the third term has limsup $\leq \alpha$. Such a point p belongs to some $G(\alpha, g_n)$, so it does not belong to $X(\alpha, g_n)$. That is, $\overline{\delta}(p, g_n) \leq \alpha$. Fix this $n = n(p)$. The third term is

$$\frac{1}{mQ} \int_{C \cap Q} f_n + \frac{1}{mQ} \int_{C \cap Q} g_n = \frac{m(C \cap Q)}{mQ} \frac{1}{m(C \cap Q)} \int_{C \cap Q} f_n + \frac{1}{mQ} \int_{C \cap Q} g_n$$

$$\leq [C : Q] \cdot n + \frac{1}{mQ} \int_Q g_n$$

As $Q \downarrow p$, $[C : Q] \to 0$ and the last term is $[g_n : Q]$ which has limsup $\overline{\delta}(p, g_n) \leq \alpha$.

All together this shows that for each $\alpha > 0$ and each k there is a set of full measure $G(\alpha, k) \subset X$ such that for all $p \in X_k \cap G(\alpha, k)$, the numbers $f(p)$, $\underline{\delta}(p, f)$, and $\overline{\delta}(p, f)$ are all contained in the interval $[k\alpha, (k+2)\alpha]$. Let G be the intersection of the sets $G(\alpha, k)$ for $k = 0, 1, 2, \ldots$ and $1/\alpha \in \mathbb{N}$. It has full measure and for all points $p \in G$ and all $\alpha > 0$, the distance between $f(p)$, $\underline{\delta}(p, f)$, and $\overline{\delta}(p, f)$ is $\leq 2\alpha$. Since α is arbitrary, this implies that for almost every p we have $f(p) = \underline{\delta}(p, f) = \overline{\delta}(p, f)$, which completes the proof. $\square$

53 Corollary *If $f : [a, b] \to \mathbb{R}$ is Lebesgue integrable and*

$$F(x) = \int_a^x f(t) \, dt$$

is its indefinite Lebesgue integral then for almost every $x \in [a, b]$ the derivative $F'(x)$ exists and equals $f(x)$.

Remark Here and below the domain of our function is $\mathbb{R}$ and we make essential use of its one-dimensionality.

Proof In dimension 1, a cube is a segment, so Theorem 51 gives

$$\frac{F(x+h) - F(x)}{h} = \fint_{[x, x+h]} f(t) \, dt \to f(x)$$

almost everywhere as $h \downarrow 0$. The same holds for $[x - h, x]$. $\square$

Corollary 53 does not characterize indefinite integrals. Mere knowledge that a continuous function G has a derivative almost everywhere and that its derivative is an integrable function f does *not* imply that G differs from the indefinite integral of f by a constant. The Devil's staircase function H is a counterexample. Its derivative exists almost everywhere, $H'(x)$ is almost everywhere equal to the integrable function $f(x) = 0$, and yet H does not differ from the indefinite integral of 0 by a constant. The missing ingredient is a subtler form of continuity.

Definition A function $G : [a, b] \to \mathbb{R}$ is **absolutely continuous** if for each $\epsilon > 0$ there exists $\delta > 0$ such that whenever $I_1, \ldots, I_n$ are disjoint intervals in $[a, b]$ we have

$$\sum_{i=1}^{n} b_i - a_i < \delta \quad \Rightarrow \quad \sum_{i=1}^{n} |G(b_i) - G(a_i)| < \epsilon.$$

54 Proposition *Every absolutely continuous function is uniformly continuous. If (I_i) is a sequence of disjoint intervals $(a_i, b_i) \subset [a, b]$ then the following are equivalent for a continuous function $G : [a, b] \to \mathbb{R}$.*

(a) $\forall \epsilon > 0 \; \exists \delta > 0$ *such that* $\displaystyle\sum_{i=1}^{n} b_i - a_i < \delta \;\Rightarrow\; \sum_{i=1}^{n} |G(b_i) - G(a_i)| < \epsilon.$

(b) $\forall \epsilon > 0 \; \exists \delta > 0$ *such that* $\displaystyle\sum_{i=1}^{\infty} b_i - a_i < \delta \;\Rightarrow\; \sum_{i=1}^{\infty} |G(b_i) - G(a_i)| < \epsilon.$

(c) $\forall \epsilon > 0 \; \exists \delta > 0$ *such that* $\displaystyle\sum_{i=1}^{n} m(I_i) < \delta \;\Rightarrow\; \sum_{i=1}^{n} m(G(I_i)) < \epsilon.$

(d) $\forall \epsilon > 0 \; \exists \delta > 0$ *such that* $\displaystyle\sum_{i=1}^{\infty} m(I_i) < \delta \;\Rightarrow\; \sum_{i=1}^{\infty} m(G(I_i)) < \epsilon.$

*Also, if G is absolutely continuous and Z is a zero set then GZ is a zero set. Finally, if G is absolutely continuous then it is **measure continuous** in the sense that given $\epsilon > 0$, there exists $\delta > 0$ such that $mE < \delta$ implies $m(GE) < \epsilon$.*

Proof Assume G is absolutely continuous. Apply the definition with just one interval (t, x). Then $|t - x| < \delta$ implies $|G(t) - G(x)| < \epsilon$, which is uniform continuity.

(a) $\Rightarrow$ (b). (a) is the definition of absolute continuity. In the definition take $\epsilon/2$ in place of ϵ. The resulting δ depends on ϵ but not on n. Thus $\sum_{i=1}^{\infty} b_i - a_i < \delta$ implies $\sum_{i=1}^{n} b_i - a_i < \delta$ implies $\sum_{i=1}^{n} |G(b_i) - G(a_i)| < \epsilon/2$ implies $\sum_{i=1}^{\infty} |G(b_i) - G(a_i)| \leq \epsilon/2 < \epsilon$, which is (b).

(b) $\Rightarrow$ (c). $m(G(I_i)) = |G(t_i) - G(s_i)|$, where $G(t_i)$ and $G(s_i)$ are the maximum and minimum of G on $[a_i, b_i]$. Let J_i be the interval between s_i and t_i. Then $J_i \subset I_i$ implies $m(J_i) \leq m(I_i)$ implies $\sum_{i=1}^{n} m(J_i) < \delta$ implies $\sum_{i=1}^{n} |G(t_i) - G(s_i)| < \epsilon$. Thus $\sum_{i=1}^{n} |G(t_i) - G(s_i)| = \sum_{i=1}^{n} m(G(J_i)) < \epsilon$, which is (c).

(c) $\Rightarrow$ (d). This is just like (a) $\Rightarrow$ (b).

(d) $\Rightarrow$ (a). Since $m(I_i) = b_i - a_i$ and $|G(b_i) - G(a_i)| \leq m(G(I_i))$ this is immediate from continuity of G.

Assume $Z \subset [a, b]$ is a zero set and G is absolutely continuous according to (d). For each $\epsilon > 0$ there exists $\delta > 0$ such that $\sum m(I_i) < \delta$ implies $\sum m(G(I_i)) < \epsilon$.

There is an open $U \subset [a, b]$ of measure $< \delta$ that contains Z. Every U is a countable disjoint union of open intervals I_i. Their total length is $mU < \delta$. Thus $GZ \subset \bigcup G(I_i)$ and by (d) we have $m(GZ) \leq \sum m(G(I_i)) < \epsilon$ so $m(GZ) = 0$.

Assume $E \subset [a, b]$ is measurable and G is absolutely continuous according to (d) with ϵ, δ as above. Regularity of Lebesgue measure implies there are compact subsets $K_n \subset E$ such that $K_n \uparrow F \subset E$, where $Z = E \setminus F$ is a zero set. (F is an F_σ-set.) Continuity implies $G(K_n)$ is compact. Since $G(K_n) \uparrow GF$, GF is measurable. Since GZ is a zero set, $GE = GF \cup GZ$ is measurable. If $mE < \delta$ then there is an open $U = \bigcup I_i \supset E$ with $mU = \sum m(I_i) < \delta$. Then $GE \subset \bigcup G(I_i)$ and by (d) we have $m(GE) \leq \sum m(G(I_i)) < \epsilon$ as desired. $\qquad\square$

Remark Here are some additional facts about absolute continuity. None is very hard to check.

(i) Disjointness of the intervals $\{I_i\}$ was assumed throughout the proof of Proposition 54. It is necessary since $\sqrt{x}$ is absolutely continuous, but repeating an interval $(0, t)$ n times produces a sum nt that can be small, despite $n\sqrt{t}$ being large. Thus, an absolutely continuous function G can have $\sum |G(b_i) - G(a_i)|$ large for *non-disjoint* intervals I_i whose total length is small.

(ii) Like $\sqrt{x}$, any continuous monotone function that is smooth except at one point is absolutely continuous.

(iii) The function $\sqrt{x} \sin(1/x)$ is continuous, measure continuous, not absolutely continuous, but sends zero sets to zero sets.

(iv) Lipschitz implies absolute continuity, but not vice versa. The example is $\sqrt{x}$ again.

55 Theorem *Let $f : [a, b] \to \mathbb{R}$ be Lebesgue integrable and let F be its indefinite integral $F(x) = \displaystyle\int_a^x f(t)\, dt$.*

(a) *For almost every x the derivative $F'(x)$ exists and equals $f(x)$.*

(b) *F is absolutely continuous.*

(c) *If G is an absolutely continuous function and $G'(x) = f(x)$ for almost every x then G differs from F by a constant.*

As we show in the next section (Corollary 62), the tacit assumption in (c) that $G'(x)$ exists is redundant. Theorem 55 then gives the following characterization of indefinite integrals. It is also called **Lebesgue's Main Theorem**.

56 Lebesgue's Antiderivative Theorem *Every indefinite integral is absolutely continuous and conversely, every absolutely continuous function has a derivative almost everywhere and up to a constant it is the indefinite integral of its derivative.*

Proof of Theorem 55 (a) This is Corollary 53.

(b) Without much loss of generality we assume $f \geq 0$. We first suppose that f is bounded, say $0 \leq f(x) \leq M$ for all x. For each $\epsilon > 0$ the choice of $\delta = \epsilon/M$ gives

$$\sum m(F(I_i)) \leq \sum M m(I_i) < \epsilon$$

whenever I_i are disjoint subintervals of $[a, b]$ having total length $< \delta$. Proposition 54 implies that F is absolutely continuous.

Now assume f is unbounded and $\epsilon > 0$ is given. Choose M so large that

$$m(\{(x, y) \in \mathcal{U}f : fx \geq M\}) < \epsilon/2.$$

Define the functions

$$g(x) = \begin{cases} fx & \text{if } fx \geq M \\ 0 & \text{otherwise} \end{cases}$$

and $f_M = f - g$. The integral of g is $< \epsilon/2$ since it is the measure of $\mathcal{U}f$ outside the rectangle $[a, b] \times [0, M]$. Let F_M and G be the indefinite integrals of f_M and g. Clearly $f = f_M + g$ implies $F = F_M + G$. See Figure 153.

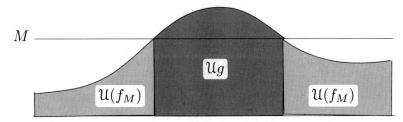

Figure 153 $\int g = m(\mathcal{U}g)$ and $\int f = \int g + \int f_M = m(\mathcal{U}g) + m(\mathcal{U}(f_M))$.

Since f_M is bounded there exists $\delta > 0$ such that

$$\sum m(I_i) < \delta \quad \Rightarrow \quad \sum m(F_M(I_i)) < \epsilon/2$$

where the I_i are disjoint intervals in $[a, b]$. Then $\sum m(I_i) < \delta$ implies

$$\sum m(F(I_i)) = \sum \int_{I_i} (f_M + g) = \sum \int_{I_i} f_M + \sum \int_{I_i} g$$
$$= \sum m(F_M(I_i)) + \sum m(G(I_i))$$
$$< \epsilon/2 + \int_a^b g < \epsilon,$$

which completes the proof that F is absolutely continuous.

(c) The Lebesgue proof resembles the Riemann proof in Chapter 3 – the Vitali Covering Lemma replaces the Lebesgue Number Lemma. We assume G is absolutely continuous and $G'(x) = f(x)$ almost everywhere. When F is the indefinite integral of f we want to show that $H = F - G$ is constant.

It is easy to see that sums and differences of absolutely continuous functions are absolutely continuous, so H is absolutely continuous and $H'(x) = 0$ almost everywhere. Fix any $x^* \in [a, b]$ and define

$$X = \{x \in [a, x^*] : H'(x) \text{ exists and } H'(x) = 0\}.$$

By assumption $mX = x^* - a$.

It is enough to show that for each $\epsilon > 0$ we have

$$|H(x^*) - H(a)| < \epsilon.$$

Absolute continuity implies there is a $\delta > 0$ such that if $I_i = [a_i, b_i] \subset [a, b]$ are disjoint intervals then

$$\sum_i b_i - a_i < \delta \quad \Rightarrow \quad \sum_i |H(b_i) - H(a_i)| < \epsilon/2.$$

Fix such a δ. Each $x \in X$ is contained in arbitrarily small intervals $[x, x+h] \subset [a, x^*]$ such that

$$\left| \frac{H(x+h) - H(x)}{h} \right| < \frac{\epsilon}{2(b-a)}.$$

These intervals form a Vitali covering $\mathcal{V}$ of X and the Vitali Covering Lemma implies that countably many of them, say $V_j = [x_j, x_j + h_j]$, disjointly cover X up to a zero set. Thus their total length is $\sum h_j = x^* - a$ and it follows that there is an N such that

$$\sum_{j=1}^{N} h_j > x^* - a - \delta.$$

Since $|H(x+h) - H(x)| < h\epsilon/2(b-a)$ on each $\mathcal{V}$-interval we have

$$\sum_{j=1}^{N} |H(x_j + h_j) - H(x_j)| < \frac{\epsilon}{2(b-a)} \sum_{j=1}^{N} h_j \leq \frac{\epsilon(x^* - a)}{2(b-a)} \leq \epsilon/2.$$

The $N+1$ intervals $I_j = [a_j, b_j]$ complementary to the (interiors of the) intervals $V_1, \ldots, V_N$ have total length $< \delta$ so $\sum_{j=0}^{N} |H(b_j) - H(a_j)| < \epsilon/2$ by absolute continuity. Thus

$$
\begin{aligned}
H(x^*) - H(a) &= \sum_{j=1}^{N} H(x_j + h_j) - H(x_j) + \sum_{j=0}^{N} H(b_j) - H(a_j) \\
&\leq \sum_{j=1}^{N} |H(x_j + h_j) - H(x_j)| + \sum_{j=0}^{N} |H(b_j) - H(a_j)| \\
&< \epsilon
\end{aligned}
$$

which completes the proof that G differs from F by a constant. $\qquad\square$

See Figure 154.

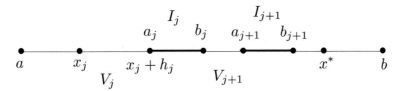

Figure 154 The complementary intervals V_j and I_j

10 Lebesgue's Last Theorem

The final theorem in Lebesgue's groundbreaking book, *Leçons sur l'intégration*, is extremely concise and quite surprising.

57 Theorem *A monotone function has a derivative almost everywhere.*

Note that no hypothesis is made about continuity of the monotone function. Considering the fact that a monotone function $[a, b] \to \mathbb{R}$ has only a countable number of discontinuities, all of jump type, this may seem reasonable, but remember – the discontinuities may be dense in $[a, b]$. If the monotone function happens to be an indefinite integral then differentiability was proved in Theorem 55.

We assume henceforth that f is nondecreasing since the nonincreasing case can be handled by looking at $-f$.

Lebesgue's proof of Theorem 57 used the full power of the machinery he had developed for his new integration theory. In contrast, the proof given below is more direct and geometric. It relies on the Vitali Covering Lemma and the following form of Chebyshev's inequality from probability theory.

The **slope** of f over $[a, b]$ is

$$s = \frac{f(b) - f(a)}{b - a}.$$

58 Chebyshev Lemma *Assume that $f : [a, b] \to \mathbb{R}$ is nondecreasing and has slope s over $I = [a, b]$. If I contains countably many disjoint subintervals I_k and the slope of f over I_k is $\geq S > s$ then*

$$\sum_k |I_k| \leq \frac{s}{S} |I|.$$

Proof Write $I_k = [a_k, b_k]$. Since f is nondecreasing we have

$$f(b) - f(a) \geq \sum_k f(b_k) - f(a_k) \geq \sum_k S(b_k - a_k).$$

Thus $s |I| \geq S \sum |I_k|$ and the lemma follows. $\square$

Remark An extreme case of this situation occurs when the slope is concentrated in the three subintervals drawn in Figure 155.

Proof of Lebesgue's Last Theorem Not only will we show that $f'(x)$ exists almost everywhere, but we will also show that $f'(x)$ is a measurable function of x and

(8) $$\int_a^b f'(x) \, dx \leq f(b) - f(a).$$

To estimate differentiability one introduces upper and lower limits of slopes called **derivates**. If $h > 0$ then $[x, x + h]$ is a "right interval" at x and $(f(x + h) - f(x))/h$ is a "right slope" at x. The lim sup of the right slopes as $h \to 0$ is called the **right maximum derivate** of f at x. It is denoted as $D^{\text{right max}} f(x)$. The lim inf of the right slopes is the **right minimum derivate** of f at x and is denoted as $D^{\text{right min}} f(x)$. Similar definitions apply to the left of x. Think of $D^{\text{right max}} f(x)$ as the steepest slope at the right of x and $D^{\text{right min}} f(x)$ as the gentlest. See Figure 156.

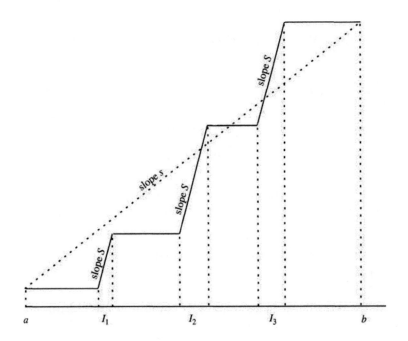

Figure 155 Chebyshev's Inequality for slopes

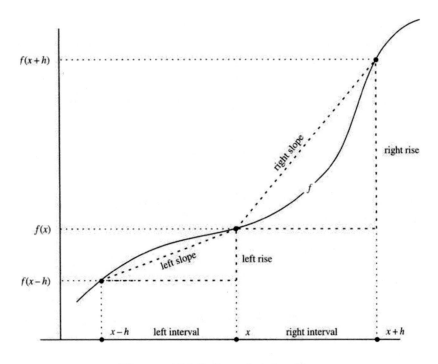

Figure 156 Left and right slopes

There are four derivates. They exist at all points of $[a, b]$ but they can take the value ∞. We first show that two are equal almost everywhere, say the left min and the right max. Fix any $s < S$ and consider the set

$$E = E_{sS} = \{x \in [a, b] : D^{\text{left min}} f(x) < s < S < D^{\text{right max}} f(x)\}.$$

We claim that

(9) $$m^* E = 0.$$

At each $x \in E$ there are arbitrarily small left intervals $[x - h, x]$ over which the slope is $< s$. These left intervals form a Vitali covering $\mathcal{L}$ of E. (Note that the point x is not the center of its $\mathcal{L}$-interval, but rather it is an endpoint. Also, we do not know a priori that E is measurable. Luckily, Vitali permits this.) Let $\epsilon > 0$ be given. By the Vitali Covering Lemma there are countably many disjoint left intervals $L_i \in \mathcal{L}$ that cover E, modulo a zero set, and they do so ϵ-efficiently. That is, if we write

$$L = \bigsqcup_i \text{int } L_i$$

then $E \setminus L$ is a zero set and $mL \leq m^* E + \epsilon$.

Every $y \in L \cap E$ has arbitrarily small right intervals $[y, y + t] \subset L$ over which the slope is $> S$. (Here it is useful that L is open.) These right intervals form a Vitali covering $\mathcal{R}$ of $L \cap E$, and by the Vitali Covering Lemma we can find a countable number of disjoint intervals $R_j \in \mathcal{R}$ that cover $L \cap E$ modulo a zero set. Since $L \cap E = E$ modulo a zero set, $R = \bigsqcup R_j$ also covers E modulo a zero set. By the Chebyshev Lemma we have

$$m^* E \leq mR = \sum_i \sum_{R_j \subset L_i} |R_j| \leq \sum_i \frac{s}{S} |L_i| \leq \frac{s}{S}(m^* E + \epsilon).$$

Since the inequality holds for all $\epsilon > 0$, it holds also with $\epsilon = 0$ which implies that $m^* E = 0$ and completes the proof of (9). Then

$$\{x : D^{\text{left min}} f(x) < D^{\text{right max}} f(x)\} = \bigcup_{\{(s, S) \in \mathbb{Q} \times \mathbb{Q} \, : \, s < S\}} E_{sS}$$

is a zero set. Symmetrically, $\{x : D^{\text{left min}} f(x) > D^{\text{right max}} f(x)\}$ is a zero set, and therefore $D^{\text{left min}} f(x) = D^{\text{right max}} f(x)$ almost everywhere. Mutual equality of the other derivates, almost everywhere, is checked in the same way. See Exercise 64.

So far we have shown that for almost every $x \in [a, b]$ the derivative of f at x exists although it may equal ∞. Infinite slope is not really acceptable and that is the purpose of (8) – for an integrable function takes on a finite value at almost every point.

The proof of (8) uses a cute trick reminiscent of the traveling secant method from Chapter 3. First extend f from $[a, b]$ to $\mathbb{R}$ by setting $f(x) = f(a)$ for $x < a$ and $f(x) = f(b)$ for $x > b$. Then define $g_n(x)$ to be the slope of the secant from $(x, f(x))$ to $(x + 1/n, f(x + 1/n))$. That is,

$$g_n(x) = \frac{f(x + 1/n) - f(x)}{1/n} = n(f(x + 1/n) - f(x)).$$

See Figure 157. Since f is almost everywhere continuous it is measurable and so is

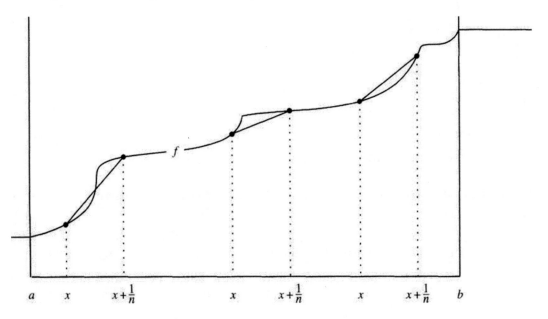

$$a \qquad x \qquad x + \frac{1}{n} \qquad\qquad x \qquad x + \frac{1}{n} \qquad\qquad x \qquad x + \frac{1}{n} \qquad b$$

Figure 157 $g_n(x)$ is the slope of the right secant at x.

g_n. For almost every x, $g_n(x)$ converges to $f'(x)$ as $n \to \infty$. Hence f' is measurable and clearly $f' \geq 0$. Fatou's Lemma gives

$$\int_a^b f'(x)\, dx = \int_a^b \liminf_{n \to \infty} g_n(x)\, dx \leq \liminf_{n \to \infty} \int_a^b g_n(x)\, dx.$$

The integral of g_n is

$$\int_a^b g_n(x)\, dx = n \int_b^{b+1/n} f(x)\, dx - n \int_a^{a+1/n} f(x)\, dx.$$

The first integral equals $f(b)$ since we set $f(x) = f(b)$ for $x > b$. The second integral is at least $f(a)$ since f is nondecreasing. Thus

$$\int_a^b g_n(x)\, dx \;\leq\; f(b) - f(a),$$

which completes the proof of (8). As remarked before, since the integral of f' is finite, $f'(x) < \infty$ for almost all x, and hence f is differentiable (with finite derivative) almost everywhere.   □

59 Corollary *A Lipschitz function is almost everywhere differentiable.*

Proof Suppose that $f : [a, b] \to \mathbb{R}$ is Lipschitz with Lipschitz constant L. Then for all $x, y \in [a, b]$ we have

$$|f(y) - f(x)| \;\leq\; L\,|y - x|.$$

The function $g(x) = f(x) + Lx$ is nondecreasing. Thus g' exists almost everywhere and so does $f' = g' - L$.   □

Remark Corollary 59 remains true for a Lipschitz function $f : \mathbb{R}^n \to \mathbb{R}$, it is **Rademacher's Theorem**, and the proof is much harder.

Definition The **variation** of a function $f : [a, b] \to \mathbb{R}$ over a partition $X : a = x_0 < \cdots < x_n = b$ is the sum $\sum_{k=1}^n |\Delta_k f|$, where $\Delta_k f = f(x_k) - f(x_{k-1})$. The supremum of the variations over all partitions X is the **total variation** of f. If the total variation of f is finite then f is said to be a function of **bounded variation**.

60 Theorem *A function of bounded variation is almost everywhere differentiable.*

Proof Up to an additive constant, a function of bounded variation can be written as the difference $f(x) = P(x) - N(x)$, where

$$P(x) \;=\; \sup\{\sum_k \Delta_k f : a = x_0 < \ldots < x_n = x \text{ and } \Delta_k f \geq 0\}$$

$$N(x) \;=\; -\inf\{\sum_k \Delta_k f : a = x_0 < \ldots < x_n = x \text{ and } \Delta_k f < 0\}.$$

See Exercise 67. The functions P and N are monotone nondecreasing, so for almost every x we have $f'(x) = P'(x) - N'(x)$ exists and is finite.   □

61 Theorem *An absolutely continuous function is of bounded variation.*

Proof Assume that $F : [a, b] \to \mathbb{R}$ is absolutely continuous and take $\epsilon = 1$. There is a $\delta > 0$ such that if (a_i, b_i) are disjoint intervals in $[a, b]$ with total length $< \delta$ then

$$\sum_i b_i - a_i < \delta \quad \Rightarrow \quad \sum_i |F(b_i) - F(a_i)| < 1.$$

Fix a partition X of $[a, b]$ with M subintervals of length $< \delta$. For any partition $Y : a = y_0 < \ldots < y_n = b$ of $[a, b]$ we claim that $\sum_k |\Delta_k f| \leq M$, where $\Delta_k f = f(y_k) - f(y_{k-1})$. We may assume that Y contains X since adding points to a partition increases the sum $\sum |\Delta_k f|$. Then

$$\sum_Y |\Delta_k F| = \sum_{Y_1} |\Delta_k F| + \ldots + \sum_{Y_M} |\Delta_k F|$$

where Y_j refers to the subintervals of Y that lie in the jth subinterval of X. The subintervals in Y_j have total length $< \delta$, so the variation of F over them is < 1 and the total variation of F is $< M$. □

62 Corollary *An absolutely continuous function is almost everywhere differentiable.*

Proof Absolute continuity implies bounded variation implies almost everywhere differentiability. □

As mentioned in Section 9, Theorem 55 plus Corollary 62 express **Lebesgue's Main Theorem,**

Indefinite integrals are absolutely continuous and

every absolutely continuous function has a derivative

almost everywhere of which it is the indefinite integral.

Appendix A   Lebesgue integrals as limits

The Riemann integral is the limit of Riemann sums. There are analogous "Lebesgue sums" of which the Lebesgue integral is the limit.

Let $f : \mathbb{R} \to [0, \infty)$ be given, take a partition $Y : 0 = y_0 < y_1 < y_2 < \ldots$ on the y-axis, and set

$$X_i = \{x \in \mathbb{R} : y_{i-1} \leq f(x) < y_i\}.$$

(We require that $y_i \to \infty$ as $i \to \infty$.) If f is measurable we define the **lower Lebesgue sum** as

$$L(f, Y) = \sum_{i=1}^{\infty} y_{i-1} \cdot mX_i.$$

L represents the measure of "Lebesgue rectangles" $X_i \times [0, y_{i-1})$ in the undergraph. If f is measurable[†] then $L \uparrow \int f$ as the Y-mesh tends to 0. It is natural to define the upper Lebesgue sum as $\sum y_i \cdot m(X_i)$ and to expect that it converges down to $\int f$ as the Y-mesh tends to 0. If $m(\{x : fx > 0\}) < \infty$ then this is true. However, if $f(x)$ is a function like e^{-x^2} then there's a problem. The first term in the upper Lebesgue sum is always ∞ even though the integral is finite. The simplest solution is to split the domain into cubes Q, work on each separately, and add the results. Then

$$L(f_Q, Y) \leq \int_Q f_Q \leq U(f_Q, Y),$$

where $L(f_Q, Y) = \sum_{i=1}^{\infty} y_{i-1} \cdot m(X_i \cap Q)$, $U(f_Q, Y) = \sum_{i=1}^{\infty} y_i \cdot m(X_i \cap Q)$, and f_Q is the restriction of f to Q. As the Y-mesh tends to 0 the lower and upper Lebesgue sums converge to the integral, just as in the Riemann case.

Upshot Lebesgue sums are like Riemann sums and Lebesgue integration is like Riemann integration, except that Lebesgue partitions the value axis and takes limits while Riemann does the same on the domain axis.

Appendix B   Nonmeasurable sets

If $t \in \mathbb{R}$ is fixed then t-translation is the mapping $x \mapsto x + t$. It is a homeomorphism $\mathbb{R} \to \mathbb{R}$. Think of the circle S^1 as $\mathbb{R}$ modulo $\mathbb{Z}$. That is, you identify any x with $x + n$ for $n \in \mathbb{Z}$. Equivalently, you take the unit interval $[0, 1]$ and you identify 1

[†]We are using the undergraph definition of measurability. Corollary 41 implies that the sets X_i are measurable so the lower Lebesgue sum makes sense.

with 0. Then t-translation becomes rotation by the angle $2\pi t$, and is denoted as $R_t : S^1 \to S^1$. If t is rational then this rotation is periodic, i.e., for some $n \geq 1$, the nth iterate of R, $R^n = R \circ \cdots \circ R$, is the identity map $S^1 \to S^1$. In fact the smallest such n is the denominator when $t = m/n$ is expressed in lowest terms. On the other hand, if t is irrational then $R = R_t$ is nonperiodic; every orbit $\mathcal{O}(x) = \{R^k(x) : k \in \mathbb{Z}\}$ is denumerable and dense in S^1.

63 Theorem *Let t be irrational and set $R = R_t$. If $P \subset S^1$ contains exactly one point of each R-orbit then P is nonmeasurable with respect to linear Lebesgue measure on S^1.*

Proof The R-orbits are disjoint sets, there are uncountably many of them, and they divide the circle as $S^1 = \bigsqcup_{n \in \mathbb{Z}} R^n(P)$. Translation is a mesisometry. It preserves outer measure, measurability, and measure. So does rotation. Can P be measurable? No, because if it is measurable with positive measure then we would get

$$m(S^1) \; = \; \sum_{n=-\infty}^{\infty} m(R^n P) \; = \; \infty,$$

a contradiction, while if $mP = 0$ then $m(S^1) = \sum_{-\infty}^{\infty} m(R^n P) = 0$, which contradicts the fact that $m[0, 1) = 1$. $\qquad\qquad\square$

But does P exist? The Axiom of Choice states that given any family of nonempty disjoint sets there exists a set that contains exactly one element from each set. So if you accept the Axiom of Choice then you apply it to the family of R-orbits and you get an example of a nonmeasurable set P, while if you don't accept the Axiom of Choice then you're out of luck.

To increase the pathology of P we next discuss translations in more depth.

64 Steinhaus' Theorem *If $E \subset \mathbb{R}$ is measurable and has positive measure then there exists a $\delta > 0$ such that for all $t \in (-\delta, \delta)$, the t-translate of E meets E.*

See also Exercise 57.

65 Lemma *If $F \subset (a, b)$ is measurable and disjoint from its t-translate then*

$$2mF \; \leq \; (b - a) + |t|.$$

Proof F and its t-translate have equal measure, so if they do not intersect then their total measure is $2mF$, and any interval that contains them must have length $\geq 2mF$. If $t > 0$ then $(a, b + t)$ contains F and its t-translate, while if $t < 0$ then $(a + t, b)$ contains them. The length of the interval in either case is $(b - a) + |t|$. $\qquad\square$

Proof of Steinhaus' Theorem By the Lebesgue Density Theorem (Theorem 47) E has lots of density points so we can find an interval (a, b) in which E has concentration $> 1/2$. Call $F = E \cap (a, b)$. Then $mF > (b - a)/2$. By Lemma 65 if $|t| < 2mF - (b - a)$ then the t-translate of F meets F, so the t-translate of E meets E, which is what the theorem asserts. $\qquad\qquad\square$

Now we return to the nonmeasurable set P discussed in Theorem 63. It contains exactly one point from each R-orbit, R being rotation by an irrational t. Set

$$A = \bigcup_{k \in \mathbb{Z}} R^{2k} P \qquad B = \bigcup_{k \in \mathbb{Z}} R^{2k+1} P.$$

The sets A, B are disjoint, their union is the circle, and R interchanges them. Since R preserves outer measure we have $m^*A = m^*B$.

The composite $R^2 = R \circ R$ is rotation by $2t$, also an irrational number. Let $\epsilon > 0$ be given. Since the orbit of 0 under R^2 is dense there is a large integer k with

$$|R^{2k}(0) - (-t)| < \epsilon.$$

For R^{2k} is the kth iterate of R^2. Thus $|R^{2k+1}(0)| < \epsilon$ so R^{2k+1} is a rotation by $< \epsilon$. Odd powers of R interchange A and B, so odd powers of R translate A and B off themselves. It follows from Steinhaus' Theorem that A and B contain no subsets of positive measure. Their inner measures are zero.

The general formula $mC = m_*A + m^*B$ in Lemma 20 implies that $m^*B = 1$. Thus we get an extreme type of nonmeasurability expressed in the next theorem.

66 Theorem *The circle, or equivalently $[0, 1)$, splits into two nonmeasurable disjoint subsets that each has inner measure zero and outer measure one.*

67 Corollary *Every measurable set $E \subset \mathbb{R}^n$ of positive measure contains a **doppelgänger** – a nonmeasurable subset N such that $m^*N = mE$, $m_*N = 0$, and N "spreads itself evenly" throughout E in the sense that if $E' \subset E$ is measurable then $m^*(N \cap E') = m(E')$.*

The proof is left to you as Exercise 50.

Appendix C   Borel versus Lebesgue

A valid criticism of Lebesgue theory as described in this chapter is that it conflicts a bit with topology, and problems arise if you try to think of Lebesgue measure theory in category terms. For example, not all homeomorphisms are meseomorphisms and composition of Lebesgue measurable functions can fail to be Lebesgue measurable. See Exercise 79.

To repair these defects Armand Borel proposed replacing the σ-algebra $\mathcal{M}$ of Lebesgue measurable sets with a smaller one, $\mathcal{B} \subset \mathcal{M}$, and restricting Lebesgue measure to it. $\mathcal{B}$ is simply the intersection of all σ-algebras that include the open sets. There is one such σ-algebra, namely $\mathcal{M}$, so $\mathcal{B}$ exists and is contained in $\mathcal{M}$. It includes all G_δ-sets (countable intersections of open sets), all $G_{\delta\sigma}$-sets (countable unions of G_δ-sets), etc. Thus $\mathcal{G}_\delta \subset \mathcal{G}_{\delta\sigma} \subset \mathcal{G}_{\delta\sigma\delta} \subset \cdots \subset \mathcal{B}$, where $\mathcal{G}_\delta$ is the collection of all G_δ-sets, $\mathcal{G}_{\delta\sigma}$ is the collection of all $G_{\delta\sigma}$-sets, etc. Likewise $\mathcal{F}_\sigma \subset \mathcal{F}_{\sigma\delta} \subset \cdots \subset \mathcal{B}$ for F_σ-sets, $F_{\sigma\delta}$-sets, etc. See Exercise 8.

A set is **Borel measurable** if it belongs to $\mathcal{B}$, and a nonnegative function is Borel measurable if its undergraph is a Borel measurable set. Equivalently a function is Borel measurable if the preimage of a Borel set is always Borel. The measure of $E \in \mathcal{B}$ is its Lebesgue measure and the integral of a Borel measurable function is its Lebesgue integral. All continuous functions are Borel measurable and the composition of Borel measurable functions is Borel measurable. That's good.

However, $\mathcal{B}$ has its own defects, the main one being that it is not complete. That is, not all subsets of a zero set are Borel measurable. (Recall that *every* subset of a zero set is Lebesgue measurable.) In the same vein, the limit of a sequence of Borel measurable functions that converge almost everywhere can fail to be Borel measurable. See Exercise 80.

I chose not to use the Borel approach in this chapter because it adds an extra layer of complication to the basic Lebesgue theory. You could not state the Monotone Convergence Theorem as "if f_n is (Borel) measurable and $f_n \uparrow f$ then $\int f_n \uparrow \int f$." No. You would also need to *assume* f is Borel measurable.

But the real reason I chose $\mathcal{M}$ over $\mathcal{B}$ is that I *like* pathology. The fact that there are ugly zero sets – zero sets carried by homeomorphisms to nonmeasurable sets – is eye-opening. I want you to see them as part of the Lebesgue picture.

Here are a couple of relevant remarks from mathoverflow in answer to the question "Why do probabilists take random variables to be Borel (and not Lebesgue) measurable?"

Yuval Peres: One reason is that probabilists often consider more than one measure on the same space, and then a negligible set for one measure (added in a completion) might be not negligible for the other. The situation becomes more acute when you consider uncountably many different measures (such as the distributions of a Markov process with different starting points.)

Terry Tao: This is also a reason why the Borel sigma algebra on the domain is often preferred in ergodic theory. (A closely related reason is because of the connection between ergodic theory and topological dynamics; a topological dynamical system has a canonical Borel sigma algebra but not a canonical Lebesgue sigma algebra.) On the other hand, a significant portion of ergodic theory is also concerned with almost everywhere convergence (wrt some reference invariant measure, of course), and then it becomes useful for the domain sigma algebra to be complete...

Appendix D   The Banach-Tarski Paradox

If the nonmeasurable examples in Appendix B do not disturb you enough, here is a much worse one. You can read about it in Stan Wagon's book, *The Banach-Tarski Paradox*. Many other paradoxes are discussed there too.

The solid unit ball in 3-space can be divided into five disjoint sets, $A_1, \ldots, A_5$, and the A_i can be moved by rigid motions to new disjoint sets A_i' whose union is two disjoint unit balls. The Axiom of Choice is fundamental in the construction, as is dimensionality greater than two. The sets A_i are nonmeasurable.

Think of this from an alchemist's point of view. A one inch gold ball can be cut into five disjoint pieces and the pieces rigidly re-assembled to make two one inch gold balls. Repeating the process would make you very rich.

Appendix E   Riemann integrals as undergraphs

The geometric description of the Lebesgue integral as the measure of the undergraph has a counterpart for Riemann integrals.

68 Theorem *A function $f : [a, b] \to [0, M]$ is Riemann integrable if and only if the topological boundary of its undergraph is a zero set, $m(\partial(\mathcal{U}f)) = 0$.*

Remark Recall from page 401 that the measure-theoretic boundary of a set A is

$$\partial_m(A) \;=\; H \setminus K$$

where H and K are the hull and kernel of A. Measurability of E is equivalent to $\partial_m(E)$ being a zero set. A function $f : [a, b] \to [0, M]$ is Lebesgue integrable if and only if it is measurable, i.e, if and only if $\mathcal{U}f$ is a measurable set, which is true if and only if $\partial_m(\mathcal{U}f)$ is a zero set. Combined with Theorem 68 this gives a nice geometric parallel between Riemann and Lebesgue integrability of bounded functions:

$$f \text{ is Riemann integrable } \iff m(\partial(\mathcal{U}f)) = 0.$$
$$f \text{ is Lebesgue integrable } \iff m(\partial_m(\mathcal{U}f)) = 0.$$

Remark Since $\partial(\mathcal{U}f) = \overline{\mathcal{U}f} \setminus \mathrm{int}(\mathcal{U}f)$, equivalent to $m(\partial(\mathcal{U}f)) = 0$ is $m(\mathrm{int}(\mathcal{U}f)) = m(\overline{\mathcal{U}f})$.

69 Lemma *If X is a metric space, $f : X \to [0, \infty)$, and*

$$\underline{f}(x) \;=\; \liminf_{t \to x} f(t) \qquad \overline{f}(x) \;=\; \limsup_{t \to x} f(t)$$

then $\mathcal{U}\underline{f} = \mathrm{int}(\mathcal{U}f)$ and $\widehat{\mathcal{U}\overline{f}} = \overline{\mathcal{U}f}$.

Proof Take any $(x, y) \in \mathcal{U}\underline{f}$. Then $y < \underline{f}(x)$ and for all (t, s) near (x, y) we have $s < f(t)$. Thus $(t, s) \in \mathcal{U}f$, $(x, y) \in \mathrm{int}(\mathcal{U}f)$, and $\mathcal{U}\underline{f} \subset \mathrm{int}(\mathcal{U}f)$. The proof of the reverse inclusion is similar, so $\mathcal{U}\underline{f} = \mathrm{int}(\mathcal{U}f)$. See Figure 158.

The proof that $\widehat{\mathcal{U}\overline{f}} = \overline{\mathcal{U}f}$ is slightly different. If $(x, y) \in \widehat{\mathcal{U}\overline{f}}$ then $y \leq \overline{f}(x)$ so there exists $t_n \to x$ such that $f(t_n) \to \overline{f}(x)$. Choose $y_n < f(t_n)$ such that $y_n \to y$. Thus $(t_n, y_n) \in \mathcal{U}f$, $(t_n, y_n) \to (x, y)$, $(x, y) \in \overline{\mathcal{U}f}$, and $\widehat{\mathcal{U}\overline{f}} \subset \overline{\mathcal{U}f}$. Conversely, if $(x, y) \in \overline{\mathcal{U}f}$ then there exists $(t_n, y_n) \in \mathcal{U}f$ such that $(t_n, y_n) \to (x, y)$. Then $y_n < f(t_n)$ and $\limsup_{n \to \infty} f(t_n) \geq \lim_{n \to \infty} y_n = y$. Thus, $y \leq \overline{f}(x)$, $(x, y) \in \widehat{\mathcal{U}\overline{f}}$, and $\overline{\mathcal{U}f} \subset \widehat{\mathcal{U}\overline{f}}$, giving equality, $\widehat{\mathcal{U}\overline{f}} = \overline{\mathcal{U}f}$. See Figure 158. $\qquad \square$

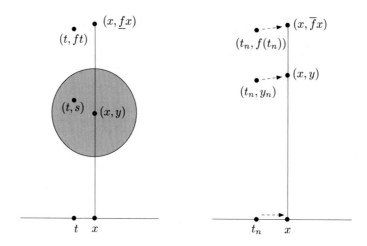

Figure 158 The shaded region is contained in the interior of $\mathcal{U}f$.

Proof of Theorem 68 Applying Lemma 69 to $f : [a, b] \to [0, M]$ gives

$$\mathcal{U}\underline{f} = \mathrm{int}(\mathcal{U}f) \quad \text{and} \quad \widehat{\mathcal{U}\overline{f}} = \overline{\mathcal{U}f}.$$

Since open sets and closed sets are measurable, this implies $\underline{f}$ and $\overline{f}$ are measurable functions. Thus

$$m(\partial(\mathcal{U}f)) \;=\; m(\overline{\mathcal{U}f} \setminus \mathrm{int}(\mathcal{U}f)) \;=\; m(\widehat{\mathcal{U}\overline{f}}) - m(\mathcal{U}\underline{f}) \;=\; \int_{[a,b]} \overline{f} - \underline{f}.$$

The integral is zero if and only if $\overline{f} = \underline{f}$ almost everywhere, i.e., if and only if f is continuous almost everywhere, i.e., by the Riemann-Lebesgue Theorem (Theorem 23 in Chapter 3) if and only if f is Riemann integrable. $\square$

70 Corollary *If f is Riemann integrable then it is Lebesgue integrable and the two integrals are equal.*

Proof Since

$$\text{interior } \mathcal{U}f \subset \mathcal{U}f \subset \text{closure } \mathcal{U}f,$$

equality of the measures of its interior and closure implies that $\mathcal{U}f$ is measurable, and it shares their common measure. Since the Lebesgue integral of f is equals $m(\mathcal{U}f)$ the proof is complete. $\square$

Remark The undergraph definition of integrals has a further expression in terms of Jordan content: The Riemann integral of a function $f : [a, b] \to [0, M]$ is the Jordan content of its undergraph, $J(\mathcal{U}f)$, provided that $J(\mathcal{U}f)$ exists. See Exercises 11 - 14. In brief, *Undergraphs lead to natural pictorial ways of dealing with integrals, both Riemann and Lebesgue.*

Appendix F   Littlewood's Three Principles

In the following excerpt from his book on complex analysis, *Lectures on the Theory of Functions*, J.E. Littlewood seeks to demystify Lebesgue theory. It owes some of its popularity to its prominence in Royden's classic text, *Real Analysis*.

> The extent of knowledge [of real analysis] required is nothing like as great as is sometimes supposed. There are three principles, roughly expressible in the following terms: Every (measurable) set is nearly a finite sum of intervals; every function (of class L^λ) is nearly continuous; and every convergent sequence of functions is nearly uniformly convergent. Most of the results of the present section are fairly intuitive applications of these ideas, and the student armed with them should be equal to most occasions when real variable theory is called for. If one of the principles would be the obvious means to settle a problem if it were "quite" true, it is natural to ask if the "nearly" is near enough, and for a problem that is actually soluble it generally is.[†]

Littlewood's First Principle expresses the regularity of Lebesgue measure (Theorem 11). Given $\epsilon > 0$, a measurable $E \subset [a, b]$ contains a compact subset covered by finitely many intervals whose union differs from E by a set of measure less than ϵ. In that sense, E is **nearly** a finite union of intervals. I like very much Littlewood's choice of the term "nearly," meaning "except for an ϵ-set," to contrast with "almost," meaning "except for a zero set."

Littlewood's Second Principle refers to "functions of class L^λ," although he might better have said "measurable functions." He means that if you have a measurable function and you are given $\epsilon > 0$ then you can discard an ϵ-set from its domain of definition and the result is a continuous function. This is **Lusin's Theorem**: *a measurable function is **nearly continuous***.

Proof of Lusin's Theorem We assume that $f : \mathbb{R}^n \to \mathbb{R}$ is measurable and $\epsilon > 0$ is given. Enumerate the set of open intervals with rational endpoints as $I_1, I_2, \ldots$. Then $E_i = f^{\mathrm{pre}}(I_i)$ is measurable, so there exists a sandwich $K_i \subset E_i \subset U_i$ where K_i is closed, U_i is open and $m(U_i \setminus K_i) < \epsilon/2^i$. Let $S_i = U_i \setminus K_i$ be this "collar" of E_i. The set $S = \bigcup S_i$ is open and has measure $< \epsilon$. Then $K = S^c$ is closed, and we claim that $f|_K$ is continuous. Let $x_k \to x$ in K. Given $\sigma > 0$, we must show there

---

[†]Reprinted from *Lectures on the Theory of Functions* by J.E. Littlewood (1994) by permission of Oxford University Press.

exists κ such that $k \geq \kappa$ implies $|f(x_k) - f(x)| < \sigma$. (To avoid abuse of notation, σ is playing the rôle of ϵ, which was already used in finding S.)

There is an I_i such that $fx \in I_i$ and the length of I_i is $< \sigma$. Thus $x \in E_i \subset U_i$. Since $x \in K$, $x \notin S_i$, i.e., $x \in K_i$. Since U_i is open and $x_k \to x$, there is a κ such that $k \geq \kappa$ implies $x_k \in U_i$. Again, since $x_k \in K$, $x_k \notin S_i$, which shows that $x_k \in K_i$ when $k \geq \kappa$. Hence $f(x_k)$ and $f(x)$ both lie in I_i, so $|f(x_k) - f(x)| < \sigma$. $\square$

Littlewood's Third Principle concerns a sequence of measurable functions $f_n : [a, b] \to \mathbb{R}$ that converges almost everywhere to a limit. Except for an ϵ-set the convergence is actually uniform, which is **Egoroff's Theorem**: *Almost everywhere convergence implies **nearly uniform convergence**.*

Proof of Egoroff's Theorem Set

$$X(k, \ell) = \{x \in [a, b] : \forall n \geq k \text{ we have } |f_n(x) - f(x)| < 1/\ell\}.$$

Fix $\ell \in \mathbb{N}$. Since $f_n(x) \to f(x)$ for almost every x we have $\bigcup_k X(k, \ell) \cup Z(\ell) = [a, b]$ where $Z(\ell)$ is a zero set.

Let $\epsilon > 0$ be given. By measure continuity $m(X(k, \ell)) \to b - a$ as $k \to \infty$. This implies we can choose $k_1 < k_2 < \ldots$ such that for $X_\ell = X(k_\ell, \ell)$ we have $m(X_\ell^c) < \epsilon/2^\ell$. Thus $m(X^c) < \epsilon$ where $X = \bigcap_\ell X_\ell$.

We claim that f_n converges uniformly on X. Given $\sigma > 0$ we choose and fix ℓ such that $1/\ell < \sigma$. For all $n \geq k_\ell$ we have

$$x \in X \quad \Rightarrow \quad x \in X_\ell = X(k_\ell, \ell) \quad \Rightarrow \quad |f_n(x) - f(x)| < 1/\ell \; < \; \sigma.$$

Hence f_n converges uniformly to f off the ϵ-set X^c. (We used σ to avoid writing ϵ with two different meanings.) $\square$

See also Exercise 83.

Appendix G Roundness

The density of a set E at p is the limit, if it exists, of the concentration of E in a ball or cube that shrinks down to p. What if you used another shape such an ellipsoid or solid torus? Would it matter? The answer is "somewhat."

Let us say that a neighborhood U of x is **K-quasi-round** if it can be sandwiched between balls $B \subset U \subset B'$ with diam $B' \leq K$ diam B. A ball is 1-quasi-round while a square is $\sqrt{2}$-quasi-round.

It is not hard to check that if x is a density point with respect to balls then it also a density point with respect to K-quasi-round neighborhoods of x, provided that K is fixed as the neighborhoods shrink to x. See Exercises 60 and 61. When the neighborhoods are not quasi-round, the density point analysis becomes marvelously complicated. See Falconer's book, *The Geometry of Fractal Sets*.

Appendix H Money

Riemann and Lebesgue walk into a room and find a table covered with hundreds of U.S. coins. (Well, ...) How much money is there?

Riemann solves the problem by taking the coins one at a time and adding their values as he goes. As he picks up a penny, a nickel, a quarter, a dime, a penny, etc., he counts: "1 cent, 6 cents, 31 cents, 41 cents, 42 cents, etc." The final number is Riemann's answer.

In contrast, Lebesgue first sorts the coins into piles of the same value (partitioning the value axis and taking preimages); he then counts each pile (applying counting measure); and he sums the six terms, "value v times number of coins with value v," and that is his answer.

Lebesgue's answer and Riemann's answer are of course the same number. It is their methods of calculating that number which differ.

Now imagine that *you* walk into the room and behold this coin-laden table. Which method would you actually use to find out how much money there is – Riemann's or Lebesgue's? This amounts to the question: Which is the "better" integration theory? As an added twist suppose you have only sixty seconds to make a good guess. What would you do then?

Exercises

1. (a) Show that the definition of linear outer measure is unaffected if we demand that the intervals I_k in the coverings be closed instead of open.
    (b) Why does this immediately imply that the middle-thirds Cantor set has linear outer measure zero?
    (c) Show that the definition of linear outer measure is unaffected if we drop all openness/closedness requirements on the intervals I_k in the coverings.
    (d) What about planar outer measure? Specifically, what if we demand that the rectangles be squares?

2. The volume of an n-dimensional box is the product of the lengths of its edges and the outer measure of $A \subset \mathbb{R}^n$ is the infimum of the total volume of countable coverings of A by open boxes.
    (a) Write out the proof of the outer measure axioms for subsets of $\mathbb{R}^n$.
    (b) Write out the proof that the outer measure of a box equals its volume.

3. A line in the plane that is parallel to one of the coordinate axes is a planar zero set because it is the Cartesian product of a point (it's a linear zero set) and $\mathbb{R}$.
    (a) What about a line that is not parallel to a coordinate axis?
    (b) What is the situation in higher dimensions?

4. The proof of Lemma 11 was done in the plane. The key insight was that a square S contains a disc Δ such that $m\Delta/mS > 1/2$. Find a corresponding inequality in n-space and write out the n-dimensional proofs of the lemma and Theorem 9 carefully.

5. Prove that every closed set in $\mathbb{R}$ or $\mathbb{R}^n$ is a G_δ-set. Does it follow at once that every open set is an F_σ-set? Why?

6. Complete the proofs of Theorems 16 and 21 in the unbounded, n-dimensional case. [Hint: How can you break an unbounded set into countably many disjoint bounded pieces?]

7. Show that inner measure is translation invariant. How does it behave under dilation? Under affine motions?

*8. Prove that $\mathbb{R} \setminus \mathbb{Q}$ is an $F_{\sigma\delta}$-set but not an F_σ-set. [Hint: Baire.] Infer that $\mathcal{F}_\sigma \neq \mathcal{F}_{\sigma\delta}$. You can google "Descriptive Set Theory" for further inequalities like this.

9. Theorem 16 implies that if E is measurable then its inner and outer measures are equal. Is the converse true? [Proof or counterexample.]

10. For an arbitrary set M define $\omega : 2^M \to [0, \infty]$ as $\omega(S) = \#(S)$, where 2^M is the power set of M (the collection of all subsets of M) and $\#(S)$ is the cardinality of S. Prove that ω is an abstract outer measure and all sets $S \subset M$ are measurable. [This is **counting measure**. It makes frequent appearances in counterexamples in abstract measure theory.]

11. The **outer Jordan content** of a bounded set $A \subset \mathbb{R}$ is the infimum of the

total lengths of *finite* coverings of A by open intervals,

$$J^*A = \inf \left\{ \sum_{k=1}^n |I_k| : \text{each } I_k \text{ is an open interval and } A \subset \bigcup_{k=1}^n I_k \right\}.$$

The corresponding definitions of outer Jordan content in the plane and n-space substitute rectangles and boxes for intervals.

(a) Show that outer Jordan content satisfies

  (i) $J^*(\emptyset) = 0$.

  (ii) If $A \subset B$ then $J^*A \leq J^*B$.

  (iii) If $A = \bigcup_{k=1}^n A_k$ then $J^*A \leq \sum_{k=1}^n J^*A_k$.

(b) (iii) is called finite subadditivity. Find an example of a set $A \subset [0,1]$ such that $A = \bigcup_{k=1}^\infty A_k$, $J^*A_k = 0$ for all k, and $J^*A = 1$, which shows that finite subadditivity does not imply countable subadditivity and that J^* is not an outer measure.

(c) Why is it clear that $m^*A \leq J^*A$, and that if A is compact then $mA = J^*A$? What about the converse?

(d) Show that the requirement that the intervals in the covering of A be open is irrelevant.

12. Prove that
$$J^*A = J^*\overline{A} = m\overline{A}$$

where $\overline{A}$ is the closure of A.

13. If A, B are compact prove that

$$J^*(A \cup B) + J^*(A \cap B) = J^*A + J^*B.$$

[Hint: Is the formula true for Lebesgue measure? Use Exercise 12.]

14. The **inner Jordan content** of a subset A of an interval I is

$$J_*A = |I| - J^*(I \setminus A).$$

(a) Show that
$$J_*A = m(\text{interior } A).$$

(b) A bounded set A with equal inner and outer Jordan content is said to have content or to be **Jordan measurable**, and we write $J_*A = JA = J^*A$, even though J is not a measure. (Is this any worse than functions with infinite integrals being nonintegrable?)

(c) Infer from Theorem 68 and the Riemann-Lebesgue Theorem that $f : [a, b] \to [0, M]$ is Riemann integrable if and only if its undergraph is Jordan measurable, and in that case its Riemann integral equals $J(\mathcal{U}f)$.

*15. Construct a Jordan curve (homeomorphic copy of the circle) in $\mathbb{R}^2$ that has positive planar measure. [Hint: Given a Cantor set in the plane, is there a Jordan curve that contains it? Is there a Cantor set in the plane with positive planar measure? (Take another look at Section 9 in Chapter 2.)]

16. Write out the proofs of Lemmas 23, 16, and 24 in the n-dimensional case.

17. Write out the proofs of the Measurable Product Theorem (Theorem 21) and the Zero Slice Theorem (Theorem 25) in the unbounded, n-dimensional case.

**18. Suppose that E is measurable and bounded.

    (a) If $E \subset \mathbb{R}$ and $\epsilon > 0$ is given, prove there exists a fat Cantor set $F \subset E$ such that $mE \leq m(F) + \epsilon$. [Hint: Review Exercise 2.151.]

    (b) Do the same in $\mathbb{R}^n$.

    (c) Do the same in $\mathbb{R}$ and $\mathbb{R}^n$ if E is nonmeasurable but $m_*E > 0$. [Hint: K_E.]

**19. Consider linear Lebesgue measure m_1 on the interval I and planar Lebesgue measure m_2 on the square I^2. Construct a mesisometry $I \to I^2$. Thus mesisometry disrespects topology: $(I, \mathcal{M}(I), m_1)$ is mesisometric to $(I^2, \mathcal{M}(I^2), m_2)$. [Hint: You might use the following outline. The inclusion $I \setminus \mathbb{Q} \to I$ is injective and preserves m_1. You can convert it to a bijection $\alpha : I \setminus \mathbb{Q} \to I$ by choosing a countable set $L \subset I \setminus \mathbb{Q}$ and then choosing any bijection $\alpha_0 : L \to L \cup (\mathbb{Q} \cap I)$. Then you can set $\alpha(x) = \alpha_0(x)$ when $x \in L$ and $\alpha(x) = x$ otherwise. Why is α is a mesisometry? (Already this shows that nonhomeomorphic spaces can have mesisometric measure spaces.) In the same way there is a mesisometry $\beta : I^2 \setminus \mathbb{Q}^2 \to I^2$. Then let $A = I \setminus \mathbb{Q}$. Express $x \in A$ as a base-2 expansion

$$x = (a_1 a_2 a_3 a_4 a_5 a_6 \ldots)$$

using the digits 0 and 1. It is unique since x is irrational. Then consider the function $\sigma : A \to I^2$,

$$\sigma(x) = (a_1 a_3 a_5 \ldots, a_2 a_4 a_6 \ldots)$$

Prove that (up to zero sets) σ sends dyadic intervals of length $1/2^n$ to dyadic squares of area $1/2^n$, and that this implies σ preserves measure. Conclude that $T = \beta \circ \sigma \circ \alpha^{-1}$ is a mesisometry $I \to I^2$.]

20. Generalize Exercise 19 with $\mathbb{R}$ in place of I and then with $\mathbb{R}^n$ in place of $\mathbb{R}$.

*21. Suppose that $U, V \subset \mathbb{R}^n$ are open. If a homeomorphism $T : U \to V$ and its inverse send Lebesgue zero sets to Lebesgue zero sets prove that it is a Lebesgue meseomorphism $(U, \mathcal{M}(U), m|_U) \to (V, \mathcal{M}(V), m|_V)$. [Note that the homeomorphism $T : \mathbb{R} \to \mathbb{R}$ which sends the fat Cantor set to the standard Cantor set sends zero sets to zero sets but T^{-1} does not.]

22. If $U, V \subset \mathbb{R}^n$ are open and $T : U \to V$ is a **Lipeomorphism** (i.e., a Lipschitz homeomorphism with Lipschitz inverse) use Exercise 21 to show that T is a meseomorphism with respect to Lebesgue measure.

23. Use Exercise 22 and the n-dimensional Mean Value Theorem to prove that a diffeomorphism $T : U \to V$ is a meseomorphism. [Pay attention to the fact that U and V are noncompact.]

24. (a) If $T : \mathbb{R} \to \mathbb{R}$ is a continuous mesisometry prove that T is rigid.
    (b) What if T is discontinuous?
    (c) Find a continuous non-affine mesisometry $T : \mathbb{R}^2 \to \mathbb{R}^2$. [Hint: Divergence.]

25. Let $f : \mathbb{R} \to [0, \infty)$ be given.
    (a) If f is measurable why is the graph of f a zero set?
    (b) If the graph of f is a zero set does it follow that f is measurable?
    **(c) Read about transfinite induction and go to stackexchange to see that there exists a nonmeasurable function $f : [a, b] \to [0, \infty)$ whose graph is non-measurable.
    (d) Infer that the measurability hypothesis in the Zero Slice Theorem (Theorem 25) is necessary since every vertical slice graph of the function in (c) is a zero set (it is just a single point) and yet the graph has positive outer measure.
    (e) Why can a graph never have positive inner measure?
    (f) How does (c) yield an example of uncountably many disjoint subsets of the plane, each with infinite outer measure?
    (g) What assertion can you make from (f) and Exercise 19?

26. Theorem 35 states that T_f is a mesisometry when $f : \mathbb{R} \to [0, \infty)$ is integrable. Prove the same thing when $f : \mathbb{R} \to [0, \infty)$ is measurable. What about a measurable function $\mathbb{R}^n \to \mathbb{R}$? [Hint: Express f as $\sum_{i,k} f_{i,k}$, where the support of $f_{i,k}$ is $[i-1, i) \cap f^{\text{pre}}([k-1, k))$. Why is $f_{i,k}$ integrable and how does this imply that T_f is a mesisometry?]

27. Using the undergraph definition, check linearity of the integral directly for two measurable characteristic functions, $f = \chi_F$ and $g = \chi_G$.

28. The **total undergraph** of $f : \mathbb{R} \to \mathbb{R}$ is $\underline{\mathcal{U}}f = \{(x, y) : y < f(x)\}$.
    (a) Using undergraph pictures, show that the total undergraph is measurable if and only if the positive and negative parts of f are measurable.
    (b) Suppose that $f : \mathbb{R} \to (0, \infty)$ is measurable. Prove that $1/f$ is measurable. [Hint: The diffeomorphism $T : (x, y) \mapsto (x, 1/y)$ sends $\mathcal{U}f$ to $(\widehat{\mathcal{U}}(1/f))^c$.]
    (c) Suppose that $f, g : \mathbb{R} \to (0, \infty)$ are measurable. Prove that $f \cdot g$ is measurable. [Hint: $T : (x, y) \mapsto (x, \log y)$ sends $\mathcal{U}f$ and $\mathcal{U}g$ to $\underline{\mathcal{U}}(\log f)$ and $\underline{\mathcal{U}}(\log g)$. How does this imply $\log fg$ is measurable, and how does use of $T^{-1} : (x, y) \mapsto (x, e^y)$ complete the proof?]
    (d) Remove the hypotheses in (a)-(c) that the domain of f, g is $\mathbb{R}$.
    (e) Generalize (c) to the case that f, g have both signs.

29. A function $f : M \to \mathbb{R}$ is **upper semicontinuous** if

$$\lim_{k \to \infty} x_k = x \quad \Rightarrow \quad \limsup_{k \to \infty} f(x_k) \leq f(x).$$

(M can be any metric space.) Equivalently, $\limsup_{y \to x} fy \leq fx$.

(a) Draw a graph of an upper semicontinuous function that is not continuous.

(b) Show that upper semicontinuity is equivalent to the requirement that for every open ray $(-\infty, a)$, the preimage $f^{\mathrm{pre}}(-\infty, a)$ is an open set.

(c) Lower semicontinuity is defined similarly. Work backward from the fact that the negative of a lower semicontinuous function is upper semicontinuous to give the definition in terms of lim infs.

30. Given a compact set $K \subset \mathbb{R} \times [0, \infty)$ define

$$g(x) = \begin{cases} \max\{y : (x,y) \in K\} & \text{if } K \cap (x \times \mathbb{R}) \neq \emptyset \\ 0 & \text{otherwise.} \end{cases}$$

Prove that g is upper semicontinuous.

31. Prove that a measurable function f is sandwiched as $u \leq f \leq v$, where u is upper semicontinuous, v is lower semicontinuous (we permit $v(x) = \infty$ at some points), and $v - u$ has small integral. [Hint: Exercise 30 and regularity.]

32. Prove Proposition 38.

33. Suppose that $f_k : [a, b] \to \mathbb{R}^n$ converges almost everywhere to f as $k \to \infty$.

(a) Verify that the Dominated Convergence Theorem fails if there is no integrable dominating function g.

(b) Verify that the inequality in Fatou's Lemma can be strict.

34. If $f_n : \mathbb{R} \to [0, \infty)$ is a sequence of integrable functions, $f_n \downarrow f$ a.e. as $n \to \infty$, and $\int f_n \downarrow 0$. Prove that $f = 0$ almost everywhere.

35. Find a sequence of integrable functions $f_k : [a, b] \to [0, 1]$ such that $\int_a^b f_k \to 0$ as $k \to \infty$ but it is not true that $f_k(x)$ converges to 0 a.e.

36. Show that the converse to the Dominated Convergence Theorem fails in the following sense: There exists a sequence of functions $f_k : [a, b] \to [0, \infty)$ such that $f_k \to 0$ almost everywhere and $\int_a^b f_k \to 0$ as $k \to \infty$, but there is no integrable dominator g. [Hint: Stare at steeples or the graph of $f(x) = 1/x$.]

37. Suppose that a sequence of integrable functions f_k converges almost everywhere to f as $k \to \infty$ and f_k takes on both positive and negative values. If there exists an integrable function g such that for almost every x we have $|f_k(x)| \leq g(x)$, prove that $\int f_k \to \int f$ as $k \to \infty$.

38. If f and g are integrable prove that their maximum and minimum are integrable.

39. Suppose that f and g are measurable and their squares are integrable. Prove

that fg is measurable, integrable, and

$$\int fg \leq \sqrt{\int f^2} \sqrt{\int g^2}.$$

[Hint: Exercise 28 helps.]

40. Find an example where Exercise 39 fails if "square integrable" is replaced with "integrable."

41. Suppose that f_k is a sequence of integrable functions and $\sum \int |f_k| < \infty$. Prove that $\sum f_k$ is integrable and

$$\int \sum_{k=1}^{\infty} f_k = \sum_{k=1}^{\infty} \int f_k.$$

*42. Prove that

$$\int_0^{\infty} e^{-x} \cos \sqrt{x} \, dx = 1 - \frac{1}{2} + \frac{2!}{4!} - \frac{3!}{6!} + \frac{4!}{8!} - \frac{5!}{10!} + \cdots$$

43. Prove that $g(y) = \int_0^{\infty} e^{-x} \sin(x+y) \, dx$ is differentiable and find $g'(y)$.

44. Write out the proof of the multidimensional Cavalieri's Principle (Theorem 39).

45. As in Corollary 41 we say that a function $f : \mathbb{R} \to \mathbb{R}$ is preimage measurable if for each $a \in \mathbb{R}$ the set $f^{\mathrm{pre}}([a, \infty)) = \{x \in \mathbb{R} : a \leq f(x)\}$ is Lebesgue measurable. *This is the standard definition for measurability of a function.* Prove that the following are equivalent conditions for preimage measurability of $f : \mathbb{R} \to \mathbb{R}$.

   (a) The preimage of every closed ray $[a, \infty)$ is measurable.
   (b) The preimage of every open ray (a, ∞) is measurable.
   (c) The preimage of every closed ray $(-\infty, a]$ is measurable.
   (d) The preimage of every open ray $(-\infty, a)$ is measurable.
   (e) The preimage of every half-open interval $[a, b)$ is measurable.
   (f) The preimage of every open interval (a, b) is measurable.
   (g) The preimage of every half-open interval $(a, b]$ is measurable.
   (h) The preimage of every closed interval $[a, b]$ is measurable.
   (i) The preimage of every open set is measurable.
   (j) The preimage of every closed set is measurable.
   (k) The preimage of every G_δ-set is measurable.
   (l) The preimage of every F_σ-set is measurable.

*46. Here is a trick question: "Are there any functions for which the Riemann integral converges but the Lebesgue integral diverges?" Corollary 70 would suggest the answer is "no." Show, however, that the improper Riemann integral $\int_0^1 f(x)\,dx$ of

$$
f(x) = \begin{cases} \dfrac{\pi}{x}\sin\dfrac{\pi}{x} & \text{if } x \neq 0 \\[2mm] 0 & \text{if } x = 0 \end{cases}
$$

exists (and is finite) while the Lebesgue integral is infinite. [Hint: Integration by parts gives

$$
\int_a^1 \frac{\pi}{x}\sin\frac{\pi}{x}\,dx = x\cos\frac{\pi}{x}\Big|_a^1 - \int_a^1 \cos\frac{\pi}{x}\,dx.
$$

Why does this converge to a limit as $a \to 0^+$? To check divergence of the Lebesgue integral, consider intervals $[1/(k+1),\, 1/k]$. On such an interval the sine of π/x is everywhere positive or everywhere negative. The cosine is $+1$ at one endpoint and -1 at the other. Now use the integration by parts formula again and the fact that the harmonic series diverges.]

*47. A nonnegative linear combination of measurable characteristic functions is a **simple function**. That is,

$$
\phi(x) = \sum_{i=1}^{n} c_i \cdot \chi_{E_i}(x)
$$

where $E_1, \ldots, E_n$ are measurable sets and $c_1, \ldots, c_n$ are nonnegative constants. We say that $\sum c_i \chi_{E_i}$ "expresses" ϕ. If the sets E_i are disjoint and the coefficients c_i are distinct and positive then the expression for ϕ is called **canonical**.

  (a) Show that a canonical expression for a simple function exists and is unique.

  (b) It is obvious that the integral of $\phi = \sum c_i \chi_{E_i}$ (the measure of its undergraph) equals $\sum c_i m(E_i)$ if the expression is the canonical one. Prove carefully that this remains true for *every* expression of a simple function.

  (c) Infer from (b) that $\int \phi + \psi = \int \phi + \int \psi$ for simple functions.

  (d) Given measurable $f, g : \mathbb{R} \to [0, \infty)$, show that there exist sequences of simple functions $\phi_n \uparrow f$ and $\psi_n \uparrow g$ as $n \to \infty$.

  (e) Combine (c) and (d) to revalidate linearity of the integral.

In fact this is often how the Lebesgue integral is developed. A "preintegral" is constructed for simple functions, and the integral of a general nonnegative measurable function is defined to be the supremum of the preintegrals of lesser simple functions.

*48. The Devil's ski slope. Recall from Chapter 3 that the Devil's staircase function $H : [0, 1] \to [0, 1]$ is continuous, nondecreasing, constant on each interval complementary to the standard Cantor set, and yet is surjective. For $n \in \mathbb{Z}$ and

$x \in [0,1]$ we define $\widehat{H}(x+n) = H(x) + n$. This extends H to a continuous surjection $\mathbb{R} \to \mathbb{R}$. Then we set

$$H_k(x) = \widehat{H}(3^k x) \quad \text{and} \quad J(x) = \sum_{k=0}^{\infty} \frac{H_k(x)}{4^k}.$$

Prove that J is continuous, strictly increasing, and yet $J' = 0$ almost everywhere. [Hint: Fix $a > 0$ and let

$$S_a = \{x : J'(x) \text{ exists}, J'(x) > a, \text{ and}$$
$$x \text{ belongs to the constancy intervals of every } H_k\}.$$

Use the Vitali Covering Lemma to prove that $m^*(S_a) = 0$.]

*49. Prove that $f : \mathbb{R} \to \mathbb{R}$ is Lebesgue measurable if and only if the preimage of every Borel set is a Lebesgue measurable. What about $f : \mathbb{R}^n \to \mathbb{R}$?

*50. (a) Prove Corollary 67: Each measurable $E \subset \mathbb{R}$ with $mE > 0$ contains a nonmeasurable set N with $m^*N = mE$, $m_*N = 0$, and for each measurable $E' \subset E$ we have $m(E') = m^*(N \cap E')$. (N is a "doppelgänger" of E.) [Hint: Try $N = P \cap E$ when $E \subset [0,1)$ and P is the nonmeasurable set from Theorem 66.]

   (b) Is N uniquely determined (modulo a zero set) by E?

51. Generalize Theorem 66 and Exercise 50 to $\mathbb{R}^n$. [Hint: Think about $P \times P$ and its complement in I^2.]

   Remark There are even worse situations. $\mathbb{R}^n$ is the disjoint union of $\#\mathbb{R}$ sets like P. This fact involves "Bernstein sets" and transfinite induction. See also Exercise 25.

52. If $T : \mathbb{R}^n \to \mathbb{R}^n$ is a meseomorphism prove that $mZ = 0$ implies $m(TZ) = 0$. [Hint: doppelgängers.]

53. Consider the function $f : \mathbb{R}^2 \to \mathbb{R}$ defined by

$$f(x,y) = \begin{cases} \dfrac{1}{y^2} & \text{if } 0 < x < y < 1 \\[2mm] \dfrac{-1}{x^2} & \text{if } 0 < y < x < 1 \\[2mm] 0 & \text{otherwise.} \end{cases}$$

   (a) Show that the iterated integrals exist and are finite (calculate them) but the double integral does not exist.

   (b) Explain why (a) does not contradict Corollary 43.

54. Do (A) or (B), but not both.
    (A) (a) State and prove Cavalieri's Principle in dimension 4.
        (b) Formulate the Fubini-Tonelli theorem for triple integrals and use (a) to prove it.
    (B) (a) State Cavalieri's Principle in dimension $n + 1$.
        (b) State the Fubini-Tonelli Theorem for multiple integrals and use (a) to prove it.
    How short can you make your answers?
55. (a) What are the densities (upper, lower, balanced, and general) of the disc in the plane and at which points do they occur?
    (b) What about the densities of the square?
    ***(c) What about the densities of the fat Cantor set?
56. Suppose that $P \subset \mathbb{R}$ has the property that for every interval $(a, b) \subset \mathbb{R}$ we have

$$\frac{m^*(P \cap (a, b))}{b - a} = \frac{1}{2}.$$

    (a) Prove that P is nonmeasurable. [Hint: This is a one-liner.]
    (b) Is there anything special about $1/2$?
57. Formulate and prove Steinhaus' Theorem (Theorem 64) in n-space.
58. The balanced density of a measurable set E at x is the limit, if exists, of the concentration of E in B where B is a ball centered at x that shrinks down to x. Write $\delta_{\text{balanced}}(x, E)$ to indicate the balanced density, and if it is 1, refer to x as a balanced density point.
    (a) Why is it immediate from the Lebesgue Density Theorem that almost every point of E is a balanced density point?
    (b) Given $\alpha \in [0, 1]$, construct an example of a measurable set $E \subset \mathbb{R}$ that contains a point x with $\delta_{\text{balanced}}(x, E) = \alpha$.
    (c) Given $\alpha \in [0, 1]$, construct an example of a measurable set $E \subset \mathbb{R}$ that contains a point x with $\delta(x, E) = \alpha$.
    **(d) Is there a single set that contains points of both types of density for all $\alpha \in [0, 1]$?
59. Prove that the density points of a measurable set are the same as its balanced density points. [Hint: Exercise 62 is relevant.]
*60. Density is defined using cubes Q that shrink down to p. What if p need not belong to Q, but its distance to Q is on the order of the edgelength ℓ of Q? That is, $d(p, Q) \leq K\ell$ for some constant K as $\ell \to 0$. (Q is a **satellite** of p.) Do we get the same set of density points?
*61. As indicated in Appendix G, $U \subset \mathbb{R}^n$ is K-quasi-round if it can be sandwiched between balls $B \subset U \subset B'$ such that $\text{diam } B' \leq K \text{ diam } B$.
    (a) Prove that in the plane, squares and equilateral triangles are (uniformly)

        quasi-round. (The same K works for all of them.)

(b) What about isosceles triangles?

(c) What about annuli of inner radius r and outer radius R such that $R/r \leq 10$, and what about balanced density for such annuli? [Hint: Draw a picture.]

(d) Formulate a Vitali Covering Lemma for a Vitali covering $\mathcal{V}$ of $A \subset \mathbb{R}^2$ by uniformly quasi-round sets instead of discs.

(e) Prove it.

(f) Generalize to $\mathbb{R}^n$.

[Hint: Review the proof of the Vitali Covering Lemma.]

*62. Consider a measure-theoretic definition of K-quasi-roundness of a measurable $W \subset \mathbb{R}^n$ as

$$\frac{\mathrm{diam}(W)^n}{mW} \leq K.$$

(a) What is the relation between the two definitions of quasi-roundness?

(b) Fix a point $p \in \mathbb{R}^n$ and let $\mathcal{W}_K$ be the family of measurable sets containing p which are K-quasi-round in the measure-theoretic sense. Prove that p is a density point of a measurable set E if and only if the concentration of E in W tends to 1 as $W \in \mathcal{W}_K$ shrinks to p. [Hint: Each W could be a fat Cantor set, but take heart from the realization that if 99% of a set is red then any 10% of it is quite pink.]

63. Let $E \subset \mathbb{R}^n$ be measurable and let x be a point of ∂E, the topological boundary of E. (That is, x lies in both the closure of E and the closure of E^c.)

(a) Is it true that if the density $\delta = \delta(x, E)$ exists then $0 < \delta < 1$? Proof or counterexample.

(b) Is it true that if $\delta = \delta(x, E)$ exists and $0 < \delta < 1$ then x lies in ∂E? Proof or counterexample.

(c) What about balanced density?

64. Choose a pair of derivates other than the right max and left min. If f is monotone write out a proof that these derivates are equal almost everywhere.

65. Exercise 3.34 asks you to prove that the set of critical values of a C^1 function $f : \mathbb{R} \to \mathbb{R}$ is a zero set. (A critical point of f is a point p such that $f'(p) = 0$ and a critical value of f is a $q \in \mathbb{R}$ such that $fp = q$ for some critical point p.) Give it another try. In (a), (b), (c), $f : [a, b] \to \mathbb{R}$ is C^1.

(a) What are the critical points and critical values of the function $\sin x$?

(b) If $f : [a, b] \to \mathbb{R}$ is C^1 why are the sets of critical points and critical values, $\mathrm{cp}(f)$ and $\mathrm{cv}(f)$, compact?

(c) How can you cover $\mathrm{cv}(f)$ with finitely many intervals of small total length? [Hint: Mean Value Theorem as an inequality.]

(d) How can you go from $[a, b]$ to $\mathbb{R}$?

66. Construct a monotone function $f : [0, 1] \to \mathbb{R}$ whose discontinuity set is exactly the set $\mathbb{Q} \cap [0, 1]$, or prove that such a function does not exist.

*67. In Section 10 the total variation of a function $f : [a, b] \to \mathbb{R}$ is defined as the supremum of all sums $\sum_{i=1}^{n} |\Delta_i f|$, where P partitions $[a, b]$ into subintervals $[x_{i-1}, x_i]$ and $\Delta_i f = f(x_i) - f(x_{i-1})$. Assume that the total variation of f is finite (i.e., f is of bounded variation) and define

$$T_a^x = \sup_P \left\{ \sum_k |\Delta_i f| \right\}$$

$$P_a^x = \sup_P \left\{ \sum_k \Delta_i f : \Delta_i f \geq 0 \right\}$$

$$N_a^x = -\inf_P \left\{ \sum_k \Delta_i f : \Delta_i f \leq 0 \right\}$$

where P ranges through all partitions of $[a, x]$. Prove that
   (a) f is bounded.
   (b) T_a^x, P_a^x, N_a^x are monotone nondecreasing functions of x.
   (c) $T_a^x = P_a^x + N_a^x$.
   (d) $f(x) = f(a) + P_a^x - N_a^x$.

*68. Assume that $f : [a, b] \to \mathbb{R}$ has bounded variation. The **Banach indicatrix** is the function

$$y \mapsto N_y = \# f^{\mathrm{pre}}(y).$$

N_y is the number of roots of $f = y$. The horizontal line $[a, b] \times y$ meets the graph of f in N_y points.
   (a) Prove that $N_y < \infty$ for almost every y.
   (b) Prove that $y \mapsto N_y$ is measurable.
   (c) Prove that

$$T_a^b = \int_c^d N_y \, dy$$

   where $c \leq \min f$ and $\max f \leq d$.

*69. (a) Assume that $A_n \uparrow A$ as $n \to \infty$ but do not assume that A_n is measurable. Prove that $m^* A_n \to m^* A$ as $n \to \infty$. (This is upward measure continuity for outer measure. [Hint: Regularity gives G_δ-sets $G_n \supset A_n$ with $m(G_n) = m^*(A_n)$. Can you make sure that G_n increases as $n \to \infty$? If so, what can you say about $G = \bigcup G_n$?])
   (b) Is upward measure continuity true for inner measure? [Proof or counterexample.]
   (c) What about downward measure continuity of inner measure? Of outer measure?

*70. Let $A \subset \mathbb{R}^n$ be arbitrary, measurable or nonmeasurable.

(a) Prove that the hull and kernel of A are unique up to zero sets.

(b) Prove that A "spreads itself evenly" through its hull in the sense that for each measurable E we have $m^*(A \cap E) = m(H_A \cap E)$.

(c) Prove the following version of the Lebesgue Density Theorem. For almost every $p \in H_A$ we have

$$\lim_{Q \downarrow p} \frac{m^*(A \cap Q)}{mQ} = 1.$$

[Hint: Review the proof of the Lebesgue Density Theorem. Taking $E = Q$ in (b) is useful in proving (c).]

71. True or false: If H_A is a hull of A then $H_A \setminus A$ is a zero set.

72. If N is a doppelgänger of a measurable set E (Corollary 67 and Exercise 50) prove that E is a measurable hull of N. (Thus N is something like a "nonmeasurable kernel of E.")

*73. Prove that the outer measure of the Cartesian product of sets which are not necessarily measurable is the product of their outer measures. [Hint: If H_A and H_B are hulls of A and B use the Zero Slice Theorem to show that their product is a hull of $A \times B$.]

*74. What about the inner measure of a product?

75. Observe that under Cartesian products, measurable and nonmeasurable sets act like odd and even integers respectively.

(a) Which theorem asserts that the product of measurable sets is measurable? (Odd times odd is odd.)

(b) Is the product of nonmeasurable sets nonmeasurable? (Even times even is even.)

(c) Is the product of a nonmeasurable set and a measurable set having nonzero measure always nonmeasurable? (Even times odd is even.)

(d) Zero sets are special. They correspond to the number zero, an odd number in this imperfect analogy. (Zero times anything is zero.)

*76. Exercise 3.18 asks you to prove that given a closed set $L \subset \mathbb{R}$, there is a C^∞ function $\beta : \mathbb{R} \to [0, \infty)$ whose zero locus $\{x : \beta(x) = 0\}$ equals L. Give it another try. Can you also do it in $\mathbb{R}^n$?

77. Suppose that $F \subset [0, 2]$ is a fat Cantor set of measure 1. Prove that there is a C^∞ homeomorphism $h : \mathbb{R} \to \mathbb{R}$ that carries $[0, 2]$ to $[0, 1]$ and sends F to a Cantor set hF of measure zero. [Hint: Use a β from Exercise 76 and a constant c to define hx as $c \int_0^x \beta(t)\, dt$. How does Exercise 3.34 help?]

78. Suppose that $f : \mathbb{R} \to [0, \infty)$ is Lebesgue measurable and $g : [0, \infty) \to [0, \infty)$ is monotone or continuous. Prove that $g \circ f$ is Lebesgue measurable. [Hint: Use the preimage definition of measurability and Exercise 45.]

79. (a) For a bijection h verify that $\chi_A = \chi_{hA} \circ h$.
    (b) Let $h : \mathbb{R} \to \mathbb{R}$ be the smooth homeomorphism supplied by Exercise 77. Why does F contain a nonmeasurable set P and why is hP measurable?
    (c) Why is the nonmeasurable function χ_P the composition $\chi_{hP} \circ h$.
    (d) Infer that *a continuous function following a Lebesgue measurable function is Lebesgue measurable (Exercise 78) but a Lebesgue measurable function following a continuous (or even smooth) function may fail to be Lebesgue measurable.*

80. Let $h : [0, 2] \to [0, 1]$ be the smooth homeomorphism supplied by Exercise 77 and let $P \subset F$ be nonmeasurable. Set $f_n(x) = 0$ for all n, x.
    (a) Is it true that the functions f_n are Borel measurable and converge almost everywhere to χ_{hP}?
    (b) Is χ_{hP} Lebesgue measurable?
    (c) Is χ_{hP} Borel measurable?
    (d) Infer that *if a sequence of Borel measurable functions converges almost everywhere to a limit function then that limit function may fail to be Borel measurable.*

81. Improve the Average Value Theorem to assert that not only is it true that for almost every p the average $\fint_Q f \, dm \to f(p)$ as $Q \downarrow p$, but actually for almost every p we have

$$\lim_{Q \downarrow p} \fint_Q |f - fp| \, dm = 0.$$

    [Hint: Apply the Average Value Theorem to each of the countably many functions $|f - r|$ where $r \in \mathbb{Q}$.]

**82. Use the Improved Average Value Theorem from Exercise 81 to give a second proof of Lusin's Theorem that does not use countable bases or preimage measurability.

83. Suppose that (f_k) is a sequence of measurable functions that converge almost everywhere to f as $k \to \infty$.
    (a) Formulate and prove Egoroff's Theorem if the functions are defined on a box in n-space.
    (b) Is Egoroff's Theorem true or false for a sequence of functions defined on an an unbounded set having finite measure?
    (c) Give an example of a sequence of functions defined on $\mathbb{R}$ for which Egoroff's Theorem fails.
    (d) Prove that if the functions are defined on $\mathbb{R}^n$ and $\epsilon > 0$ is given then there is an ϵ-set $S \subset \mathbb{R}^n$ such that for each compact $K \subset \mathbb{R}^n$, the sequence of functions restricted to $K \cap S^c$ converges uniformly.

84. Why does Lusin's Theorem imply that if $f : B \to \mathbb{R}$ is measurable and $B \subset \mathbb{R}^n$ is bounded then f is nearly *uniformly* continuous? What if B is unbounded but has finite measure?

*85. Show that nearly uniform convergence is transitive in the following sense. Assume that f_n converges nearly uniformly to f as $n \to \infty$, and that for each fixed n there is a sequence $f_{n,k}$ which converges nearly uniformly to f_n as $k \to \infty$. (All the functions are measurable and defined on $[ab]$.)

  (a) Show that there is a sequence $k(n) \to \infty$ as $n \to \infty$ such that $f_{n,k(n)}$ converges nearly uniformly to f as $n \to \infty$. In symbols

$$\operatorname*{nulim}_{n\to\infty} \operatorname*{nulim}_{k\to\infty} f_{n,k} = f \quad \Rightarrow \quad \operatorname*{nulim}_{n\to\infty} f_{n,k(n)} = f.$$

  (b) Why does (a) remain true when almost everywhere convergence replaces nearly uniform convergence? [Hint: The answer is one word.]

  (c) Is (a) true when $\mathbb{R}$ replaces $[a, b]$?

  (d) Is (b) true when $\mathbb{R}$ replaces $[a, b]$?

86. Consider the continuous functions

$$f_{n,k}(x) = (\cos(\pi n! x))^k$$

for $k, n \in \mathbb{N}$ and $x \in \mathbb{R}$.

  (a) Show that for each $x \in \mathbb{R}$,

$$\lim_{n\to\infty} \lim_{k\to\infty} f_{n,k}(x) = \chi_{\mathbb{Q}}(x),$$

    the characteristic function of the rationals.

  (b) Infer from Exercise 24 in Chapter 3 that there can not exist a sequence $f_{n,k(n)}$ converging everywhere as $n \to \infty$.

  (c) Interpret (b) to say that everywhere convergence can not replace almost everywhere convergence or nearly uniform convergence in Exercise 85.

87. (a) Prove that, up to a zero set, the measure-theoretic boundary of a measurable set E is contained in its topological boundary, $\partial_m(E) \subset \partial E$.

  (b) Construct an example of a continuous function $f : [a, b] \to [0, M]$ such that $\partial(\mathcal{U}f) \neq \partial_m(\mathcal{U}f)$. [Hint: A picture is worth a thousand formulas.]

88. Generalize Theorem 68 to functions of several variables. That is, prove that a bounded nonnegative function defined on a box in n-space is Riemann integrable if and only if the topological boundary of its undergraph is a zero set.

89. The L^1-norm of the integrable function $f : [a, b] \to \mathbb{R}$ is $\|f\| = \int |f|$. This gives a metric on the set $\mathcal{L}$ of integrable functions $[a, b] \to \mathbb{R}$ as $d_{L^1}(f, g) = \|f - g\|$. We say that $f_n \to g$ L^1-converges** to g if $\|f_n - g\| \to 0$.

(a) Prove that $\mathcal{L}$ is a complete metric space.

(b) Prove that $\mathcal{R}$ is dense in $\mathcal{L}$ where $\mathcal{R}$ is the set of Riemann integrable functions.

(c) Infer that $\mathcal{L}$ is the completion of $\mathcal{R}$ with respect to the L^1-metric. (This constructs Lebesgue integrals with minimal reference to Lebesgue measure.)

(d) What happens if we replace $[a, b]$ with a box in $\mathbb{R}^n$?

90. A theory of integration more general than Lebesgue's is due to Arnaud Denjoy. Rediscovered by Ralph Henstock and Jaroslav Kurzweil, it is described in Robert McLeod's book, *The Generalized Riemann Integral*. The definition is deceptively simple. Let $f : [a, b] \to \mathbb{R}$ be given. The **Denjoy integral of f, if it exists, is a real number I such that for each $\epsilon > 0$ there is a function $\delta : [a, b] \to (0, \infty)$ and

$$\left| \sum_{k=1}^{n} f(t_k) \Delta x_k - I \right| < \epsilon$$

for all Riemann sums with $\Delta x_k < \delta(t_k)$, $k = 1, \ldots, n$. (McLeod refers to the function δ as a "gauge" and to the intermediate points t_k as "tags".)

(a) Verify that if we require the gauge $\delta(t)$ to be continuous then the Denjoy integral reduces to the Riemann integral.

(b) Verify that the function

$$f(x) = \begin{cases} \dfrac{1}{\sqrt{x}} & \text{if } 0 < x \le 1 \\ 100 & \text{if } x = 0 \end{cases}$$

has Denjoy integral 2. [Hint: Construct gauges $\delta(t)$ such that $\delta(0) > 0$ but $\lim_{t \to 0^+} \delta(t) = 0$.]

(c) Generalize (b) to include all functions defined on $[a, b]$ for which the improper Riemann integral is finite.

(d) Infer from (c) and Exercise 46 that some functions are Denjoy integrable but not Lebesgue integrable.

(e) Read McLeod's book to verify that every nonnegative Denjoy integrable function is Lebesgue integrable and the integrals are equal; and every Lebesgue integrable function is Denjoy integrable and the integrals are equal. Infer that the difference between Lebesgue and Denjoy corresponds to the difference between absolutely and conditionally convergent series – if f is Lebesgue integrable, so is $|f|$, but this is not true for Denjoy integrals.

**91. Four types of convergence of a sequence of measurable functions (f_n) are: Almost everywhere convergence, L^1 convergence, nearly uniform convergence, and convergence in measure. This last type of convergence requires that for each $\epsilon > 0$ we have

$$m(\{x : |f_n(x) - g(x)| > \epsilon\}) \to 0$$

as $n \to \infty$. Consulting the tetrahedron in Figure 159, decide which oriented edges represent implications for sequences of functions defined on $[a, b]$, on $\mathbb{R}$, or represent implications on neither $[a, b]$ nor $\mathbb{R}$.

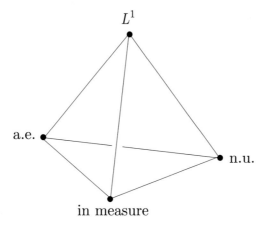

Figure 159 You might label an edge that represents implication only for functions defined on $[a, b]$ with a single arrow, but use a double arrow if the implication holds for functions defined on $\mathbb{R}$. For example, how should you label the edge from a.e. to n.u.?

92. Assume that the (unbalanced) density of E exists at every point of $\mathbb{R}$, not merely at almost all of them. Prove that up to a zero set, $E = \mathbb{R}$, or $E = \emptyset$. (This is a kind of **measure-theoretic connectedness. Topological connectedness of $\mathbb{R}$ is useful in the proof.) Is this also true in $\mathbb{R}^n$?

***93. [Speculative] Density seems to be a first-order concept. To say that the density of E at x is 1 means that the concentration of E in a ball B containing x tends to 1 as $B \downarrow x$. That is,

$$\frac{m(B) - m(E \cap B)}{mB} \to 0.$$

But how fast can we hope it tends to 0? We could call x a **double density point** if the ratio still tends to 0 when we square the denominator. Interior points of E are double density points. Are such points common or scarce in a measurable set? What about balanced density points? What about fractional powers of the denominator?

Suggested Reading

There are many books on more advanced analysis and topology. Among my favorites in the "not too advanced" category are these.

1. Kenneth Falconer, *The Geometry of Fractal Sets*.
   Here you should read about the Kakeya problem: How much area is needed to reverse the position of a unit needle in the plane by a continuous motion? Falconer also has a couple of later books on fractals that are good.

2. Thomas Hawkins, *Lebesgue's Theory of Integration*.
   You will learn a great deal about the history of Lebesgue integration and analysis around the turn of the last century from this book, including the fact that many standard attributions are incorrect. For instance, the Cantor set should be called the Smith set; Vitali had many of the ideas credited solely to Lebesgue, etc. Hawkins' book is a real gem.

3. John Milnor, *Topology from the Differentiable Viewpoint*.
   Milnor is one of the clearest mathematics writers and thinkers of the twentieth century. This is his most elementary book, and it is only seventy-six pages long.

4. James Munkres, *Topology, a First Course*.
   This is a first-year graduate text that deals with some of the same material you have been studying.

5. Robert Devaney, *An Introduction to Chaotic Dynamical Systems*.
   This is the book you should read to begin studying mathematical dynamics. It is first rate.

One thing you will observe about all these books – they use pictures to convey the mathematical ideas. Beware of books that don't.

© Springer International Publishing Switzerland 2015
C.C. Pugh, *Real Mathematical Analysis*, Undergraduate Texts
in Mathematics, DOI 10.1007/978-3-319-17771-7

Bibliography

1. Ralph Boas, *A Primer of Real Functions*, The Mathematical Association of America, Washington DC, 1981.
2. Andrew Bruckner, *Differentiation of Real Functions*, Lecture Notes in Mathematics, Springer-Verlag, New York, 1978.
3. John Burkill, *The Lebesgue Integral*, Cambridge University Press, London, 1958.
4. Paul Cohen, *Set Theory and the Continuum Hypothesis*, Benjamin, New York, 1966.
5. Robert Devaney, *An Introduction to Chaotic Dynamical Systems*, Benjamin Cummings, Menlo Park, CA, 1986.
6. Jean Dieudonné, *Foundations of Analysis*, Academic Press, New York, 1960.
7. Kenneth Falconer, *The Geometry of Fractal Sets*, Cambridge University Press, London, 1985.
8. Russell Gordon, *The Integrals of Lebesgue, Denjoy, Perron, and Henstock*, The American Mathematical Society, Providence, RI, 1994.
9. Fernando Gouvêa, *p-adic Numbers*, Springer-Verlag, Berlin, 1997.
10. Thomas Hawkins, *Lebesgue's Theory of Integration*, Chelsea, New York, 1975.
11. George Lakoff, *Where Mathematics Comes From*, Basic Books, New York, 2000.
12. Edmund Landau, *Foundations of Analysis*, Chelsea, New York, 1951.
13. Henri Lebesgue, *Leçons sur l'intégration et la recherche des fonctions primitives*, Gauthiers-Villars, Paris, 1904.
14. John Littlewood, *Lectures on the Theory of Functions*, Oxford University Press, Oxford, 1944.
15. Ib Madsen and Jørgen Tornehave, *From Calculus to Cohomology*, Cambridge University Press, Cambridge, 1997.

© Springer International Publishing Switzerland 2015
C.C. Pugh, *Real Mathematical Analysis*, Undergraduate Texts
in Mathematics, DOI 10.1007/978-3-319-17771-7

16. Jerrold Marsden and Alan Weinstein, *Calculus III*, Springer-Verlag, New York, 1998.

17. Robert McLeod, *The Generalized Riemann Integral*, The Mathematical Association of America, Washington DC, 1980.

18. John Milnor, *Topology from the Differentiable Viewpoint*, Princeton University Press, Princeton, 1997.

19. Edwin Moise, *Geometric Topology in Dimensions 2 and 3*, Springer-Verlag, New York, 1977.

20. James Munkres, *Topology, a First Course*, Prentice Hall, Englewood Cliffs, NJ, 1975.

21. Murray Protter and Charles Morrey, *A First Course in Real Analysis*, Springer-Verlag, New York, 1991.

22. Dale Rolfsen, *Knots and Links*, Publish or Perish, Berkeley, 1976.

23. Halsey Royden, *Real Analysis*, Prentice-Hall, Englewood Cliffs, NJ, 1988.

24. Walter Rudin, *Principles of Mathematical Analysis*, McGraw-Hill, New York, 1976.

25. James Stewart, *Calculus with Early Transcendentals*, Brooks Cole, New York, 1999.

26. Arnoud van Rooij and Wilhemus Schikhof, *A Second Course on Real Functions*, Cambridge University Press, London, 1982.

Index

© Springer International Publishing Switzerland 2015
C.C. Pugh, *Real Mathematical Analysis*, Undergraduate Texts
in Mathematics, DOI 10.1007/978-3-319-17771-7